GEOTECHNICAL SPECIAL PUBLICATION NO. 99

ADVANCES IN UNSATURATED GEOTECHNICS

PROCEEDINGS OF SESSIONS OF GEO-DENVER 2000

SPONSORED BY
The Geo-Institute of the American Society of Civil Engineers

August 5-8, 2000
Denver, Colorado

EDITED BY
Charles D. Shackelford
Sandra L. Houston
Nien-Yin Chang

GEO INSTITUTE

ASCE American Society of Civil Engineers
1801 ALEXANDER BELL DRIVE
RESTON, VIRGINIA 20191–4400

Abstract: This publication includes papers that advance our understanding of measuring, characterizing, and predicting the properties, behavior, and performance of unsaturated materials typically encountered or used in geotechnical engineering practice. The materials addressed include coarse-grained soils, compacted soils, geotextiles, loess, residual soils, shales, spoil fill, and swelling and expansive soils. These unsaturated materials are characterized with respect to soil–water characteristic curves, hydraulic conductivity, shear strength, and/or stress–strain properties. Both static and dynamic loading conditions are considered. Advances in the state of the art with respect to both laboratory and field measurements of unsaturated soil properties are included. Finally, predictions of unsaturated soil properties and their behavior with respect to specific applications, such as landfill liners and covers, heave of slabs on grade, and the bearing capacity of foundations on or in unsaturated soils, are addressed.

Library of Congress Cataloging-in-Publication Data

Geo-Denver 2000 (2000 : Denver, Colo.)
Advances in unsaturated geotechnics : proceedings of sessions of Geo-Denver 2000 : August 5-8, 2000, Denver, Colorado / sponsored by the Geo-Institute of the American Society of Civil Engineers ; edited by Charles D. Shackelford, Sandra L. Houston, Nien-Yin Chang.
p. cm.—(Geotechnical special publication ; no. 99)
Includes bibliographical references and index.
ISBN 0-7844-0510-7
1. Soil mechanics—Congresses. 2. Soil moisture—Congresses. I Shackelford, Charles D. II. Houston, Sandra L. III. Chang, Nien-Yin. IV. American Society of Civil Engineers. Geo-Institute. V. Title. VI. Series.

TA710.A1 G34 2000
624.1'513--dc21 00-040141

Library of Congress Catalog Card No: 00-040141
ISBN 0-7844-0510-7
Manufactured in the United States of America.

Geotechnical Special Publications

1 *Terzaghi Lectures*
2 *Geotechnical Aspects of Stiff and Hard Clays*
3 *Landslide Dams: Processes, Risk, and Mitigation*
4 *Tiebacks for Bulkheads*
5 *Settlement of Shallow Foundation on Cohesionless Soils: Design and Performance*
6 *Use of In Situ Tests in Geotechnical Engineering*
7 *Timber Bulkheads*
8 *Foundations for Transmission Line Towers*
9 *Foundations & Excavations in Decomposed Rock of the Piedmont Province*
10 *Engineering Aspects of Soil Erosion, Dispersive Clays and Loess*
11 *Dynamic Response of Pile Foundations–Experiment, Analysis and Observation*
12 *Soil Improvement: A Ten Year Update*
13 *Geotechnical Practice for Solid Waste Disposal '87*
14 *Geotechnical Aspects of Karst Terrains*
15 *Measured Performance Shallow Foundations*
16 *Special Topics in Foundations*
17 *Soil Properties Evaluation from Centrifugal Models*
18 *Geosynthetics for Soil Improvement*
19 *Mine Induced Subsidence: Effects on Engineered Structures*
20 *Earthquake Engineering & Soil Dynamics II*
21 *Hydraulic Fill Structures*
22 *Foundation Engineering*
23 *Predicted and Observed Axial Behavior of Piles*
24 *Resilient Moduli of Soils: Laboratory Conditions*
25 *Design and Performance of Earth Retaining Structures*
26 *Waste Containment Systems: Construction, Regulation, and Performance*
27 *Geotechnical Engineering Congress*
28 *Detection of and Construction at the Soil/Rock Interface*
29 *Recent Advances in Instrumentation, Data Acquisition and Testing in Soil Dynamics*
30 *Grouting, Soil Improvement and Geosynthetics*
31 *Stability and Performance of Slopes and Embankments II*
32 *Embankment of Dams–James L. Sherard Contributions*
33 *Excavation and Support for the Urban Infrastructure*
34 *Piles Under Dynamic Loads*
35 *Geotechnical Practice in Dam Rehabilitation*
36 *Fly Ash for Soil Improvement*
37 *Advances in Site Characterization: Data Acquisition, Data Management and Data Interpretation*
38 *Design and Performance of Deep Foundations: Piles and Piers in Soil and Soft Rock*
39 *Unsaturated Soils*
40 *Vertical and Horizontal Deformations of Foundations and Embankments*
41 *Predicted and Measured Behavior of Five Spread Footings on Sand*
42 *Serviceability of Earth Retaining Structures*
43 *Fracture Mechanics Applied to Geotechnical Engineering*
44 *Ground Failures Under Seismic Conditions*
45 *In Situ Deep Soil Improvement*
46 *Geoenvironment 2000*
47 *Geo-Environmental Issues Facing the Americas*
48 *Soil Suction Applications in Geotechnical Engineering*
49 *Soil Improvement for Earthquake Hazard Mitigation*
50 *Foundation Upgrading and Repair for Infrastructure Improvement*
51 *Performance of Deep Foundations Under Seismic Loading*
52 *Landslides Under Static and Dynamic Conditions–Analysis, Monitoring, and Mitigation*
53 *Landfill Closures–Environmental Protection and Land Recovery*
54 *Earthquake Design and Performance of Solid Waste Landfills*
55 *Earthquake-Induced Movements and Seismic Remediation of Existing Foundations and Abutments*
56 *Static and Dynamic Properties of Gravelly Soils*
57 *Verification of Geotechnical Grouting*
58 *Uncertainty in the Geologic Environment*
59 *Engineered Contaminated Soils and Interaction of Soil Geomembranes*
60 *Analysis and Design of Retaining Structures Against Earthquakes*

61 *Measuring and Modeling Time Dependent Soil Behavior*
62 *Case Histories of Geophysics Applied to Civil Engineering and Public Policy*
63 *Design with Residual Materials: Geotechnical and Construction Considerations*
64 *Observation and Modeling in Numerical Analysis and Model Tests in Dynamic Soil-Structure Interaction Problems*
65 *Dredging and Management of Dredged Material*
66 *Grouting: Compaction, Remediation and Testing*
67 *Spatial Analysis in Soil Dynamics and Earthquake Engineering*
68 *Unsaturated Soil Engineering Practice*
69 *Ground Improvement, Ground Reinforcement, Ground Treatment: Developments 1987-1997*
70 *Seismic Analysis and Design for Soil-Pile-Structure Interactions*
71 *In Situ Remediation of the Geoenvironment*
72 *Degradation of Natural Building Stone*
73 *Innovative Design and Construction for Foundations and Substructures Subject to Freezing and Frost*
74 *Guidelines of Engineering Practice for Braced and Tied-Back Excavations*
75 *Geotechnical Earthquake Engineering and Soil Dynamics III*
76 *Geosynthetics in Foundation Reinforcement and Erosion Control Systems*
77 *Stability of Natural Slopes in the Coastal Plain*
78 *Filtration and Drainage in Geotechnical/Geoenvironmental Engineering*
79 *Recycled Materials in Geotechnical Applications*
80 *Grouts and Grouting: A Potpourri of Projects*
81 *Soil Improvement for Big Digs*
82 *Risk-Based Corrective Action and Brownfields Restorations*
83 *Design and Construction of Earth Retaining Systems*
84 *Effects of Construction on Structures*
85 *Application of Geotechnical Principles in Pavement Engineering*
86 *Big Digs Around the World*
87 *Jacked Tunnel Design and Construction*
88 *Analysis, Design, Construction, and Testing of Deep Foundations*
89 *Recent Advances in the Characterization of Transportation Geo-Materials*
90 *Geo-Engineering for Underground Facilities*
91 *Special Geotechnical Testing: Central Artery/Tunnel Project in Boston, Massachusetts*
92 *Behavioral Characteristics of Residual Soils*
93 *National Geotechnical Experimentation Sites*
94 *Performance Confirmation of Constructed Geotechnical Facilities*
95 *Soil-Cement and Other Construction Practices in Geotechnical Engineering*
96 *Numerical Methods in Geotechnical Engineering: Recent Developments*
97 *Innovations and Applications in Geotechnical Site Characterization*
98 *Pavement Subgrade, Unbound Materials, and Nondestructive Testing*
99 *Advances in Unsaturated Geotechnics*

Preface

Traditional geotechnical engineering practice has been based primarily on consideration of the limiting conditions represented by either completely dry or completely saturated soils. However, a significant portion, if not the majority, of the problems encountered in geotechnical engineering practice involve unsaturated soils and materials. The assumption of completely dry or completely saturated conditions often results in conservative analyses and designs in many of the applications involving unsaturated materials. However, this assumption also can result in unconservative analyses and designs in some applications, such as in evaluating moisture and gas migration through soil barriers used in waste containment applications. Also, the assumption of saturated soil conditions simply is not relevant in some applications, such as in evaluating the heave of foundations on swelling or expansive soils. Finally, rising costs coupled with technological advancements provide motivation to reduce some of the conservatism currently inherent in geotechnical engineering practice by improving our ability to measure, characterize, and predict the properties, behavior, and performance of unsaturated materials.

Consideration of these factors has resulted in this publication on *Advances in Unsaturated Geotechnics* that comprises the papers presented at the Unsaturated Soils sessions at Geo-Institute's *GeoDenver 2000* Conference held in Denver, Colorado, August 3-8, 2000. These sessions were organized and co-sponsored by the *Subcommittee on Unsaturated Soils* (chaired by Charles D. Shackelford) of the Geo-Institute's *Soil Properties Committee* (chaired by Sandra L. Houston), and Technical Committee No. 6 (TC6) on *Unsaturated Soils* of the *International Society for Soil Mechanics and Geotechnical Engineering* (ISSMGE).

The papers presented in this publication advance our current understanding of unsaturated materials typically encountered or used in geotechnical engineering practice. The papers pertain to the measurement, characterization, and prediction of the properties, behavior, and performance of a wide range of materials, including coarse-grained soils, compacted soils, geotextiles, loess, residual soils, shales, spoil fill, and swelling and expansive soils. These materials are characterized for a wide range of properties, including the soil–water characteristic curves, hydraulic conductivity, shear strength, and/or stress–strain properties, and include consideration of both static and dynamic loading conditions. Advances in the state of the art with respect to both laboratory and field measurements of unsaturated soil properties are included, and predictions associated with specific applications, such as landfill liners and covers, heave of slabs on grade, and the bearing capacity of foundations on or in unsaturated soils, are addressed.

Each of the papers included in this publication received at least one positive peer review. All of the papers are eligible for discussion in the ASCE/G-I *Journal of Geotechnical and Geoenvironmental Engineering,* and all of the papers are eligible for ASCE awards. The Editors express their sincere appreciation to all of the reviewers, with special thanks to Linda Hinshaw (Colorado State University) and Maria Marshall (Arizona State University) who assisted in the coordination of the reviews and final organization of the papers.

Charles D. Shackelford, Professor, Department of Civil Engineering, Colorado State University, Fort Collins, CO, USA

Sandra L. Houston, Professor and Chair, Department of Civil Engineering, Arizona State University, Tempe, AZ, USA

Nien-Yin Chang, Professor, Department of Civil Engineering, University of Colorado, Denver, CO, USA

Contents

Part V: Water and Gas Migration

Indexes

Use of a New Thermal Conductivity Sensor to Measure Soil Suction

F. Shuai[1] and D.G. Fredlund[2]

Abstract

Soil suction (i.e., negative pore-water pressure) measurements are essential to predicting and verifying the behavior of an unsaturated soil. Some difficulties were encountered with soil suction measurements made using earlier versions of thermal conductivity soil suction sensors. An improved thermal conductivity sensor has been developed at the University of Saskatchewan, Saskatoon, Canada. The durability and the accuracy of the new sensor has been significantly improved by using a specially designed ceramic and improved electronics.

Introduction

The measurement of negative pore-water pressure is essential to understanding the behavior of an unsaturated soil. The negative pore-water pressure (or soil suction) affects water flow, water storage, volume change and shear strength. As a result, there is need for a reliable technique to measure soil suction in an unsaturated soil. Various techniques are now available for measuring soil suction. Of these technique, the thermal conductivity sensor has been proven to be the most promising device for the in situ measurement of soil suction. The attractiveness of the thermal conductivity soil suction sensor lies primarily in its ability to produce a reasonably reliable measurement of soil suction over a relatively wide range of suctions and the measurements are essentially unaffected by the salt content of the soil (Lee and Fredlund, 1984 and Fredlund and Wong, 1989). Another advantage of thermal conductivity sensors is their versatility and ability to be connected to a data acquisition system for continuous and remote monitoring.

---

[1] Postdoctoral Fellow, Department of Civil Engineering, University of Saskatchewan, Saskatoon, SK., Canada, S7N 5A9

[2] Professor, Department of Civil Engineering, University of Saskatchewan, Saskatoon, SK., Canada, S7N 5A9

Thermal conductivity suction sensors consist of a temperature sensing element and a miniature heater embedded in a porous ceramic block (Fig. 1). The sensor is heated at a standard rate for a fixed heating period. During the heating period, part of the heat is dissipated and the undissipated heat results in a temperature increase at the center of the ceramic block. The amount of heat dissipation is controlled by the presence of water in the ceramic block since water is a better thermal conductor than air. The water content of the ceramic block is dependent upon the soil suction applied to the ceramic block by the surrounding soil. Thus, the change in the temperature at the center of the ceramic block is closely related to the change in matric suction of the surrounding soil. In other words, the temperature increase at the center of the sensor due to the heat pulse is a function of the soil suction of the surrounding soil being tested.

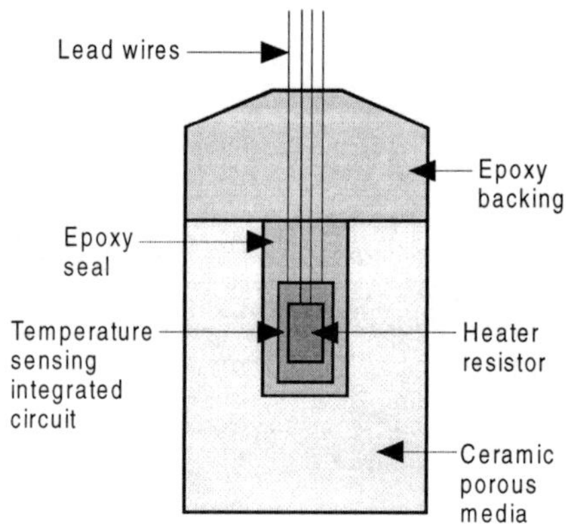

Figure 1 Cross-section diagram of a thermal conductivity soil suction sensor.

There have been numerous difficulties and shortcomings experienced with previously developed versions of thermal conductivity suction sensors. These difficulties and shortcomings can be identified as: 1.) low strength and poor durability of the ceramic tip, 2.) insensitivity and inaccuracy particularly in the higher range of suctions, and 3.) poor stability of the electronic signal. In order to solve the difficulties associated with soil suction measurements, an improved thermal conductivity suction sensor has been developed at the University of Saskatchewan, Saskatoon, Canada.

In this paper, the special design considerations associated with the new thermal conductivity sensor, both in the ceramic and in the electronic aspects, are discussed. The calibration of the new thermal conductivity sensor is described. Some field and laboratory data obtained using the new thermal conductivity suction sensor are presented.

Special Design Considerations Associated with the New Thermal Conductivity Sensor

In order to ensure good performance of the new sensor, past problems needed to be addressed during the development of the new sensor. These considerations include the quality of the ceramic tip, the stability of the output signals and the sensitivity and accuracy of the new sensor.

Characteristic of the Ceramic Tip

The key component of the sensor is the ceramic tip. The ceramic used for soil suction sensors needs to have characteristics considerably different from regular ceramics. In general, regular ceramics possess a low porosity with a relatively uniform pore size distribution, while a high porosity with a wide distribution of pore sizes is required for the soil suction sensor tip in order to ensure a consistent accuracy that extends to the highly negative pore-water pressures. Due to the extremely wide range of possible soil suctions, the desired range of pore sizes for an acceptable ceramic tip should range from 0.1 to 0.0001 mm which corresponds to soil suctions ranging from 1 to 1500 kPa (Fredlund etc., 1994). A ceramic porosity higher than 50% is desired.

The ceramics must have high strength characteristics for use in geotechnical applications. In other words, the ceramic must be sufficiently robust to withstand stresses associated with drilling or driving the sensor into position in the field. It must also be durable to resist the effects of freeze-thaw cycles and drying-wetting cycles in harsh environments. In general, ceramics manufactured with a low porosity and a narrow range in pore size distribution have high strength characteristics whereas those with a high porosity and a wide pore size distribution have been found to be soft, friable and of relatively low strength.

A series of laboratory tests was performed to make a suitable ceramic for a new thermal conductivity suction sensor. The new ceramic tip with a high porosity (i.e., more than 60%) and a wide range of pore sizes has been developed. The new ceramic tip has pore sizes that range from 0.05 mm to less than 0.0001 mm and can be used to measure suctions from approximately 5 kPa to 1500 kPa. In addition, there is a near-linear relationship between water content and logarithm of soil suction between 5 kPa and 500 kPa. This ensures a reasonably accurate measurement of matric suction over a wide range of soil suction. Figure 2 shows a typical soil-water characteristic curve for the ceramic.

The strength of the new ceramic tip has also been significantly improved. The compressive strength of the ceramic has been increased to approximately 2100 kPa. The average splitting tensile strength is about 600 kPa. The greater strength helps prevent cracking and crumbling during installation and prolong operation in the field.

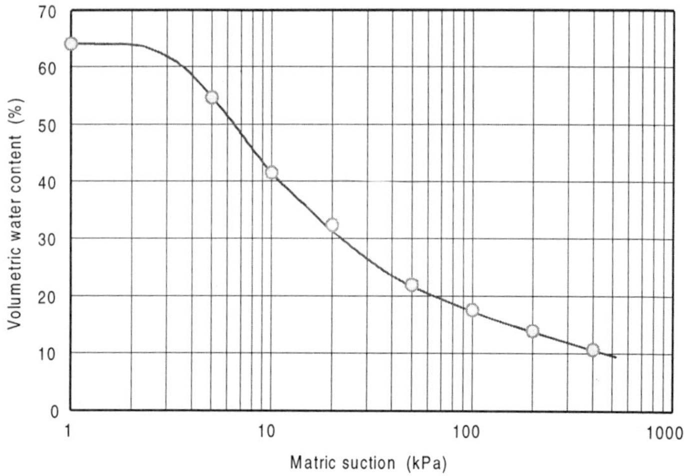

Figure 2. A typical soil-water characteristic curve for the new ceramic.

Characteristic of Electronic Design

The electronic design of the sensor mainly consists of three parts; namely, the temperature sensor selection, the signal conditioning design and the constant current sink design.

Temperature sensor

Two kinds of temperature sensors have been used to measure temperature changes inside the ceramic; namely, thermocouple and integrated circuit (i.e., IC).

A thermocouple consists of a junction of two different conductors (e.g., copper and constantan) when subjected to a temperature gradient generates a voltage. The device is sensitive to temperature gradient, but the voltage generated is very low, usually in the range of tens of microvolts, and is non-linear. This makes thermocouples less desirable for use in a thermal conductivity sensor.

An integrated circuit uses a silicon semi-conductor thermistor to measure temperature. Thermistors are resistors with a high rate of resistance change with temperature. The output voltage from the silicon semi-conductor thermistor can be amplified in the integrated circuit in order to provide improved resolution. The main advantages of the integrated circuit sensor is its higher resolution and its linear relationship between output and absolute temperature.

An integrated circuit, IC, was selected for the new sensor due to its higher accuracy. Another advantage with the integrated circuitry sensor is its ability to measure the soil temperature as well as the soil suction.

Signal conditioning

In order to increase the resolution and accuracy of suction measurements, the quality of output signal from the temperature sensor should be improved and unwanted noise should be removed before the signal is sent to the data logger. This process is called signal conditioning.

Three approaches were used for signal conditioning; namely, amplification, isolation and filtering. The output signal from the integrated circuit was first amplified using an amplifier to the full range of the data logger. The amplified signal was isolated from outside noise sources and unwanted noises were filtered out using a filter prior to entering the data logger.

Signal conditioning significantly improves the quality of the output signal from the temperature sensor. A comparison of heating curves measured before and after improvement is presented in Fig. 3.

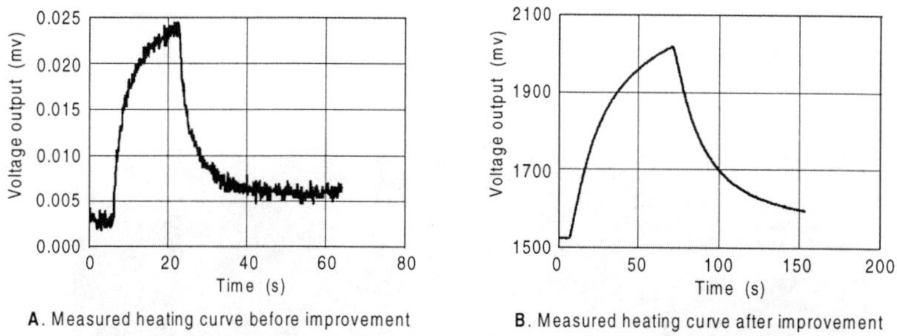

Figure 3 The heating curves measured before and after improvement.

Constant current sink

A precisely controlled heating voltage across the heating element is necessary if reading are to be accurate and reproducible. There are many factors that influence the heating voltage across the heating element, such as cable length, environmental temperature and voltage vibration of power source, to name only a few. In order to eliminate these influences, a constant current sink was designed and manufactured at the University of Saskatchewan.

The constant current sink is able to maintain a constant current of 200 mA through the heater resistor. The constant current compensates for differences in resistance when different lengths of extension wires are used. It also compensates for small changes in the heating resistance caused by a change in environmental temperature. This constant current sink, along with the amplifier mentioned previously, were installed on the same PC board and called a constant current sink and amplifier. A picture of this device is shown in Fig. 4.

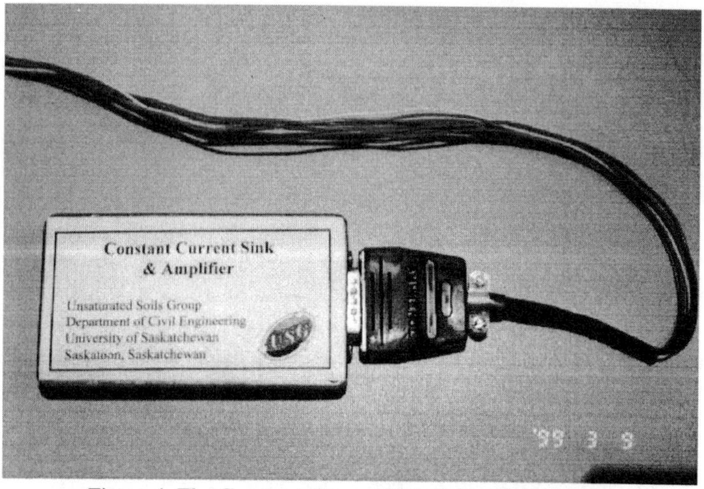

Figure 4 The Constant Current Sink & Amplifier device

Figure 5 The thermal conductivity soil suction sensor developed at the University of Saskatchewan, Saskatoon, Canada.

Characteristic of New Thermal Conductivity Sensor

With the above design considerations, an improved thermal conductivity soil suction sensor was developed at the University of Saskatchewan, Saskatoon, Canada (Fig. 5). The durability and accuracy of the new sensors have been significantly improved over that of earlier versions. The new thermal conductivity sensor has been found to be quite sensitive and accurate in measuring soil suction in the range from 5 to 1500 kPa. This is the range of greatest interest for geotechnical and geo-environmental engineering. The main technical specifications for the new sensor are listed in Table 1.

Table 1 Technical specification

Measurement parameters	Soil suction Soil temperature
Measurement range	Soil suction 5 to 1500 kPa Temperature -40 °C to 110 °C
Accuracy	Less than ± 5% for suction measurement ± 0.5 °C for temperature measurement
Resolution (using CR10X)	0.33 mV
Soil types	Suitable for all soil types
Protection	Suitable for long-term burial
Temperature	0 to 40 °C for suction measurement (no damage when used in frozen soils, but suction reading will be incorrect)
Power supply	12V ~ 15V DC, 250 mA
Size	Diameter: 28 mm, Length: 38 mm
Cable length	Standard: 8m, Maximum: 100m

Calibration of the New Sensor

The accuracy of the soil suction measurements obtained when using the thermal conductivity suction sensor is dependent upon their calibration. The calibration of the new sensor was performed by applying a range of matric suction values to the sensor. The sensor was embedded in a soil that was placed on a Tempe cell (Fig. 6). The soil in the Tempe cell provided continuity between the water phase in the porous ceramic tip and in the high air entry plate. The matric suction was applied by increasing the air pressure in the Tempe cell while maintaining the water pressure below the high air entry disk. The change in voltage output from the sensor was monitored periodically until suction equilibrium was achieved. The above procedure was repeated for various applied suctions ranging from 0 to 400 kPa so that a calibration curve could be obtained.

The calibration curve obtained from the calibration process is non-linear. In order to use the calibration curve to calculate the soil suction from the output voltage of the sensor, it is important to have a reasonably accurate characterization of the calibration curve. The following equation is proposed to fit the relationship between output voltage, ΔV, and soil suction, ψ.

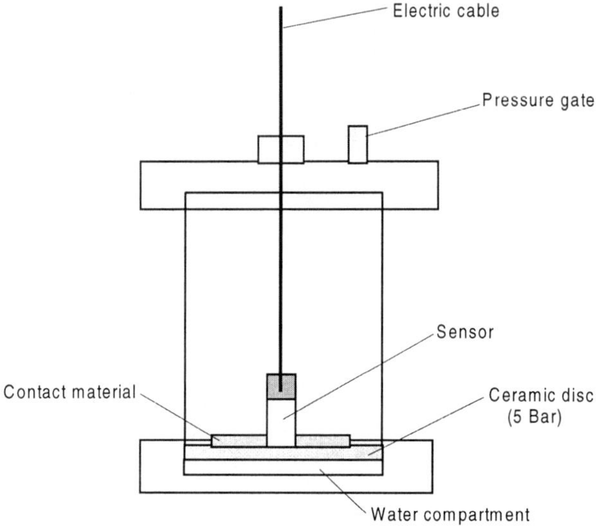

Figure 6 The physical layout of the test apparatus for calibration of sensors.

$$\psi = \left[\frac{b \cdot (\Delta V - a)}{c - \Delta V} \right]^{d} \qquad [1]$$

where:

a = parameter designating the output voltage under saturated conditions,
c = parameter designating the output voltage under dry conditions,
d = parameter designating the slope of the calibration curve, and
b = parameter related to the inflection point on the calibration curve.

A typical calibration curve for a sensor is shown in Fig 7 along with its parameter values. Since there are only four parameters in Eq. 1, only five calibration points are required to establish the calibration curve. As a result, the calibration process is simplified and the time required for calibration is significantly reduced. Equation 1 facilitates the calculation of the soil suction from the output voltage of the sensor and increases the accuracy of the suction measurement.

Typical results of Soil Suction Measurements

Tests were undertaken to verify the performance of the new sensors. Those tests were divided into two parts, laboratory tests and field tests. Typical results are presented in this section.

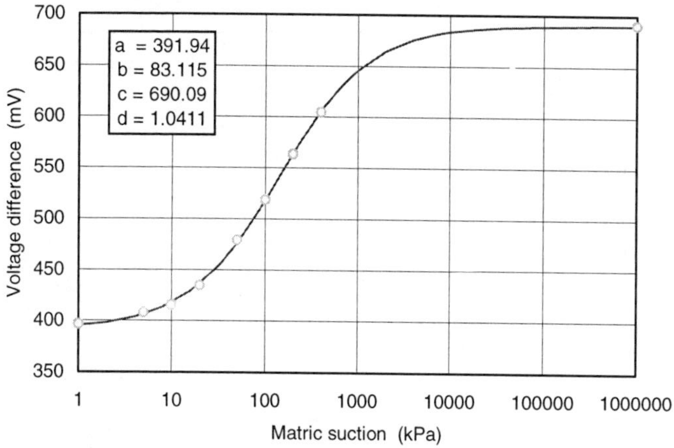

Figure 7 A typical calibration curve for a thermal conductivity sensor

Figure 8 presents soil suction results for laboratory measurements on an Indian Head till specimen. The soil specimen was compacted with a initial water content of 16%. Laboratory measurements were carried out using two sensors inserted from each end of a soil sample. One sensor was initially saturated while the other was initially air-dried. The sensors were inserted into predrilled holes in the sample. The sample with the installed sensors were wrapped in aluminum foil and sealed by duck tape to minimize moisture loss. The response of both sensors were monitored. The readings from both sensors were stable and showed that the same suction can be measured provided the effect of hysteresis of the ceramic is taken into account. The results indicate that the time required for the initially dry sensor to come to equilibrium with the soil specimen is less than the equilibrium time for the initially saturated sensor.

The new sensors have also been installed in the test cover site in the Key Lake area, Saskatchewan, Canada. The cover consisted of 0.7 m sandy till underlain by 0.6 m outwash sand.

A total of 22 new sensors were installed at the site. Twelve sensors were installed in a trench and another 10 sensors were installed in a observation shaft. The layout of sensor installation in the cover is shown in Fig. 9.

Data from sensors located at 14 cm, 35 cm, 58 cm and 115 cm from September 12 to September 18, 1997, are presented in Fig. 10 along with precipitation data. Soil suction data obtained from the sensors indicates that soil suction usually decreased after a heavy rain and then increased until the next rain event. The increase in soil suction was caused by evaporation from the cover surface. Therefore, the fluctuations in soil suction were most significant at the top of the cover and became less significant with depth (Fig. 10).

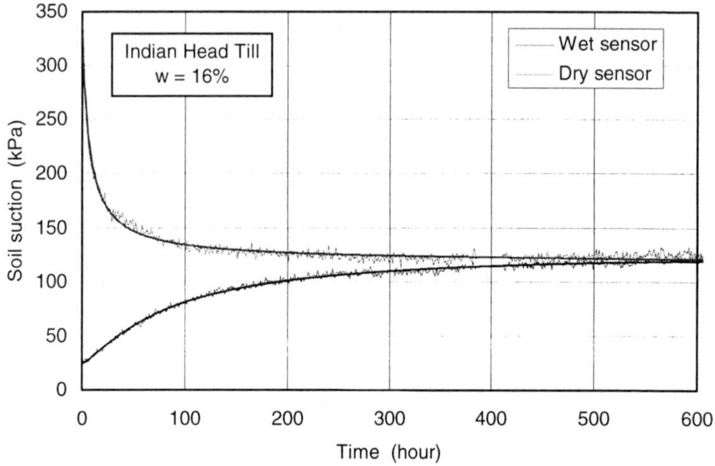

Figure 8 Laboratory measurements of soil suction on Indian Head till

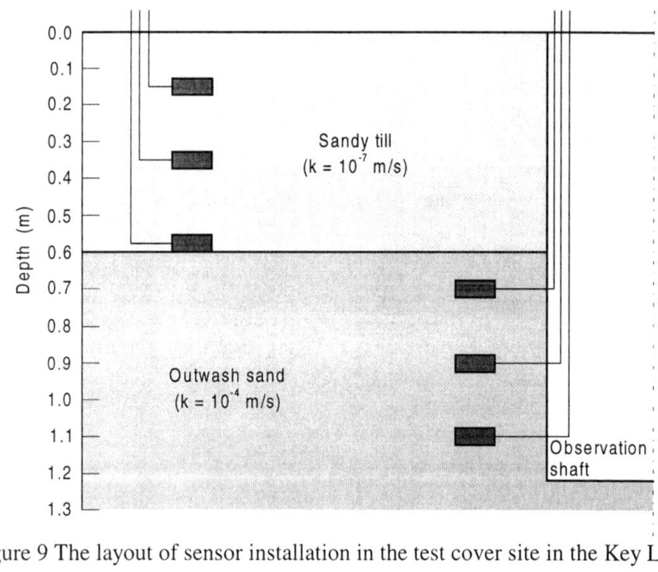

Figure 9 The layout of sensor installation in the test cover site in the Key Lake.

Conclusions

An improved thermal conductivity soil suction sensor was developed to eliminate the difficulties encountered with early versions of the device. A specially designed ceramic has been used as the ceramic tip for the new sensor. The ceramic has a high porosity and a wide range of pore-size distribution. The strength, durability and the range of suction measurement of the new sensor have also been significantly improved.

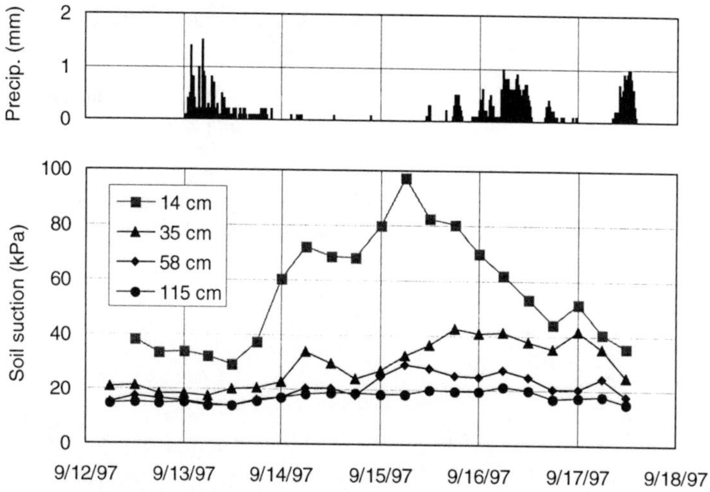

Figure 10 Soil suction measured in a test cover site

The stability, sensitivity and accuracy of the new sensor have been enhanced through use of superior integrated circuitry for the temperature sensing device, advanced signal conditioning, and a specially designed constant current sink device.

A laboratory procedure for the purpose of calibration has been set up. A calibration equation was proposed to facilitate in the calculation of soil suction from the output voltage.

The new thermal conductivity sensor has been found to be quite sensitive and accurate in measuring soil suction in the range from 5 to 1500 kPa with a coefficient of variation less than ±5% over the range. To-date, its performance has been satisfactory.

Acknowledgments

The authors wish to acknowledge the financial support from the Natural Sciences and Engineering Research Council of Canada (NSERC) and other seven organizations, Placer Dome Canada Ltd., Placer Dome Inc., Westmin Resources Ltd., Cominco Ltd., Cambior Inc., Rio Algom Ltd., and Saskatchewan Highways and Transportation.

References

Fredlund, D.G., Gan, J. K-M., and Li, W-X. 1994. Thermal conductivity suction sensors - design consideration, *Proceeding of the 13th International Conference on Soil Mechanics and Foundation Engineering*, New Delhi, India, pp. 291-296

Fredlund, D.G. and Rahardjo, H. 1993. *Soil Mechanics for Unsaturated Soils*, New York, NY; John Wiley & Sons, 486 pp.

Fredlund, D.G. and Wong, D.K.H. 1989. Calibration of thermal conductivity sensors for measuring soil suction, *ASTM Geotechnical Testing Journal*, vol. 12, no. 3, pp. 188-194.

Lee, R.K.C. and Fredlund, D.G. 1984. Measurement of soil suction using the MCS 6000 gauge, *Proc. 5th Int. Conf. Expansive Soils*, Inst. of Eng., Adelaide, Australia, pp. 50-54.

Evaluation of a Moisture Probe for Monitoring Soil Suction in Residual Soils

James B. Nevels, Jr.[1]
Member ASCE

Abstract:

The study of matrix and total suction versus depth is important in the analysis and design of slabs on grade such as light structure building slabs and highway pavements. This paper quantifies the seasonal moisture profiles with depth in two typical expansive Oklahoma clays. These soils can be characterized as being medium to highly plastic, blocky structured, with a medium stiff consistency, shallow in depth and residual. Following a detailed field investigation and laboratory soil analysis, the project involves some variation in typical access tube installation technique and moisture monitoring with a capacitance probe. Adjustments to the access tube installation appear to work well. The paper presents characteristic curve equations for different soil horizons which are used along with the monitored moisture profiles to produce suction versus depth changes with time in the field. To date, some change in suction within the active zone has been recorded.

1: Introduction:

The study of matrix and total suction versus depth is important in the analysis and design of slabs on grade such as a light structure building slab or highway pavement. Free field heave is a term applied to describe the heave of a slab due to change in soil suction only with no change in total stress. The displacement of slabs under this condition would occur just due to environmental effects upon the slab with no load applied to the surface. This condition is assumed to vary linearly from a maximum value of heave at the surface to zero at the depth of the active zone. The prediction of heave in a slab can be made by summing the volumetric strains due to the change in moisture

[1] Soils and Foundations Engineer, Oklahoma Department of Transportation / Materials Division, 200 NE 21st Street, Oklahoma City, OK 73105.

in the soil for layers through the depth of the active zone such as used in the clod test procedure (Nelson and Miller 1992). This change in soil suction used for the free field heave case can be made by studying the variation of in situ moisture with depth.

This paper studies the changes in in situ moisture with depth over time for two typical shrink-swell residual Oklahoma clay soils. By doing so, soil suction variability and ranges in different layers can be estimated. This will be beneficial in clarifying suction versus depth in the active zone. In addition, this study addresses the effects of access tube installation on moisture measurement and monitoring with a capacitance moisture probe.

2. *SITE CHARACTERIZATION*

The two soils selected for this study are identified for the purposes of discussion by their pedological soil series names, Huska and Doolin respectively. The Huska soil is located in southern Payne County and the Doolin in central Cleveland County. Both soils are located in central Oklahoma along I-35 in relatively undisturbed areas within the right of way line. In each case the ground condition consists of one percent sloped ground with short grass in place. They are located on high stream divides on uplands in the Central Red Rolling Prairie physicographic region of Oklahoma. Each soil has similar geographic settings in that the mean annual temperature is 16.7°C, the mean annual precipitation is 838.2 mm and the long term mean annual Thornthwaite moisture index is 5. (Soil Survey of Payne County, Oklahoma, 1987) and (Soil Survey of Cleveland County, Oklahoma, 1987).

2.1 Pedological and Geological Description

The Huska series consists of deep moderately well drained, very slowly permeable soil that formed in interbedded shale and sandstone. The Doolin series is identical except that it forms from shale and generally is the deeper residual soil. Ground slopes can range from 1 to 5 percent for the Huska and 0 to 3 percent for the Doolin. On the very flattest ground slope for the Doolin series the upper part of the soil profile may form in old alluvium. Both soils can collectively be characterized as brown to reddish brown mottled, medium to high plasticity, blocky structured, medium stiff to hard residual clays with a high shrink-swell potential. Both series are known for seasonal perched watertables. The taxonomic class for the Huska and Doolin soils are respectively Fine mixed superactive, thermic Mollic Natrustalfs; and Fine, smectitic, thermic Typic Natrustolls. Implications of these taxonomic classifications are as follows:

c) Fine, mixed, superactive, thermic Mollic Natrustalfs: This soil is an Alfisol. The mean annual soil temperature is 15°C or higher but lower than 22°C. The moisture regime is ustic, and the moisture control section is dry in some or all parts for 90 or more cumulative days in normal years. The clay mineralogy is mixed and has a ratio of cation exchange capacity to clay of 0.6 by weight. All 'B' subhorizons are natric. The Btnz1 and Btnz2 have an accumulation of salts more soluble than gypsum.

d) Fine, smectitic, thermic Typic Natrustolls: This soil is a Mollisol. The mean annual soil temperature is 15°C or higher but lower than 22°C. The moisture regime is ustic, and the moisture control section is dry in some or all parts for 90 or more cumulative days in normal years. The clay mineralogy is smectitic. All 'B' subhorizons are natric.

The water state class based on the Soil Survey for both soils is moist which has three subdivisions and a soil suction ranging from 1 to 1500 kPa. As indicated at times theses soils are at a dry water state class which also has three subdivisions and a soil suction greater than 1500 kPa. The exchangeable sodium (percent) in the B subhorizons varies from 22 to 53 and from 15 to 18 respectively for Huska and Doolin soils.

Underlying the Huska soil at this particular site is the alternating siltstone and shale of the Wellington-Admire geologic unit and for the Doolin soil is underlying shale of the Hennessey geologic unit. Both geologies are of Permian age. The blocky and/or prismatic soil structure transitions into a platy structure for the weathered siltstone and shale. The appropriate reference for the soil and shallow geology is the (Soil Survey of Payne County, Oklahoma, 1987) and (Soil Survey of Cleveland County, Oklahoma, 1987)

2.2 Borings

For this study a continuous Standard Penetration Test (SPT) boring was made at each site by air rotary drilling using a CME 75 drill rig with a 88.9 mm stepped wing bit according to ASTM D 1586-84 standard. Soil samples were broken into close intervals for Atterberg limit and moisture content tests and stored in 0.15mm thick plastic bags. These SPT borings were terminated at refusal according to ASTM D 1586-84 subsection rule 7.2.2, 50 blows in any one 152.4 increment. At the end of drilling and at 24 hours no watertable was observed in either SPT boring. However, during the course of this study wet walls in boring were observed at these locations. The SPT borings served as the focal point for the location and reference of all remaining borings and in situ tests. In two separate 1.52 m offset borings at each site from the respective SPT borings, 75 mm O.D. continuous thin-walled tube samples were taken for further laboratory tests according to ASTM D 1587-94 standard. The site stratigraphy and index properties for the Huska and Doolin soils are presented in Figures 1 and 2 respectively. It was noted during the boring that a soil structure change occurs at approximately 1.49 m from a blocky to platy condition for the Huska soil. Also Figure 1 indicates at approximately 1.4 m the zone of seasonal fluctuation may stabilize. For the Doolin soil there is a very abrupt change in soil structure from a blocky to a weak coarse prismatic condition at approximately 0.8 m.

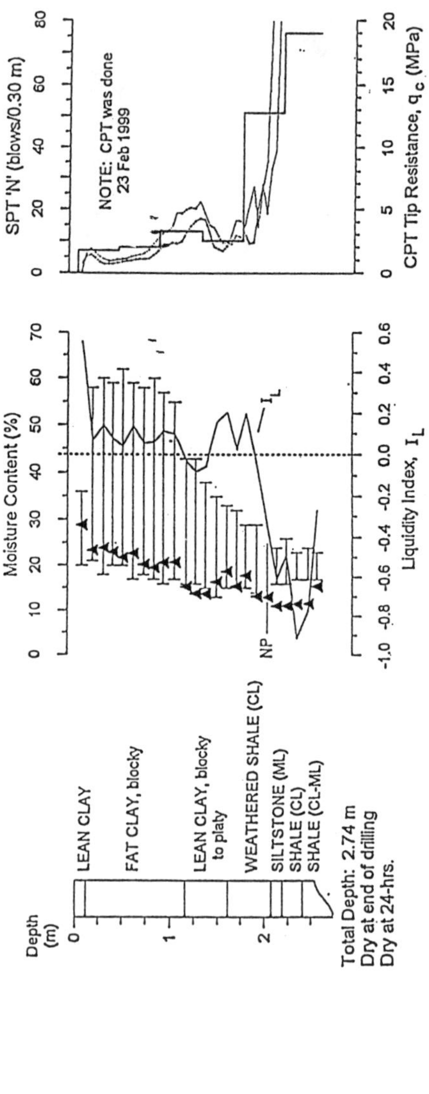

Figure 1. Stratigraphy of the Huska series: Boring No. 1, 9 April 1998

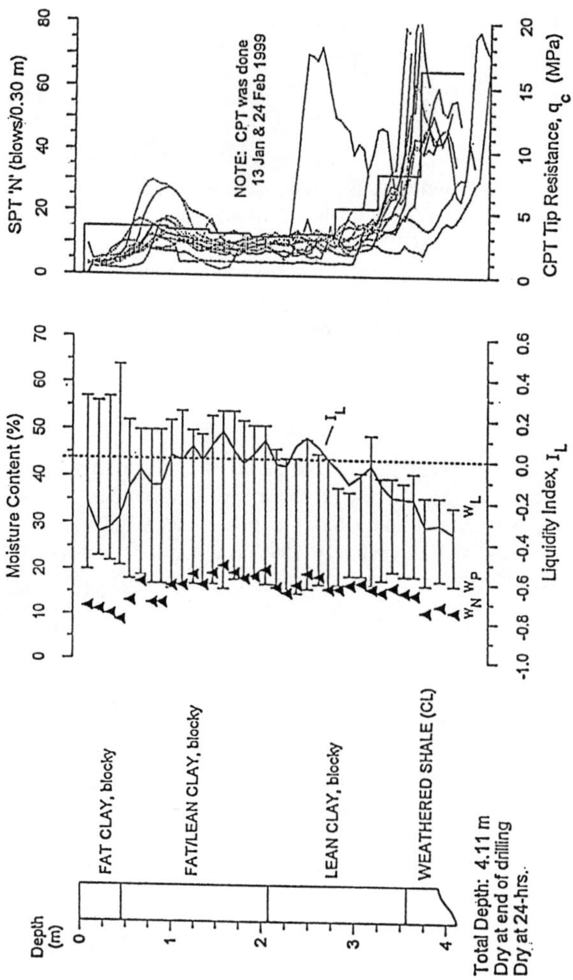

Figure 2. Stratigraphy of the Doolin series: Boring No. 9, 23 July 1998

The zone of seasonal fluctuation tends to stabilize at approximately 2.2 m. For the auger borings used to install the PVC access tubes (to be discussed in more detail in a later section), they were conducted according to ASTM D 1452-80 standard. Again the auger borings along with offset pushtube borings are arranged in a grid pattern around the SPT boring with appropriate consideration for sampling interference. The reference for all ASTM field sampling and testing procedures is (ASTM Volume 04.08, 1998).

2.3 In Situ Tests

The location of the Huska and Doolin soils were selected within very large mapped areas for each soil series. Still it is considered necessary to perform some in situ testing to verify the consistency of the soil profile laterally and with depth. The electric cone penetration test (CPT) without pore pressure measurement conducted according to ASTM D 3441-94 standard was used to accomplish this task. Figures 1 and 2 present the range of tip resistance (q_c) profiles with depth. The reference for all ASTM in situ testing procedures is (ASTM Volume 04.08, 1998).

3. LABORATORY ANALYSIS

Laboratory analysis consisted of two parts. The first part was to document the index properties and in situ density for these two residual soils. The second part was to study the in situ total and matrix suction. The reference for all ASTM laboratory test was (ASTM Volume 04.08, 1998) and that for the AASHTO test was (AASHTO Part II, 19th Edition, 1998).

3.1 Index Tests

The site stratigraphy and index properties are presented graphically in Figures 1 and 2 with the index and moisture test conducted according to ASTM D 4318-93, D 422-63(1990) and D 2216-96 standards respectively. Also shown is the liquidity index calculated for each site suggesting a preconsolidated stress history which is documented for Oklahoma residual soils derived from Permian geology primarily due to desication (Means and Parcher, 1968). The in situ wet density was determined with depth at each site from thin-walled tube samples according to AASHTO T 233-86 (1996), reference Figure 3a and 3b. From the wet density and associated moisture, a volumetric moisture (θ) can be calculated which is needed in the correlation between the frequency count (D reading) measured by the capacitance probe to be discussed in a later section.

3.2 Suction Tests

The in situ total and matrix suction was measured for these residual soils using the filter paper method according to ASTM D 5298-94 test standard. The

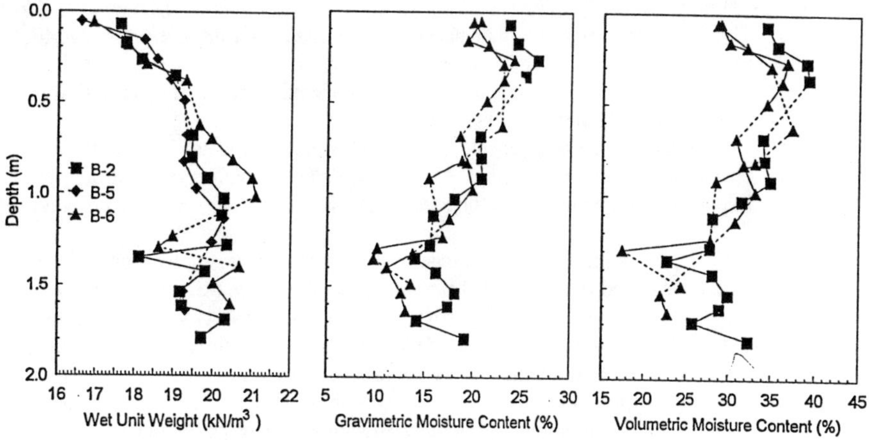

Figure 3a. Unit weight and moisture profiles at Huska series site

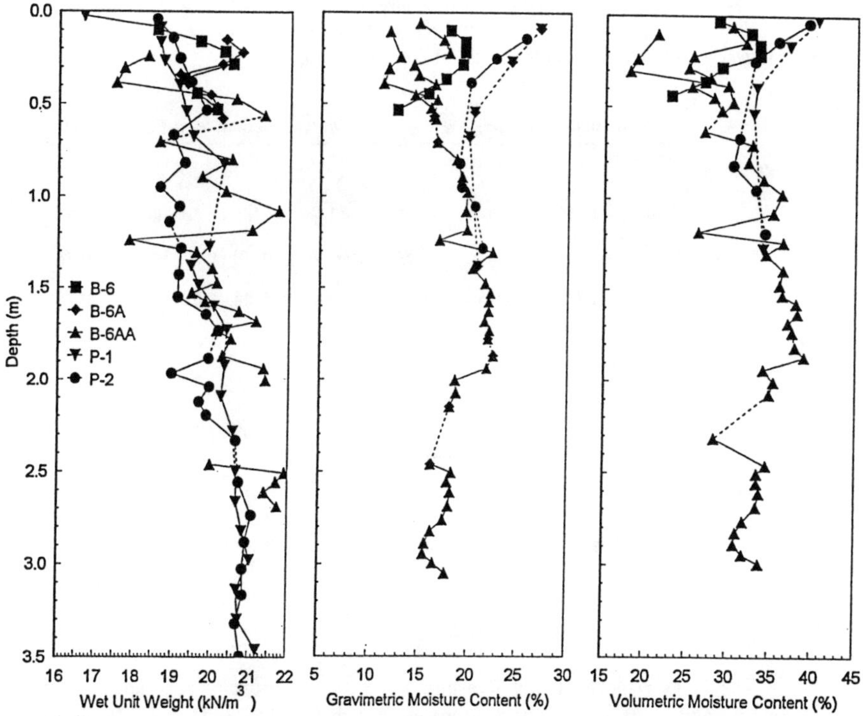

Figure 3b. Unit weight and moisture profiles at Doolin series site

procedure used deviated from the D 5298-94 apparatus requirements in four ways (Nevels 1996):

1. The Fisherbrand 9-790A filter papers used in the tests were not pretreated to prevent organism growth or biological decomposition.
2. A 230-ml polyethylene specimen container having a screw-top lid seal was used instead of a metal or glass container.
3. Two wraps of masking tape 12.7 mm wide were used instead of the flexible plastic electrical tape to further seal the outside lid / container connection.
4. Expandable polyethylene netting was used instead of screen wire or brass discs.

All ASTM D 5298-94 procedure requirements were met. The in situ total and matrix suctions were measured on thin-walled tube samples taken from the offset borings to the Huska and Doolin SPT borings. These particular suction tests were carried out on approximately 45-mm thick X 75 mm OD samples split in half with three stacked filter papers between the sample halves to measure the matrix suction and two stacked filter paper placed on the top sample half all sealed in a 230-ml container. The filter paper moisture contents following equilibrium were converted to suction measurements using the Schleicher and Schuell No. 589 calibration curve adopted by the ASTM standard. A moisture characteristic curve was measured for samples at several different depths from the 'A' and 'B' horizons for the Huska soil and for the 'B' horizon for the Doolin soil, see Table 1 for equations.

	Total Suction, h_T (kPa)	**R^2**	**Matric Suction, h_M (kPa)**	**R^2**
Huska Series				
"A" horizon	$\log(h_T) = 4.9572 - 0.1344w$	0.87	$\log(h_M) = 5.6485 - 0.2042w$	0.81
"B" horizon	$\log(h_T) = 5.6565 - 0.1237w$	0.86	$\log(h_M) = 6.0370 - 0.1615w$	0.88
"C" horizon	$\log(h_T) = 4.2191 - 0.0748w$	0.79	$\log(h_M) = 6.5524 - 0.3731w$	0.75
Doolin Series				
"B" Horizon	$\log(h_T) = 5.4465 - 0.1072w$	0.92	$\log(h_M) = 5.8380 - 0.1573w$	0.96

R^2: Correlation Coefficient
w: gravimetric moisture content, (%)

Table 1. Soil-moisture characteristic equations

4. *MOISTURE PROFILE MEASUREMENT*

The measurement of soil moisture has been traditionally accomplished by gravimetric and neutron probe techniques as specified in ASTM D 2216-96 and D 6031- 96 respectively (ASTM Volumes 04.08 and 04.09, 1998). The gravimetric method is most common and accurate but is less desirable because of the time and labor in physically removing and testing samples. The neutron probe is the best known indirect method of in situ moisture measurement. Its main fault besides being nuclear is that the probe averages moisture over a large sphere of influence at least 200 mm around the source of neutrons. The applicability of using an electrical property such as the dielectric constant of a soil and water matrix to measure field moisture content is a more recent indirect technique. The advantages are that it is non-nuclear, portable, has consistent quick readings and improved measurement resolution.

4.1 Capacitance Probe

The dielectric constant of soil material is proportional to the capacitance of the soil and is a measure of its electric dipole moment per unit volume. For most common solid constituents of soil the dielectric constant ranges from 2 to 4. The dielectric constant of free water on the other hand ranges from 78 to 81. Since water has a much higher dielectric constant (or capacitance) than the solid constituents of soil, a measure of the capacitance serves as a measure of the moisture content of the soil. The capacitance of soil increases with an increase in the number of soil water molecules which are free to relax as their electric dipoles respond to the field reversal of the probe electrodes (R.M. Mean et al., 1994) The relationship of soil moisture to the dielectric property (capacitance) is presented in (Selig and Mansukhani, 1975). The Troxler Sentry Model 200 AP capacitance probe was used in this study (Sentry 200AP User's Manual, 1992). This unit uses frequency domain reflectometry to make soil capacitance measurements. A pair of electrodes is inserted into the soil. The soil acts as the dielectric completing a capacitance circuit. As high frequency radio waves are pulsed through the capacitance circuitry, a natural resonant frequency is established which is dependent upon the soil capacitance. The instrument measures a frequency difference or shift in resonant frequency (D reading) which is related to the dielectric constant by the geometry of the electric field established around the probe electrodes. The frequency difference (D) determines the soil moisture according to the Sentry 200-AP User's Manual using an exponential equation in the form of Equation 1 shown below.

[1] $$m = 1/F_1 \ln(DF_2)/F_o$$

where F_1, F_1, F_2 are factory calibration constants, D is frequency difference and in the emitted and received frequency between the two electrodes m represents the volumetric soil moisture. A linear calibration may also be used according to the Manual. The procedure for the use of this probe is given by (Mead et al., 1994).

4.2 Access Tube Installation

The capacitance probe uses an access tube similar to that of the neutron probe technique for measuring volumetric soil water content with depth. The probe is lowered into the access tube and the frequency difference between the emitted and received frequency is measured. A non-metallic access tube is required by the capacitance probe used in the study. A PVC pipe with the following tolerances was used: I.D. 51.82-52.83 mm, O.D. 59.69-60.96 mm. It has been stressed in all reported literature that careful access tube installation is required for successful operation of this operation of this technique. Minimal soil disturbance and a snug fit of the PVC casing in the soil is required. The method of installation recommended is the hand augering through the inside diameter of the access tube and then driving the tube in a few millimeters at a time. To further facilitate installation, a specially arranged and anchored tube guide and beveling of the access tube inward is recommended. As noted in Figures 1 and 2, these soils are medium stiff to hard with depth based on the SPT 'N' resistance. It was decided to experiment with the installation technique. The method used first in this study was to first take continuous 50.8 mm O.D. thin-walled samples with the hydraulics of the drill rig and then push hydraulically with the drill rig the PVC access tube (mean O.D. 60.3 mm) into the smaller 50.8 mm O.D. hole squeezing the soil outward. The second method of access tube installation was to hand auger a 95.3 mm hole and insert the 60.3 mm PVC and place a neat cement grout (1:1 ratio of cement to water) in the annulus. Figure 4 shows the effect on the capacitance probe D readings based on installation technique. Also indicated in Figure 4, a deep installation at the Doolin site was purposely made to show the effect of a small void space around the annulus of the access tube. First a hollow thick-walled 3.05 m long by 60.3 mm diameter N-size drill rod was hydraulically pushed into the ground. Following this operation, the PVC access tube was carefully inserted in the hole. The result was a reasonably snug fit at the bottom with variable annular void space up toward the ground surface. All access tube ends were sealed with O-ring tips. Each access tube installation was protected at the ground surface with a leak-resistant manhole cover.

4.3 Moisture Monitoring and Field Calibration

Moisture content monitoring was made at 121.92 mm intervals with the capacitance probe on a random basis at each site as time would permit. It was found convenient to develop the data base for the moisture monitoring in terms of the 'D' readings, see Figures 5 to 8. There were instances where rainwater infiltrated through the manhole coverseal and filled the access tube with water, and it was handled by swabbing out the access tube with a sponge attached to 19.05 mm diameter PVC pipe. As mentioned in section 4.1, the exponential and linear correlations between the volumetric moisture and the D readings are recommended by and already programmed in the equipment by the manufacturer. Field calibrations were made at each site between paired data (calculated volumetric moisture from Figure 3 and initial D readings) for various depths to improve the accuracy of the calibration in converting the D reading to a gravimetric moisture content. These calibrations were based on the 50.8mm I.D. thin walled samples and the calibrations were then applied

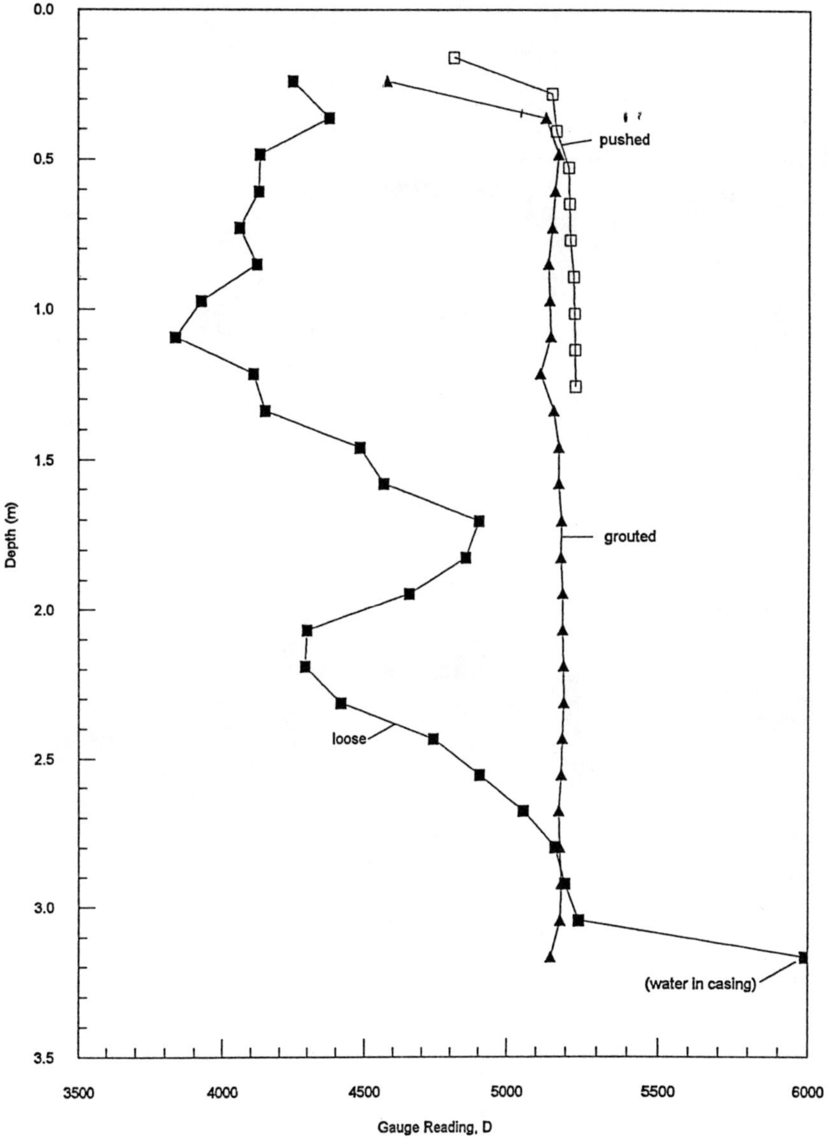

Figure 4. Comparison of installation methods, Doolin series site, 18 December 1998

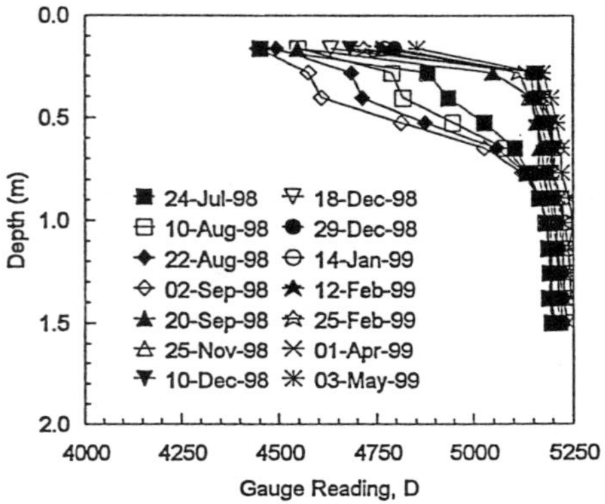

Figure 5. Gauge reading vs. depth, pushed tube at Doolin series site

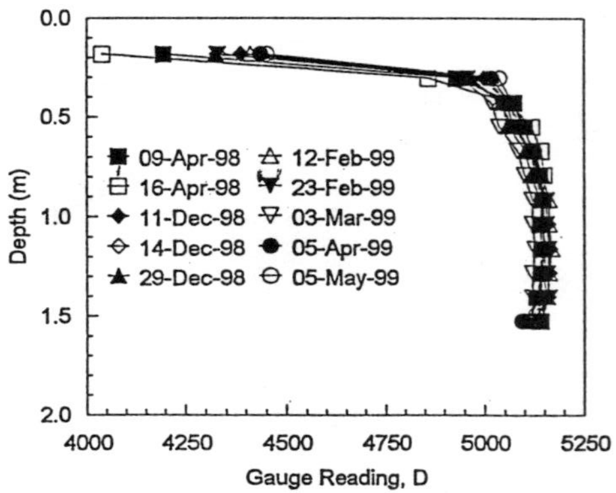

Figure 6. Gauge reading vs. depth, pushed tube at Huska series site

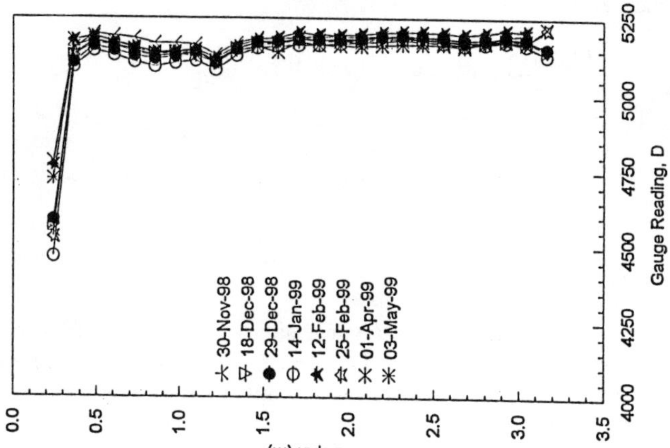

Figure 7. Gauge reading vs. depth, grouted tube at Doolin series site

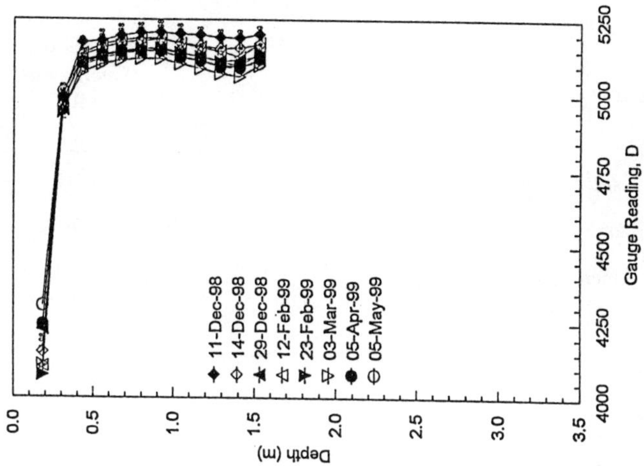

Figure 8. Gauge reading vs. depth, grouted tube at Huska series site

to the pushed, hand-augered and grouted installations. The correlation coefficient (r^2) did not compare favorably with what has been reported in the literature (Jooste and Scullion, 1995)(Yoder et al., 1997). More advanced statistical analyses in the form of quadratic and multilinear regression were tried without success, see Table 2.

Huska Series		**Correlation Coefficient R^2**
"A" horizon		
Linear:	$D = 6335 - 42.15\theta$	0.94
"B" horizon		
Linear:	$D = 5468 - 10.53\theta$	0.47
Doolin Series		
"B" horizon		
Linear:	$D = 4754 + 12.39\theta$	0.51
Exponential:	$D = -5227e^{(-0.14\theta)} + 5225$	0.67

D: moisture gauge reading
θ: volumetric moisture content, (%)

Table 2: Calibration of moisture probe.

5. *DISCUSSION*

A pedological soil survey and a conventional geotechnical site assessment indicate varying consistency with depth for these soils with regard to surface soil horizons, engineering index properties, in situ density, and SPT N-resistance, and CPT q_c resistance. Both soils normally exist in an unsaturated condition. The water state class is generally moist but can also be dry depending on the seasonal wet periods. The moisture monitoring in terms of the capacitance probe D reading appears to be consistent to date with little change. The theory of the capacitance suggests that one can use this particular probe to convert the D readings to volumetric moisture. The results shown in Table 2 from our study on these shallow residual soils do not support the use of this approach in making precise moisture readings with depth based on in situ calibration attempts. The D reading although not precise does show some qualitative change in moisture with depth from the time of installation. Preceding the Doolin soil access tube installation a very wet season had occurred. From July 24, 1998, the ground was in a drying out phase until mid September when some rain again occurred, see moisture content profile in Figure 2 and the July 24, 1998, to September 2, 1998, D reading profiles in Figure 5. At the Huska site quite the opposite had occurred. The moisture profile in Figure 1 shows a wetter in situ condition near the ground surface matched by D reading profiles showing essentially no change. Only marginally light rainfall occurred during this period, see Figures 1 and 6. The range of the D readings in the grouted access tube installations shown in Figures 7-8 appear to match the snug fit access tube installation D readings. Finally the D readings do seem to indicate the approximate depth of seasonal moisture fluctuation in Figure 5. The problem of calibration of the probe may be due primarily to the variation in computed volumetric moisture content as indicated by the range in in situ bulk density seen in Figure 3 (Snethen, Pers. Comm). The effect of bulk density with depth can be eliminated by calibration when the density remains constant (Kuráž 1982). Selig and Mansukhani (1975) pointed out the following: a) The relationship of a composite dielectric constant of a soil sample to moisture per unit volume will be nonlinear and soil type dependent, b) the dielectric constant decreases with increasing measurement frequency, thus decreasing the sensitivity to moisture change c) the measurement of a frequency change of an oscillator circuit may be simpler and more reliable for field monitoring than a radio frequency bridge, and d) an unsaturated in situ clay can be modeled as a network of many parallel - series combinations of resistors and capacitors such that the apparent capacitance of the soil will change with the measurement frequency. This is considered the case even if the dielectric for each element remains constant.

The use of volumetric water applied primarily in agriculture and the geo-sciences is based on total volume. Since some soils shrink and swell with wetting and drying, the total volume may change. Another variable in moisture measurement is the size of sample. The capacitance probe measures a disk shaped volume of soil 101.6 mm high by 254 mm in diameter according to the manufacturer. Substracting out the access tube, yields a soil sample of about 8 kg compared to approximately a 200 g sample from the bulk density specimen (McKeen, Pers. Comm.). It appears to date that the calibration of this particular probe has been restricted to repacked samples (Mead et al., 1994)(Yoder et al., 1997)(Jooste and Scullion, 1995). Jooste and Scullion (1975) also report that a laboratory calibration is required if volumetric moisture content range dies not have at least a 15 percent variation then the data will not fit a regression line.

6. *SUCTION VERSUS DEPTH*

To predict suction versus depth over time, the approach presented in this paper consisted of the following steps:

a. Calculate the volumetric moisture from the calibration curves shown in Table 2 for each D regarding depth in the respective 'A' and 'B' horizons

b. Convert these volumetric moisture contents to gravimetric moisture (w) through the following relation derived by fitting a curve to measured gravimetric and calculated moisture contents:

$$2. \quad w = \frac{2.38 \log \Theta}{10.54 \log \Theta}$$

c. Use the gravimetric moisture in the appropriate characteristic curve equation.

As an example, the total suction versus depth for the Doolin soil presented in Figure 5 using the exponential calibration equation from Table 2 for both 'A' and 'B' horizons is given in Figure 9. By consequence the SPT moisture profiles for the Huska and Doolin soils represent the practical range in moisture variation seen for residual soils of Permian geology in Central Oklahoma. The zone of seasonal moisture fluctuation was estimated at 1.4 m for the Huska soil and 2.2 m for the Doolin soil. Applying the moisture versus depth profiles to the Doolin characteristic curve from Table 2 shows a probable boundary in total suctioin versus depth.

7. *CONCLUSIONS*

There appears to be an inadequate correlation between the frequency shift (D reading) and the volumetric moisture for this probe based on field collected samples and calibration for these residual soils. From the data collected the variability in bulk density in the field and resulting volumetric moisture content is apparently a major contributing factor for this. The behavior of unsaturated clay is very complex with regard to dielectric permittivity. The positive aspects of this probe are that it is relatively easy to use, portable and the frequency differences (D readings) appear to be consistent over time. The study of D readings versus depth may provide an effective way of monitoring the zone of seasonal fluctuation.

The major issue for clay soils in geotechnical engineering applications is in developing a calibration technique that takes into account the non-linearity over a wide moisture range and one that improves the sensitivity to measurement. For easier application of a capacitance probe it would be more appropriate to develop the characteristic curve in terms of volumetric moisture versus suction. At this stage there appears to be some qualitative value in using this probe. Determining changes in in situ soil suction based on probe D readings using a laboratory calibration needs further study.

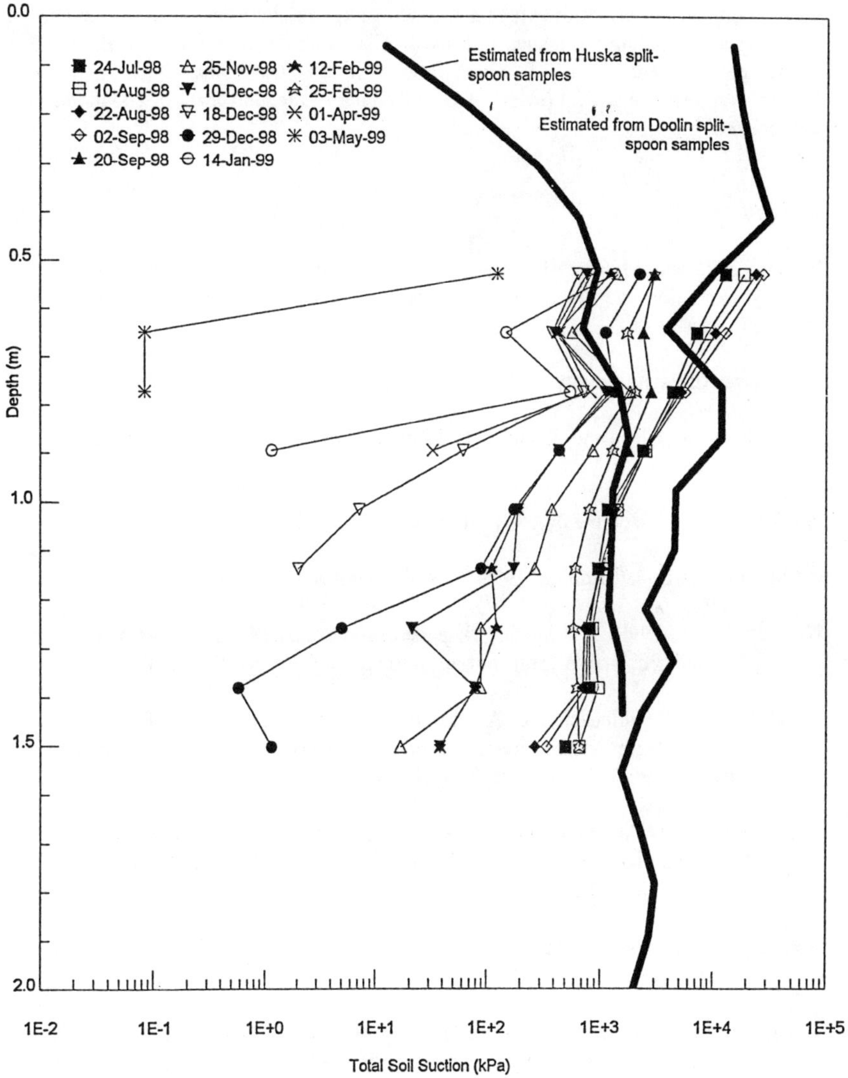

Figure 9. Estimated total suction profiles in the Doolin series, "B" horizon

ACKNOWLEDGMENTS

The author wishes to extend a special thanks to Shawn Garza for his patience in typing and proofing the manuscript and to Christopher Clarke for calculation and figure development, both of whom are with the Oklahoma Department of Transportation. The critical review of Gerald Miller of the University of Oklahoma is appreciated

REFERENCES

Nelson, J.D. and Miller D.J., 1992. Expansive Soils Problems and Practice in Foundation and Pavement Engineering. New York: John Wiley & Sons, Inc. 259 pp.

Soil Survey of Payne County, Oklahoma, 1987. USDA, Soil Conservation Service.

Soil Survey of Cleveland County, Oklahoma, 1987. USDA, Soils Conservation Service.

ASTM, Section 4, Volume 04.08, 1998. Soil and Rock (I).

AASHTO, Part II, 19th Edition, 1998. Test Procedures.

Nevels, Jr., J.B., 1995. Soil Suction Dispersion and Statistical Analysis with Depth at the Medford Airport, Transportation Research Record 1596, 76-83.

Mead, R.M., I.C. Paltineanu, J.E. Ayars and J. Liu, 1994. Capacitance Probe Use in Soil Moisture Measurement, American Society of Agriculture Engineers Paper No. 94-2122, St. Joseph, Michigan, ASAE.

Selig, E.T. and A.M. Mansukhani, 1975. Relationship of Soil Moisture to the Dielectric Property, Journal of the Geotechnical Engineering Division, ASCE, Vol. 101, No. GT8, 755-770.

Sentry 200-AP User's Manual, 2nd Ed., March 1992. Troxler Electronics Laboratories.

Jooste, J. and T. Scullion, 1995. Monitoring of Vertical Moisture Barriers Using Troxler Sentry 200-AP Device, Transportation Research Record 1481, 10-20.

Yoder, R.E., D.L. Johnston, J.B. Wilkerson, and D.C. Yoder, 1997. Soil Water Sensor Performance, Soil and Water Division, ASAE, November, 121-133.

Snethen. Personal Communication, 1999.

Kuráž, V., 1981. Testing of a Field Dielectric Soil Moisture Meter, ASTM Geotechnical Testing Journal, Vol. 4 No.3, 111-116.

McKeen. Personal Communication, 1999.

Utilization of SPT-T, CPT and DMT Tests to Predict The Ultimate Bearing Capacity of Precast Concrete Pile in Brazilian Unsaturated Residual Soil

Anna S. P. Peixoto[1], Unicamp, Paulo J. R. de Albuquerque[2], USP, and David de Carvalho[3], Unicamp.

Abstract

In situ tests, mainly the Standard Penetration Test, SPT, are commonly used in Brazilian foundation practice. This paper presents the estimating of ultimate bearing capacity of a 14.0m deep instrumented precast concrete pile with 0.18m diameter, that was driven in an experimental research site at the University of Campinas - Unicamp. The soil of the site is characterized as residual of Diabase. The soil is unsaturated up to 18.0m depth and its geotechnical characterization obtained by laboratory and field tests is presented here. The skin friction and point resistance were obtained by an instrumented load test. The load test results were compared with results obtained by methods to predict the bearing capacity of the foundation element using the Standard Penetration Test with torque measurement, SPT-T, Cone Penetration Test, CPT, and Flat Dilatometer Test, DMT, results.

Introduction

To calculate the bearing capacity of foundations, field tests like SPT, CPT and DMT are commonly performed. In Brazilian practice, mainly the SPT and the CPT have been used over the last decades. Nowadays the DMT and SPT-T are also used by some engineers.

[1] Civil Engineer, M.Sc., Unicamp, Feagri, DConru, Caixa Postal 6011, Campinas/SP, Brazil, CEP 13083-970. annasil@agr.unicamp.br

[2] Civil Engineer, M.Sc., POLI-USP, Oito de Maio,90, Sorocaba/SP, Brazil, CEP 18090-140. pjrocha@facens.br

[3] Professor, Dr, Unicamp, Feagri, DConru, Caixa Postal 6011, Campinas/SP, Brazil, CEP 13083-970. david@agr.unicamp.br

The SPT-T test, Standard Penetration Test with torque measurement, has been used in Brazil since 1991. The test provides a static resistance parameter in an unexpensive way. The test has been done as follows: after penetration of the sampler with the count of blows, an adapter is placed which allows the joining of the torquemeter. Afterwards, rotation on the group rod-sampler is applied by using the torquemeter. The maximum torque is measured and the rotation continues without interruption until the torque stands constant, when the residual torque is obtained.

This paper presents the analysis of the applicability of some formulas using these tests to calculate the ultimate bearing load of precast concrete piles driven in a residual unsaturated soil. For the comparison of the results a static load test was carried out in a 14.0m long precast concrete pile with a diameter of 0.18m. The pile shaft was instrumented with strain-gages, which allowed the obtainment of the point load and the skin friction during the load test. The Diabase residual soil of the site presents a 6.0m thick upper layer of sandy silty clay, over a sandy clayey silty.

Subsoil

The subsoil in the place is characterized by two layers of soil over the saprolitic material from the diabase rocks from the Serra Geral group. The 6.0m thick upper layer is composed by highly porous, lateritic sandy silty clay. The second layer is composed of an 18.0m sandy clayey silty. The water table is located at 18m depth. The two layers of lateritic and saprolitic soils have differences in the mechanical behavior that depend on the degree of weathering. Both represent the called Tropical Soils.

Figure 1 presents the data from the SPT and CPT tests. Table 1 shows the results of a Cone Penetration Test (CPT) and a Standard Penetration Test with torque measurement (SPT-T). These tests are carried out 2.0m from the pile center, before it is driven. The q_C and f_C means the cone tip resistance and the cone friction. The "N_{SPT}" means the numbers of blows, "T", the torque measurement, both of the SPT-T test. The f_T means the adhesion soil-sampler obtained by Ranzini (1994), as the following equation:

$$f_T = \frac{T}{(41.336h - 0{,}032)}$$

where f_T (kPa), adhesion soil-sampler; T (m.kN), torque measurement; h (m), length of penetration of sampler.

Table 2 shows the geotechnical parameters obtained from laboratory tests performed in undisturbed samples obtained from a pit excavated to 16.0m. The "γ_{NAT}" means bulk density; "γ_S", density of the solid particles; "w", the natural moisture content; "e", voids ratio; "n", porosity; "c", cohesion and "ϕ", angle of shear resistance, both obtained by consolidated undrained test.

Table 3 shows the results from the DMT test, in which "ID" means material index; "Kd", horizontal stress index and "ED", dilatometer modulus.

Table 1. Results from SPT-T and CPT tests.

Soil	Depth (m)	q_C (kPa)	f_C (kPa)	N_{SPT}	T (N.m)	f_T (kPa)
Sandy Silty Clay	1.0	*	*	1/25	20	10,98
	2.0	250	14.0	1/25	20	10,98
	3.0	633	16.3	2	20	10,98
	4.0	1300	33.3	2	20	10,98
	5.0	1500	77.3	5	40	21,97
	6.0	3767	80.0	8	80	43,93
Sandy Clayey Silt	7.0	2375	148.0	4	70	38,44
	8.0	2400	157.3	6	80	43,93
	9.0	2067	173.7	7	100	54,92
	10.0	2300	190.0	7	90	49,42
	11.0	2533	218.0	9	90	49,42
	12.0	2733	265.3	10	130	71,39
	13.0	2750	278.5	8	160	87,86
	14.0	2633	281.7	8	140	76,88
	15.0	2433	220.3	8	140	76,88

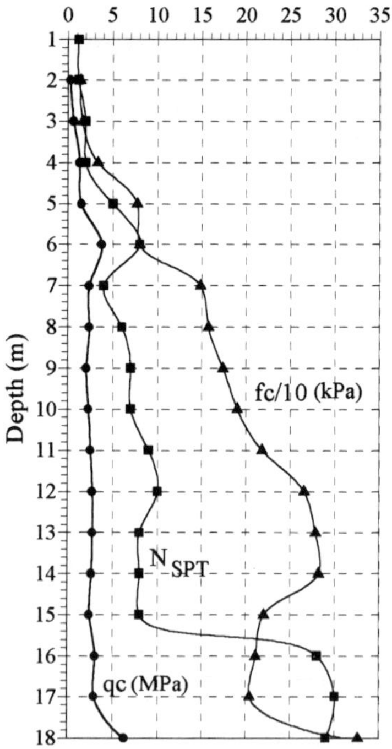

Fig. 1. Results from SPT and CPT.

Table 2. Geothecnical Characteristics of the Subsoil

Depth (m)	γNAT (kN/m^3)	γ_s (kN/m^3)	w (%)	e	n (%)	c (kPa)	ϕ (degrees)
1.0	13.4	29.7	24.3	1.77	63.8	5	31.5
2.0	13.0	29.1	23.4	1.76	63.7	11	31.5
3.0	13.0	29.5	22.8	1.79	64.1	2	30.5
4.0	13.0	30.1	23.7	1.86	65.0	0	26.5
6.0	15.4	30.1	24.6	1.44	59.0	18	18.5
7.0	15.4	29.1	26.3	1.40	58.2	31	22.5
8.0	14.8	29.5	28.1	1.56	60.1	18	25.5
9.0	15.0	30.1	29.9	1.60	61.5	60	20.1
10.0	15.1	30.1	30.5	1.60	61.6	78	22.8
12.0	16.1	29.6	33.8	1.46	59.4	87	18.3
14.0	16.4	30.6	32.8	1.48	59.7	76	19.1
16.0	16.7	30.1	39.2	1.51	60.1	55	22.0

Table 3. Results from DMT test.

Depth (m)	ID	Kd	ED (bar)
1	0.38	2.0	12
2	0.45	1.6	16
3	0.38	1.6	16
4	0.91	0.9	25
5	0.74	1.1	30
6	1.08	1.0	28
7	1.00	0.7	48
8	1.03	0.7	68
9	1.68	0.7	165
10	1.77	1.4	201
11	1.17	2.4	176
12	1.19	1.9	187
13	0.95	2.3	141
14	0.74	2.0	134
15	0.88	1.5	99
16	0.61	1.5	99
17	0.50	1.8	110
18	0.49	2.9	110
19	0.45	1.1	114

Load Test

The load test was of the Slow Maintained Load Type, following the determinations of the Brazilian code NBR12131/91 (ABNT, 1991). The instrumentation comprised strain-gages installed in full bridge in steel bars. The concrete pile was built with a steel pipe inside of 40.0mm diameter, in which the instrumented steel bars were installed after the pile was driven. The 14.0m pile had two 8.0m and 6.0m segments which were welded. To drive the pile, a

1,650Kg mass falling from a 0.50m height was used. The load-settlement curve and the load-transference curves obtained are presented in figure 2 and figure 3.

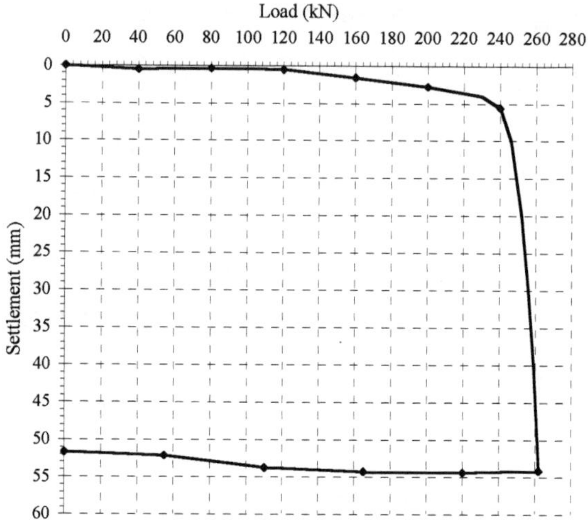

Fig. 2. Load-Settlement Curve.

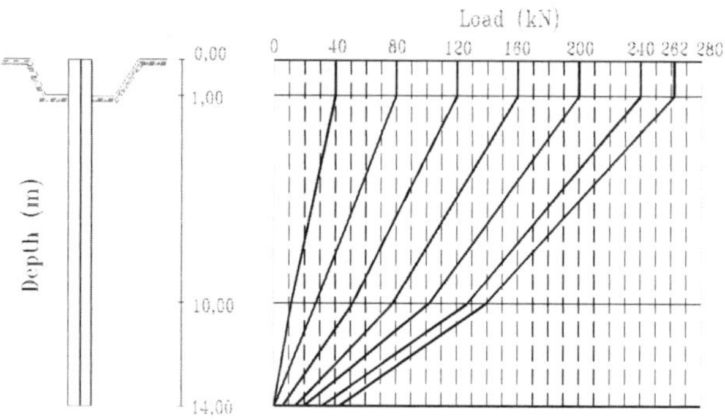

Fig. 3. Load-Transference Curve.

Analysis

The results of the ultimate tip load, "Q_{utl}", and shaft load, "Q_{usl}", obtained in the load test were compared with results of ultimate loads predicted by methods based on data from SPT, SPT-T, CPT and DMT tests. The parameter "Q_{us}" refers to ultimate lateral load, "Q_{ut}" refers to ultimate point load , and "Q_u" means ultimate load . The results obtained are presented in table 4.

Table 4. Results of ultimate loads predicted and their comparison with the ultimate load result from the load test.

	Methods	Q_{us} (kN)	Q_{ut} (kN)	Q_u (kN)	Q_{us}/Q_{usl}	Q_{ut}/Q_{utl}	Q_u/Q_{ul}
SPT	Aoki & Velloso	101	29	130	0,46	0,67	0,50
	Décourt & Quaresma	226	41	267	1.03	0.95	1.02
	Meyerhof	87	61	148	0.40	1.42	0.56
CPT	Aoki & Velloso	312	38	350	1.43	0.90	1.33
	Philipponat	321	30	351	1.47	0.70	1.33
SPT-T	Décourt & Quaresma	233	52	285	1.06	1.21	1.09
	Alonso	219	41	260	1.00	0.95	0.99
DMT	FHWA-AS-91-044	158	4	162	0.73	0.09	0.62

Out of the methods based on data from the SPT test, that of Decourt-Quaresma (1978) was the one that showed the best results to predict shaft load and point load. The ultimate load predicted was 2% above the one obtained in the static load test. The other methods did not attain such precision. They only reached values 50% below the ultimate load obtained in the load test.

Methods that use data from SPT-T provide good results varying by 10% above, Decourt (1996), and 1% below, Alonso (1996), the value obtained in the load test. This shows the potential of this test to predict the bearing capacity of deep foundations.

The methods of Aoki&Velloso (1975) and Philipponat (1981) which use data from CPT have provided values about 33% above the ultimate load obtained in the load test. This is particularly due to the value of skin friction which reached values 40% above those obtained in the load test.

The recommended procedure described in FHWA-AS-91-044 (1992) which uses data from DMT have provided values about 40% below the ultimate load obtained in the load test. This is particularly due to the value of point load which reached values 12 times below those obtained in the load test.

Conclusions

The results obtained from the analyses conducted show that:
- The SPT-T test is an advancement of the SPT test since it supplies a skin friction value at a low cost. Its use has led to good predictions of the skin friction of piles. The number of blows obtained in the SPT test, when used with empirical formulas set from the experience with determined types of soils, can lead to good results in predicting the ultimate load of a pile.
- Formulas that use CPT results provide values above those of the load test. This points to the need to use reducing factors for the soil being analyzed.
- In this paper, the result obtained with the use of DMT test was not near the point ultimate of obtained in the load test. This shows the need to continue performing the DMT test in tropical soils (lateritic and saprolitic soils), so that more precise ratios can be determined between the data supplied by the test and soil parameters.

Acknowledgements

The authors gratefully acknowledge the FAPESP (São Paulo State Research Support Foundation) for the financial support to the first writer.

References

A.B.N.T. (1991) *Prova de Carga Estática*: NBR 12131/91. Rio de Janeiro: ABNT, 1991. 4p.

ALBUQUERQUE, P.J.R. (1996) *Análise do comportamento de estaca pré-moldada de pequeno diâmetro, instrumentada, em solo residual de diabásio da região de Campinas.* Campinas-SP, FEAGRI-UNICAMP, 1996. 154p. Dissertação de Mestrado.

ALONSO, U. R. (1996) Estimative of adhesion in piles from lateral friction measured with torque in SPT-T tests. *Soils & Rocks, Latin-American Geotechnical Journal*, ABMS-ABGE, 19 (1), p.81-84. April, 1996

AOKI, N. & VELLOSO, D. (1975) Um método aproximado para estimativa da capacidade de carga em estacas. In: *Congresso Panamericano de Mecânica dos Solos e Engenharia de Fundações*, V, 1975, Buenos Aires, Anais, p.367-376.

DECOURT, L. & QUARESMA, A. R. (1978) Capacidade de Carga de Estacas a Partir de Valores de SPT. In: *Congresso Brasileiro de Mecânica dos Solos e Engenharia de Fundações*, VI, Rio de Janeiro, Anais, 1978, v.1, p.45-53.

FHWA-AS-91-044 (1992) *The Flat Dilatometer Test.* U.S. Department of Transportation. Federal Highway Administration

MEYERHOF, G.G. (1976) Bearing capacity and settlement of pile foundation. *Journal of the soils mechanics and foundation division.* ASCE, 99(3), p.197-228.1976.

PHILIPPONAT, G. (1978) Méthode pratique de calcul des pieux à L'aide du pénétrométre statique. *Informations techniques bulletin*, Paris, Soletanche Enterprise, 21p. 1978.

Interpretation of Pressuremeter Tests in Unsaturated Soil

Gerald A. Miller[1] and Kanthasamy K. Muraleetharan[1]

Abstract

This paper examines cylindrical cavity expansion in unsaturated soil and its application to the interpretation of pressuremeter test results. An analytical solution for cylindrical cavity expansion in unsaturated soil is presented and used to examine the influence of unsaturated soil characteristics on the pressuremeter response. The pressuremeter response is characterized by the limit pressure and pressuremeter modulus. Effects of matric suction and initial degree of saturation are examined in light of unsaturated soil mechanics theories. Application of the unsaturated theory to the design of foundations is discussed in comparison to existing methods of pressuremeter interpretation. A methodology for predicting pressuremeter response under different moisture conditions is presented along with an example.

Introduction

The pressuremeter is a useful tool for assessing in situ stress-strain behavior of soils. The pressuremeter modulus and limit pressure can be used directly in settlement and bearing capacity predictions of shallow foundations, respectively. Results can also be used to estimate soil shear strength by employing cylindrical cavity expansion theory and assumptions of ideal soil behavior. Typically, for fine-grained soil, undrained behavior is assumed while coarse-grained soils are assumed to behave in a drained fashion. Current methods of interpretation do not account for the effects of partial saturation. This paper examines the effects of partial saturation on the pressuremeter response via a cylindrical cavity expansion theory developed for unsaturated soils. The theory is used to interpret results of field and laboratory tests and to present a preliminary framework for interpreting pressuremeter test results from unsaturated soils.

[1] Assistant Professor, School of Civil Engineering and Environmental Science, University of Oklahoma, 202 West Boyd Street, Room 334, Norman, OK 73072

Cylindrical Cavity Expansion

Vesic (1972) presented equations for analyzing the expansion of a cylindrical cavity in a soil characterized by elastic properties E and ν (Young's modulus and Poisson's ratio), and strength parameters c and ϕ (cohesion and friction angle). A cross-section through the cavity and surrounding soil is presented in Fig. 1, where the plastic zone is defined by R_u and R_p and corresponds to the ultimate pressure p_u acting in the cavity. The soil in the plastic zone is assumed to behave as a compressible plastic solid characterized by c, ϕ, and volumetric strain, ε_v. Soil beyond the plastic zone behaves as a linear elastic solid described by E and ν. By considering equilibrium conditions at the interface between the plastic and elastic zones, Vesic (1972) arrived at the expression for the ultimate cavity pressure given as:

$$p_u = cF_c + pF_p \tag{1}$$

$$F_p = (1+\sin\phi)(I_{rr}\sec\phi)^{\sin\phi/(1+\sin\phi)} \tag{2}$$

$$F_c = (F_p - 1)\cot\phi \tag{3}$$

$$I_{rr} = \frac{I_r}{\left(1 + I_r \varepsilon_v \sec\phi\right)} \tag{4}$$

$$I_r = \frac{E}{2(1+\nu)(c + p\tan\phi)} \tag{5}$$

where: F_c and F_p are dimensionless cylindrical cavity expansion factors, I_r is the rigidity index, and I_{rr} is the reduced rigidity index. The rigidity index is the ratio of shear modulus to the soil shear strength for an initial isotropic stress, p. Reduced rigidity index results from volume change in the plastic zone. Vesic (1972) presented a methodology for estimating volume change in the plastic zone based on soil mechanics principals and laboratory testing. A parametric study presented by Vesic showed that the principal parameters affecting the ultimate cavity pressure were the initial stress state of the soil, strength, volume change characteristics and rigidity index of the soil. The size of the plastic zone is defined by the rigidity index as follows:

$$R_p/R_u = (I_{rr}\sec\phi)^{0.5} \tag{6}$$

Examination of Eqs. 1-6 reveals that the size of the plastic zone and the ultimate cavity pressure depends on reduced rigidity index and shear strength. Vesic showed that for stiff soils with large shear strength the ultimate cavity pressure is quite sensitive to the volumetric strain.

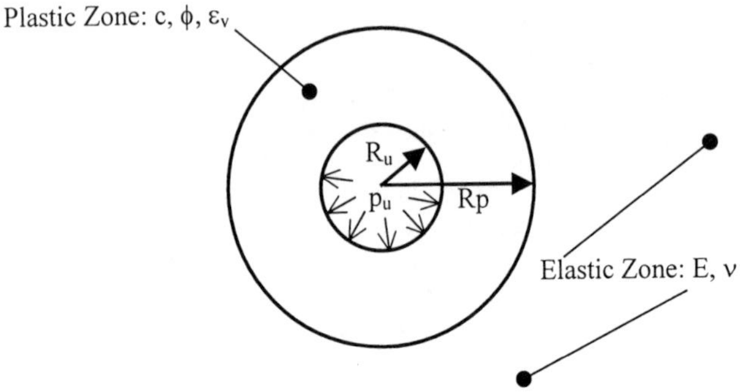

Figure 1. Expansion of a Cylindrical Cavity (After Vesic 1972)

Equations 1-6 can be used to estimate the undrained shear strength of purely cohesive soils (ϕ=0, ε_v=0) from pressuremeter test results with the assumption that the pressuremeter expansion can be treated as the expansion of an infinitely long cylindrical cavity. This is generally a satisfactory assumption provided the length to diameter ratio of the inflatable portion of the probe exceeds 6.5 (Briaud 1992). For this purpose the limit pressure, p_L, from the pressuremeter is equal to the ultimate cavity pressure, p_u. If the volumetric strain in the plastic zone can be estimated, then for a purely frictional material (c=0) pressuremeter results can be used to predict the friction angle (Vesic 1972). Methods for estimating undrained shear strength for cohesive soils and friction angle of cohesionless soils are summarized by Briaud (1992).

It is clear that the expressions developed by Vesic (1972) and methods of interpretation of pressuremeter test results hinge on the assumption of soil behavior defined by completely drained or undrained behavior. Implicit in these assumptions is that the soil behavior depends on a single independent stress-state variable, i.e., the effective stress. For unsaturated soil, it is now widely accepted that the soil behavior depends on two independent stress-state variables, the net normal stress, defined as the difference between the total stress and pore air pressure (σ_n-u_a) and the matric suction, defined as the difference between the pore air and pore water pressure (u_a-u_w). To accurately describe the expansion of a cylindrical cavity in unsaturated soil, these two independent stress-state variables must be incorporated into the derivation of the ultimate cavity pressure. These expressions are presented in a subsequent section.

Bearing Capacity and Settlement from Pressuremeter Test Results

While the pressuremeter can be used to estimate soil shear strength, methods are available for predicting the bearing capacity and settlement of foundations using the limit pressure and pressuremeter modulus directly. The bearing capacity for a shallow foundation is given by (Briaud 1992):

$$q_u = kP_{Le}+q_o \tag{7}$$

where: q_u = the ultimate bearing capacity,
P_{Le} = the equivalent net limit pressure near the foundation level,
$P_{Le} = P_L - p_{oH}$,
P_L = limit pressure,
p_{oH} = total horizontal stress at rest,
k = empirically determined bearing capacity factor,
q_o = total vertical overburden stress at the foundation level.

The equivalent net limit pressure is the geometric mean of the net limit pressure within a distance of 1.5 foundation widths above and below the foundation level. Net limit pressures can be estimated using p_{oH} determined from the initial portion of the pressuremeter curve. Recommended values of k are in the range of 0.8 to 2 for sand, 0.8 to 1.5 for silt and 0.8 to 1.2 for clay (Briaud 1992). Higher values of k correspond to equivalent depths of embedment equal to 3 times the foundation width, while lower values correspond to zero embedment.

Settlement of foundations can be estimated using an elasticity-based method and incorporating a pressuremeter modulus. One example, is the method presented by Janbu et al. (1956) where the settlement, s, is given by

$$s=\mu_0\mu_1(1 - \nu^2)qB/E \tag{8}$$

where: μ_0, μ_1 are correction factors to account for depth of embedment and depth to a rigid stratum, respectively,
ν = Poisson's ratio,
q = bearing pressure,
B = footing width,
E = pressuremeter modulus within the zone of influence.

The pressuremeter-based method has been shown to give excellent predictions of bearing capacity relative to traditional bearing capacity equations, particularly for sands (Briaud 1992). Settlement equations also provide reasonable predictions (Briaud 1992).

Field and Laboratory Testing in Unsaturated Soil

Pressuremeter equations presented previously were developed with saturated soil mechanics in mind and do not account for the influence of matric suction on the limit pressures or moduli obtained from test results in unsaturated soils. Consider the data shown in Fig. 2, obtained from prebored pressuremeter tests conducted in an unsaturated silt during wet and dry periods. The site was instrumented with

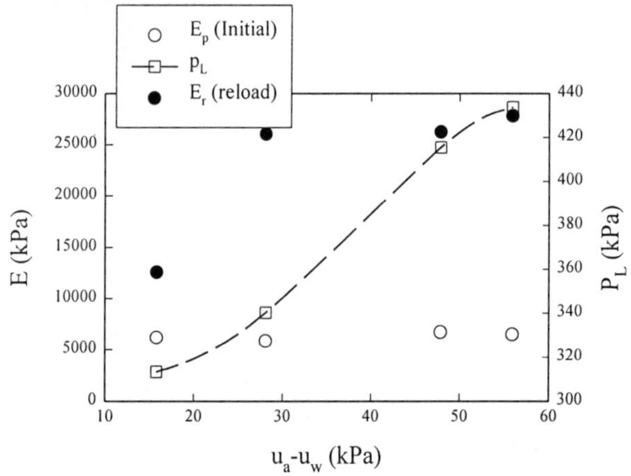

Figure 2. Pressuremeter Field Test Results

tensiometers to monitor matric suction. The limited data obtained thus far, shows that the limit pressure was sensitive to the matric suction at the time of testing. A similar variation in pressuremeter modulus is also shown in Fig. 2. Figure 3 contains data collected from small-diameter pressuremeter tests conducted in the same silt compacted in a rigid wall calibration chamber (Miller and Muraleetharan 1998).

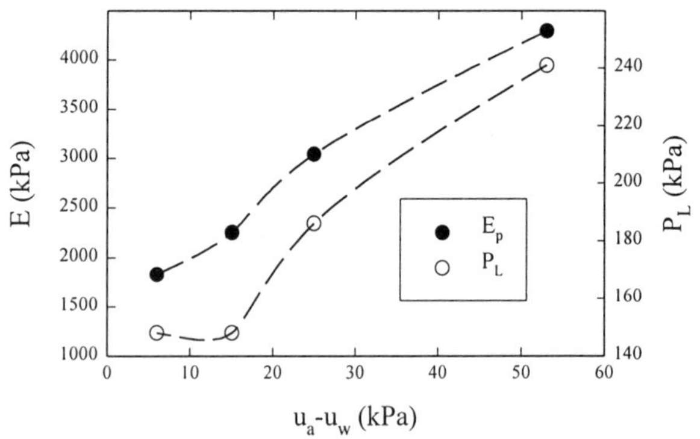

Figure 3. Pressuremeter Calibration Chamber Test Results

Data shown in Fig. 3 further demonstrates the influence of matric suction on the PMT limit pressure and modulus. Thus, bearing capacities and settlements predicted with Eqs. 7 and 8 will vary as a function of matric suction as well. If it is assumed that the basic form of Eqs. 7 and 8 are valid for unsaturated soils, then to account for varying matric suction it would be beneficial to be able to predict how the limit pressure and pressuremeter modulus will change as moisture conditions change. The remainder of this paper describes how a cylindrical cavity expansion theory for unsaturated soils can be used to predict limit pressures given: results from a single pressuremeter test, some basic soil information, and knowledge of the soil-water characteristic behavior.

Cylindrical Cavity Expansion Theory for Unsaturated Soils

It is now widely accepted that the stress state for an unsaturated soil can be adequately described by two independent stress state variables, the net normal stress and the matric suction. Two independent stress tensors can be used to fully describe the state of stress at a point within a soil mass as presented in Fredlund and Rahardjo (1993).

Extending Vesic's (1972) original cavity expansion theory, Muraleetharan et al. (1998) derived the following expression for the ultimate cavity pressure (p_u) of an expanding cylindrical cavity in an unsaturated soil,

$$p_u = F_p\left[(p - u_{af}) - \frac{E}{H(1-2\nu)}\Delta(u_a - u_w) + \Delta u_a\right] + F_c c + u_{af} \tag{9}$$

where: F_p, F_c = dimensionless cylindrical cavity expansion factors given by,

$$F_p = (1+\sin\phi) I_{rr}^{\sin\phi/(1+\sin\phi)} \tag{10}$$

$$F_c = (F_p - 1)\cot\phi \tag{11}$$

p = initial (prior to the expansion) total mean normal stress,
u_{af} = final pore air pressure in the plastic zone,
c = cohesion,
ϕ = internal angle of friction,
I_{rr} = reduced rigidity index,

$$I_{rr} = \frac{1+\varepsilon_v}{\dfrac{f_2}{I_r} - \dfrac{2(1+\nu)\Delta(u_a - u_w)}{H(1-2\nu)}\sin\phi + \dfrac{2(1+\nu)\Delta u_a}{E}\sin\phi + \varepsilon_v} \tag{12}$$

$$f_2 = \cos\phi \tag{13}$$

ε_v = volumetric strain in the plastic zone at the limit pressure
I_r = rigidity index

$$I_r = \frac{E}{2(1+\nu)[(p-u_a)\tan\phi + c]} \tag{14}$$

$\Delta(u_a\text{-}u_w)$ = change in matric suction during the cavity expansion,
Δu_a = change in pore air pressure during the cavity expansion,
E = Young's modulus,
ν = Poisson's ratio, and
H = Elastic modulus with respect to the matric suction (Fredlund and Rahardjo 1993).

In deriving Eq. (9) it is assumed that at the limit pressure the radius of the cavity is R_u and this cavity is surrounded by a plastic zone of radius R_p and beyond that the soil behaves as a linear elastic material. The values of R_p and R_u are related by the following equation.

$$R_p/R_u = I_{rr}^{0.5} \tag{15}$$

Furthermore, it is assumed that the soil within the plastic zone is a compressible plastic solid defined by a Mohr-Coulomb failure criterion for unsaturated soil. The stress state variables can be used to describe the shear strength of unsaturated soil using a Mohr-Coulomb type equation with the following form:

$$\tau = c' + (\sigma_n - u_a)\tan\phi + (u_a - u_w)\tan\phi^b \tag{16}$$

where: τ = shear stress at failure on a failure plane,
c' = effective cohesion,
σ_n = total normal stress on the failure plane,
u_a = pore air pressure at failure,
ϕ = angle of internal friction associated with the net normal stress,
u_w = pore water pressure at failure, and
ϕ^b = angle of internal friction associated with the matric suction.

Note that for a saturated soil $u_a = u_w$ and c' and ϕ become effective-stress strength parameters. Eq. (16) describes a plane on a three dimensional plot with shear strength on the vertical axis and matric suction and net normal stress on the horizontal axes. For the situation where ϕ is constant, Eq. (16) predicts that increasing matric suction will result in failure envelopes with similar slope when projected on the plane bounded by the $(\sigma_n - u_a)$ and τ axes. The slopes of the failure envelopes in this plane are given by ϕ and the cohesion intercept, c, is given by,

$$c = c' + (u_a - u_w)\tan\phi^b \tag{17}$$

and thus, Eq. (16) can be simplified to,

$$\tau = c + (\sigma_n - u_a) \tan \phi \qquad (18)$$

Development of a Method for Interpreting Pressuremeter Tests in Unsaturated Soil

In a previous study (Miller and Muraleetharan 1998), pressuremeter data was obtained with a small diameter probe in a rigid wall calibration chamber containing unsaturated silt. This data is used to show how the cylindrical cavity expansion theory for unsaturated soil may be used for interpreting pressuremeter test results from unsaturated soil. Development of a framework for interpreting the PMT in unsaturated soil is the focus of continued research at the University of Oklahoma involving: pressuremeter testing in a large flexible wall calibration chamber, laboratory testing, analytical and numerical modeling. The method outlined here is one approach that holds promise.

The basic notion of the proposed method of interpretation is that given the limit pressure and moisture condition relevant to a given pressuremeter test, it is desired to predict the limit pressure that will exist if the moisture conditions change. Of particular interest would be the case where the soil in question is expected to experience an increase in moisture content, and hence a reduction in strength.

Examination of Eq. 9 indicates that several variables must be known in order to predict the limit pressure as a function of changing moisture conditions. The use of this equation is complicated by the fact that many of the variables are dependent on moisture content. To begin implementing the theory into a practical framework of interpretation, several assumptions were made as follows:

1) The soil-water characteristic curve is available or can be reasonably well estimated. In this paper, the soil-water characteristic behavior is expressed in terms of the degree of saturation. Furthermore, it is assumed that the soil-water characteristic curve is not altered by the expansion of the pressuremeter probe.
2) The soil shear strength parameters, c', ϕ, and ϕ^b for the extended Mohr-Coulomb envelope for unsaturated soil are known or can be reasonably well estimated.
3) Elastic soil properties E, H, and ν, can be estimated with reasonable certainty based on pressuremeter test results. Furthermore, the change in these properties due to a change in moisture conditions has a small effect on predicted limit pressures resulting from changes in moisture conditions (preliminary sensitivity analyses using Eq. 9 show this is a reasonable assumption).
4) Pore air pressures remain equal to zero throughout the pressuremeter probe expansion.
5) The initial total mean normal stress can be estimated with reasonable certainty.

Provided items 1 through 5 are satisfied, what are needed to predict the limit pressure using Eq. 9 are volumetric strain in the plastic zone, ε_v, and the change in matric suction, $\Delta(u_a\text{-}u_w)$, expected during expansion for a given initial moisture condition. It is expected that the outcome of the current research will include the development of families of curves that relate ε_v and $\Delta(u_a-u_w)$ to the initial moisture condition for various soil types under various states (i.e. density and stress level).

To explore this approach further, the rigid-wall calibration chamber test results shown in Fig. 3, in combination with the soil water characteristic curve shown in Fig. 4, were used to calculate values of ε_v and $\Delta(u_a-u_w)$ necessary to match actual PMT limit pressures using Eq. 9. The soil properties that were used in the calculations are shown in Table 1. An iterative process was used in the calculations, whereby a value of $\Delta(u_a-u_w)$ was assumed and ε_v was computed such that p_u from Eq. 9 equaled p_L from a given PMT test. Next, the value of ε_v was used to calculate the volume change in the plastic zone around the probe and the corresponding change in degree of saturation. Based on the change in degree of saturation, a corresponding change in matric suction was determined using the equation for the soil-water characteristic curve shown in Fig. 4. This new value of $\Delta(u_a-u_w)$ was then used to calculate another ε_v. The process was repeated until ε_v and $\Delta(u_a-u_w)$ were consistent with the soil-water characteristic curve, as demonstrated in Fig. 5 where the predicted changes in matric suction and degree of saturation are presented. The resulting values of ε_v and $\Delta(u_a-u_w)$ for the chamber PMT tests are shown in Figs. 6 and 7 as a function of initial degree of saturation (S_i) and initial matric suction, $(u_a-u_w)_i$. Figures similar to these and representative of different soil types can be developed to assist in the interpretation of PMT results in unsaturated soils, as illustrated in the following example.

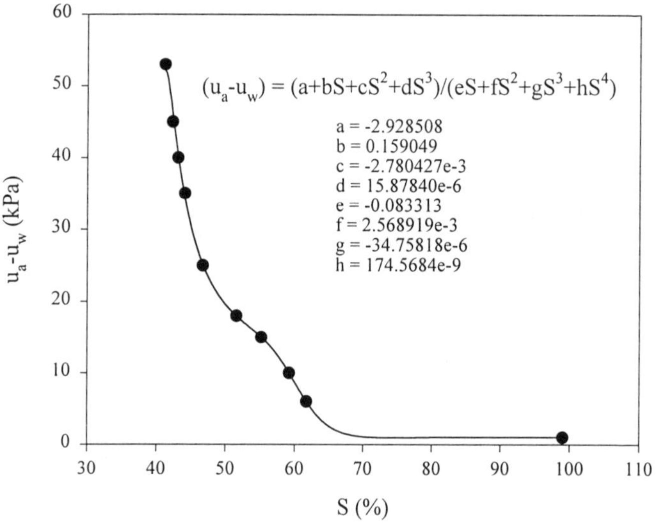

Figure 4. Soil-Water Characteristic Curve for Calibration Chamber Soil (Minco Silt)

Table 1. Soil Properties used for and Results of Backcalculation for Calibration Chamber PMT Data

Test No.	$(u_a-u_w)_i$ (kPa)	$\Delta(u_a-u_w)$ (kPa)	ε_v	H (kPa)	E (kPa)	p (kPa)	S (%)	e
1	6	-4.99	0.07429	9135	1827	39	61.7	0.804
2	15	-13.53	0.08083	11270	2254	39	55.2	0.814
3	25	-9.59	0.06716	15200	3040	39	46.7	0.854
4	53	-23.29	0.04408	21470	4294	39	41.2	0.905

Note: shear strength parameters used in backcalculation with Eq. 9 include: ϕ=31, ϕ^b=27.5, and c′=5 kPa, based on direct shear tests.

Consider the field test results shown in Fig. 2. Given the pressuremeter data for the test corresponding to an initial matric suction of 56 kPa, it desired to estimate the limit pressure at different initial matric suctions. For comparison purposes, initial matric suctions of 16, 28, and 48 kPa corresponding to the other PMT tests at the site are selected. The calculations proceed as follows:

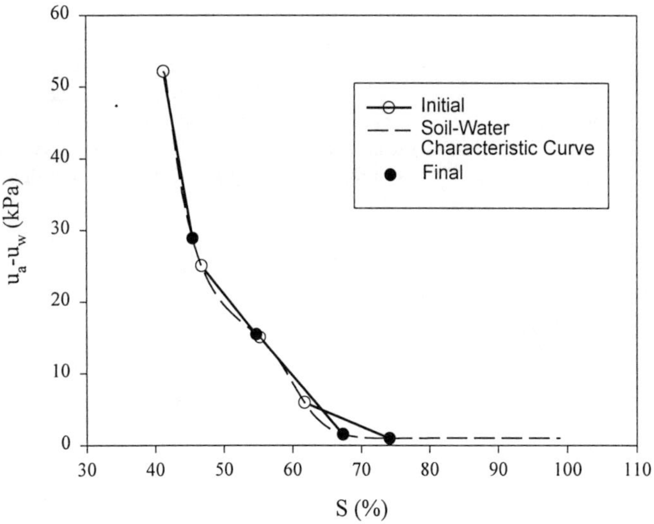

Figure 5. Calculated Changes in Degree of Saturation and Matric Suction Compared to the Soil-Water Characteristic Curve for Calibration Chamber Soil (Minco Silt)

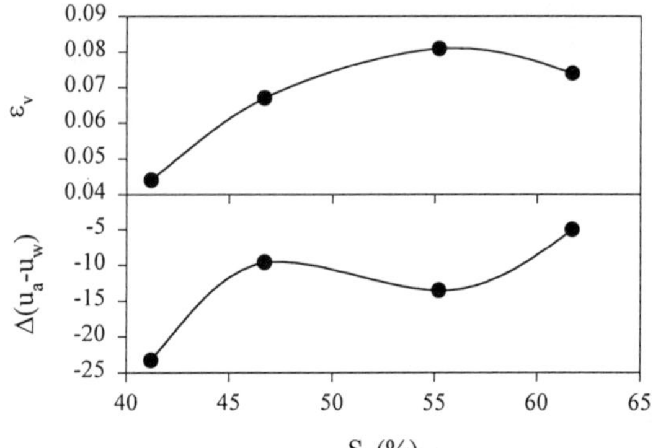

Figure 6. Initial Degree of Saturation versus Backcalculated Change in Matric Suction and Volumetric Strain in the Plastic Zone for Chamber PMT Tests

1) Use Fig. 7 to estimate $\Delta(u_a-u_w)$ = -25 kPa for initial matric suction $(u_a-u_w)_i$ equal to 56 kPa. It is assumed that since the soil used in the chamber tests came from the field site in question, that trends observed in Fig. 7 are similar for field site soils.
2) Using Eq. 9 and $\Delta(u_a-u_w)$ = -25 kPa, determine ε_v necessary to achieve agreement between the predicted limit pressure, p_u, and the actual PMT limit pressure, p_L = 434 kPa. Modulus value E was taken as the PMT reload modulus, and based on values of some typical soils (Fredlund and Rahardjo 1993) H was assumed equal to 5E. Initial mean normal stress, p, was taken equal to the vertical total stress at the PMT test depth (1 m). Table 2 summarizes the input values used in the calculations.

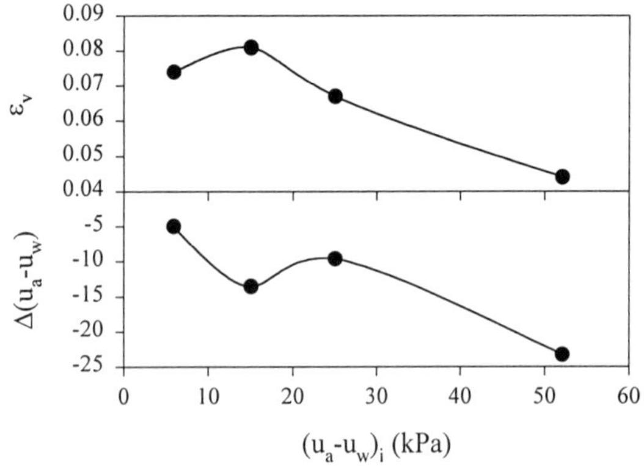

Figure 7. Initial Matric Suction versus Backcalculated Change in Matric Suction and Volumetric Strain in the Plastic Zone for Chamber PMT Tests

3) Determine the ratio of ε_v calculated in step 2) to the value of ε_v corresponding to $(u_a-u_w)_i = 56$ kPa in Fig. 7.
4) Assuming that the general trends in Fig. 7 are similar for the soil at the field site, determine values of ε_v and $\Delta(u_a-u_w)$ corresponding to $(u_a-u_w)_i$ values of 16, 28, and 48 kPa.
5) Multiply values of ε_v from step 4) by the ratio determined in step 3). The ε_v can vary widely depending on soil density, elastic parameters, and other properties; however, it is assumed that the general trend of ε_v in Fig. 7 will be similar for similar soil types.
6) Use Eq. 9, $\Delta(u_a-u_w)$ from step 4), and modified ε_v from step 5) to estimate limit pressures corresponding to $(u_a-u_w)_i$ values of 16, 28, and 48 kPa.

Results of the calculations and predictions are shown in Table 2 and Fig. 8. As demonstrated in Fig. 8, this preliminary approach holds promise, as predicted values are within 25% of the actual values. Furthermore, predictions are conservative relative to actual PMT limit pressures.

Table 2. Soil Properties used for and Results of Backcalculation and Predictions based on Field PMT Data

Test No.	$(u_a-u_w)_i$ (kPa)	$\Delta(u_a-u_w)$ (kPa)	ε_v	H (kPa)	E (kPa)	p (kPa)	p_u (kPa)	p_L (kPa)	p_u/p_L
1	16	-13.2	0.00567	137800	27560	20	235	310	0.757
2	28	-10.5	0.00448	137800	27560	20	334	338	0.989
3	48	-21	0.00329	137800	27560	20	403	413	0.976
4	56	-25	0.00280	137800	27560	20	434	434	1.000

Note: shear strength parameters used with Eq. 9 include: ϕ=31, ϕ^b=27.5, and c′=5 kPa, based on direct shear tests.

The importance of accurately predicting the effect of matric suction on the limit pressure is demonstrated in Fig. 9. In Fig. 9, the bearing capacity predicted using Eq. 7, is plotted as a function of initial matric suction using both predicted and actual limit pressures. As shown, a significant change in bearing capacity occurs with changes in initial matric suction. Hence, it is very important to be aware of moisture conditions corresponding to the time a pressuremeter test is conducted if the results are to be used in geotechnical predictions.

Summary and Conclusions

Laboratory and field pressuremeter test data in unsaturated soil demonstrate that the matric suction at the time of testing has a significant influence on the resulting limit pressure and modulus. Furthermore, the bearing capacity of shallow foundations predicted using PMT limit pressures is similarly affected by matric suction. An unsaturated soil mechanics theory for cylindrical cavity expansion can be applied to the interpretation of pressuremeter tests in unsaturated soil. A

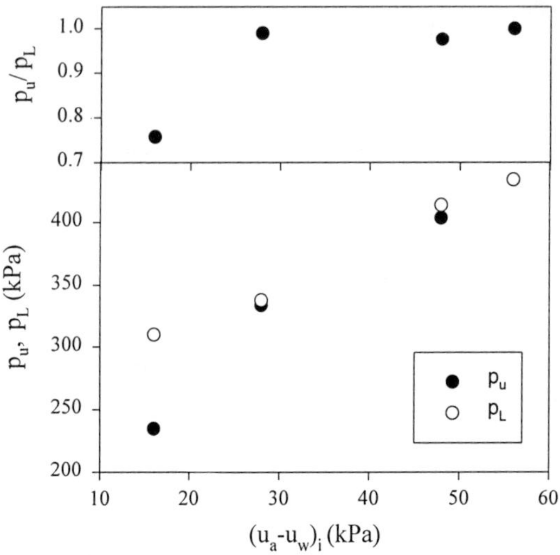

Figure 8. Initial Matric Suction versus Predicted (p_u) and Actual (p_L) Limit Pressures for Field Tests

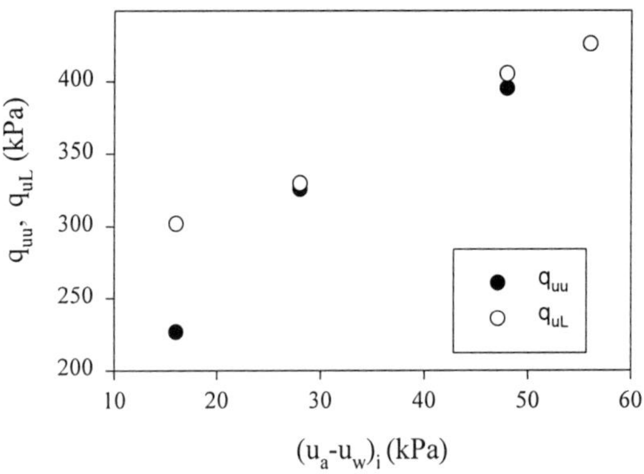

Figure 9. Initial Matric Suction versus Bearing Capacity using Predicted (q_{uu}) and Actual (q_{uL}) Limit Pressures

preliminary approach for estimating the limit pressure for different values of initial matric suction based on a single PMT test (corresponding to one value of initial matric suction) was introduced in this paper. Predictions using the preliminary method for interpreting pressuremeter test results from unsaturated soil were within 25% of actual values, and show that the suggested approach holds promise.

Extensive work is being carried out at the University of Oklahoma to build upon the preliminary interpretive framework presented in this paper.

References

Briaud, J.L. (1992), *The Pressuremeter*, A.A. Balkema, Rotterdam.

Janbu, N., Bjerrum, L. and Kjaernsli, B. (1956), *Guidance in the Solution of Foundation Problems*, Norwegian Geotechnical Institute Publication No. 16.

Miller, G.A. and Muraleetharan, K.K. (1998), "In Situ Testing in Unsaturated Soil," *Proc. of UnSat'98, 2nd International Conference on Unsaturated Soils*, Beijing, China, Aug. 27-30.

Fredlund, D.G. and Rahardjo, H. (1993), *Soil Mechanics for Unsaturated Soils*, John Wiley and Sons, Inc., New York.

Muraleetharan, K.K., Salehipour, S.A., Dhavala, D. (1998), *Cavity Expansion in Unsaturated Soils: Theory and Application*, Technical Report, University of Oklahoma School of Civil Engineering and Environmental Science, Norman.

Vesic, A.S. (1972), "Expansion of Cavities in an Infinite Soil Mass," *Journal of the Soil Mechanics and Foundation Division*, ASCE, 98(3), pp. 265-290.

Ultrasonic Testing for Compacted Clayey Soils

Nazli Yesiller[1], Associate Member, ASCE, Gokhan Inci[2], Student Member, ASCE, and Carol J. Miller[3], Member, ASCE

Abstract

In this study, tests were conducted to investigate the use of ultrasonic methods to determine compaction characteristics of clayey soils. In particular, through-transmission test method was used to determine P-wave velocities in compacted clayey soils. Effects of soil type and compaction conditions on velocity were investigated. Tests were conducted on three soils that had low, medium, and high plasticity. Soils were prepared at water contents ranging from dry to wet of optimum using standard and modified compaction efforts. It was observed that velocity increased with increasing compactive effort and decreasing plasticity and clay content. Moreover, the variation of ultrasonic velocity with water content was similar to the variation of dry density with water content. Shapes of the water content versus velocity plots correlated well with the shapes of compaction plots. Access to two opposite surfaces of samples was required to perform through-transmission measurements. Additional tests were conducted to determine the feasibility of using surface-transmission with access required only to top surface of soils to measure velocity. Velocities that were similar to through-transmission velocities were obtained with surface measurements using correction factors. Through-transmission can be used in the laboratory; whereas surface-transmission can potentially be used in the field to determine ultrasonic velocity of compacted clayey soils. This method shows promise for determination of compaction properties of clayey soils in the field.

Introduction

Clay soils are compacted for construction of various structures and facilities. Compacted clay soils are commonly used as liners for waste containment facilities; as

---

[1] Assistant Professor, Department of Civil and Environmental Engineering, Wayne State University, Detroit, MI 48202.

[2] Graduate Research Assistant, Department of Civil and Environmental Engineering, Wayne State University, Detroit, MI 48202.

[3] Professor, Department of Civil and Environmental Engineering, Wayne State University, Detroit, MI 48202.

embankments, subgrades, bases, and backfills for foundations and transportation facilities; and as hydraulic barriers for various facilities. Verifying the properties and condition of compacted clay soils is a significant component of design and construction with these materials. Compaction characteristics of soils are required for design, material selection, and quality control purposes.

Compaction characteristics of soils are determined by analyzing the relationship between water content and dry density (or unit weight) of soils. Proctor compaction tests are commonly used in the laboratory to determine the variation of dry density with water content. The relationship between dry density and water content of soils is demonstrated using a compaction curve. Compaction properties of field soils are compared with the compaction properties of the soils determined in the laboratory to verify the effectiveness of construction procedures. Destructive and nondestructive methods are used to determine density and water content of field soils.

Destructive test methods that are used most commonly include sand cone and rubber balloon methods. These methods involve excavating a hole in the compacted soil. The volume of the hole and the weight of the excavated soil are determined to calculate the density of the soil. Care must be exercised in determination of the weight and volume of the soil. Part of the excavated soil is used to determine the water content of the soil. The total density and water content are used to calculate the dry density of the soil. These methods can both be time consuming. The predominant nondestructive method is nuclear density meter. Both the density and water content of a compacted soil can be determined in a short period of time using this method. The density and water content of a soil are inferred from transmission properties of gamma rays in the soil. Water content is determined with transmission on the ground surface. The gamma source in the density meter is placed in a small hole excavated in the soil to obtain a representative density for a lift of soil. Presence of radioactive material in the density meter requires special handling procedures and licensed personnel. Agreement has not been fully reached on the reliability of nuclear methods (Brown 1996). Various other destructive, nondestructive, and empirical methods are available for determining compaction characteristics of soils. Extensive background information is available on determination of compaction characteristics of soils in the laboratory and in the field and also on verification of field compaction procedures (Winterkorn and Fang 1975, Brown 1996).

Ultrasonic testing can provide a fast and simple approach for determining characteristics of compacted clayey soils. This nondestructive method can be used as an alternative to existing methods to analyze laboratory or field compacted soils. This investigation was conducted to assess the feasibility of using ultrasonic testing (in this case ultrasonic velocity measurements) to determine compaction characteristics of clayey soils. Variation of ultrasonic velocity with water content and density of compacted soils was analyzed. Effects of soil type and compaction conditions on the velocity were analyzed. Transducer arrangements that would facilitate use of the method in the field were investigated.

Ultrasonic Testing for Compacted Soils

Ultrasonic testing is used for nondestructive evaluation of materials and structures. Ultrasonic waves are stress waves with frequencies higher than 20 kHz that propagate in mass media. Propagation of ultrasonic waves in a material is affected by the properties and condition of the material. Transmission of waves in a material is quantified generally using two parameters: velocity and attenuation. Ultrasonic velocity can be correlated to elastic constants and mechanical properties of a material, whereas ultrasonic attenuation can be correlated to microstructural properties of a material (McIntire 1991).

An extensive investigation of the use of ultrasonic testing for compacted soils was reported in an early study by Sheeran et al. (1967). Velocities of P-waves were determined on three types of soils. It was observed that peak velocities and maximum dry densities occurred within ± 0.5% water content for laboratory compacted soils. The ultrasonic velocities increased with increasing dry density until the optimum water content. However, at water contents higher than the optimum water content, the velocities decreased with increasing dry density. Field tests were conducted on a low plasticity sandy silt. Velocities for laboratory compacted samples (Proctor compaction) of the soil varied between 300 m/sec and 1400 m/sec. Field velocities for the soil were within the range of laboratory velocities.

Significant progress has been made in ultrasonic testing equipment, data acquisition, and data analysis since the study reported by Sheeran et al. (1967). However, ultrasonic testing has been used on compacted soils to a limited extent in recent studies. Wang et al. (1991) conducted a simple study to determine the variation of ultrasonic velocity with water content and pressure for a silt loam compacted under static pressure. The compaction pressure and water content ranged from 143 kPa to 430 kPa and 11% to 20%, respectively. The ultrasonic wave velocities in the soil were between 360 m/sec and 560 m/sec. The velocities increased with increasing compaction pressure. In general, the velocities increased with decreasing water content with some exceptions. The increase in the wave velocity at low water contents was attributed to the high bulk modulus of the soil at low water contents. Sologyan (1990) provided a brief summary of applications of ultrasonic testing to determine density of soils in the field. The information provided was related to agricultural applications. It was suggested that ultrasonic testing was used effectively to determine density, water content, and microstructural properties of soils. However, specific details were not provided.

Materials

Tests were conducted on three soils with varying plasticities, with the intent to use local soils. Soils 1 and 2 were obtained from two landfills located in Southeast Michigan. However, a high plasticity soil could not be located in this area. Therefore, Soil 2 was mixed with 25% bentonite (by weight) to make Soil 3. Index properties of the soils are presented in Table 1.

Table 1 – Soil Properties

Property		Soil 1	Soil 2	Soil 3
Classification (USCS)		SC	CL	CH
Particle Size (%)	Sand	56	3	2
	Silt	27	38	34
	Clay	17	59	64
Atterberg Limits	Liquid Limit	16	40	83
	Plasticity Index	7	17	60
Specific Gravity		2.68	2.68	2.69

Equipment

The electronic equipment used for velocity measurements consists of three units: (1) P-wave transducers, (2) pulser-receiver, and (3) data acquisition system (Fig. 1). Two transducers are used for measurements, one transducer for transmitting ultrasonic waves, the other for receiving ultrasonic waves that travel through the test sample. The transducers are 50-kHz-center frequency narrowband transducers with diameters of 44.5 mm. The transducers are actuated by a pulser-receiver which is connected to a computer for digitization of data (Fig. 1). The 10-MHz-bandwidth broadband pulser-receiver has an adjustable repetition rate in the range of 20 Hz to 2 kHz, with a 1V output for synchronous triggering of the signal acquisition board during data acquisition. The data acquisition system includes a computer equipped with an A/D (analog to digital converter) board with 100 MHz sampling rate and a digital oscilloscope software package which is used for viewing waveforms and for adjusting data acquisition parameters.

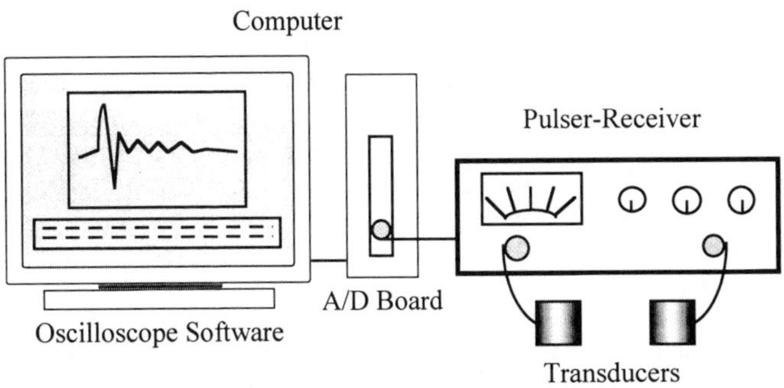

Figure 1. Electronic Equipment

Ultrasonic Test Procedure

The ultrasonic tests conducted for this study consisted of determination of velocity of P-waves in the test soils using two transducer arrangements. The test configurations used in the study are presented in Fig. 2. The tests were conducted using through-transmission and surface-transmission methods. The through-transmission test technique is also referred to as direct transmission. In this method, ultrasonic waves are sent from one surface of a material with a transmitting transducer and the transmitted waves are received by a receiving transducer placed on the opposite surface of the material. This method is used mainly for testing highly attenuating materials such as soils. In surface measurements, the transducers are placed on the same surface of a material and waves are transmitted from the transmitting transducer to the receiving transducer along the surface of the material.

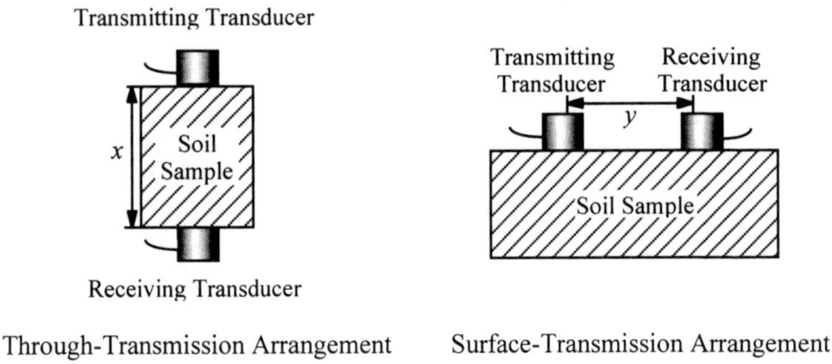

Figure 2. Test Configurations

The velocity of the P-waves is obtained as the ratio of the travel path to the travel time of the waves. In the through-transmission test technique, the travel path is the thickness of the samples (indicated by x in Fig. 2). In the surface measurements, the travel path is the distance between the transducers along the surface of the soil (indicated by y in Fig. 2). The travel path of the waves for both test configurations is determined using mechanical measurements (i.e. using a pair of calipers with an accuracy of 0.02 mm).

The travel time for both of the test methods is obtained from ultrasonic measurements as the first arrival time of the waves at the receiving transducer. The first arrival time was calculated as the difference between the time of application of the pulse by the transmitting transducer and the arrival time of the waveform in the receiving transducer. The A/D board was synchronized with the pulser-receiver such that data collection started at the time of pulse application. Therefore, there was no uncertainty in the determination of the pulse application time which was equal to 0 μsec. Wave averaging and calibration procedures were used to minimize random and systematic errors associated with determination of the first arrival time of the waves at the receiving transducer. A single waveform recorded in the computer consisted of an average of 32 waveforms. For each measurement, four such waveforms were recorded. The average of these four waveforms was further processed using a computer code to determine the first arrival time. A schematic description of the procedure is presented in Fig 3. Initially, a baseline was determined on the early portion of the waveform using a statistical procedure. Also, the waveform was rectified to facilitate the analysis of the waves. A threshold was then determined as twice the amplitude of maximum variation (representative of noise in the waveform) in the baseline. Finally, the first arrival time was determined as the intersection point of the waveform with the threshold. Similar procedures are described in literature for the determination of first arrival time (Jones and Facaoaru 1969, Komlos et al. 1996).

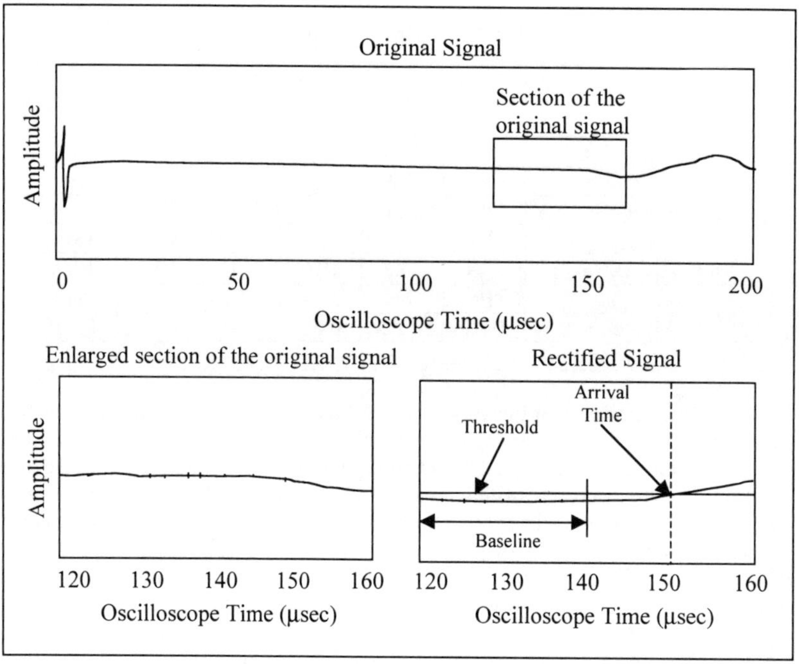

Figure 3. Signal Processing

Testing Program

Laboratory tests consisted of determination of P-wave velocity on compacted soils using the through-transmission and surface-transmission methods. The soils were compacted using standard and modified Proctor efforts. Effects of soil properties and compaction conditions on velocity were analyzed. Comparisons were made between traditional compaction plots (dry density vs. water content) and compaction plots obtained with ultrasonic measurements (velocity vs. water content). Comparisons were also made between direct and surface measurements.

In the tests using the through-transmission test technique, samples were used for determination of density and water content as well as P-wave velocity. Velocity was determined on each sample that was used for determination of compaction characteristics of the soils. In most cases, three samples were prepared at each target water content. Compaction plots were generated using dry density vs. water content and velocity vs. water content (Figs. 4 - 6). In addition, plots of total density versus water content and void ratio versus water content are presented in Figs. 4 – 6. Maximum dry density, maximum velocity, maximum total density, and minimum void ratio with corresponding water contents are presented in Table 2.

Table 2 –Properties of Compacted Soils

Property	Soil 1		Soil 2		Soil 3	
	STD	MOD	STD	MOD	STD	MOD
Max. $\gamma_{(dry)}$ (kN/m^3)	20.5	21.6	15.9	18.2	15.3	16.7
w_{opt} (%)	9	7.2	22	15	25	19
Max. Velocity (m/sec)	1350	1550	980	1470	920	1040
$w_{vel\ max}$	8.3	7	22	16.5	25	18.5
Max. $\gamma_{(total)}$ (kN/m^3)	22.4	23.2	19.6	21	19.1	19.9
$w_{\gamma t\ max}$ (%)	10	7.6	23.2	16.2	25	21
Min. e	0.28	0.22	0.65	0.45	0.73	0.58
$w_{e\ min}$	9	7.2	22	15	24.5	18

It was observed that the variation of ultrasonic velocities with water content was similar to the variation of dry density with water content (Figs. 4 - 6). Velocity increased with increasing water contents up to the optimum water content. Beyond optimum, velocity decreased with increasing water contents. The maximum velocity was obtained approximately at the optimum water content. The highest variation with respect to maximum dry density, 1.5%, was for Soil 2 compacted with modified effort (Table 2). The variation of velocity with water content was also similar to the variation of total density with water content. In addition, the variation of velocity with water content was similar to inverse of the variation of void ratio with water content. The maximum velocity was obtained approximately at maximum total density and minimum void ratio (Table 2). The highest variation with respect to total density, 2.5%, was for Soil 3 compacted with modified effort. The highest variation with respect to void ratio, 1.5%, was for Soil 2 compacted with modified effort. For Soil 3, variation of velocity with water content was closer to variation of total density with water content than variation of dry density with water content. The location of maximum dry density was not clear for Soil 3 (LL = 83). Similar compaction plots were reported in the literature for soils with LL>70 (Brown 1996).

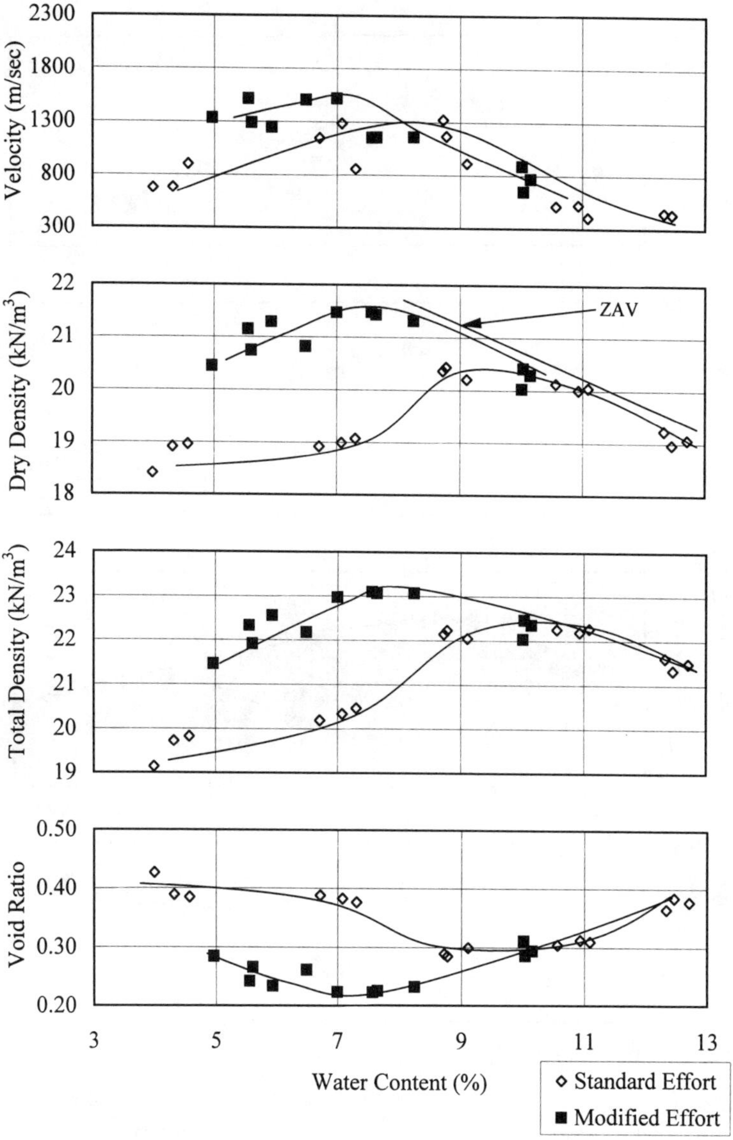

Figure 4. Compaction Characteristics of Soil 1

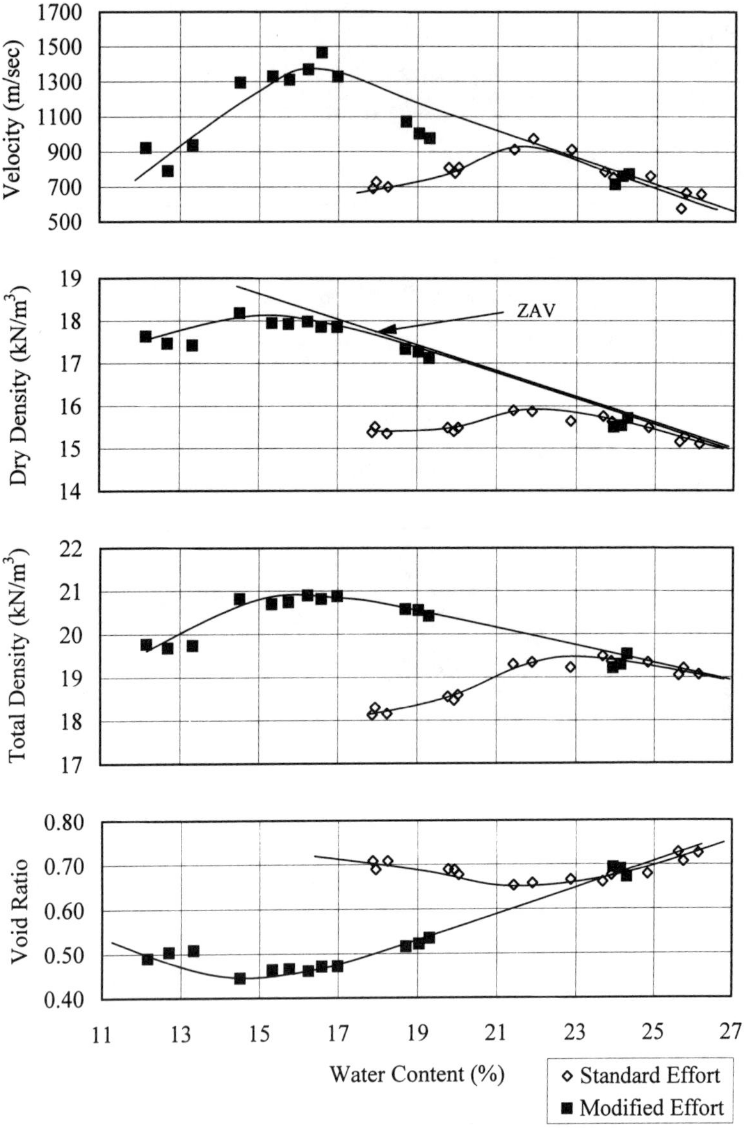

Figure 5. Compaction Characteristics of Soil 2

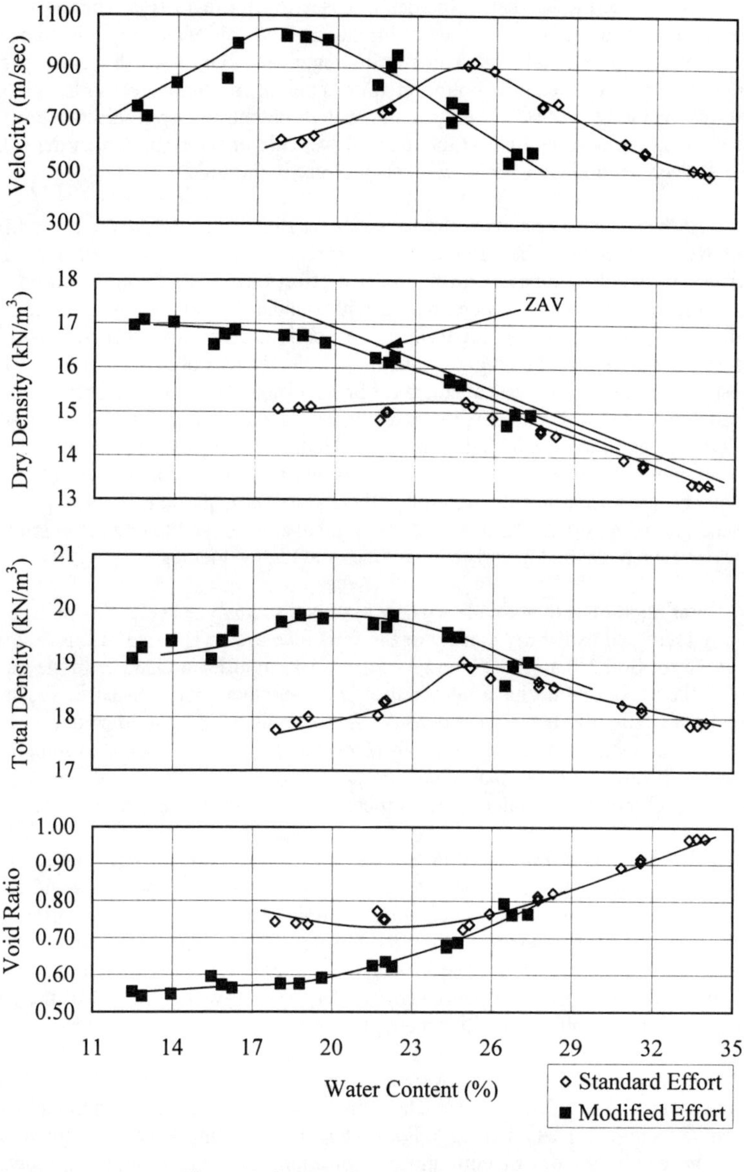

Figure 6. Compaction Characteristics of Soil 3

In a three-phase system such as a compacted soil, wave transmission occurs through all of the phases. In general, velocities in solids are higher than velocities in liquids, which are higher than velocities in gases (McIntire 1991). Therefore, it is expected that high velocities occur at high solid and low water and air contents. Highest solid and lowest void contents indicated by maximum dry density and minimum void ratio occur at optimum water content in compacted soils. Hence, maximum velocity occurs at the optimum water content. The velocities are lower than the maximum value both at the dry and wet side of the optimum as the solids contents decrease and water or air contents increase in the soils.

At the same water content the velocities for samples compacted with modified effort were higher than the velocities for samples compacted with standard effort (Figs. 4 - 6). Soils compacted with modified effort have higher solids content and lower voids content than soils compacted with standard effort. High solids content results in high velocities. Velocity in the test soils decreased with increasing clay content and plasticity. The highest maximum velocity was obtained for Soil 1 and lowest maximum velocity was obtained for Soil 3 (Table 2). As the plasticity and the clay content of the soils increase, the porosity also increases resulting in a decrease in the densities and increase in water contents. High densities and low water contents result in high velocities, whereas low densities and high water contents result in low velocities. In addition, it is expected that there were some particle contacts between the sand grains in Soil 1. Particle contacts could also have facilitated transmission of the waves which resulted in early arrival times and high velocities.

Variation of velocity with dry density was also analyzed (Fig. 7). In general, velocity increased as the dry density of the soils increased. The rate of this increase was more for Soil 1 than Soils 2 and 3. For Soil 1, significant changes in the travel path of the waves could have occurred as more contacts between particles (sand grains) occurred with increases in density. Increases in solid-to-solid contacts could increase velocity more than decreases in voids content (without increases in particle contacts). It must be noted that microstructural analyses were not conducted in this study to verify potential changes in the structure of the soils during compaction.

Use of through-transmission method in the field requires obtaining samples for the ultrasonic tests. However, access only to top surface of soils is required in surface-transmission without any need for sampling. Additional laboratory tests were conducted to compare velocities determined using through-transmission and surface-transmission measurements. The objective of these tests was to determine the feasibility of using surface measurements to determine P-wave velocity of the soils for potential field application of this method.

Surface transmission tests were conducted on samples prepared in 45.7-cm-diameter and 15.7-cm-deep molds. The compaction energy used was similar to the standard Proctor compaction energy. For each soil, three samples were prepared: one at approximately 5% dry of optimum, one at approximately optimum, and one at approximately 5% wet of optimum water content. The transducers were placed on the surface of the samples. Measurements were conducted at varying transducer separations. Velocity was determined as the ratio of the path length of the waves (distance between transducers) to the first arrival time of the waves. Through-transmission tests were also conducted.

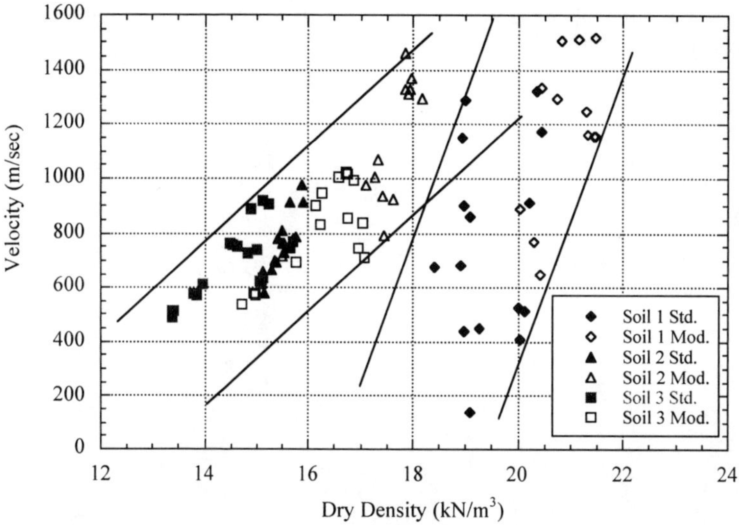

Figure 7. Variation of Velocity with Dry Density

There are uncertainties in the determination of the path length of the waves in surface-transmission. The path length can be affected by the non-uniform distribution of excitation over the cross sectional area of the piezoelectric crystal in a transducer (Krautkramer and Krautkramer 1983). The path length depends on the position of the portion of the crystal that undergoes maximum deformation. The velocity can vary depending on the assumed path length, three possible variations are presented below:

$$v = \frac{L_{c-c}}{t_{arr}}, \quad v = \frac{L_{ir-ir}}{t_{arr}}, \quad v = \frac{L_{or-or}}{t_{arr}} \qquad (1)$$

where:
v is the velocity,
$L_{c\text{-}c}$ is the distance between the center of the transducers,
$L_{ir\text{-}ir}$ is the distance between the inner rim of the transducers,
$L_{or\text{-}or}$ is the distance between the outer rim of the transducers.

The center-to-center transducer separations were varied between 5 cm and 25 cm at 5-cm increments. Velocities were determined using the three path lengths for each particular transducer separation. In addition, velocities were determined using through-transmission measurements. It was observed that the direct velocities (Fig. 8) were similar to the direct velocities obtained in the previous set of tests (Figs. 4 - 6). For Soil 1, the density of the soil compacted at the optimum water content was lower than the density obtained in the previous tests. As a result the velocity for the sample was lower than the velocity expected to be obtained at the optimum water content (Fig. 4 and Fig. 8).

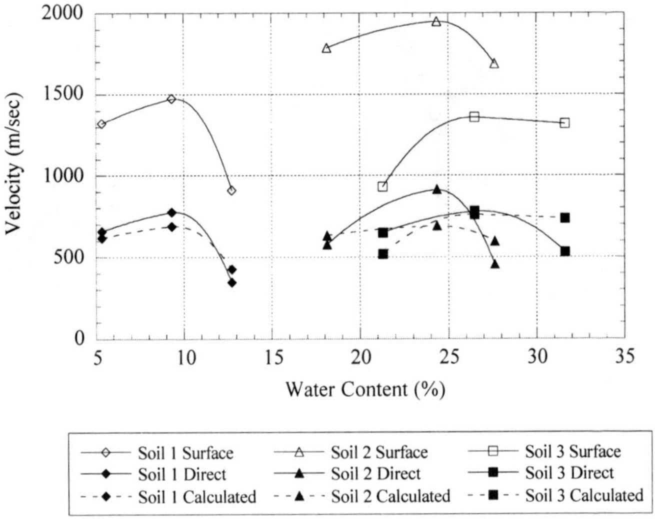

Figure 8. Direct and Surface Measurements

In most cases, velocities obtained from the surface measurements were not similar to direct velocities. There were differences in the magnitude of the velocities as well as differences in the variation of the velocities with water content (i.e. surface and direct measurements indicated dissimilar trends). For all of the soils, the variation of direct velocities with water content followed similar trends with the variation of 5-cm center-to-center surface velocities with water content (illustrated on Fig. 8). However, the magnitudes of these surface velocities were not similar to direct velocities. Variations in surface and direct velocity measurements were observed in ultrasonic testing of concrete (Komlos et al. 1996). A correction factor was suggested to be used in European standards to convert surface velocities to direct velocities. A similar approach was used in this study. A correction factor was calculated for each soil using the average of the ratios of the direct velocity to indirect velocity. The correction factors were 0.47, 0.35, and 0.56 for Soils 1, 2, and 3, respectively. The variation between measured and calculated (using the correction factor) direct velocities were between 2.6% to 38% with a mean variation of 20% (with 11% average deviation from the mean). The calculated direct velocities are also presented on Fig. 8. It is believed that an average variation of 20% between direct and surface measurements can be acceptable for compacted clayey soils.

Conclusions

Tests were conducted to analyze the use of ultrasonic methods to investigate compaction characteristics of clayey soils. Ultrasonic tests were conducted to determine P-wave velocity in the soils. It was observed that variation of velocity with water content was similar to the variation of density with water content. Maximum velocities and dry densities were within ± 1.5% water content for the test soils. Velocities increased with increasing compactive efforts and decreasing plasticities. These results were explained with the distribution of various phases in a compacted soil mass. High velocities occur at high solids contents and low water and air contents and conversely, low velocities occur at low solids contents and high water and air contents. The magnitude and variation of velocities obtained in this study correlated well with the distribution of the phases in the soil masses. In addition, it was observed that velocities increased with increasing dry density. The rate of velocity increases with density was higher for Soil 1 than Soils 2 and 3. Significant changes in the travel path of the waves could have occurred for Soil 1 as more particles (sand grains) contacted with increasing densities. Particle contacts could have facilitated transmission of the waves which resulted in early arrival times and high velocities. Potential changes in the structure of the soils during compaction were not verified by microstructural analysis in the study.

Through-transmission (with access required to opposite surfaces of soils) and surface-transmission (with access required only to top surface of soils) arrangements were used to determine velocity. Through-transmission arrangement was used to obtain baseline velocity measurements. Velocities that were on average within 20% of through-transmission velocities could be obtained with surface-transmission using soil-dependent correction factors.

In summary, it is believed that ultrasonic testing can be used effectively to determine compaction characteristics of soils both in the laboratory and in the field. Through-transmission tests can be conducted in the laboratory, whereas surface-transmission can be used in the field to determine velocity of compacted clayey soils. This method shows promise for determination of compaction properties of clayey soils in the field. The method can be used as an alternative to existing field methods. However, field verification is required to complete the development of the method.

Acknowledgement

Funding for this study was provided by the National Science Foundation under Grant CMS-9713922. Ms. Tarhonda Rhodes and Mr. Gnanatilake Gamage assisted with laboratory tests and data analysis. Their efforts are greatly appreciated.

References

Brown, R. W., Editor, (1996). *Practical Foundation Engineering Handbook*, McGraw-Hill, New York, NY.

Jones, I., and Facaoaru, I., (1969). "Recommendations for Testing Concrete by the Ultrasonic Pulse Method," *Materials and Structures*, RILEM, 2 (10), pp. 275-284.

Komlos, K., Popovics, S., Nurnbergerova, T., Babal, B., and Popovics, J. S., (1996). "Comparison of Five Standards on Ultrasonic Pulse Velocity Testing of Concrete," *Cement, Concrete, and Aggregates,* CCAGDP, 18 (1), pp. 42-48.

Krautkramer, J., and Krautkramer, H., (1983). *Ultrasonic Testing of Materials*, Springer-Verlag, New York, NY.

McIntire, P., (1991). *Nondestructive Testing Handbook, Volume Seven Ultrasonic Testing*, American Society for Nondestructive Testing, Columbus, OH.

Sheeran, D. E., Baker, W. H. and Krizek, R. J., (1967). "Experimental Study of Pulse Velocities in Compacted Soils," *Highway Research Record No. 177*, Highway Research Board, pp. 226-238.

Sologyan, A. I., (1990). "Survey of Methods and Means for Determining Soil Density in the Field," *Soviet Journal of Nondestructive Testing,"* Plenum Publishing Corp., 25 (7), pp. 480-486.

Wang, R., Haibo, G., Ay, C., Schuler, R., Gunasekaran, S., and Shinners, K., (1991). "Ultrasonic Method to Evaluate Soil Moisture and Compaction," presented at 1991 International Winter Meeting, Paper No. 91-1522, ASAE, St. Joseph, MI.

Winterkorn, H. E., and Fang, H. Y., Editors, (1975). *Foundation Engineering Handbook*, Van Nostrand Reinhold, Princeton, NJ.

Use of Grain-Size Functions in Unsaturated Soil Mechanics

Murray D. Fredlund[1]
G. Ward Wilson[2]
Delwyn G. Fredlund[2]

Abstract

The grain-size distribution is commonly used for soil classification; however, there is also potential to use the grain-size distribution as a basis for estimating unsaturated soil behavior. Mathematically representing the grain-size distribution provides several benefits to soil mechanics. An example is the estimation of the soil-water characteristic curve from the grain-size distribution. Much emphasis has recently been placed on the estimation of unsaturated soil-property functions from the soil-water characteristic curve. Several methods have been proposed that use the grain-size distribution as the basic information for the estimation of the soil-water characteristic curve.

Two mathematical forms are presented to represent grain-size distribution curves; namely, a unimodal form and a bimodal form. The equations presented in this paper can provide a close representation of a wide variety of grain-size distribution for different soil types.

1. Introduction

The grain-size distribution is a simple, yet informative classification test routinely performed in soil mechanics. Valuable information regarding the amount of each particle size can be determined in the laboratory through the use of a series of sieves and hydrometer analysis. Recent research has made use of the grain-size distribution as the basis for the estimation of soil properties such as the soil-water characteristic curve (Gupta and Larson, 1979; Arya and Paris, 1981; Haverkamp and Parlange, 1986). It is of value to be able to mathematically represent the grain-size distribution curve as a continuous function that will allow further analysis to be performed. The

[1] Graduate student ,Department of Civil Engineering, University of Saskatchewan, Saskatoon, Sask., S7N 5A9
[2] Professor of Civil Engineering, Department of Civil Engineering, University of Saskatchewan, Saskatoon, Sask., S7N 5A9

specific objective of this paper is to develop a mathematical description for the grain-size distribution of any given soil. This forms the foundation step for the general procedure to determine related soil property functions and constitutive relations.

Mathematically representing the grain-size distribution provides several benefits to soil mechanics. Firstly, the soil can be classified using the best-fit parameters. Secondly, the mathematical equation can be used as the basis for further soil analysis and description. An example is the estimation of the soil-water characteristic curve from the grain-size distribution (Fredlund et al, 1997). Thirdly, a mathematical equation can provide a method of representing the entire curve between measured data points.

Two models to fit grain-size data are proposed in this paper. These models consist of a unimodal and a bimodal mathematical function. The two new equations provide great flexibility for fitting a wide variety of soils.

2. *Definition of variables*

The grain-size distribution for a soil is defined as the relationship between percent passing (by mass) and the particle size. It has also been called the mass-based aggregate size distribution or the ASD. The *particle size* represents the size of particles that can pass a particular sieve mesh. The *percent passing* represents the mass percentage of particles passing a particular sieve size.

3. *Background*

ASTM D422-54T (1958) presented a standard for determining the grain-size distribution. Standard sieve sizes, reporting methods, and methods for performing a hydrometer analysis are presented. The sieve analysis allowed points on the grain-size distribution to be determined for particle sizes greater than the #200 sieve or 0.074 mm. The hydrometer analysis presented by ASTM, standardizes a method for determining the grain-size distribution for particles smaller than the #200 sieve. Interpretation of the grain-size distribution is typically carried out manually.

Gardner (1956) used a two-parameter, log-normal distribution to fit grain-size distribution data. Kemper and Chepil (1965) further confirmed the work of Gardner (loc cit.). The log-normal distribution often failed to provide a close fit of the grain-size distribution at the extremes of the curve (Gardner, 1956; Hagen et al, 1987). Wagner and Ding (1994) later improved upon the log-normal equation by presenting three and four parameter log-normal equations.

Campbell (1985) presented a classification diagram based on the assumption that the particle-size distribution is approximately log-normal. This assumption led to the particle-size distribution being approximated with a Gaussian distribution function. With this assumption, any combination of sand, silt, and clay can be represented by a geometric (or log) mean particle diameter and a geometric standard deviation. Values were summarized in a modified USDA textural classification chart by Shirizi and Boersma (1984).

The first limitation associated with using a log-normal type of equation is the assumption that the grain-size distribution is symmetric. In reality, the grain-size distribution is often non-symmetric and can be better fit by a different type of equation. Secondly, a method for fitting soils that are bimodal or gap-graded is often of value and the four-parameter log-normal equations have not been found to be satisfactory for fitting gap-graded grain-size distributions.

There are three primary types of grain-size distributions. These three types of distributions are known as *well-graded* soils, *uniform* soils, and *gap-graded* soils. Figure 1 illustrates each type of grain-size distribution. This paper focuses on these three categories of grain-size distributions and provides equations to fit the experimental data for each category. Well-graded soils and uniform soils are examined and a unimodal method of fitting an equation is developed. Then a mathematical method of representing a gap-graded soil is subsequently presented.

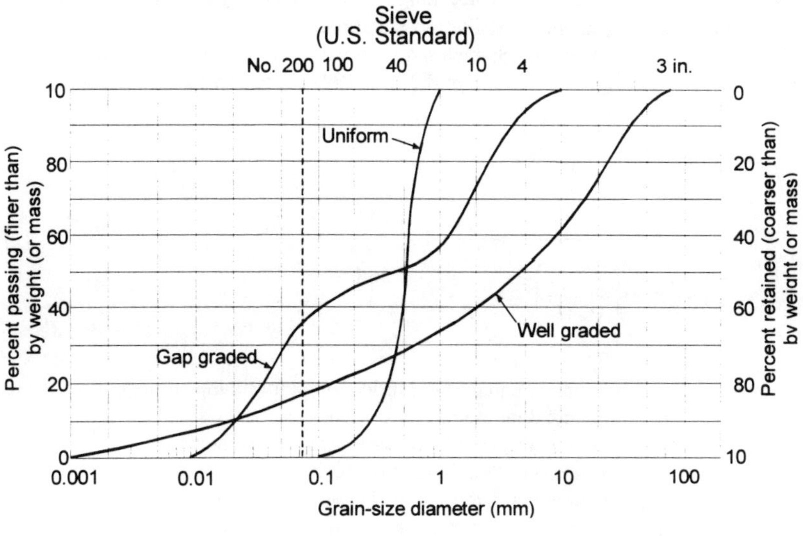

Fig. 1 Three primary types of grain-size distribution curves (Holtz and Kovacs, 1981)

4. *Unimodal Equation for the Grain-size Distribution*

The selection of an appropriate, mathematical equation involves a review of a variety of equations that could be used to fit soils data. It has been observed that the soil-water characteristic curve possessed a shape similar to that of the grain-size distribution. This is to be expected since the soil-water characteristic curve describes the void distribution in a soil while the grain size curve provides information on the distribution of the solid phase of the soil. Since the solids plus the voids add up to the total soil volume, it is to be expected that the distribution of the solids phase (i.e., grain-size distribution) would tend to bear an inverse type relationship to the distribution of voids (i.e., soil-water characteristic curve), and vice versa.

Equations used to fit the soil-water characteristic curve have been proposed by Brooks and Corey, 1964; Gardner, 1974; van Genuchten, 1980; Burdine, 1953; Mualem, 1976; Fredlund and Xing, 1994. Brooks and Corey (1964) and Gardner (1974) presented three parameter equations while van Genuchten (1953) and Fredlund and Xing (1994) presented four parameter equations. It would appear that a similar forms of equations could be used to represent the grain-size distribution.

An accurate representation of the clay fraction of the grain-size distribution was considered necessary in order to complete the mathematical function. The Fredlund and Xing (1994) equation allows independent control over the lower end of the curve (i.e., the fine particle size range), and was selected as the basis for the development of a grain-size distribution equation. The reversed scale of the grain-size distribution as well as characteristics unique to the grain-size distribution, required that the original Fredlund and Xing (1994) equation to be modified to the form shown below:

$$P_p(d) = \frac{1}{\left[ln\left[exp(1)+\left(\frac{a_{gr}}{d}\right)^{n_{gr}}\right]\right]^{m_{gr}}}\left[1-\left[\frac{ln\left(1+\frac{d_{rgr}}{d}\right)}{ln\left(1+\frac{d_{rgr}}{d_m}\right)}\right]^7\right] \qquad [1]$$

where: a_{gr} = parameter equal to the inflection point on the curve and related to the initial breaking point of the curve,
n_{gr} = parameter related to the steepest slope of the curve,
m_{gr} = parameter related to the shape of the curve,
d_{rgr} = parameter related to the diameter of the fines in a soil,
d = diameter of any particle size under consideration, and
d_m = diameter of the minimum allowable size particle.

Equation [1] is referred to as the unimodal equation and can be used to fit a wide variety of soils as shown in Figs. 2, and 3. A quasi-Newton least squares regression algorithm was used to adjust three of the five parameters to fit the equation to each soil. The algorithm progressively minimizes the squared differences between the equation and experimental data. The best-fit particle size distribution function can be plotted over the grain-size distribution data, typically on a logarithmic scale.

The unimodal equation provides significant improvements in the fit of grain-size data over previous mathematical representations (i.e., log-normal distribution) of particles present in a soil. Another form can also be used to visualize the distribution of particle sizes by differentiating the particle size distribution curve. The differentiation produces a particle-size probability density function (PDF). The differentiated form of the unimodal grain-size equation can be seen in Eq. [2]. The parameters presented in the particle-size probability density function, PDF, are the same as defined in Eq. [1].

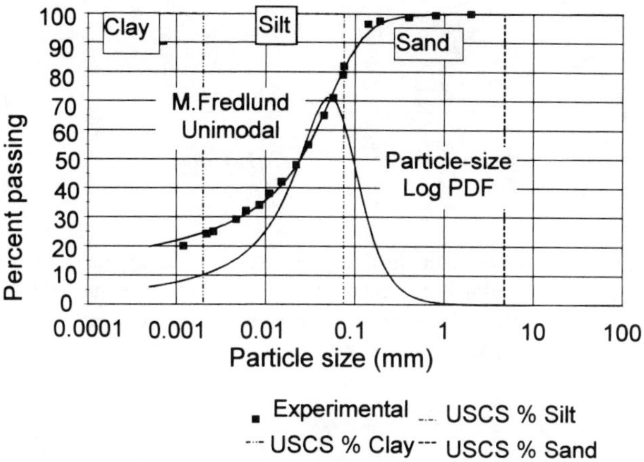

Fig. 2 Logarithmic probability density function for uniform silt from the Pilot Butte area of Saskatchewan (optimum compression) best fit with the unimodal equation.

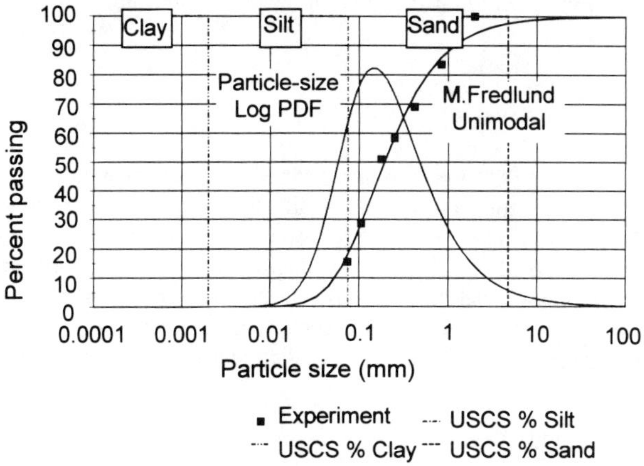

Fig. 3 Logarithmic probability density function for Rubicon Sandy Loam best fit with the unimodal equation.

$$\frac{dp_p}{dd}=\frac{1}{\left[ln\left[exp(1)+\left(\frac{a_{gr}}{d}\right)^{n_{gr}}\right]\right]^{m}}\left[1-\frac{ln\left(1+\frac{d_{rgr}}{d}\right)^{7}}{ln\left(1+\frac{d_{rgr}}{d_m}\right)^{7}}\right]m_{gr}\left(\frac{a_{gr}}{d}\right)^{n_{gr}}\frac{n_{gr}}{d\left[\left[exp(1)+\left(\frac{a_{gr}}{d}\right)^{n_{gr}}\right]ln\left[exp(1)+\left(\frac{a_{gr}}{d}\right)^{n_{gr}}\right]\right]}$$

$$+\frac{7}{\left[ln\left[exp(1)+\left(\frac{a_{gr}}{d}\right)^{n_{gr}}\right]\right]^{m_{gr}}}\frac{ln\left(1+\frac{d_{rgr}}{d}\right)^{6}}{ln\left(1+\frac{d_{rgr}}{d_m}\right)^{7}}\frac{d_{rgr}}{\left[d^{2}\left(1+\frac{d_{rgr}}{d}\right)\right]} \quad [2]$$

The particle size distributions presented in this paper are calculated using Eq. [2]. The highest point on the PDF plot is the mode or the most frequent particle size. Since Eq. [2] is a PDF, the natural laws of probability hold and the area under the differentiated curve must equal 1 as shown below.

$$\int_{-\infty}^{+\infty}\frac{dP_p}{dd}dx=1 \quad [3]$$

where: x = particle-size diameter.

Equation [2] can also be used to calculate probabilities. Equation [4] shows how to calculate the probability that a soil particle diameter will fall in a certain range. Equation [2] can be arithmetically integrated between specified particle diameter sizes and the probability can be determined by the following relationship.

$$probability(d_1<d<d_2)=\int_{x=d_1}^{x=d_2}p(x)dx \quad [4]$$

It is convenient to represent the PDF function in a different manner when plotting on a logarithmic scale. The arithmetic PDF function will often appear distorted when plotted on a logarithmic scale. The peak of Eq. [4] will not represent the most frequent particle size because of the logarithmic distribution of the particle-size scale. To overcome this limitation, the PDF function is often represented as shown in Eq. [5]. Taking the log of particle size and differentiating the grain-size equation produces a PDF that appears more physically realistic. The peak of Eq. [5] will represent the most frequent particle size.

$$p_l(d)=\frac{dP_p}{d\log(d)}=\frac{dP_p}{dd}\ln(10)\cdot d \quad [5]$$

where: $p_l(d)$ = logarithmic probability density function.

The probability of the logarithmic PDF can be calculated as follows.

$$probability(d_1 < d < d_2) = \int_{x=log(d_1)}^{x=log(d_2)} p_l(x)dx \qquad [6]$$

The probability density function for various grain-size curves can be seen in Figs. 2, and 3.

5. Parametric Study of the Proposed Grain-size Distribution Equation

A parametric study of the proposed unimodal equation shows behavior similar to that of the original Fredlund and Xing (1994) equation. The a_{gr} parameter is related to the initial break of the equation and its effect on the grain-size distribution curve can be seen in Fig. 4 where a_{gr} is varied and the other equation parameters are held constant. The a_{gr} parameter provides an indication of the largest particle sizes.

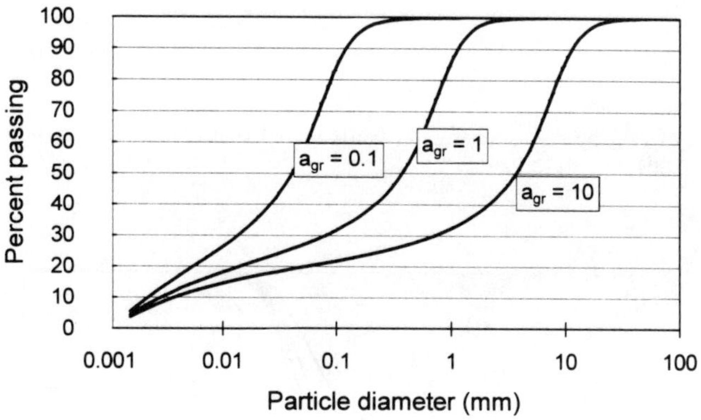

Fig. 4 Effect of varying the a_{gr} parameter while n_{gr} = 4.0, m_{gr} = 0.5, dr_{gr} = 1000, and d_m = 0.001.

Figure 5 shows how the parameter n_{gr} influences the slope of the grain-size distribution. The point of maximum slope of the grain-size distribution indicates the gradation of the particle sizes (i.e., on a logarithm scale) in the soil as seen in Fig. 5. The parameter m_g controls the break onto the finer particle size of the sample. The effect of the m_g parameter can be seen in Fig. 6. The parameter, d_{rgr}, affects the shape of this fine particle size of the curve. However, the amount of variation produced on the curve is quite minimal as shown in Fig. 7. In some cases the d_{rgr} can be modified to improve the fit of the overall equation. It was found that a value of 0.001 for d_{rgr} provided a reasonable fit in most cases.

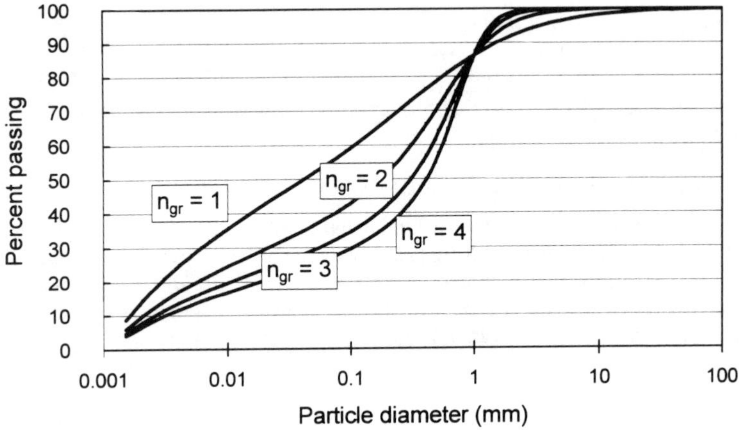

Fig. 5 Effect of varying the n_{gr} parameter while a_{gr} = 1.0, m_{gr} = 0.5, dr_{gr} = 1000, and dm = 0.001

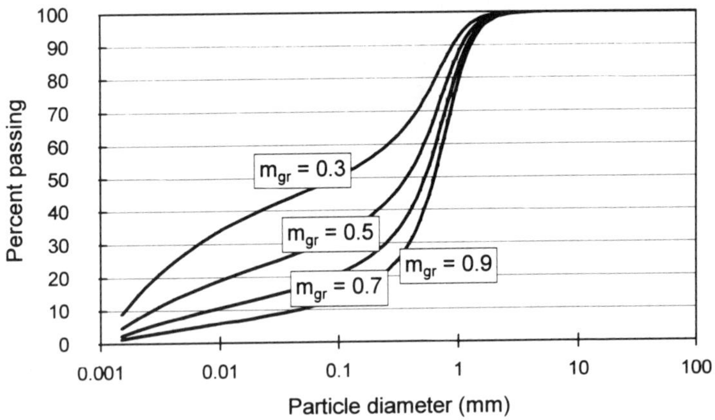

Fig. 6 Effect of varying the m_{gr} parameter while a_g = 1.0, n_{gr} = 4.0, dr_{gr} = 1000, and d_m = 0.001

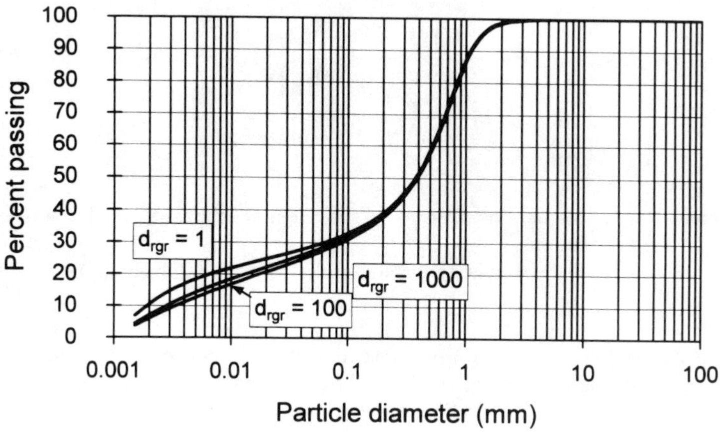

Fig. 7 Effect of varying the d_{rgr} parameter while a_{gr} = 1.0, n_{gr} = 4.0, m_{gr} = 0.5, and d_m = 0.001

The unimodal equation (Eq. [1]) appears to have versatility in handling a wide variety of soil types.

6. Bimodal Equation for the Grain-size Distribution Curve

There is a limitation in using the unimodal equation (i.e., Eq. [1]) when the soils are gap-graded. In this case, it is necessary to consider the use of a bimodal, best-fit. Soils frequently have particle size distributions that are not consistent with a unimodal distribution and as a result, attempts to fit the unimodal equation to certain data sets can often lead to unsatisfactory results.

The characteristic shape of a bimodal or gap graded soil is the double "hump" seen in the experimental data. These anomolies indicate that the particles are concentrated around two separate particle size ranges. From a mathematical standpoint, a gap-graded soil can be viewed as a combination of two or more separate soils (Durner, 1994). This allows for the "stacking" of more than one unimodal equation.

$$P_p(d)=\left[\sum_{i=1}^{k} w_i\left[\frac{1}{\left[ln\left[exp(1)+\left(\frac{a_{gr}}{d}\right)^{n_{gr}}\right]\right]^{m_{gr}}}\right]\right]\left[1-\left[\frac{ln\left(1+\frac{d_r}{d}\right)}{ln\left(1+\frac{d_r}{d_m}\right)}\right]^7\right] \qquad [7]$$

where: k = the number of "subsystems" for the total particle-size distribution,
w_i = the weighting factors for the subcurves, subject to $0 < w_i < 1$ and
$\Sigma w_i = 1$.

For a bimodal curve, k would be equal to 2 and the number of parameters to be determined would be 4 times [k + (k-1)]. The unimodal equation is used as the basis for the bimodal equation. The final equation for a bimodal curve is shown below in its extended form.

$$P_p(d)=\left\{w\left[\frac{1}{\left(\ln\left(\exp(1)+\left(\frac{a_{bi}}{d}\right)^{n_{bi}}\right)\right)^{m_{bi}}}\right]+(1-w)\left[\frac{1}{\left(\ln\left(\exp(1)+\left(\frac{j_{bi}}{d}\right)^{k_{bi}}\right)\right)^{l_{bi}}}\right]\right\}\left[1-\left(\frac{\ln\left(1+\frac{dr_{bi}}{d}\right)}{\ln\left(1+\frac{dr_{bi}}{d_m}\right)}\right)^7\right] \qquad [8]$$

where: a_{bi} = parameter related to the initial breaking points on the curve,
n_{bi} = parameter related to the steepest slope on a portion of the curve,
m_{bi} = parameter related to the shape of the curve,
j_{bi} = parameter related to the second breaking point along the curve,
k_{bi} = parameter related to the second steep slope along the curve,
l_{bi} = parameter related to the second shape of the curve,
dr_{bi} = parameter related to the amount of fines in a soil,
d = diameter of any particle size under consideration, and
d_m = diameter of the minimum allowable size particle.

The bimodal data sets can be closely fit using the bimodal best-fit equation (Fig. 8). However, the bimodal fit provided only an *adequate* fit of unimodal data sets. In other words, unimodal data sets were better fit using the unimodal equation. The results of fitting the bimodal curve to several different soils can be seen in Figs. 8 to 10.

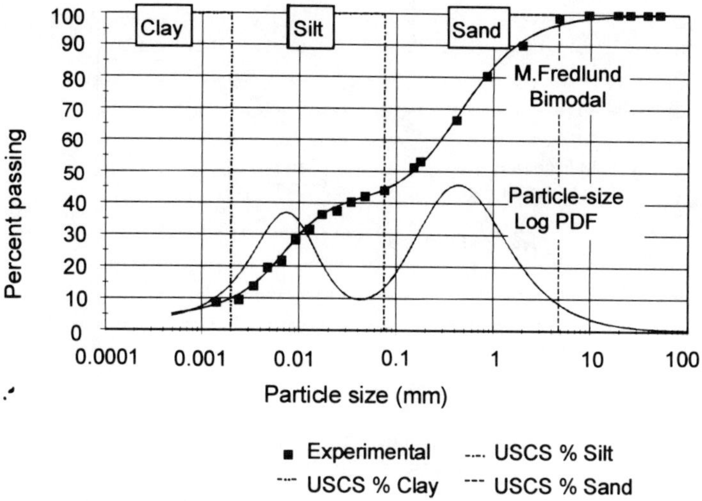

Fig. 8 Logarithmic probability density function fitted with a bimodal equation, for a gap-graded Saprolitic Soil tested at the University of Saskatchewan

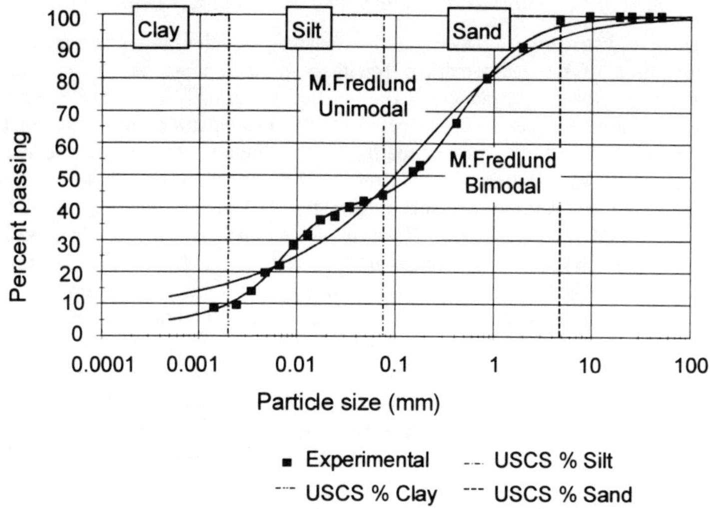

Fig. 9 Example of a bimodal fit on a gap-graded Saprolitic Soil tested at the University of Saskatchewan ($R^2 = 0.999$)

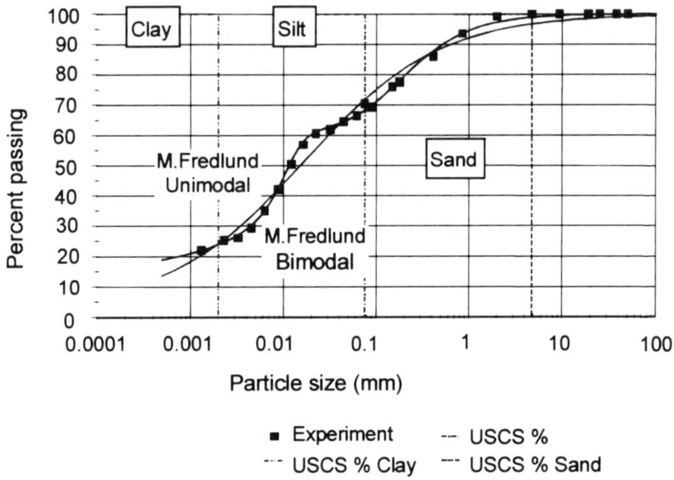

Fig. 10 Example of a bimodal fit of a gap-graded Saprolitic Soil tested at the University of Saskatchewan ($R^2 = 0.999$)

7. Application of the Mathematical Function for the Grain-size Distribution

The grain size distribution has been used extensively for the classification of soils. The application of the mathematical equations in this paper can be applied to geotechnical engineering practice. The use of equations to fit the grain-size distribution provides several advantages. Firstly, the equations presented in this paper provide a method for estimating a continuous function. Secondly, quantification of soils based on their grain size distribution is possible when equations are fit to datasets of soils information. Thirdly, equations provide a consistent method for determining physical indices such as percent clay, percent sand, percent silt, and particle diameter variables such as d_{10}, d_{20}, d_{30}, d_{50}, and d_{60}.

It has also been found that the grain size distribution is central to most methods of estimating the soil-water characteristic curve (Gupta and Larson, 1979; Arya and Paris, 1981; Haverkamp and Parlange, 1986, Ranjitkar and Sunder, 1989). An accurate representation of the soil particle sizes is essential when the grain-size distribution curve is used as the basis for the estimation of the soil-water characteristic curve. The equations presented in this paper appear to provide an excellent basis for the estimation of the soil-water characteristic curve (Fredlund et. al, 1997).

7.1 Parameters of the grain-size distribution equations

The unimodal fit of the grain-size distribution has been fit to many experimentally measured grain-size data sets extracted from research papers. The unimodal fit performed well with the exception of soils exhibiting bimodal behavior. The parameters of the unimodal equation vary in a manner similar to the parameters in the Fredlund and Xing (1994) soil-water characteristic curve equation. This study also investigated whether equation parameters could be grouped according to soil textural classifications. For example, is there a range of the n_{gr} parameter typical for silty sands? The results of this research indicate that *general* parameter groups can be identified but *specific* parameter groupings cannot be identified. The influence of equation parameters on each other does not allow for *specific* groupings. It was found that grouping soils is more successful when parameters with physical significance are selected. Successful groupings of soil properties has been achieved by combining soils according to physical parameters such as percent clay, percent silt, and percent sand and through the use of variables such as d_{10}, d_{20}, d_{30}, d_{50}, and d_{60}.

7.2 Determining physical parameters from the grain-size distribution equation

One of the benefits of the two grain-size equations presented in this paper is that conventional physical variables can be computed from the curves. The most commonly used variables are percent clay, percent sand, and percent silt. Also used are diameter variables such d_{10}, d_{20}, d_{30}, d_{50}, and d_{60}. The equations presented are of the form, $P_p(d)$ where d is particle diameter (mm). The percent clay, percent silt, and percent sand can therefore be a computed by substituting in the appropriate diameters. The diameters used depend upon the criteria associated with the various classification methods. The divisions can be determined for any classification method by substituting into the equations the appropriate diameters as shown in Fig. 11.

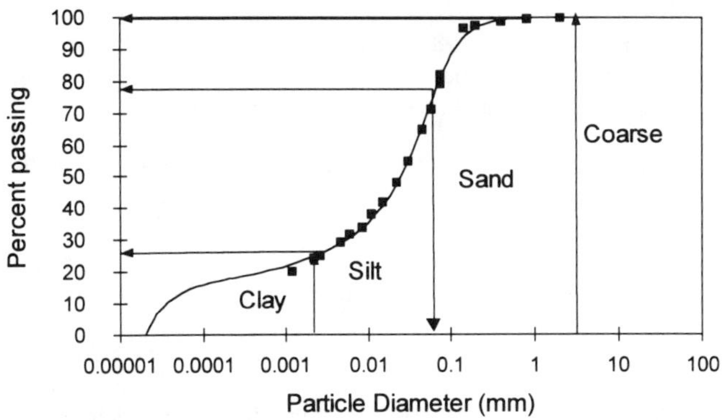

Fig. 11 Determination of the soil fractions (i.e., % clay, % silt, and % sand) when using the unimodal equation

The diameter variables must be computed in an inverse manner. The particle size diameter answers to the question,"What particle diameter has 10 percent of the total mass smaller than this size?" Taking the inverse of either the unimodal or bimodal equation is difficult. A half-length algorithm was therefore used to compute diameters from the grain size curve. An initial guess diameter was selected and the correction distance was progressively halved until the iteration process yielded a minimal error.

8. *Conclusions*

Unimodal and bimodal equations are presented to fit essentially any grain-size distribution dataset. The unimodal equation was found to provide a good fit of a variety of soils. The extremes of the grain-size distribution were also well-fit by the equation.

Gap-graded soils can be fit using a bimodal equation. The bimodal equation allows for a mathematical representation of any grain-size distribution where the sample contains two distinctly different, but dominant particle size groups.

Mathematical representation of the grain-size distribution provides numerous benefits. Curves can be identified and categorized. Likewise, the grain-size curves can be located in a data base using searching techniquies. Grain-size variables (i.e., *% clay*, d_{10}, d_{60}, etc.) can be mathematically determined from the equation. The unimodal and bimodal equations provide a method for fitting the three primary types of well-graded soils, uniform soils, and gap-graded soils.

The proposed continuous mathematical function for the grain-size curve sets the stage for further analysis to estimate the soil-water characteristic curve of a soil.

9. *References*

Arya, L. M., and Paris J. F., 1981, A physicoempirical model to predict the soil moisture characteristic from particle-size distribution and bulk density data, Soil Science Society of America Journal, Vol. 45, pp. 1023-1030.

Brooks R. H. and Corey A.T., 1964, Hydraulic Properties of Porous Media, Colorado State Univ. Hydrol. Paper, No. 3, 27, pp. March 1964.

Burdine, N. T., 1953, Relative permeability calculations from pore size distribution data, Journal of Petroleum Technology, Vol. 5, No. 3, pp. 71-78.

Campbell, G. S., 1985, Soil Physics with Basic, Elsevier, New York.

Fredlund, D. G., and Xing, A., 1994, Equations for the soil-water characteristic curve, Canadian Geotechnical Journal, Vol. 31, No. 3, pp. 521-532.

Fredlund, M. D., Fredlund, D. G., and Wilson, G. W., 1997, Prediction of the Soil-Water Characteristic Curve from Grain-Size Distribution and Volume-Mass

Properties, 3rd Brazilian Symposium on Unsaturated Soils, Rio de Janeiro, April 22-25.

Gardner, W. R., 1974, The permeability problem, Soil Science, 117, 243-249.

Gupta, S. C., and W. E. Larson, 1979, Estimating soil-water retention characteristics from particle size distribution, organic matter percent, and bulk density, Water Resources Research Journal, Vol. 15, No. 6, pp. 1633-1635.

Hagen, L. J., Skidmore, E. L. and Fryrear, D.W., 1987, Using two sieves to characterize dry soil aggregate size distribution, Transactions of the ASAE, 30(1), 162-165.

Haverkamp, R., and Parlange, J. Y., (1986), Predicting the water-retention curve from a particle-size distribution: 1. Sandy soils without organic matter, Soil Science, Vol. 142, No. 6, pp. 325-339.

Holtz, R. D. and Kovacs, W. D., 1981, An introduction to geotechnical engineering, Prentice-Hall, Inc., Englewood Cliffs, New Jersey.

Kemper, W. D. and Chepil, W. S., 1956, Size distribution of aggregates. In Methods of Soil Analysis, part 1. ed. C.A. Black Agronomy 9:499-510.

Kohnke, H., 1968, Soil Physics, McGraw-Hill Book Company, New York.

Mualem, Y., 1976, A new model for predicting the hydraulic conductivity of unsaturated porous media, Water Resources Res., 12, 513-522.

Ranjitkar S., and Sunder B., 1989, Prediction of hydraulic properties of unsaturated granular soils based on grain size data, Ph.D Thesis, University of Massachusetts, 75-131.

Shirizi M. A, and Boersma L., 1984, A unifying quantitative analysis of soil texture, Soil Science Society of America Journal, 48, 142-147.

van Genuchten, M. T., 1980, A closed form equation for predicting the hydraulic conductivity of unsaturated soils, Soil Science Society America Journal, pp. 892-890.

Wagner, L. E., and Ding, D., 1994, Representing aggregate size distributions as modified lognormal distributions, American Society of Agricultural Engineers, Vol. 37, No. 3, pp. 815-821.

Soil-Water Characteristic Curve Variability

Claudia E. Zapata[1], William N. Houston[2], Sandra L. Houston[2], and Kenneth D. Walsh[3]

Abstract

The most widely-used constitutive relations for unsaturated soils use matric soil suction as a state variable. The soil-water characteristic curve (SWCC), the relationship between soil suction and some measure of the water content, can be measured or predicted based on soil index properties such as the grain-size distribution (GSD) function. Estimation based on index properties is highly desirable due to its simplicity and low cost and would be the path of choice to the SWCC, provided the accuracy of the estimate were adequate. Whether measured or estimated, there is variability and uncertainty associated with the SWCC that in turn will directly impact any model for the unsaturated soil behavior that makes use of this relationship. The variability in the SWCC associated with direct suction measurements as well as the variability associated with the prediction of the SWCC based on GSD was investigated. Three different soils that cover a typical range of soils encountered in practice were used in this study. The investigated sources of variability related to direct suction measurements included the equations used to fit the suction data, the different methods available to measure suction, the operator, the number of data points used to define the SWCC, and the range of suction covered in the measurements. The variability observed in the SWCC when different predictive algorithms are used is also reported. Finally, a new model for predicting the SWCC based on soil index properties is presented, using a database of 190 soils. Results showed that the operator as well as the range of suction covered in the measurements produced significant variability in the measured SWCC. Surprisingly, the variability in the SWCC as predicted by soil index properties and/or GSD-based algorithms was found to be as small or smaller than that associated with the operator (person(s) measuring the SWCC).

Introduction

The soil-water characteristic curve (SWCC) is the relationship between soil water suction and the water content of the soil. The form of water content most often used when soil water suction is initially measured is gravimetric water content. However, when published, the most commonly used form of water content is volumetric water content (Houston et al., 1999). The degree of saturation, S, is also used sometimes as a measure of water content for the SWCC.

The suction used for the SWCC is usually the matric suction, u_a-u_w, but total suction is occasionally used as well. The SWCC directly relates to the shear strength, volume change, and fluid flow characteristics of unsaturated soils. It is now generally

[1] Faculty Research Associate, Arizona State University, Department of Civil and Environmental Eng., P.O Box 875306, Tempe, AZ 85287-5306

[2] Professor, Arizona State University, Department of Civil and Environmental Eng., P.O Box 875306, Tempe, AZ 85287-5306

[3] Associate Professor, Arizona State University, Del E. Webb School of Construction, Tempe, AZ, 85287

accepted that the behavior of an unsaturated soil cannot be related to a single effective stress variable. Rather, the total normal stress, σ_n, the pore-air pressure, u_a, and the pore-water pressure, u_w, must be combined in two independent stress state parameters. The two stress variables usually selected are the net normal stress $(\sigma_n - u_a)$, and the matric suction $(u_a - u_w)$ (Fredlund and Morgenstern,1977). The SWCC is commonly used together with the water content to estimate soil suction for use in the calculation of the state variable. Therefore, uncertainty in the SWCC relationship leads to variability in predicting unsaturated soil behavior. Despite the well-recognized importance of suction, it has not become routinely measured in geotechnical laboratories or in the field. An investigation of practice throughout the United States, conducted as a part of the present study, showed that less than 20% of commercial geotechnical laboratories performed suction measurements on a regular basis (Zapata, C., 1999). It also appears that much less than 20% of the laboratories of practicing geotechnical consultants perform suction measurements on a routine basis. This fact suggests that the geotechnical engineering profession would rely on empirically derived methods for estimating the SWCC, if the estimations were of satisfactory accuracy.

To encourage geotechnical engineers to implement unsaturated soil mechanics theory in routine practice, it is necessary to evaluate to what degree of accuracy the SWCC can be ascertained using various approaches. To quantify the variability associated with the determination of the SWCC, it is necessary to also quantify the uncertainty associated with direct suction measurements. This uncertainty will serve as a benchmark to decide whether or not the variability in any given prediction is acceptable. If, for example, the uncertainty associated with the predicted SWCC were comparable to that associated with the directly measured SWCC, then one might be prone to embrace the predictive algorithm.

The objectives of the studies described in this paper were:

1) To evaluate the variability in the SWCC.
2) To develop a new relationship between the SWCC and soil index properties.

Background

The development of a SWCC for a particular soil from suction measurements can require several tests and can take some time to obtain all the necessary data. In order to facilitate the efficient determination of the SWCC, several mathematical models have been developed to describe the SWCC for a particular soil from just a few points. The second important reason why we fit functions to the experimental SWCC raw data is that many applications of the SWCC require that it be differentiated or integrated and be continuous. Therefore, a set of data for a SWCC without a function fitted to it is of limited use. Some of the fitted functions that have been proposed are summarized in Table 1. Generally, the equations in Table 1 have been validated for certain soils and ranges in suction. The equations with three and four parameters seem to be the more suitable to represent the SWCC (Leong and Rahardjo, 1996). However, at this point it is not possible to state that a particular equation is capable of best-fitting every soil encountered. The process of fitting experimental suction data to one of the proposed equations requires a minimum number of experimentally obtained suction measurements, depending upon the number of unknown parameters in the chosen function.

Table 1. Soil-Water Characteristic Curve Equations Used to Fit Experimental Data

Reference	Equation	Parameter Description
Fredlund and Xing(1994) (F&X)	$$\theta_w = C(h)\times\left[\frac{\theta_s}{\left[\ln\left[\exp(1)+\left(\frac{h}{a}\right)^b\right]\right]^c}\right] \quad ...(1)$$ $$C(h)=\left[1-\frac{\ln\left(1+\frac{h}{h_r}\right)}{\ln\left(1+\frac{10^6}{h_r}\right)}\right] \quad ....(2)$$	θ_w volumetric water content a = a soil parameter which is primarily a function of the air entry value of the soil in kPa. b = a soil parameter which is primarily a function of the rate of water extraction from the soil, once the air entry value has been exceeded. c = a soil parameter which is primarily a function of the residual water content. h_r = a soil parameter which is primarily a function of the suction at which residual water content occurs in kPa.
van Genuchten (1980) (van G.)	$$\theta_w = \theta r + \frac{\theta_s - \theta r}{\left[1+\left(\frac{h}{a}\right)^b\right]^c} \quad ....(3)$$	θ_r = residual volumetric water content. a = a soil parameter which is primarily a function of the air entry value of the soil in kPa. b = a soil parameter which is primarily a function of the rate of water extraction from the soil, once the air entry value has been exceeded. c = a soil parameter which is primarily a function of the residual water content.
McKee and Bumb (1987)	$$\theta_w = \theta r + \frac{\theta s - \theta r}{1+\exp(1)^{\left[\frac{(h-a)}{b}\right]}} \quad ....(4)$$	θ_r = residual volumetric water content. a = curve-fitting parameter b = curve-fitting parameter
van Genuchten and Mualem (1980) (van G.&M)	$$\theta_w = \theta r + \frac{\theta_s - \theta r}{\left[1+\left(\frac{h}{a}\right)^{bm}\right]^{\left(1-\frac{1}{bm}\right)}} \quad ....(5)$$	θ_r = residual volumetric water content. a = a soil parameter which is primarily a function of the air entry value of the soil in kPa. b_m = a soil parameter which controls the slope at the inflection point in the soil-water characteristic curve.
van Genuchten and Burdine (1980) (van G.&B)	$$\theta_w = \theta r + \frac{\theta_s - \theta r}{\left[1+\left(\frac{h}{a}\right)^{b}\right]^{\left(1-\frac{2}{b}\right)}} \quad ....(6)$$	θ_r = residual volumetric water content. a = a soil parameter which is primarily a function of the air entry value of the soil in kPa. b = a soil parameter which is primarily a function of the rate of water extraction from the soil, once the air entry value has been exceeded.

Table 1. Continued

Reference	Equation	Parameter Description
Gardner (1958)	$\theta_w = \theta_r + \dfrac{\theta_s - \theta_r}{1+\left(\dfrac{h}{a}\right)^b}$ ....(7)	θ_r = residual volumetric water content. a = a soil parameter which is primarily a function of the air entry value of the soil in kPa. b = a soil parameter which is primarily a function of the rate of water extraction from the soil, once the air entry value has been exceeded.
Brooks and Corey (1964) (B&C)	$\theta_w = \theta_r + (\theta_s - \theta_r)\left(\dfrac{a_b}{h}\right)^{b_b}$ ....(8)	θ_r = residual volumetric water content. a_b = bubbling pressure in kPa. b_b = pore size index.
Williams et al. (1983)	$\ln \Theta_e = A + B \ln h$ ....(9)	A = fitting parameter B = fitting parameter
Farrel and Larson (1972)	$h = (u_a - u_w)_b \exp[\alpha(\theta_s - \theta_w)]$ ....(10)	α = empirical constant $(u_a - u_w)_b$ = air-entry value
Assouline et al. (1998)	$\theta_w = \theta_L + (\theta_s - \theta_L)\left[1 - \exp\left[-\xi\left(\dfrac{1}{\psi} - \dfrac{1}{\psi_L}\right)^{\eta}\right]\right]$ ....(11)	Ψ = capillary head Ψ_L = capillary head that corresponds to a very low water content, at which the hydraulic conductivity is negligible. θ_L = volumetric water content at capillary head Ψ_L. η = fitting parameter ξ = fitting parameter

Soil suction measurement methods can be generally classified as either direct or indirect. Direct methods include pressure plates, pressure membranes, and tensiometers. These methods measure the pore pressure in the soil or impose a known pressure to the soil and allow the water content to come to equilibrium with the imposed pressure. Indirect methods include filter paper, porous blocks, and heat dissipation sensors. These methods use measurements or indicators of water content or a physical property that is sensitive to a change in water content (e.g., water content in the filter paper method, electrical resistance, and thermal conductivity). Before suction measurements are taken, a baseline calibration run is made and all suction measurements are based on this calibration. Thorough descriptions of methods to measure soil suction can be found in Ridley and Wray (1995), and Lee and Wray (1995).

Another option for obtaining the SWCC is to predict it, based on the GSD and other soil properties. The vast majority of the proposed predictive algorithms can be classified into three major approaches. The first of these approaches is based upon statistical estimation of water contents at selected matric suction values. These water contents, at each suction value, are correlated to soil properties. This process generally requires a regression analysis followed by a curve fitting procedure. The final product may or may not involve an equation describing the SWCC. This approach has been followed by several researchers including Visser (1969), Gupta and Larson (1979), Rawls et al. (1982), Cassel, Ratliff, and Ritchie (1983), Hutson and Cass (1987), Rawls, Gish, and Brakensiek (1991), Rajkai, K. and Varallyay, Gy. (1992), Rawls and Brakensiek (1982), Gregson et al. (1987), Williams et al. (1992), Livneh, M., Kinsky, J., and Zaslavsky, D. (1970), Reddi, L. and Poduri, R. (1997), Aubertin, M., Ricard, J., and Chapuis, R. (1998), and Mbagwu, J. and Mbah, C. (1998).

The second approach includes those methods that correlate, by regression analysis, soil properties with the fitting parameters of an analytical equation that represents the SWCC. This statistical approach has been followed by Ghosh, R. (1980), Williams et al. (1983), Ahuja et al. (1985), Rawls, W., Ahuja, L., and Brakensiek, D. (1992), Cresswell, H. and Paydar, Z. (1996), and Tomasella, J. and Hodnett, M. (1998), as well as others.

Although the first two approaches both use a statistical correlation between soil texture and other soil data and water contents, they differ in the parameters which are correlated with soil properties. van Genuchten and Leij (1992) suggest that the first of these two approaches should be preferred over the second for reasons of simplicity.

The third approach includes the methods that estimate the SWCC using a physics-based conceptual model. It involves physical models based upon the conversion of the GSD (textural information) into a pore-size distribution, which in turn is related to a distribution of water contents and associated pore pressures. This approach was first presented by Arya, L. and Paris, J. in 1981, and followed by Haverkamp, R. and Parlange, J. (1986); Mishra, S., Parker, J. C., and Singhal, N. (1989); Arya and Dierolf (1992); Smettem, K. and Gregory, P. (1996); Basile, A. and D'Urso, G. (1997), and Fredlund, M. et al. (1997), among others. This approach has generated considerable interest, although there is still a need for improvement in the representation of the pore space.

For those models that use statistical correlation (approaches one and two above) the quality of the results obtained obviously depends heavily on the database use to fit and test the equations. Those models using a physical conceptual base are generally limited to soils with large pores, namely sandy soils, and to low matric suction ranges. The predictions for fine-grained soils remain rather unreliable. It seems that the representation of the pore-size distribution (PSD) is the most critical factor in formulating relationships between pore and particle radii, which in turn are used to estimate the SWCC.

Comparison of the different models can be found in van Genuchten and Leij (1992); Williams, R. D. and Ahuja, L. R. (1992); Kern, J. S. (1995); and Nandagiri, L. and Prasad, R. (1997), among others.

Variability in the SWCC

The first research objective is subdivided into two parts. Part 1 focuses on factors that are believed to cause variability in the SWCC when experimental measurements of suction are used. Part 2 is directed toward assessing the variability in the SWCC when it is predicted from correlations with soil index properties or when first principles are used to predict its position. The approaches used under Part 2 are based on either empirical relationships or surface tension and calculated radii of menisci.

Part 1: Variability in the SWCC when Estimated from Direct Suction Measurements

Variability should be expected when direct measurement of matric suction is used to define the SWCC because it is very difficult to measure suction with precision. Several factors may contribute to this variability, including:

1) The equation used to fit the experimental data
2) The operators' ability and experience
3) The number of data points used to define the SWCC
4) The range in suction covered by the actual measurements
5) The method used to acquire matric suction data

Soil Characterization

In an effort to quantify the variability in SWCC, a soil characterization program was conducted for three different soils. The soils were selected to cover a typical range of soils frequently encountered in engineering practice. The three test soils are referred to as El Paso sand, Price Club silt, and Fountain Hills clay. Each soil was thoroughly mixed and split in such a way that all sub-samples were representative and very close to identical.

Soil characterization involved sieve analysis, hydrometer analysis, Atterberg limits, and specific gravity of solids. Grain-size distribution curves for the three soils are shown in Figure 1. The soils were classified as a poorly graded sand, SP (El Paso), a sandy-low plasticity silt, ML (Price Club), and a silty-highly plastic clay, CH (Fountain Hills). The results from the soil characterization tests are summarized in Table 2.

Matric Suction Data Acquisition

About 80 commercial geotechnical and agricultural laboratories and educational institutions that performed soil testing on a regular basis were contacted and surveyed. Survey results showed that 14 institutions were willing to perform soil matric suction measurements for a SWCC, using at least one method. The most commonly used methods were the pressure plate, the pressure membrane, and the buried filter paper.

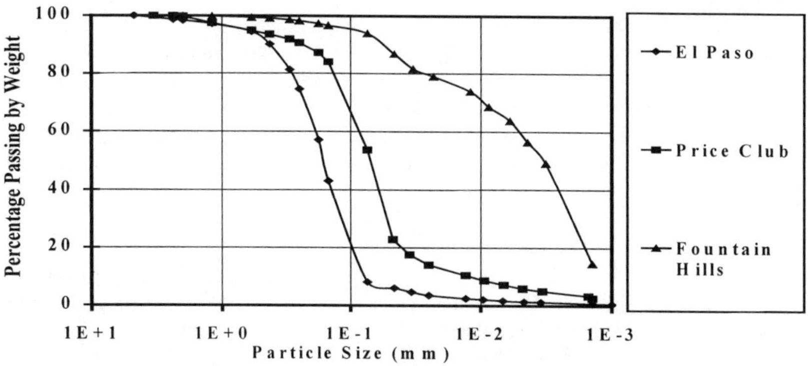

Figure 1. Grain-Size Distribution Curves

Table 2. Soil Characterization

Characteristics	El Paso	Price Club	Fountain Hills
Grain Size Distribution			
Percent sand	91.9	46.4	6.1
Percent silt	6.7	47.6	34.7
Percent clay	1.4	6.0	59.2
Smallest Particle Size (mm)	1e-4	1e-5	1e-5
Atterberg Limits			
Liquid Limit, LL (%)	---	22	70
Plastic Limit, PL (%)	NP	18	29
Plasticity Index, PI (%)	---	4	41
Specific Gravity of Solids, G_s	2.65	2.73	2.77
Dry Density, ρ_d (gm/cm^3)	1.39	1.33	1.14
USCS Classification	SP	ML	CH

Those facilities that used pressure plate extractors exclusively had limitations in the range of suction they could provide. Generally, pressure plate and pressure membrane measurements were limited to 1500 kPa due to the air-entry value of the ceramic plates or membranes used. Thus, the full range of suction was not covered when pressure plates and membranes were used exclusively. On the other hand, laboratories that measured matric suction by means of buried filter paper could give a wider range of measured values.

The three soils were sent to nine different laboratories chosen from those contacted to acquire the matric suction information. In addition, two more data sets were acquired in our laboratory by different operators using different suction determination methods. Table 3 is a summary of the data reported by each laboratory. Relationships between matric suction and volumetric water content reported by the different laboratories are shown in Figures 2, 3, and 4 for El Paso sand, Price Club silt, and Fountain Hills clay, respectively.

A preliminary data analysis was performed in order to identify outlying data. Based on statistical considerations and visual inspection, the data provided by laboratories 2, 10, and 11 for El Paso sand were considered outliers and were eliminated from further consideration. For Price Club silt, data from laboratories 2 and 10 were also eliminated from the analysis, while for the Fountain Hills clay, the statistical analysis did not identify any particular outlier. The reduction of the database resulted in retention of 43, 40, and 43 data points, for the sand, silt, and clay, respectively.

Best-Estimate SWCC and its Confidence Band

To perform variability studies, it was deemed necessary to fit the suction data with a continuous function. Based on previous studies (e.g., Leong and Rahardjo, 1996), seven equation were chosen to fit the data from those listed in Table 1. The equations are:

1) Equations (1) and (2) by Fredlund and Xing (F&X)
2) Equation (1) by Fredlund and Xing with C(h) = 1
3) Equation (3) by van Genuchten (van G.)

4) Equation (5) by van Genuchten and Mualem (van G.&M)
5) Equation (6) by van Genuchten and Burdine (van G.&B)
6) Equation (7) by Gardner, and
7) Equation (8) by Brooks and Corey (B&C)

Table 3. Data Collected from Each Laboratory

Lab #	Number of Suction Points			Dry Density (gm/cm^3)			Method
	El Paso	Price Club	Fountain Hills	El Paso	Price Club	Fountain Hills	
Lab 1	4	4	4	1.39	1.33	1.11 1.02	pp
Lab 2	2	2	2	1.38	1.39	1.19	pp
Lab 3	3	3	3	1.31	1.43	1.01	pp
Lab 4	1	3	4	1.38	1.28	1.18	fp
Lab 5	12	10	7	1.38	1.39 1.28	1.18	fp
Lab 6	13	7	6	1.38	1.28	1.18	pm
Lab 7	3	3	4	1.36 1.23 1.20	1.37 1.10 1.32	0.92 0.94 1.19 1.21	fp
Lab 8	4	4	5	1.25 1.45 1.57 1.60	1.39 1.43 1.50 1.73	1.25 1.16 1.13 1.26 1.17	fp
Lab 9	3	3	3	1.54	1.48	1.12	pp
Lab 10	5	5	5	1.38	1.28	1.18	fp
Lab 11	3	3	---	1.38	1.28	---	fp

pp = pressure plate; pm = pressure membrane; fp = filter paper

The fit to the experimental data was performed by an iteration process that minimized the sum of the squared residuals (SS_E). Once the best-estimate SWCC was found for each soil, its 95% confidence band (95% c.b.) was determined. The confidence band is associated with the fitting process and represents the uncertainty in the fitting parameters of the equations used to represent the data. Those interested in the details of the procedure for finding the confidence band for the response function at any abscissa are referred to Bates and Watts (1988), Mishra et al. (1989), and Huet et al. (1996).

The best-estimate SWCCs and their confidence bands are depicted in Figures 5, 6, and 7, for the sand, the silt, and the clay, respectively. The equation proposed by Fredlund and Xing (eq. 1) was found to best fit the data for the sand, while the van Genuchten (eq. 3) was found to be the best fit to the experimental data for the silt and the clay, although the Fredlund and Xing equation was a fairly close second for these soils as well.

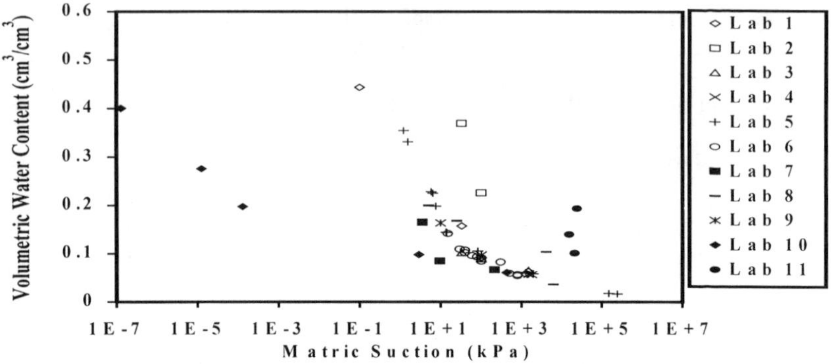

Figure 2. Matric Suction Experimental Results for El Paso Sand

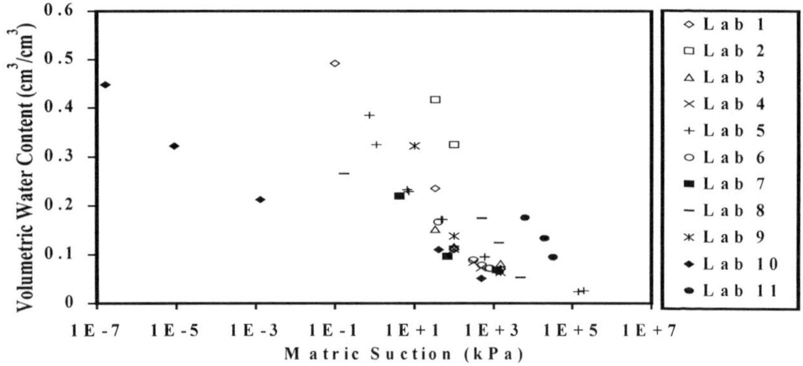

Figure 3. Matric Suction Experimental Results for Price Club Silt

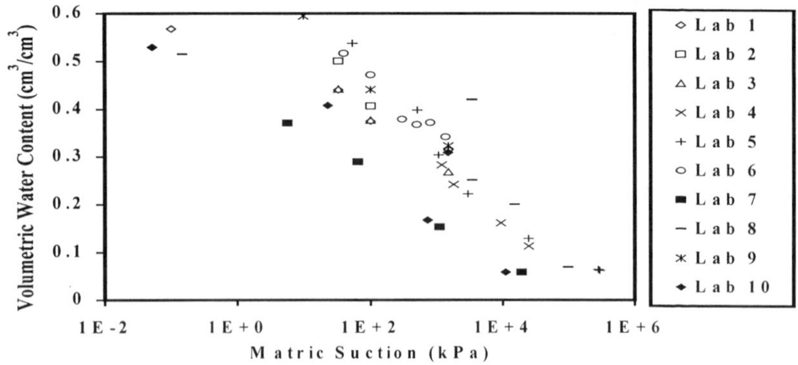

Figure 4. Matric Suction Experimental Results for Fountain Hills Clay

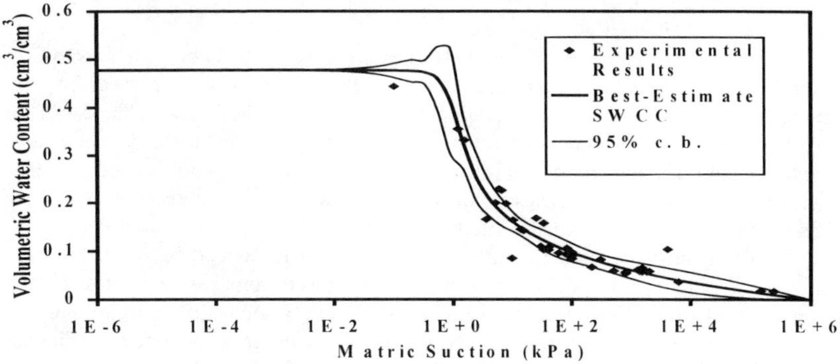

Figure 5. Best-Estimate SWCC and Confidence Band - El Paso Sand

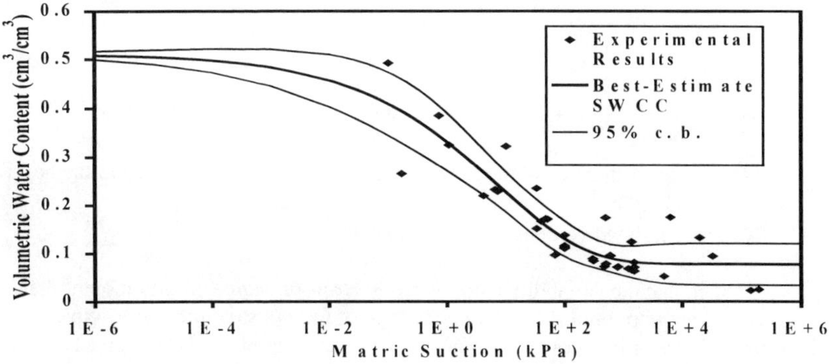

Figure 6. Best-Estimate SWCC and Confidence Band - Price Club Silt

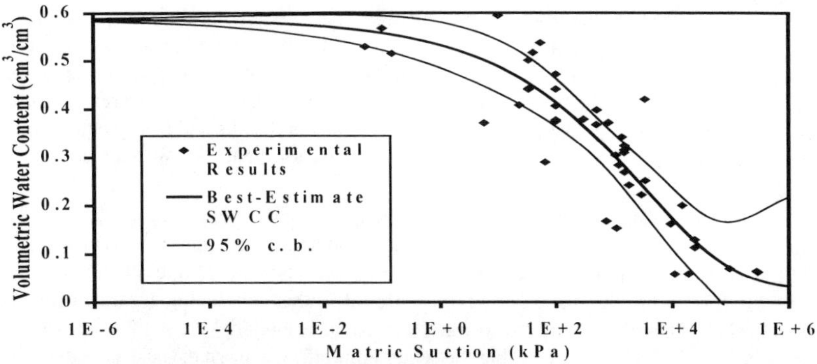

Figure 7. Best-Estimate SWCC and Confidence Band – Fountain Hills Clay

Variability in the SWCC as a Function of the Type of Equation Used to Fit Experimental Data

In practice, an engineer is unlikely to examine seven different equations to fit experimental suction data. Therefore, it may be useful to know what error can be expected from choosing a single equation that may not give the *best* fit to the data. For this purpose, the seven equations previously discussed were fitted to the data obtained for the three soils. The range of variability in water content determination related with the type of equation used to fit the experimental data is depicted in Figures 8, 9, and 10 for El Paso sand, Price Club silt, and Fountain Hills clay, respectively.

To quantify the variability around the best estimate value, the predicted volumetric water contents obtained by each equation were compared with the ordinates from the best-fit equation, at each level of suction. The absolute difference of those two values was plotted as a percentage of the saturated volumetric water content in Figures 11, 12, and 13, corresponding to El Paso sand, Price Club silt, and Fountain Hills clay, respectively. The variability was computed as follows:

$$\%\ \text{Variability} = \frac{\left|\hat{\theta}_i - \bar{\theta}_i\right|}{\theta_{sat}} \times 100\% \qquad (12)$$

Where:

$\hat{\theta}_i$ = predicted volumetric water content at i suction level, for the equation being tested

$\bar{\theta}_i$ = best estimate volumetric water content at i suction level, from the best fit equation

θ_{sat} = saturated volumetric water content

Notice that the variability is measured in terms of degree of saturation of the soil, based on the assumption that the volume change of the soil structure is negligible. This definition of variability allows for comparison of different materials, as well as for extrapolation to materials similar to the ones in this study. To evaluate the magnitude of the variability, the 95% confidence band for the best-estimate SWCC was plotted in each figure, also as a percentage of the saturated volumetric water content.

A summary of the results is presented in Table 4. Among the four-parameter equations, the one proposed by van Genuchten had the best overall performance for the range of soils studied, while the Fredlund and Xing equation, using a correction factor equal to 1, performed the best among the three-parameter equations. The rest of the equations had an acceptable performance except for the Brooks and Corey equation, which deviated substantially near the air-entry region for the three soils (see bold numbers in Table 4).

For El Paso sand, the predicted θ_w given by the different equations deviated from the one given by Fredlund and Xing. The Fredlund and Xing equation provided the best fit for the El Paso sand and it forces the curve to zero volumetric water content at 1,000,000 kPa. Fredlund and Xing (1994) provide rather convincing arguments that all curves should have essentially zero water content at 1,000,000 kPa, perhaps the most convincing of which is the fact that the relative humidity at oven dryness is extremely low and corresponds approximately to

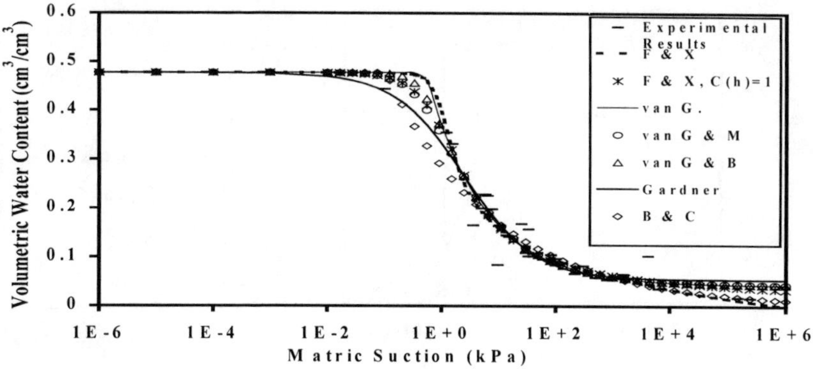

Figure 8. El Paso Sand Data Fitted with Different Equations

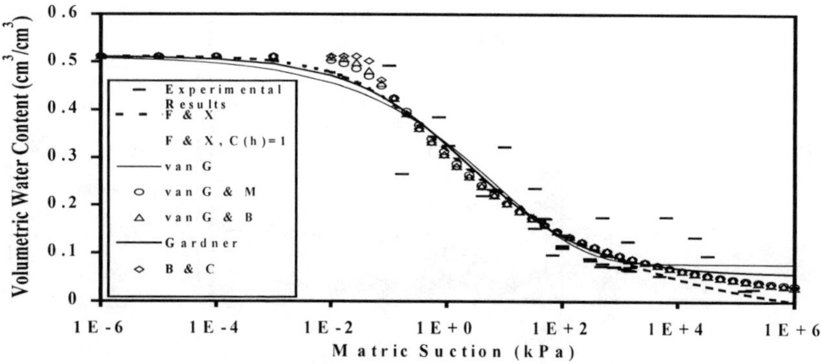

Figure 9. Price Club Silt Data Fitted with Different Equations

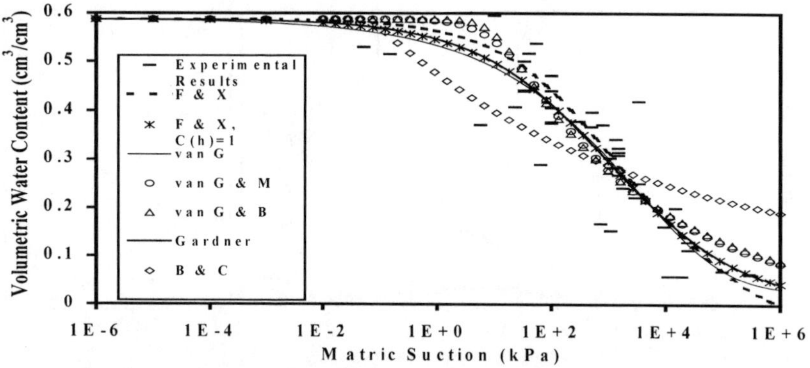

Figure 10. Fountain Hills Clay Data Fitted with Different Equations

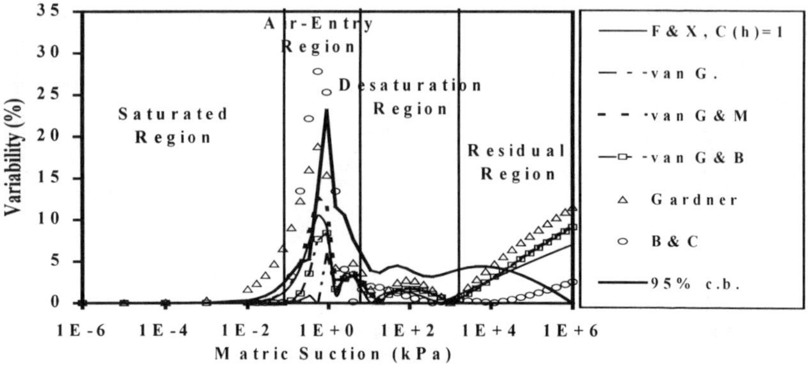

Figure 11. Variability due to Different Equations - El Paso Sand

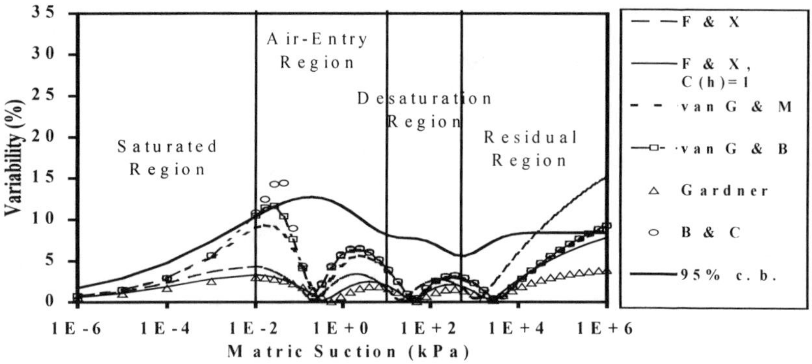

Figure 12. Variability due to Different Equations – Price Club Silt

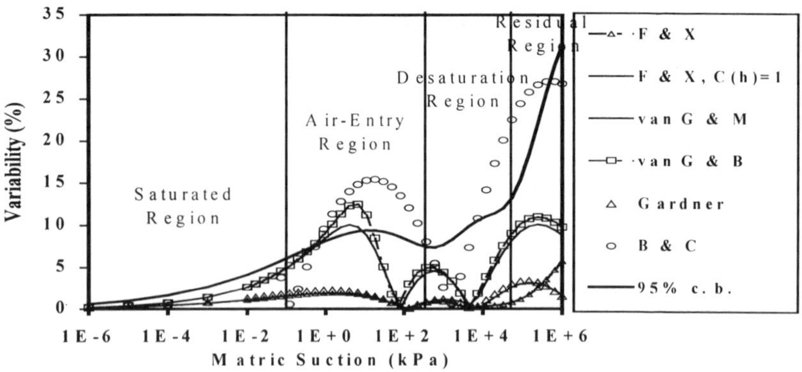

Figure 13. Variability due to Different Equations – Fountain Hills Clay

Table 4. Maximum Percentage of Variability due to Different Equations

Equation	Saturated Region	Air-Entry Region	Desaturation Region	Residual Region
		El Paso Sand		
95% Confidence Band	2	23	6	4
Fredlund & Xing	0	0	0	0
Fredlund &. Xing, C(h)=1	1	11	2	**7**
van Genuchten	0	6	2	**9**
van Genuchten & Mualem	1	9	3	**9**
van Genuchten & Burdine	0	8	2	**9**
Gardner	**7**	19	**7**	**12**
Brooks and Corey	0	**28**	2	3
		Price Club Silt		
95% Confidence Band	10	13	8	8
Fredlund & Xing	4	4	3	**15**
Fredlund &. Xing, C(h)=1	3	3	2	8
van Genuchten	0	0	0	0
van Genuchten & Mualem	9	9	4	**9**
van Genuchten & Burdine	10	12	4	**9**
Gardner	3	3	2	4
Brooks and Corey	**11**	**14**	4	**9**
		Fountain Hills Clay		
95% Confidence Band	6	9	12	31
Fredlund & Xing	2	2	1	6
Fredlund &. Xing, C(h)=1	2	2	2	3
van Genuchten	0	0	0	0
van Genuchten & Mualem	5	**10**	7	10
van Genuchten & Burdine	5	**12**	8	11
Gardner	2	2	2	3
Brooks and Corey	4	**15**	**20**	27

1,000,000 kPa. Studies described in the latter part of this paper show that SWCCs passing through 1,000,000 kPa at zero water content fit very well for essentially all sands and fairly well for most silts and clays.

Variability in the SWCC due to Different Operators

The variability in the SWCC due to different operators was analyzed by fitting a SWCC to each set of data collected from the different laboratories. Each of these SWCCs was then compared to the best fit equation for each soil (i.e., Fredlund and Xing for the sand, van Genuchten for the silt and the clay). The results are depicted in Figures 14, 15, and 16, after excluding outliers and laboratories reporting less than 3 data points due to the minimum requirement when fitting a nonlinear equation. The variability was accounted the same way as it was for the previous analysis. Results are shown in Figures 17, 18, and 19 for the sand, the silt and the clay, respectively. A summary of the results is presented in Table 5.

El Paso sand results show that the function fitted to data obtained from laboratory 8 gives a misrepresentation of the SWCC with a variability as high as 47% in the air-entry region (see bold numbers in Table 5). Laboratories 3, 5, and 6 provided what appears to be the best representation of the SWCC throughout the entire range of suction measurements. For the Price Club silt, data reported by laboratory 11 showed high deviation, as high as 55%, in the desaturation region. Note that the SWCC produced by Laboratory 11 is skewed to the right of the best-estimate SWCC by at least two orders of magnitude (Figure 15). Data from laboratories 3, 4, and 5 gave a good fit compared to the best-estimate SWCC for this soil. Fountain Hills clay data show laboratories 4, 8, and 10 as the best predictors. The rest of the laboratories failed to produce SWCCs which matched well throughout the entire range of suction (labs 3, 6, 7, 9, and 10) or in a particular region, showing variability in the results as high as 54% (laboratory 6).

Variability in the SWCC as Related to the Number of Data Points Reported

An attempt was made to correlate the number of data points and the range in suction measurements reported, with the performance of each laboratory. The results are shown in Figure 20. For El Paso sand, those laboratories that reported 12 and 13 data points (laboratories 5 and 6, respectively) provided a good estimate of the SWCC. However, data from lab 3 also provided a good estimate of the SWCC with only three data points. Therefore, some correlation was found for the sand, but there is not very clear evidence that the quality of the resulting SWCC increases substantially with number of points.

For the silt and the clay, poor correlation was found between the number of reported data points and the maximum variability. It seems that three data points are enough to make a good representation of the entire SWCC, provided of course that all three are relatively reliable. Comparisons within the overall data set (sand, silt, and clay together) support this conclusion.

Variability in the SWCC as Related to the Range in Suction Measurements Reported and The Method Used for Measurement

The range in suction provided by each lab is determined by the method used to acquire the data. Therefore, the effects of these factors were considered together. The method employed by each laboratory for determining the matric suction of the soil is listed in Table 3. Out of eleven, six laboratories used the buried filter paper method, four used the pressure plate method, and one used the pressure membrane method. It is reasonable to expect the pressure plate and the pressure membrane methods to produce comparable results, because both of them give a desorption curve. It is not clear whether the filter paper method gives a desorption or an adsorption curve, as this depends on the method used to calibrate the filter paper.

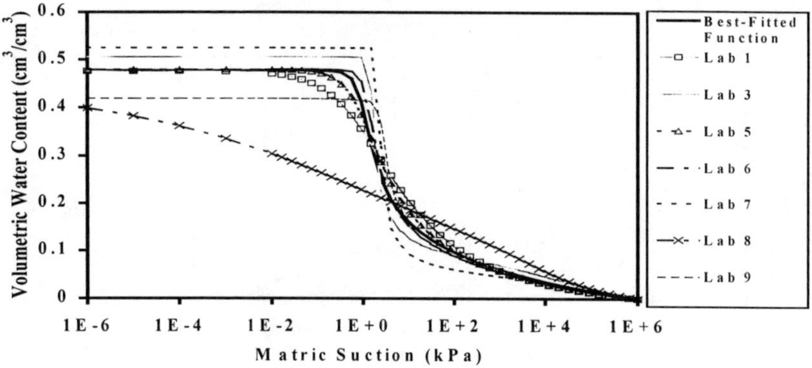

Figure 14. El Paso Sand SWCCs by Different Operators

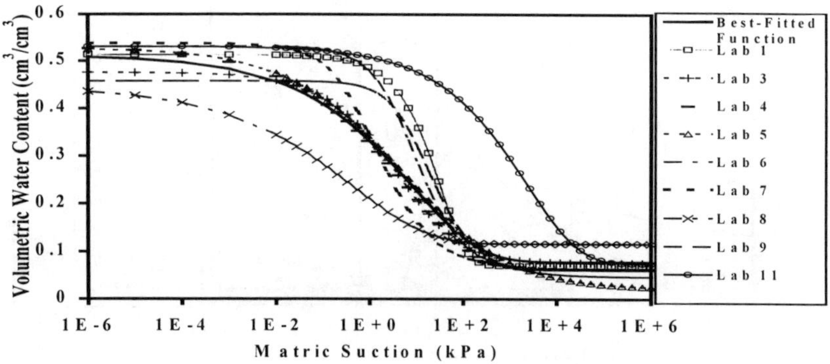

Figure 15. Price Club Silt SWCCs by Different Operators

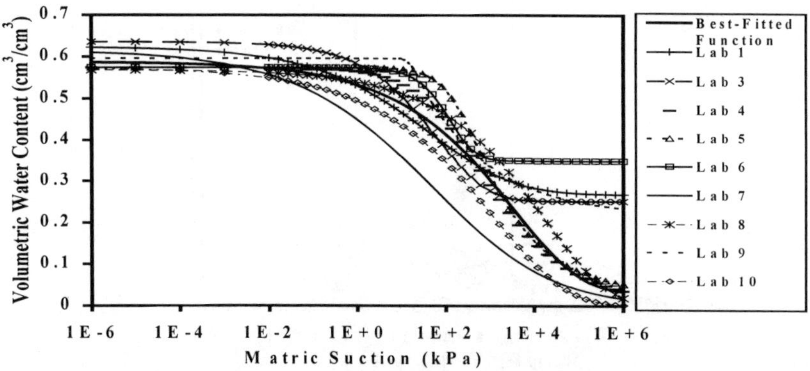

Figure 16. Fountain Hills Clay SWCCs by Different Operators

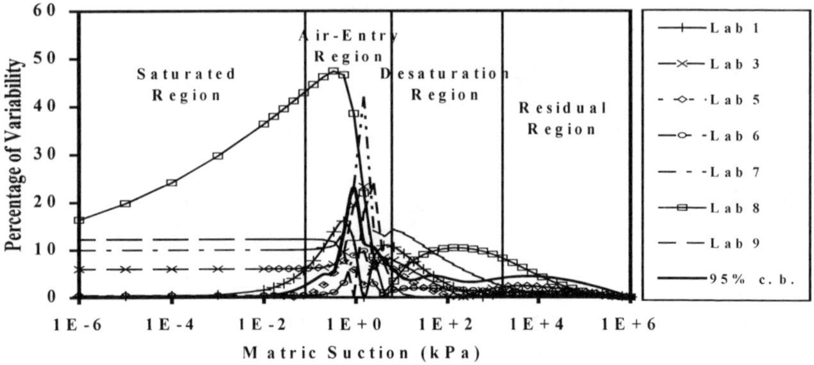

Figure 17. Variability due to Different Operators – El Paso Sand

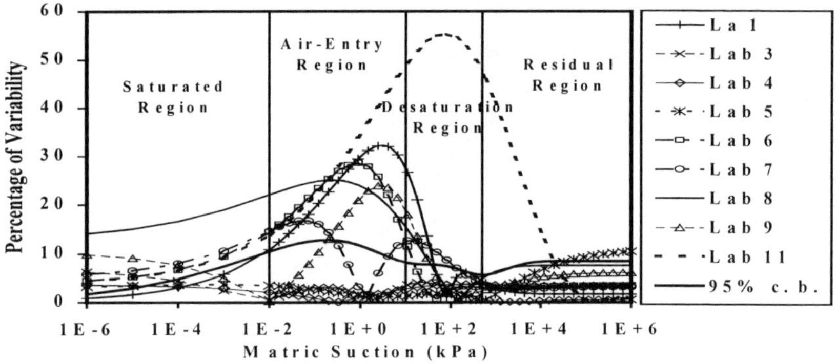

Figure 18. Variability due to Different Operators – Price Club Silt

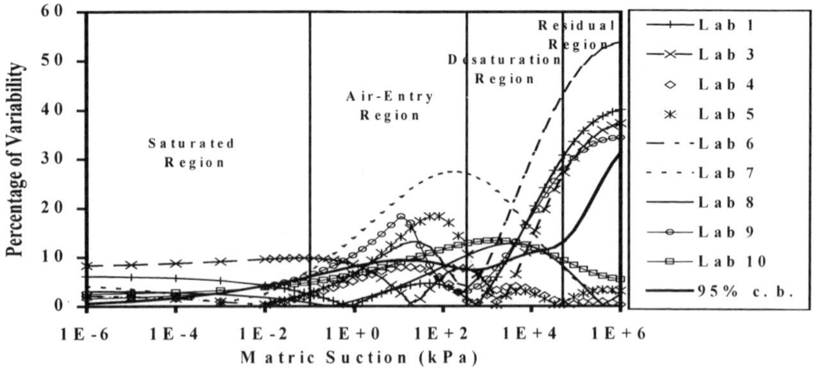

Figure 19. Variability due to Different Operators – Fountain Hills Clay

Table 5. Maximum Percentage of Variability due to Different Operators

Laboratory	Saturated Region	Air-Entry Region	Desaturation Region	Residual Region
		El Paso Sand		
Conf. Band	2	23	6	4
1	**6**	11	**11**	1
3	**6**	19	**8**	2
5	0	8	6	1
6	0	10	2	1
7	**10**	**42**	**14**	2
8	**43**	**47**	**10**	**7**
9	**12**	**24**	2	0
		Price Club Silt		
Conf. Band	10	13	8	8
1	**11**	**32**	**27**	4
3	6	3	4	1
4	4	2	4	4
5	4	3	2	**10**
6	**14**	**29**	**11**	3
7	**15**	**17**	**13**	4
8	**22**	**25**	**15**	8
9	10	**24**	**18**	6
11	**14**	**47**	**55**	**46**
		Fountain Hills Clay		
Conf. Band	6	9	12	31
1	6	5	**28**	**40**
3	**10**	**10**	**24**	**37**
4	2	8	4	1
5	2	**18**	11	3
6	2	**13**	**40**	**54**
7	**7**	**27**	**27**	7
8	3	**10**	**13**	8
9	6	**18**	**25**	**34**
10	5	**12**	**13**	9

Numerical comparisons are limited due to the range of suction covered by each method. The range of measurements obtained with the pressure plate and pressure membrane is limited by the air-entry value of the ceramic stone or membrane used, which is about 1,500 kPa. On the other hand, the filter paper can give a wider range of suction measurements. Thus, filter paper suction values cannot be compared with the other two methods in the extreme ranges of suction. For this reason, the data obtained by the different methods were compared: 1) throughout the entire range in suction, by fitting the data obtained with each method with a SWCC equation, and 2) in a limited range of suction where data were available from all three methods, by comparing the absolute residuals of the experimental data.

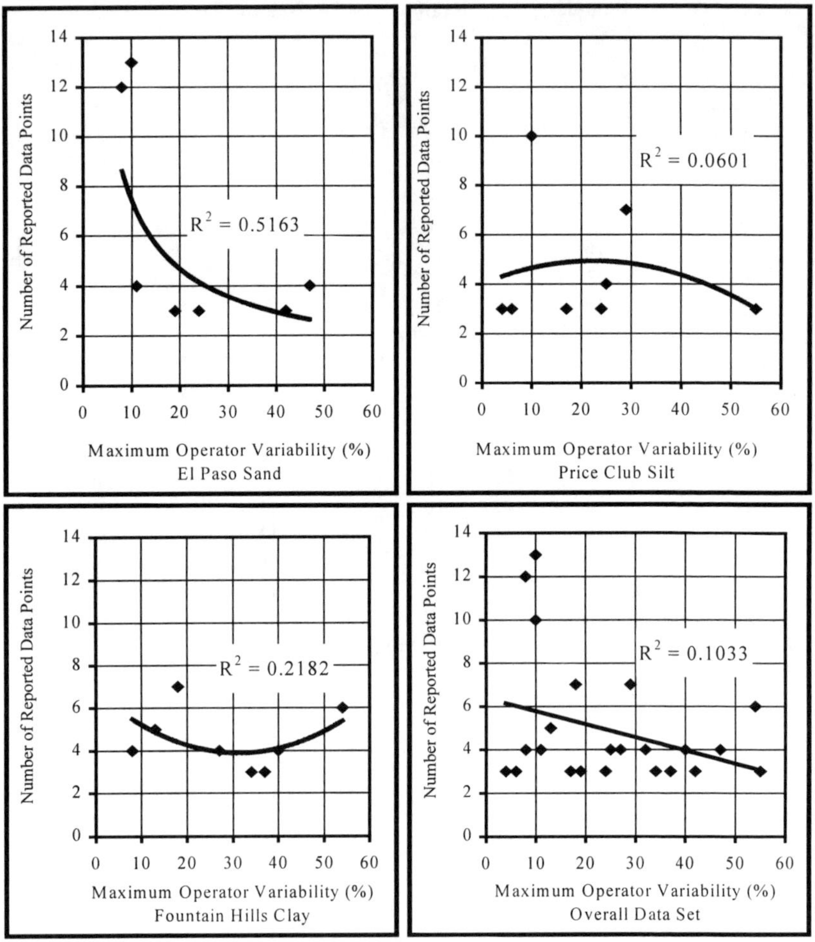

Figure 20. Correlation between Number of Data Points and Maximum Operator Variability

Variability in the SWCC throughout the Entire Range of Suction due to the Method Used to Measure Suction. To find the θ_w variability over the full range of suction, the experimental measurements were divided by method used. Each data set for each method of measurement was fitted to an equation. Then the SWCC for each test method was compared to the best-estimate SWCC. Figures 21 through 23 show the results corresponding to El Paso sand, Price Club silt, and Fountain Hills clay, respectively.

The variability in the predicted θ_w among methods was compared with the best-predicted θ_w at each level of suction. This difference was plotted as a percentage of the saturated volumetric water content in figures 24 through 26. For comparison, the 95% confidence band of the best-estimate SWCC was plotted in the same figure, also as a percentage of the saturated volumetric water content.

Results showed that none of the methods used produced higher variability than that corresponding to the 95% confidence band of the best-estimate SWCC. Although the variability increased generally as the fines content increased, no strong trend was seen. The air-entry region appears to be the most critical region for sand, and the desaturation region the most critical for the clay. The silt data showed a more or less uniform variability throughout the entire range of suction measurements.

Variability in a Limited Suction Range of the SWCC due to Method Used to Measure Suction. For comparison in the range where all three methods produced data, the absolute residuals were plotted. Absolute residuals, e, are obtained from the difference between the measured volumetric water content and its corresponding fitted value from the best-estimate SWCC:

$$e = \left|\theta_i - \hat{\theta}_i\right| \qquad (13)$$

Figure 27 shows the absolute residuals, in terms of volumetric water content, for the three soils analyzed. The horizontal lines represent the data mean or average.

Three statistical methods were used to check the scatter of the residuals: Analysis of Variance or ANOVA test, Duncan's multiple range test, and the Barlett test. Information about these tests can be found in Neter et al. (1985) and Hines and Montgomery (1990).

The ANOVA, or Analysis of Variance, determines whether or not the scatter of the measurements obtained by each method affected the overall fit of the data to the chosen function (Fredlund and Xing in the case of the sand, van Genuchten in the cases of the silt and the clay). The ANOVA indicated with 95% confidence, that there was not enough evidence to prove that having different test methods significantly affected the overall fit in any of the soils analyzed. This result is in agreement with the analysis performed in the previous section.

The second test performed on the data was the Barlett test. This test was used to check if the scatter of the residuals within each method statistically differed from each other. The variance of the data was used as the measure of the scatter. The Barlett test

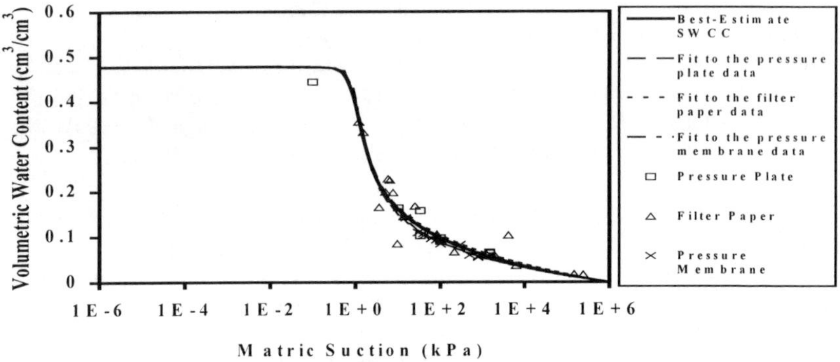

Figure 21. SWCC Obtained by Different Suction Methods - El Paso Sand

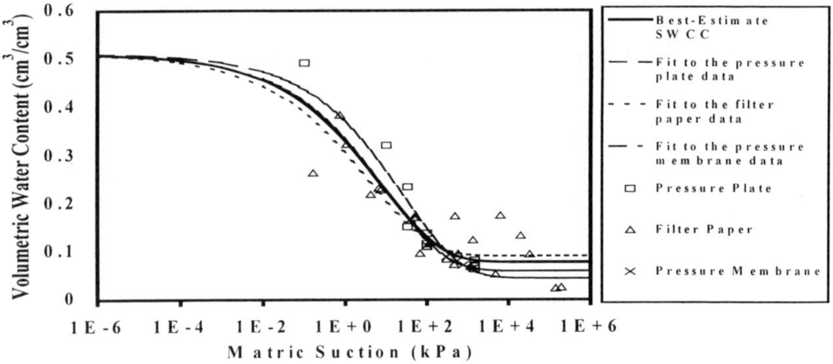

Figure 22. SWCC Obtained by Different Suction Methods – Price Club Silt

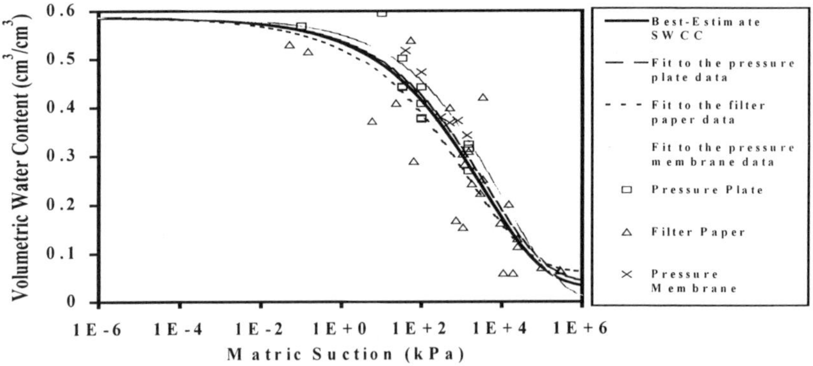

Figure 23. SWCC Obtained by Different Suction Methods – Fountain Hills Clay

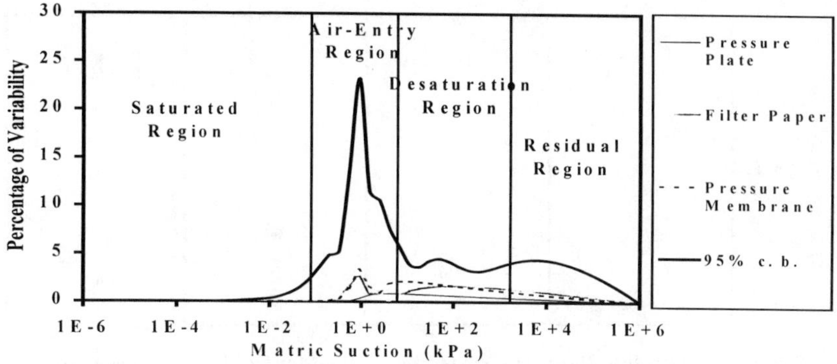

Figure 24. Variability Encountered due to Different Methods - El Paso Sand

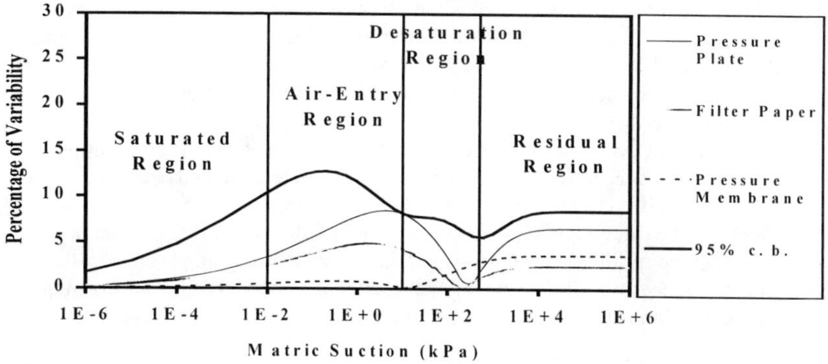

Figure 25. Variability Encountered due to Different Methods - Price Club Silt

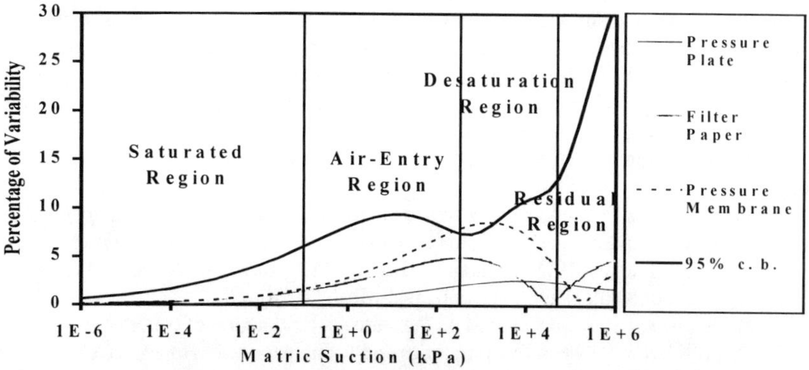

Figure 26. Variability Encountered due to Different Methods - Fountain Hills Clay

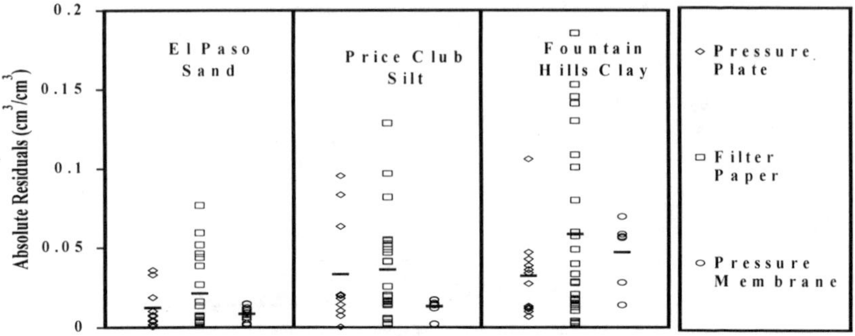

Figure 27. Absolute Residuals Sorted by Method for the Three Types of Soil

showed, with 95% confidence, that the variances were significantly different for the three soils analyzed. For the sand and the clay, the variance of the residuals obtained by the filter paper technique was higher than the variance of the data obtained by the other two methods. For the silt, the highest scatter or variance of the data residuals was for the data obtained by the pressure plate method. Pressure membrane residuals had the lowest variance for the three soils.

The third test performed on the absolute residuals was the Duncan's multiple range test. This test was performed to establish if the different methods' residual means were significantly different from each other. The Duncan test showed with 95% confidence, that the filter paper residual's mean was significantly different from the pressure membrane residual's mean for two of the soils: the sand and the silt. No significant differences were found among the residual means for the clay. Pressure plate and pressure membrane means were not significantly different from each other for the three types of soil.

Part 2: Variability in the SWCC when Predicted from Soil Index Properties

Several reported models used to predict the SWCC from the grain-size distribution (GSD) and soil index properties were reviewed. From those models, 16 were implemented to estimate the SWCC for the three soils previously described. These models were chosen because (i) they have remained popular throughout the years and several researchers have used them, (ii) they use commonly measured soil properties, and (iii) they are relatively easy to implement.

The following models were used to predict the SWCC for the three soils of this study:

1) *Approach 1*: Models using statistical estimation of the water content at a specified matric suction from GSD and/or volume-mass soil properties: Gupta-Larson (1979), Rajkai-Varallyay (1992), Rawls-Brakensiek (1982), Aubertin et al. (1998), and Mbagwu-Mbah (1998).
2) *Approach 2*: Models based on the statistical estimation of water content by correlation between soil properties and fitting parameters of a SWCC

equation: Ghosh (1980), Williams et al. (1983), Rawls et al. (1992), Cresswell-Paydar (1996), and Tomasella-Hodnett (1998).

3) *Approach 3*: Models that based the prediction of the SWCC on a physics based conceptual model using textural information: Arya-Paris (1981), Haverkamp-Parlange (1986), Mishra et al. (1989), Arya-Dierolf (1992). Smettem-Gregory (1996), and Fredlund, M. et al. (1997).

Figures 28 through 30 show the SWCCs predicted using Approach 1, along with the best-estimate SWCC (experimental) for El Paso sand, Price Club silt, and Fountain Hills clay respectively. The Rawls-Brakensiek model provided the best prediction of all the models. Nevertheless, it consistently produced over-predicted values of water content within the air-entry region for all three soils and under-predicted values at higher levels of suction for the clay.

Figures 31 through 33 show the SWCCs predicted using Approach 2, along with the best-fit SWCCs for the three soils. The model by Rawls et al. estimated the Brooks and Corey (BC) and the van Genuchten (vG) parameters. The Williams et al. model yielded the best prediction for the three soils even though the function presents a discontinuity at the air-entry value.

Figures 34 through 36 show the SWCCs predicted using Approach 3, along with the best-fit SWCCs for the three soils. The best predictive equation was that presented by Fredlund, M., even though under-prediction of the θ_w was observed for the three soils at high suction levels.

From the aforementioned 16 models, the best three predictive models for each soil were compared to the experimentally determined SWCC to obtain the variability in the different regions of the SWCC. To statistically determine the best predictive models of the SWCC for the three soils analyzed the experimentally obtained volumetric water content, θ_e, was compared to the predicted volumetric water content, θ_p, by means of the mean squared error. The mean squared error is given by:

$$MSE = \frac{SSE}{n} \qquad (14)$$

Where:
MSE = mean squared error
SSE = sum of the squared error
n = number of compared data points

Table 6 shows the MSE for the models in each of the approaches for the three soils analyzed. The MSE for the best three predictive models is shown in bold.

For El Paso sand, the best predictive models were those proposed by Rawls and Brakensiek (1982), Williams et al. (1983), and Fredlund, M. et al. (1997). These models use Approaches 1, 2, and 3 respectively, indicating that any of the three approaches can be used to predict an acceptable SWCC for this type of soil. For the Price Club silt, the best predictors were the models proposed by Rawls and Brakensiek (1982), Williams et al. (1983), and Tomasella and Hodnett (1998). The Rawls-Brakensiek model uses Approach 1, while the remaining two models use Approach 2. Lastly, the best predictive models for the Fountain Hills clay were Rawls and Brakensiek (1982), Ghosh (1980), and

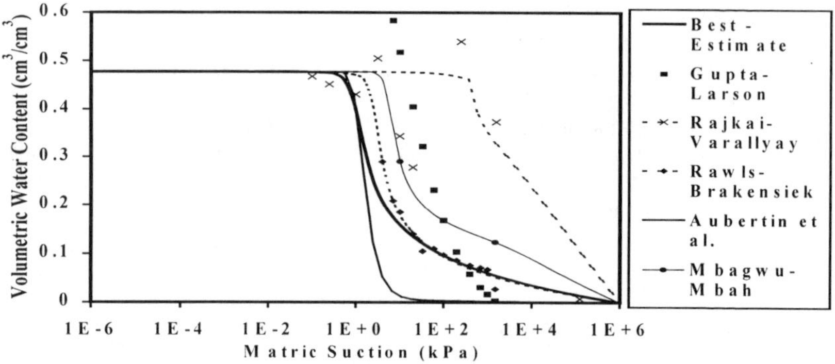

Figure 28. Predicted SWCCs Using Approach 1 - El Paso Sand

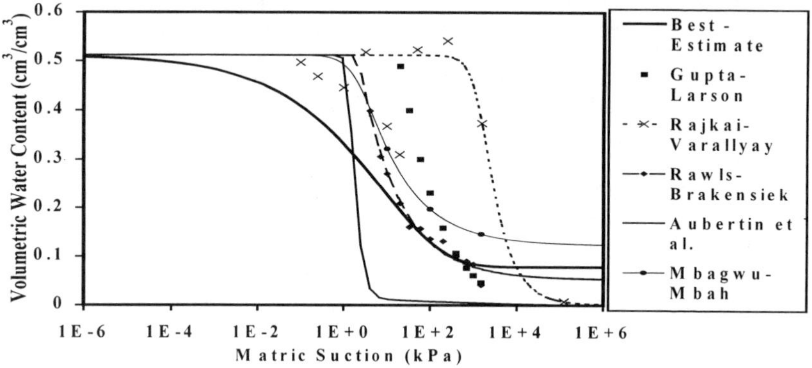

Figure 29. Predicted SWCCs Using Approach 1 – Price Club Silt

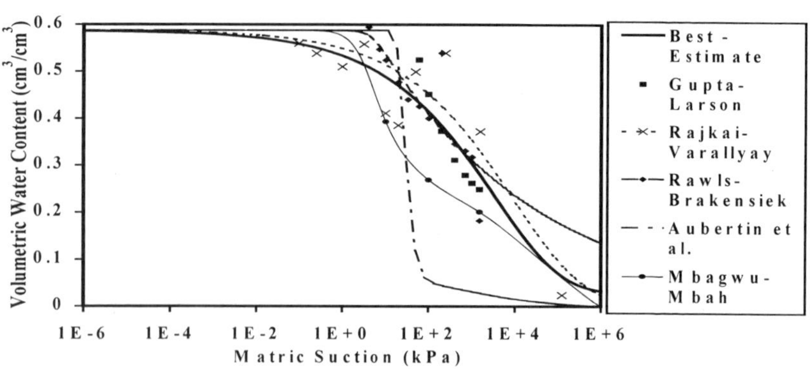

Figure 30. Predicted SWCCs Using Approach 1 – Fountain Hills Clay

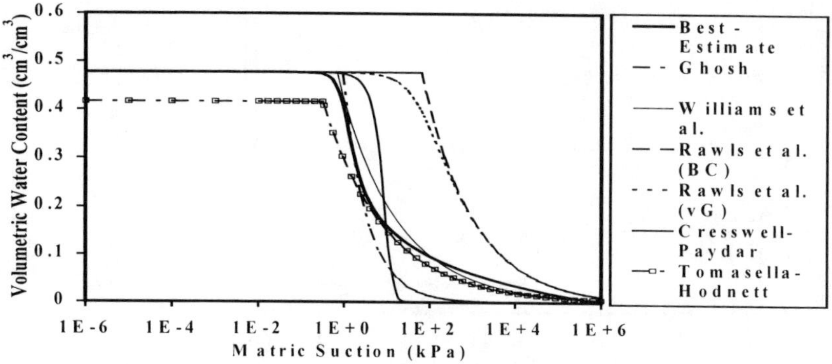

Figure 31. Predicted SWCCs Using Approach 2 – El Paso Sand

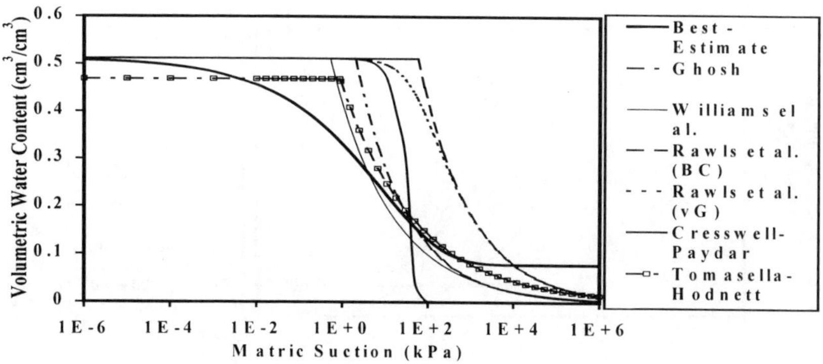

Figure 32. Predicted SWCCs Using Approach 2 – Price Club Silt

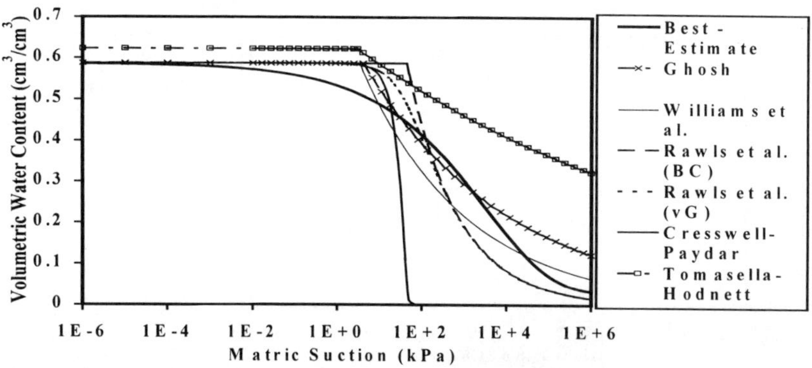

Figure 33. Predicted SWCCs Using Approach 2 – Fountain Hills Clay

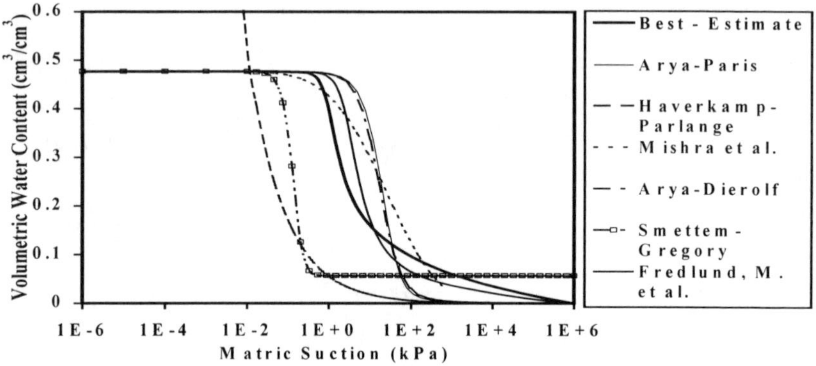

Figure 34. Predicted SWCCs Using Approach 3 – El Paso Sand

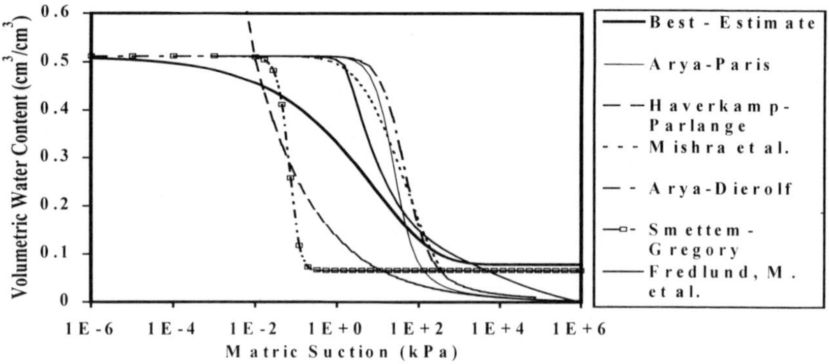

Figure 35. Predicted SWCCs Using Approach 3 – Price Club Silt

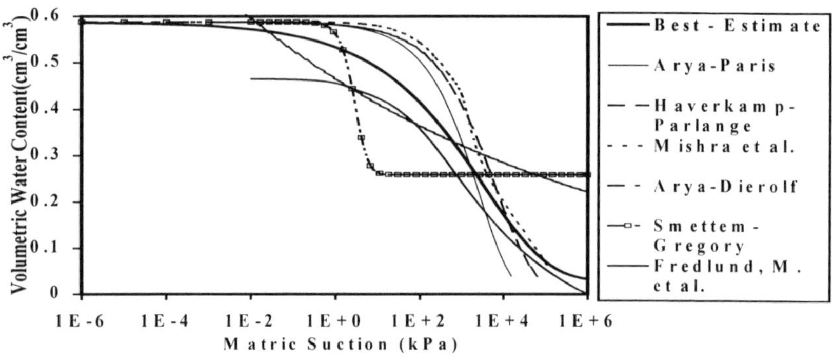

Figure 36. Predicted SWCCs Using Approach 3 – Fountain Hills Clay

Table 6. Mean Squared Error for the Predicted SWCC

Predicted Model	El Paso Sand	Price Club Silt	Fountain Hills Clay
Approach 1			
Gupta-Larson	0.058030	0.065569	0.049108
Rajkai-Varallyay	0.074219	0.053200	0.005673
Rawls-Brakensiek	**0.001302**	**0.005494**	**0.003205**
Aubertin et al.	0.004274	0.012960	0.024997
Mbagwu-Mbah	0.006422	0.006927	0.004629
Approach 2			
Ghosh	0.002123	0.007725	**0.002639**
Williams et al.	**0.000431**	**0.004605**	**0.002581**
Rawls et al. (BC)	0.027459	0.027726	0.004927
Rawls et al. (vG)	0.022665	0.022811	0.003554
Cresswell-Paydar	0.005351	0.014693	0.032291
Tomasella-Hodnett	0.001910	**0.002912**	0.027930
Approach 3			
Arya-Paris	0.007221	0.012544	0.003431
Haverkamp-Parlange	53.28111	1.300468	0.015440
Mishra et al.	0.003489	0.009895	0.004753
Arya-Dierolf	0.006248	0.013939	0.004347
Smetten-Gregory	0.019319	0.017434	0.016242
Fredlund, M. et al.	**0.001440**	0.006290	0.004268

Williams et al. (1983). The first model uses Approach 1, while the other two are based on Approach 2.

The variability, as previously defined, associated with the prediction of the SWCC is depicted in figures 37 through 39 for El Paso sand, Price Club silt, and Fountain Hills clay, respectively. Included on these figures is the 95% confidence band for the best estimate of the experimentally determined SWCC, shown for comparison.

The results showed that for the sand and the silt, the highest variability was found at the air-entry region, whereas for the clay, the variability at the residual region was the highest. Table 7 is a summary of the maximum variability encountered within each region of the SWCC for the three soils. It can be observed that although the magnitude of the variability in the results is considerable, it was lower than the variability found when different operators experimentally determined the SWCC (compare Table 7 with Table 5).

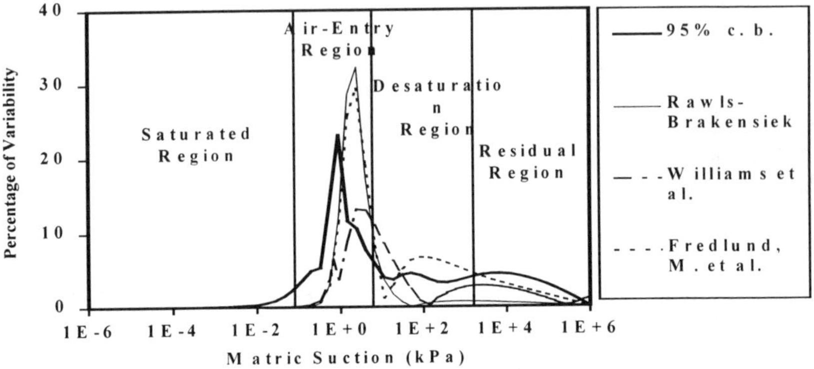

Figure 37. Variability in Predicted SWCC - El Paso Sand

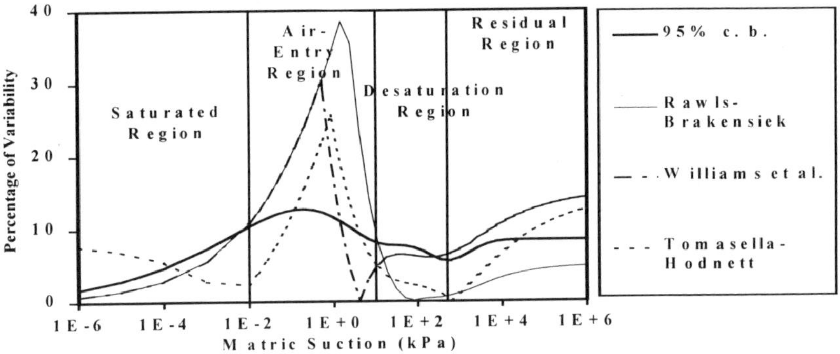

Figure 38. Variability in Predicted SWCC – Price Club Silt

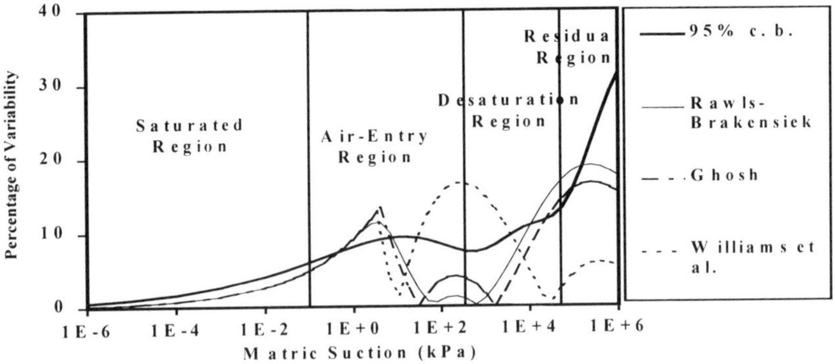

Figure 39. Variability in Predicted SWCC – Fountain Hills Clay

Table 7. Maximum Percentage of Variability when SWCC is Predicted from Soil Index Properties

Predicted Model	Saturated Region	Air-Entry Region	Desaturation Region	Residual Region
			El Paso Sand	
95% Confidence Band	2	23	6	4
Rawls-Brakensiek	0	**32**	**8**	1
Williams et al.	0	13	**11**	3
Fredlund	0	**29**	**8**	4
			Price Club Silt	
95% Confidence Band	10	13	8	8
Rawls-Brakensiek	10	**39**	8	5
Williams et al.	10	**30**	6	**14**
Tomasella-Hodnett	8	**26**	5	**13**
			Fountain Hills Clay	
95% Confidence Band	6	9	12	31
Rawls-Brakensiek	5	**11**	**15**	19
Ghosh	5	**13**	**13**	17
Williams et al.	5	**17**	**17**	6

Given the high variability that is still present in the experimental determination of the SWCC compared with the variability found for the investigated predictive models, a new predictive model will be presented. This new model is based on statistical correlations of very simple and easy to measure soil index properties with the fitting parameters of the SWCC function proposed by Fredlund and Xing (equation 1). This approach was selected because it is clear from Table 6 that Approach 2 was clearly the most succesful.

A New Empirical Model for Predicting the SWCC from Index Properties

A database characterizing approximately 190 soils was assembled from research papers and a knowledge-based program developed by Soilvision Systems Ltd (Dunn and Palmer, 1994; Escario and Juca, 1989; Fredlund, 1995; Fredlund et al., 1995; Gan et al., 1988; Ghosh, 1980; Haverkamp and Parlange, 1986; Houston et al., 1999; Krahn and Fredlund, 1972; Livneh et al., 1970; Marinho and Stuermer, 1998; Oberg and Sallfors, 1997; Rahardjo et al., 1995; Rohm and Vilar, 1995; Sabbagh, 1995; Vanapalli et al., 1998, SoilVision, 1997). The soils were divided into two categories: soils having a Plasticity Index (PI) greater than zero and soils having a PI equal to zero. Data for approximately 70 soils with PI greater than zero and 120 soils with PI equal to zero were collected.

The data assembled for the soils with PI greater than zero included the percentage passing #200 sieve and the Atterberg Limits, particularly the Plasticity Index. For soils with PI equal to zero (non-plastic soils), the diameter D_{60} was gathered. Included in the collected data was a measured and fairly well-defined SWCC.

For the soils with PI greater than zero, the product of the percentage passing the #200 sieve, as a decimal, was multiplied by the PI as a percentage, to form the weighted PI. This value was designated as *wPI*, and used as the main soil property for correlation. The reasoning behind this choice is as follows. The equilibrium soil suction at a given degree of saturation was expected to be proportional to the specific surface area of the soil. The PI is a fair indicator of surface area and the use of PI alone was considered. However, a soil with a small percentage of highly active clay would have a high PI but only a moderate specific surface area. Therefore, the weighted PI, wPI, was considered a better indicator of soil particle surface area available for water adsorption and retention.

For non-plastic soils, several functions of gradation parameters were tried. It was decided finally that correlation with D_{60} worked as well as any of the other correlations. Therefore, the D_{60} was the main soil property for correlation for the soils with PI equal to zero. For simplicity, it was decided to use the Fredlund and Xing equation for all soils, regardless of PI.

For Soils with PI > 0

The Fredlund and Xing equation fitting parameters in equation 1 (parameters a, b, c, and h_r, defined in Table 1), were correlated with the new *wPI* parameter. The equations found are the following:

$$a = 0.00364(\text{wPI})^{3.35} + 4(\text{wPI}) + 11 \qquad (15)$$

$$\frac{b}{c} = -2.313(\text{wPI})^{0.14} + 5 \qquad (16)$$

$$c = 0.0514(\text{wPI})^{0.465} + 0.5 \qquad (17)$$

$$\frac{h_r}{a} = 32.44e^{0.0186(\text{wPI})} \qquad (18)$$

The wPI parameter in equations 15 through 18 is defined as:

$$\text{wPI} = \text{Passing \#200} \times \text{PI} \qquad (19)$$

Where:

Passing #200 = Material passing the #200 U.S. Standard Sieve expressed as a decimal

PI = Plasticity Index (%) = Liquid Limit – Plastic Limit

In those cases where the saturated volumetric water content, θ_{sat}, is unknown, the user can make use of the following correlation:

$$\theta_{sat} = 0.0143(\text{wPI})^{0.75} + 0.36 \qquad (20)$$

However, although equation (20) produces a more or less unbiased estimate of the θ_{sat}, the scatter is very considerable and it is highly desirable to have direct measurements of density, or better yet, density and specific gravity, G_s, so that θ_{sat} can be calculated from direct measurements. Equation (21) for estimating G_s can be used with only small to moderate error when directly measured G_s values are not available:

$$G_s = 0.041(wPI)^{0.29} + 2.65 \qquad (21)$$

For Soils with PI = 0

For granular soils with Plasticity Index equal to zero, the parameter used to relate to the SWCC was the Diameter D_{60} from the grain-size distribution (GSD) curve. The correlations found are the following:

$$a = 0.8627(D_{60})^{-0.751} \qquad (22)$$

$$\bar{b} = 7.5 \qquad (23)$$

$$c = 0.1772\ln(D_{60}) + 0.7734 \qquad (24)$$

$$\frac{h_r}{a} = \frac{1}{D_{60} + 9.7e^{-4}} \qquad (25)$$

Where:

D_{60} = Grain diameter corresponding to 60% passing by weight or mass (mm)

$\bar{b}$ = Average value of fitting parameter b

No correlation between the 'b' parameter and D_{60} was found. Therefore, a constant average b value is suggested. In those cases where the θ_{sat} is unknown, the following average value is recommended for soils with PI equal to zero:

$$\bar{\theta}_{sat} = 0.36 \qquad (26)$$

The individual fitting parameters for the Fredlund and Xing SWCC equation, a, b, c, and h_r, did not correlate very well with the index properties wPI and D_{60}. Perhaps not surprisingly, the R^2 values were quite low. It has been experience of the authors that when a complex function that requires three or four parameters for fitting is used, the fitting parameters themselves rarely correlate well with soil properties. For the case in point, the SWCC, this result is probably related to the fact that numerous sets of fitting parameters, which differed substantially, could be found which would produce SWCCs which were nearly the same. In the final analysis, what the user wants is a predictive model that will predict the SWCC with acceptable accuracy. Therefore, based on this experience, it was decided to delay the judgment of goodness of fit until the final predicted SWCCs could be compared to measured SWCCs. This comparison is made in the following section.

Comparison of Measured SWCCs and SWCCs Predicted by the New Model

Figures 40 through 46 shows the results obtained. Figures 40 through 43 show the SWCCs for the soils with wPI greater than zero. Figures 44 and 45 show the SWCCs for the soils with wPI equal to zero. In figures 40 through 45, the solid curves represent the "predicted" band corresponding to the wPI and D_{60} values indicated. The prediction is derived from the correlations obtained from the database (190 soils), using equations 15 through 25. The data points shown in figures 40 through 45 represent the actual, measured SWCCs, after some smoothing. Each symbol type represents an experimental curve. The goodness of the fit can be judged by observing the extent to which the "predicted" band is centered on and envelops the experimental data. For each figure, the experimental data subset represents the same range in wPI (or D_{60}), as does the predicted band given by the solid curves. Figure 46 summarizes the results obtained for both groups of soils.

Data for the 190 soils used in this study and hundreds of other SWCCs examined by the authors show clearly that the SWCC moves gradually to the right with increasing plasticity. The correlations and algorithms developed for this study provide a smooth transition across the spectrum of soils with no plasticity to those with high plasticity. Although Figures 40 through 45 show that the predicted SWCC curves match the measured SWCC fairly well, and probably about as well as can be done at this time, there is nevertheless considerable scatter. Data presented in the earlier sections of this paper show that if a single soil is sent out to a dozen laboratories across the country for SWCC measurement, the results show a variability greater than that of the experimental data in Figure 42, for example. Likewise, if a single laboratory is asked to reproduce the SWCC for a single soil, i. e., make multiple measurements of the SWCC for a single soil, the variability can be as great as the difference between the wPI = 10 curve and the wPI = 30 curve in Figure 42.

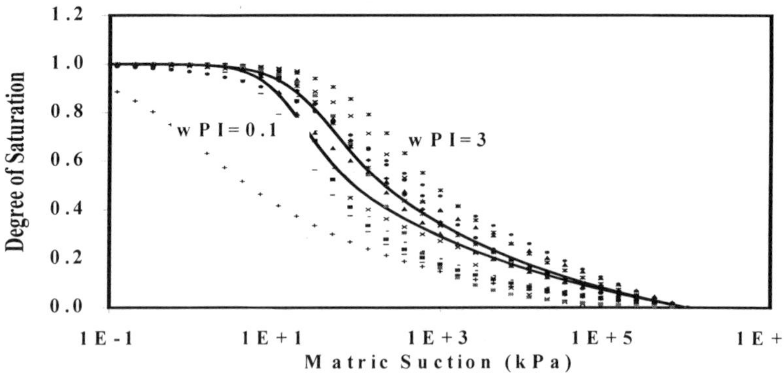

Figure 40. Range of SWCCs for 14 Soils with wPI between 0.1 and 3

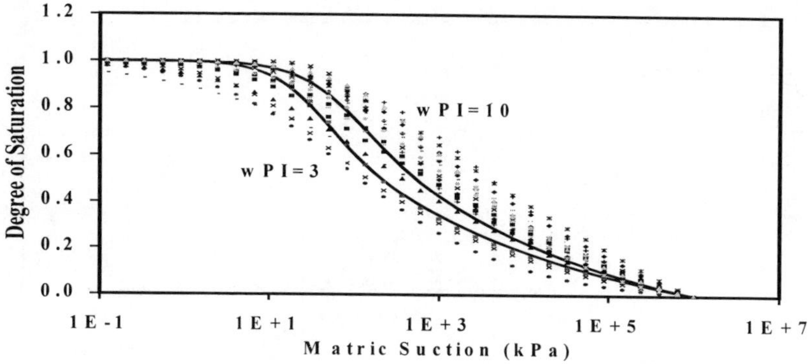

Figure 41. Range of SWCCs for 21 Soils with wPI between 3 and 10

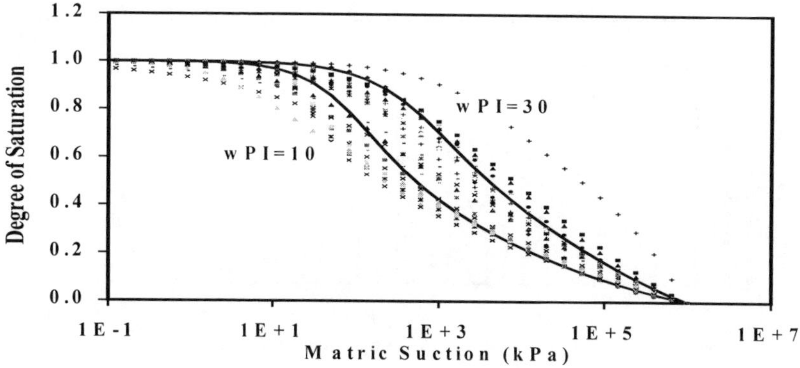

Figure 42. Range of SWCCs for 24 Soils with wPI between 10 and 30

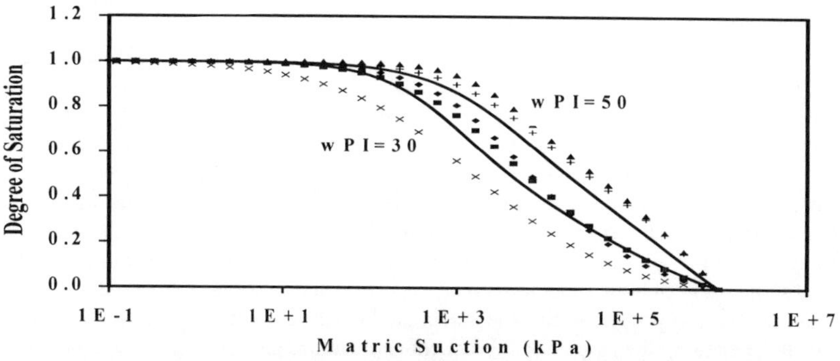

Figure 43. Range of SWCCs for 5 Soils with wPI between 30 and 50

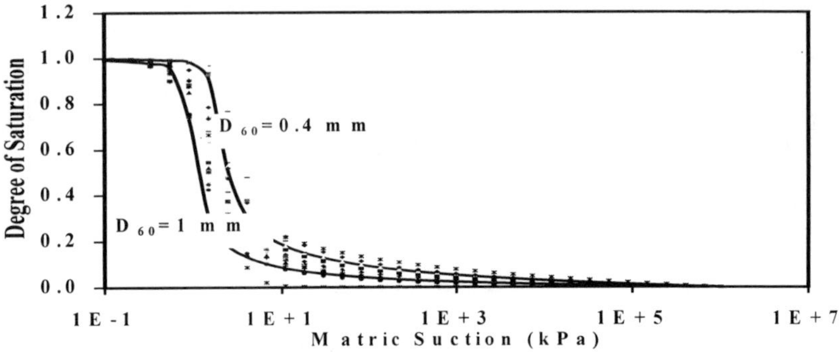

Figure 44. Range of SWCCs for 18 Soils with D_{60} between 1 and 0.4 mm.

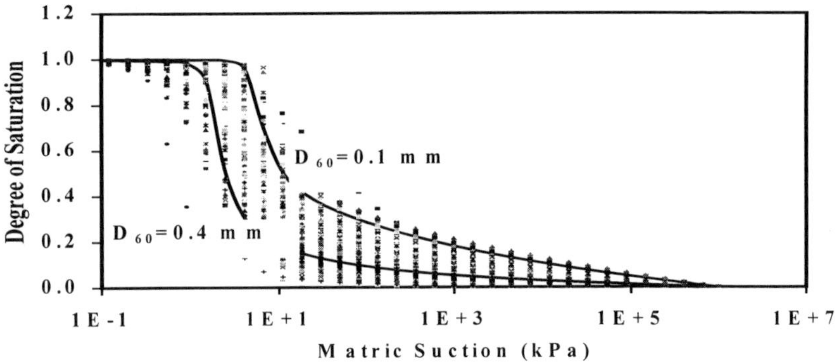

Figure 45. Range of SWCCs for 108 Soils with D_{60} between 0.4 and 0.1 mm.

The Case for Compensating Errors

The use of models to predict engineering behavior of materials dates back to the inception of engineering itself. Even the creator of each model, however, would not claim that his/her model is perfect. The imperfections may lie in the fitting parameters, which characterize the model, or in the functional form of the model itself. Because imperfections are the norm rather than the exception, it has long been an important tenet of engineering practice to check, and if possible calibrate, a predictive model against actual performance under prototype conditions. This calibration normally takes the form of adjustment of the fitting parameters or coefficients in the model. To the extent that the calibration has been done well, any remaining errors in the coefficients or the form of the model tend to compensate so that the model predicts the correct, observed performance. The engineer using the model for subsequent applications checks to see that the new conditions are sufficiently similar to the calibration conditions that good results can be expected.

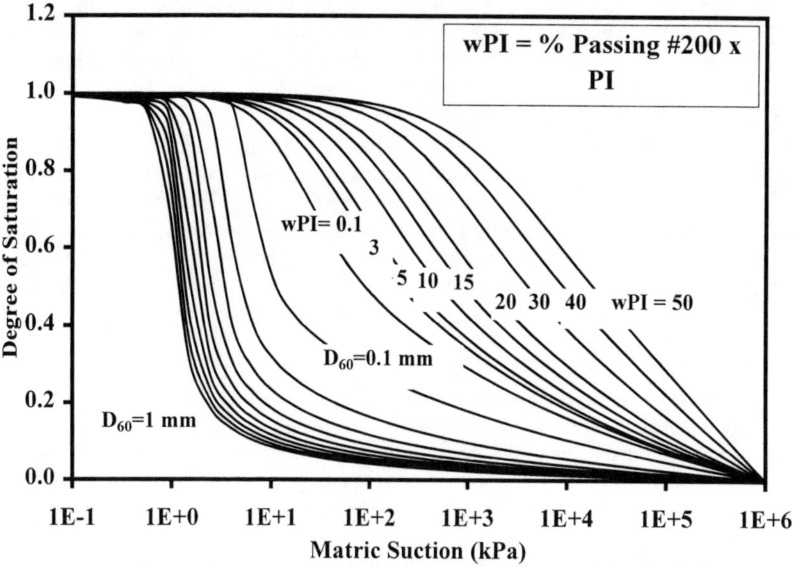

Figure 46. Predicted SWCC based on D_{60} and wPI

This utilization of compensating errors is common in all branches of engineering and science and is the feature that allows us to get satisfactorily good predictions with imperfect models. Obviously, the better the model and the smaller the compensating errors, the more confidently we can extrapolate to field conditions differing somewhat from calibration conditions.

The utilization of compensating errors in connection with SWCCs can be postulated as follows. Suppose the shear strength, τ_f, is to be estimated from Equation (27).

$$\tau_f = c' + (\sigma - u_a)\tan\phi' + (u_a - u_w)\tan\phi^b \quad (27)$$

Normally c' and ϕ' would be determined by a lab test program on saturated soil, but tan ϕ^b must be assessed from the testing of unsaturated soil. It is well known that tan ϕ^b approaches tan ϕ' as S approaches 100%, but it is also well known that tan ϕ^b decreases as S decreases, usually non-linearly. Thus, a lab test program on unsaturated specimens is needed to establish the functions given by (28a) or (28b).

$$\tan\phi^b = f(S) \quad (28a)$$

or

$$\tan\phi^b = f(u_a - u_w) \quad (28b)$$

The values of $u_a - u_w$ for use in Equation (27) and (28b) can be measured directly or estimated from a SWCC, such as those depicted by Figure 46. Suppose the SWCC from Figure 46 is adopted and the wPI=30. As discussed at length in this paper, the SWCC for wPI=30 will have some error associated with it. Assume the actual SWCC for the soil lies to the right of the wPI=30 curve. This means that u_a - u_w will be under-estimated for each S value, both in Equation (27) and (28b). This under-estimation of u_a - u_w produces a corresponding error in the tan ϕ^b function.

Now, when the results of the lab test program are applied to the field, the same wPI=30 curve is used to estimate field values of u_a - u_w, which are again under-estimated. When these values of u_a - u_w are used in Equations (28a) and (27), compensating errors can be expected to produce a fairly good estimate of shear strength, τ_f. Note that compensating errors should not be expected if $u_a - u_w$ were directly measured for the lab test program, but the SWCC from Figure 46 were used to estimate u_a - u_w for the field.

Conclusions

The high variability encountered in the experimentally obtained suction measurements and the numerous sources of error in the measured values leads to the conclusion that even some of the most experienced researchers have difficulties in getting a unique SWCC for a soil. It appears that using a large database in conjunction with a knowledge-based system is perhaps the best way to achieve the best fit and to select the optimal SWCC.

These observations have led the authors to conclude that soil suction and SWCCs simply cannot be measured with great precision at the present time. Researchers and practitioners dealing with unsaturated soils need to recognize and acknowledge this condition. There are two corollaries to this conclusion: The first is that the SWCC can probably be estimated from D_{60} or wPI (see Figure 46) about as accurately as it can be measured, unless the laboratory or person making the measurement is highly experienced. Secondly, it is difficult to develop a predictive model for SWCCs that is consistent with all of the SWCCs reported in the literature because of the fairly high probability that any given measured SWCC has significant experimental error associated with it. It therefore follows that the agreement between predicted and measured results depicted in Figures 40 through 45 could be improved if the experimental error from the measured results could be removed. Of course, this latter conclusion is optimistic and rests on the contention that the experimental errors tend to be both positive and negative.

Acknowledgement

A part of this work is based upon research supported by the National Science Foundation under grant no. CMS9612073. This support is gratefully acknowledged.

References

Ahuja, L. R., Naney, J. W. and Williams, R. D. (1985). " Estimating Soil Water Characteristics from Simpler Properties or Limited Data". *Soil Science Society of America Journal,* Vol. 49, No. 5, pp. 1100-1105.

Arya, L. M. and Paris, J. F. (1981). "A Physicoempirical Model to Predict the Soil Moisture Characteristic from Particle-Size Distribution and Bulk Density". *Soil Science Society of America Journal,* Vol. 45, No. 6, pp. 1023-1030.

Arya, L. M. and Dierolf, T. S. (1992). "Predicting Soil Moisture Characteristics from Particle-Size Distributions: An Improved Method to Calculate Pore Radii from Particle Radii". *Proceedings of the International Workshop on Indirect Methods for Estimating the Hydraulic Properties of Unsaturated Soils.* van Genuchten, M. Th., Leij, F. J., and Lund, L. J., editors. University of California. Riverside, CA. pp. 115-124.

Aubertin M., Richard J.F., and Chapuis Robert P. (1998). "A Predictive Model for the Water Retention Curve: Application to Tailings from Hard-Rock Mines". *Canadian Geotechnical Journal.* Vol. 35, No. 1, pp. 55-69.

Basile A. and D'Urso G. (1997). "Experimental Corrections of Simplified Methods for Predicting Water Retention Curves in Clay–Loamy Soils from Particle–Size Determination". *Soil Technology.* Vol. 10, No. 3. pp. 261 –272.

Bates, D. and Watts, D. (1988). *Nonlinear Regression Analysis and its Applications*. John Wiley & Sons.

Cassel, D. K., Ratliff, L. F., and Ritchie, J. T. (1983) "Models for Estimating In-situ Potential Extractable Water Using Soil Physical and Chemical Properties". *Soil Sci. Soc. Am.* J., 47(4), 764-769.

Cresswell, H. P. and Paydar, Z. (1996). "Water Retention in Australian Soils. 1. Description and Prediction Using Parametric Functions". *Australian Journal of Soil Research.* Vol. 34, No. 2. pp. 195–212.

Dunn, R. J. and Palmer, B. S. (1994). "Lessons Learned from the Application of Standard Test Methods for Field and Laboratory Hydraulic Conductivity Measurement". *Hydraulic Conductivity and Waste Contaminant Transport in Soil, ASTM STP 1142.* D. E. Daniel and S. J. Trautwein, editors. American Society for Testing and Materials, Philadelphia.

Escario, V. and Juca, J. (1989). "Strength and Deformation of Partly Saturated Soils". *Proceedings of the Twelfth International Conference on Soil Mechanics and Foundation Engineering*, Rio de Janeiro, Vol. 3, pp. 43-46.

Fredlund, D. G. and Morgenstern, N. R. (1977). "Stress State Variables for Unsaturated Soils". *Journal of the Geotechnical Engineering Division*, ASCE, GT5, Vol. 103, pp. 447-466.

Fredlund, D. G., (1995), "Prediction of Unsaturated Soil Functions Using the Soil-Water Characteristic Curve", *Proceedings of the Bength B. Broms Symposium on Geotechnical Engineering,* Singapore, 13-15 December, pp. 113-133.

Fredlund, D. G., Vanapalli, S. K., Xing, A., and Pufalel, D. E., (1995), "Predicting the Shear Strength Function for Unsaturated Soils Using the Soil-Water Characteristic Curve", In: *Unsaturated Soils*, Alonso and Delage, editors.

Fredlund, M., Fredlund, D., and Wilson, G. (1997). "Prediction of the Soil-Water Characteristic Curve from Grain-Size Distribution and Volume-Mass Properties". *Proceedings of the Third Brazilian Symposium on Unsaturated Soils,* NONSAT'97, Rio de Janeiro, Brazil.

Gan, J. K. M, Fredlund, D. G., and Rahardjo, H. (1988). "Determination of the Shear Strength Parameters of an Unsaturated Soil Using the Direct Shear Test". *Canadian Geotechnical Journal.* Vol. 25, pp. 500-510.

Ghosh, R. K. (1980). "Estimation of Soil-Moisture Characteristics from Mechanical Properties of Soils". *Soil Science,* Vol. 130, No. 2, pp. 60-63.

Gregson, K. Hector, D. J., and McGowan, M. (1987). "A One-parameter Model for the Soil Water Characteristic" *Soil Science.* 38, 483 – 486.

Gupta, S. C. and Larson, W. E. (1979). "Estimating Soil Water Retention Characteristics from Particle Size Distribution, Organic Matter Percent, and Bulk Density". *Water Resources Research,* Vol. 15, No. 6, pp. 325-339.

Haverkamp, R. and Parlange, J. Y. (1986). "Predicting the Water-Retention Curve from Particle-Size Distribution: 1. Sandy Soils without Organic Matter". *Soil Science,* Vol. 142, No. 6, pp. 325-339.

Hines, W. W. and Montgomery, D. C. (1990). *Probability and Statistics in Engineering and Management Science.* (3rd ed.). John Wiley & Sons, Inc.

Houston, W. N., Houston, S. L., Zapata, C. E., Manepally, C., and Lawrence, C. (1999). "Influence of Compressibility on Use and Interpretation of Soil Water Characteristic Curves. *Proceedings of the XI Panamerican Conference on Soil Mechanics and Geotechnical Engineering.* Foz do Iguassu, Brazil. pp. 947-954.

Huet, S., Bouvier, A., Gruet, M., and Jolivet, E. (1996). *Statistical Tools for Nonlinear Regression: A Practical Guide with S-PLUS Examples.* Springer series in statistics. Springer-Verlag New York, Inc.

Hutson, J. L., and Cass, A. (1987). "A Retentivity Function for Use in Soil-Water Simulation Models". *Soil Science Journal.* 38(1), 105 – 113.

Kern, J. S. (1995). "Evaluation of Soil–Water Retention Models Based on Basic Soil Physical Properties". *Soil Science Society of American Journal.* Vol. 59, No. 4, pp. 1134–1141.

Krahn, J., and Fredlund, D. G. (1972). "On Total, Matric and Osmotic Suction". *Soil Science.* Vol. 114, No. 5, pp. 339 - 347.

Lee, H. C. and Wray, W. K. (1995). "Techniques to Evaluate Soil Suction – A Vital Unsaturated Soil Water Variable". In: *Unsaturated Soils.* Alonso E. E. and Delage, P., editors. Proceedings of the First International Conference on Unsaturated Soils, UNSAT'95. 6-8 September, Paris, France. pp. 615-622.

Leong, E. C., and Rahardjo, H. (1996). "A Review on Soil-Water Characteristic Curve Equations". Geotechnical Research Report NTU/GT/96-5, Nanyang Technological University, NTU-PWD Geotechnical Research Centre, Singapore.

Livneh, M., Kinsky, J. and Zaslavsky, D. (1970). "Correlation of Suction Curves with the Plasticity Index of Soils". *Journal of Materials,* JMLSA, Vol. 5, No. 1, March 1970, pp. 209-220.

Marinho, F. A. M. and Stuermer, M. M. M. (1998). "Aspects of the Storage Capacity of a Compacted Residual Soil". *Proceedings of the Second International Conference on Unsaturated Soils.* Volume 1. Beijing, China. August, 27-30.

Mbagwu, J. S. C. and Mbah, C. N. (1998). "Estimating Water Retention and Availability in Nigerian Soils from Their Saturation Percentage". *Communications in Soil Science and Plant Analysis.* Vol. 29, No. 7/8. pp. 913-922.

Mishra, S., Parker, L. C., and Singhal, N. (1989). "Estimation of Soil Hydraulic Properties and their Uncertainty from Particle Size Distribution Data". *Journal of Hydrology*, Vol. 108. pp. 1-18.

Nandagiri L. and Prasad R. (1997). "Relative Performances of Textural Models in Estimating Soil Moisture Characteristic". *Journal of Irrigation and Drainage Engineering–ASCE.* Vol. 123, No. 3. pp. 211–214.

Neter, J., Wasserman, W., and Kutner, M. H. (1985). *Applied Linear Statistical Models. Regression, Analysis of Variance, and Experimental Designs.* (2nd ed.). Richard D. Irwin, Inc.

Oberg AL. and Sallfors G. (1997). "Determination of Shear Strength Parameters of Unsaturated Silts and Sands Based on the Water Retention Curve". *Geotechnical Testing Journal.* Vol. 20, No. 1. pp. 40–48.

Rahardjo, H., Chang, H. M. F. and Lim, T. T. (1995). "Shear Strength and in Situ Matric Suction of a Residual Soil". In *Unsaturated Soils.* Alonso & Delage, editors. pp. 637-643.

Rajkai, K. and Varallyay, Gy. (1992). "Estimating Soil Water Retention from Simpler Properties by Regression Techniques". *Proceedings of the International Workshop on Indirect Methods for Estimating the Hydraulic Properties of Unsaturated Soils.* van Genuchten, M. Th., Leij, F. J., and Lund, L. J., editors. University of California. Riverside, CA. pp. 417-426.

Rawls, W. J. and Brakensiek, D. L. (1982). "Estimating Soil Water Retention from Soil Properties". *Journal of the Irrigation and Drainage Division.* ASCE, Vol. 108, No. IR2, pp. 166–171.

Rawls, W. J., Brakensiek, D. L., and Saxton, K. E. (1982). "Estimation of Soil Water Properties". *Trans. ASAE.* 25, 1316-1320 and 1328.

Rawls, W. J., Gish, T. J., and Brakensiek, D. L. (1991). "Estimating Soil Water Retention from Soil Physical Properties and Characteristics". *Adv. in Soil Sci.* 16, 213 – 234.

Rawls, W. J., Ahuja, L. R., and Brakensiek, D. L. (1992). "Estimating Soil Hydraulic Properties from Soils Data". *Proceedings of the International Workshop on Indirect Methods for Estimating the Hydraulic Properties of Unsaturated Soils.* van Genuchten, M. Th., Leij, F. J., and Lund, L. J., editors. University of California. Riverside, CA. pp. 329-340.

Reddi L.N. and Poduri R. (1997). "Use of Liquid Limit State to Generalize Water Retention Properties of Fine-Grained Soils". *Geotechnique.* Vol 47, No. 5, pp.1043-1049.

Ridley, A. M. and Wray, W. K.. (1995). "Suction Measurement: A Review of Current Theory and Practices". In: *Unsaturated Soils.* Alonso E. E. and Delage, P., editors. Proceedings of the First International Conference on Unsaturated Soils, UNSAT'95. 6-8 September, Paris, France. pp. 1293-1322.

Rohm, S. A. and Vilar, O. M. (1995), "Shear Strength of an Unsaturated Sandy Soil". In *Unsaturated Soils.* Alonso & Delage, editors. pp. 189-193.

Sabbagh, A. (1995). "Prediction of Volume Change in Unsaturated Clays". In *Unsaturated Soils.* Alonso & Delage, editors. pp. 791-796.

Smettem, K. R. J. and Gregory, P. J. (1996). "The Relation between Soil Water Retention and Particle Size Distribution Parameters for Some Predominantly Sandy Western

Australian Soils". *Australian Journal of Soil Research.* Vol. 34, No. 5. pp. 695 – 708.

SoilVision User's Guide. (1997). Version 1.2 [Computer software]. SoilVision Systems, Ltd. Saskatoon, Saskatchewan, Canada.

Tomasella, J. and Hodnett, M. G. (1998). "Estimating Soil Water Retention Characteristics from Limited Data in Brazilian Amazonia". *Soil Science.* Vol. 163, No. 3, pp. 190–202.

Vanapalli, S K., Sillers, W. S. and Fredlund, M. D. (1998). "The Meaning and Relevance of Residual State to Unsaturated Soils". *51st Canadian Geotechnical Conference.* Edmonton, Albert, October 4-7, pp. 1-8.

van Genuchten, M. Th. and Leij, F. J. (1992). "On Estimating the Hydraulic Properties of Unsaturated Soils". *Proceedings of the International Workshop on Indirect Methods for Estimating the Hydraulic Properties of Unsaturated Soils.* van Genuchten, M. Th., Leij, F. J., and Lund, L. J., editors. University of California. Riverside, CA. pp. 1-14.

Visser, W. C. (1969). "An Empirical Expression for the Desorption Curve". *Water in the Unsaturated Zone.* pp. 329 – 335. (Proc. UNESCO IASH Symp: Wageningen, Netherlands).

Williams, J., Prebble, R., Williams, W., and Hignett, C. (1983). "The Influence of Texture, Structure and Clay Mineralogy on the Soil Moisture Characteristic". *Australian Journal of Soil Research*, Vol. 21, pp. 15 – 32.

Williams, R. D. and Ahuja, L. R. (1992). "Estimating Soil Water Characteristics Using Measured Physical Properties and Limited Data". *Proceedings of the International Workshop on Indirect Methods for Estimating the Hydraulic Properties of Unsaturated Soils.* van Genuchten, M. Th., Leij, F. J., and Lund, L. J., editors. University of California. Riverside, CA. pp. 405-416.

Zapata, C. E. (1999). "Uncertainty in Soil-Water Characteristic Curve and Impacts on Unsaturated Shear Strength Predictions", Ph.D. Dissertation, Arizona State University, Tempe, United States.

The Influence of the compaction energy on the SWCC of a residual soil

Fernando A. M. Marinho[1] and Mônica M. Stuermer[2]

Abstract

The proliferation of the use of the soil water characteristic curve (SWCC) to obtain the unsaturated soil parameters is unavoidable and in some cases desired. However, the complex interrelation between soil water content and soil water suction makes the adequate indirect determination of the unsaturated soil parameters a difficult task. There are many aspect to be considered when using SWCC for predicting mechanical or hydraulic behaviour of the soil. The paper describes a study involving a residual soil of Gneiss, compacted using three different compaction energies. Soil water characteristic curves (SWCC) are obtained with volume measurement during the drying process and the soil suction was obtained using the filter paper method. The filter paper was placed in direct contact with the soil sample. Discussions on the relationship between the compaction energy, SWCC shape and the drying behaviour of the soil samples are given.

Introduction

The soil water characteristic curves (SWCC) has been used to obtain unsaturated soil parameters (mechanical and hydraulic) via some model. The use of models to obtain the SWCC itself is also common practice. Any model for obtaining the SWCC should consider the effect of the soil type and its stress history. The soil type may be considered explicitly (a soil specific model) or implicitly (determining few experimental points). Generally speaking the models of the first type do not consider the structure of the soil or may consider it through an empirical parameter

[1] University of São Paulo (USP) - Brazil - fmarinho@usp.br

[2] University Mackenzie - Brazil - Former MSc Student at USP - stuermer@mackenzie.com.br

basically related to the density of the sample. The second type of model may have the structure indirectly considered, since it uses some actual measurements. In addition to this particular aspect of the use of the SWCC for predicting the unsaturated soil parameters there is the effect of the stress history. In natural soils this may be a reflection of the overconsolidation under saturated state or due to a process of dissecation or yet a compaction procedure. In compacted soils, the compaction energy and the initial water content may control the SWCC shape at some level of suction.

The soil water characteristic curves relates the amount of water that can be retained into the pores of a porous material with the soil water suction. This amount of water can be quantified in gravimetric or volumetric way. It is useful to make use of other parameters such as void ratio to have a better visualisation of the phenomenon involved.

Any porous material may have a range of pore sizes and these pores can store water. One can say that the material has a capacity for storing water. The water is stored in small spaces where the surface effect is significant. In this way some energy has to be applied to withdraw the water from the pores. The relation between this energy and the amount of water withdraw is known as soil water characteristic curve. The bigger the size of the pore the smaller the energy required to empty that pore. It should be pointed out that the energy required to empty a pore is the same that may be used to shrink the soil sample. If shrinkage of the soil is not possible, due to the proximity of the particles, the soil will desaturate.

It is found in the literature that the relation between gravimetric water content and suction is, in some cases, fairly independent of the initial dry unit weight (e.g. Olsen and Langfelder, 1965; Marinho and Chandler, 1993, among others). However, considering the enormous and broad use of the SWCC, it may be useful to look in more detail the influence of the initial dry unit weight on the SWCC.

Olson and Langfelder (1965) carried out a series of tests using five different soils where the suction at the compaction water content was determined using the axis translation technique. The results are presented in figure 1 in terms of gravimetric water content. Each soil was compacted statically using a different pressure and a range of water contents, which resulted in different initial void ratio for each sample. The symbols shown in figure 1 do not represent experimental points; they are plotted just to facilitate the identification of the soils tested. The connections between the points indicate the same compaction pressure. These curves do not represent the SWCC. By plotting the data in this way one can say that the relation between water content and suction is practically independent of the compaction pressure, for the pressures used and for the soils tested.

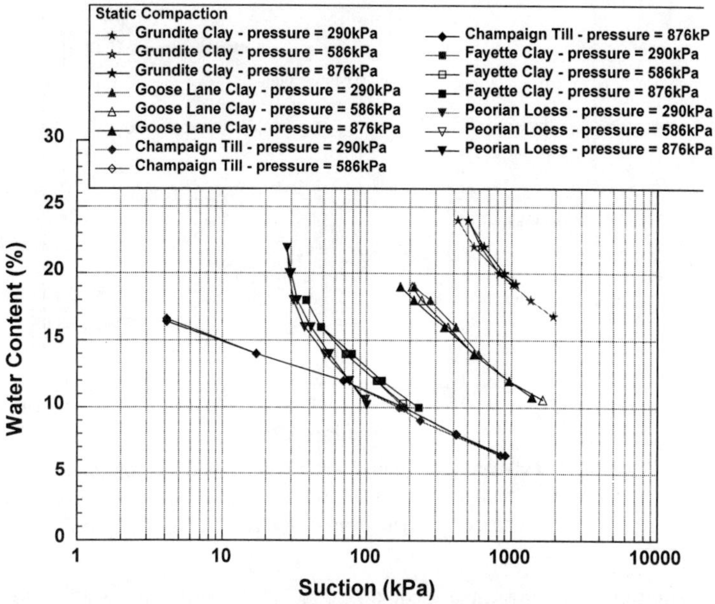

Figure 1 - Relation between gravimetric water content and suction for five different soils compacted at different pressures (data from Olson & Langfelder, 1965).

Olson and Langfelder (1965) also investigated the effect of the compaction procedure on the suction of two of the soils shown in figure 1. The Goose Lane clay and the Champaign till were compacted using both, kneading compaction and static compaction. The results are presented in figure 2. The compaction curves obtained using the kneading compaction on the Goose Lane clay were similar to those obtained using static compaction. For the Champaign till the difference was considerable. It can be seen that the kneading compaction lead to values of suction different from the one obtained by static compaction. This may suggest a potential effect of the structure on a SWCC. The tendency of the curves to coincide for higher values of soil suction is likely due to the fact that the water are located into pores of similar size/shape.

Although the results showed an influence of the compaction procedure on the value of suction for the same water content, the soil water characteristic curve may present some other differences during a drying path. If a drying test is performed the results may not follow the lines drawn in figure 1 and 2, in other words, the suction capacity ($C = \dfrac{\Delta w}{\Delta \log(suction)}$) of each sample may be different just after compaction.

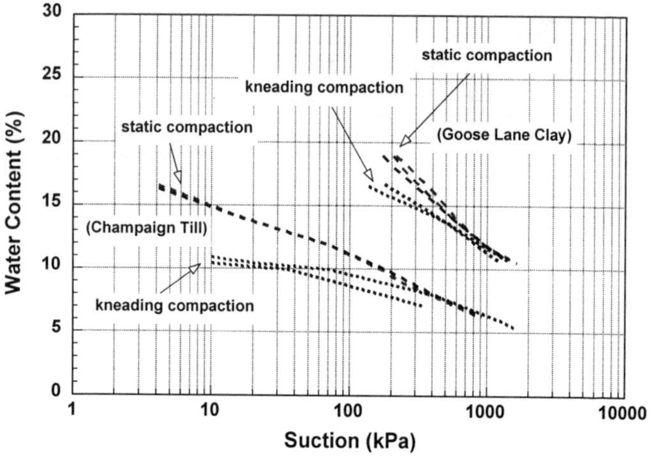

Figure 2 - Relation between gravimetric water content and suction for Goose Lane Clay and Champaign till under static compaction and under kneading compaction (data from Olson & Langfelder, 1965).

From the data presented it could be concluded that the influence of the initial dry unit weight on the SWCC is negligible and the type of mechanical compression (e.g. static or kneading compaction) may give different relation between gravimetric water content and suction. However, the data presented by Olson and Langfelder (1965) did not allow for the SWCC to be obtained. The results are for different samples and not for the same sample on drying.

In order to try to identify the effect of stress history on the SWCC, data from literature were analysed. The result is presented in figure 3, in terms of suction capacity (between 100kPa and 1000kPa) versus liquid limit of the corresponding soil. After the data were plotted it was possible to identify an upper and lower limits of that relation. The slurry line refers to samples prepared using water content above the liquid limit. The line related to stiff soils is almost horizontal, indicating the high energy required to extract water from the pores of the soils.

Comparing the results from literature with those obtained from the data of Olson & Langfelder (1965), it can be observed that the suction capacity obtained is higher than the expected values in some cases. The lower the liquid limit the smaller will be that effect. In figure 3 these results are identified by its corresponding compaction procedure (static compaction, SC or kneading compaction, KC).

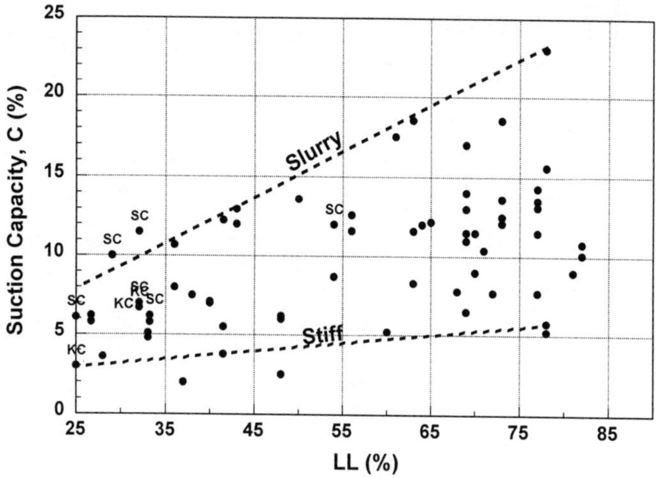

Figure 3 - Relation between the suction capacity (C) and liquid limit (LL).

This paper presents the results of a series of drying tests carried out on compacted residual soil of gneiss under different energy of compaction. The soils suction was measured using the filter paper technique. The filter paper was placed in direct contact with the soil sample. At higher soil water contents this procedure will measure matric suction, and , as the sample dries, there will be a gradual transition to a measurement of total suction The results also show the influence of the energy on the shrinkage process.

The soil tested

The soil used is a residual soil of gneiss from the top layer of the profile. The liquid and plastic limits of the material tested are 48% and 29%, respectively. The clay content (%<2μm) and specific gravity of the soil are 45% and 2.7, respectively. The soil was compacted using three different energies: Standard Proctor (SP), Modified Proctor (MP) and a non-standard lower energy (NE). The compaction curves are presented in Figure 4 together with lines of iso-degree of saturation. The values of maximum dry unit weight and optimum water content are presented in Table 1. All the compacted samples are named according to the code indicated in figure 4. Some of these samples were used to obtain the drying path of the soil water characteristic curves.

Compaction Energy	γ_{dmax} (kN/m³)	w_{opt} (%)
Standard Proctor (SP)	15,4	25
Modified Proctor (MP)	16,7	19
Non Standard Energy (NE)	14,3	28

Table 1: Compaction parameters.

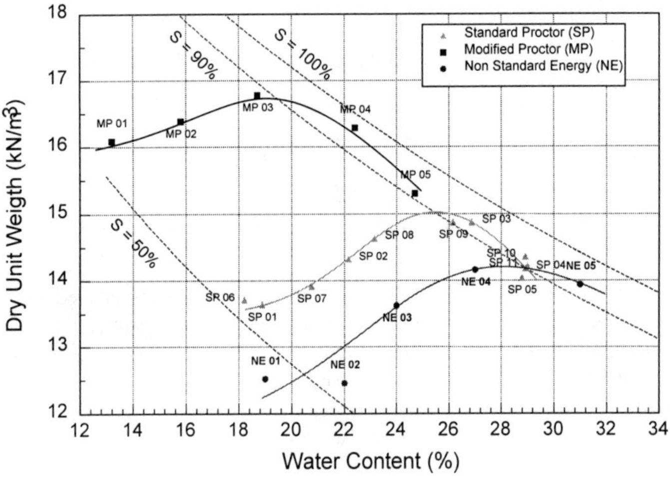

Figure 4: Compaction curves for three different energies.

Suction measurements

The suction measurements were obtained with the filter paper method using Whatman nº 42, quantitative filter paper. The calibration curve used was the one suggested by Chandler et al (1992).

Twenty-one samples were used to obtain the soil suction and the soil water characteristic curves. Five samples using the modified proctor (MP), eleven samples using the standard proctor (SP) and five samples with the non-standard energy (NE). All samples were weighted and some of them had their volume determined by measuring the external dimension of the sample, using a vernier calliper reading to 0.01mm. After determining the mass and volume, five filter papers were placed in each sample after compaction. The filter paper were placed in direct contact with the soil sample. This procedure will give matric suction for higher soil water content and, as the soil dries, there is a tendency for the measurement of total suction. This is due to the reduction of contacts between the filter paper and the soil water. In this case the transference of water from the soil to the filter paper start to be via vapour. The sample and the filter papers were wrapped in two layers of cling film and left for seven days inside a polystyrene box. The temperature variation inside the box during the process of equilibrium, between the filter paper and the soil, was less than 2°C. The filter papers were weighted using an electronic balance reading to 0.0001g. The use of electronic balance together with plastic bags should be made with care. The plastic may induce some static that affects the readings. It is recommended to use a spacer between the balance plate and the plastic bag.

Initial Characteristics of the Samples

The determination of the characteristic curves started at the compaction water content. This procedure does not allow for the determination of the SWCC at very low values of suction. The suction measured after compaction and the volumetric characteristics of the sample are presented in figures 5 to 7, according to the energy of compaction. It is important to emphasize that the results presented in this way do not represent the soil water characteristic curves for the soil.

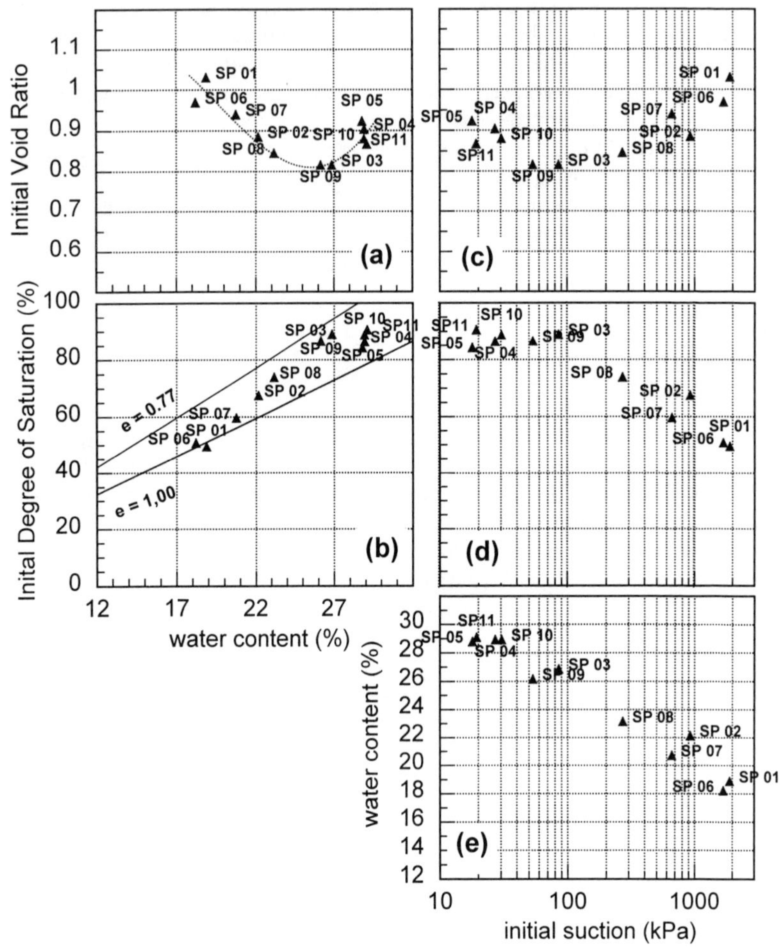

Figure 5: Initial conditions of the samples compacted using the standard proctor energy.

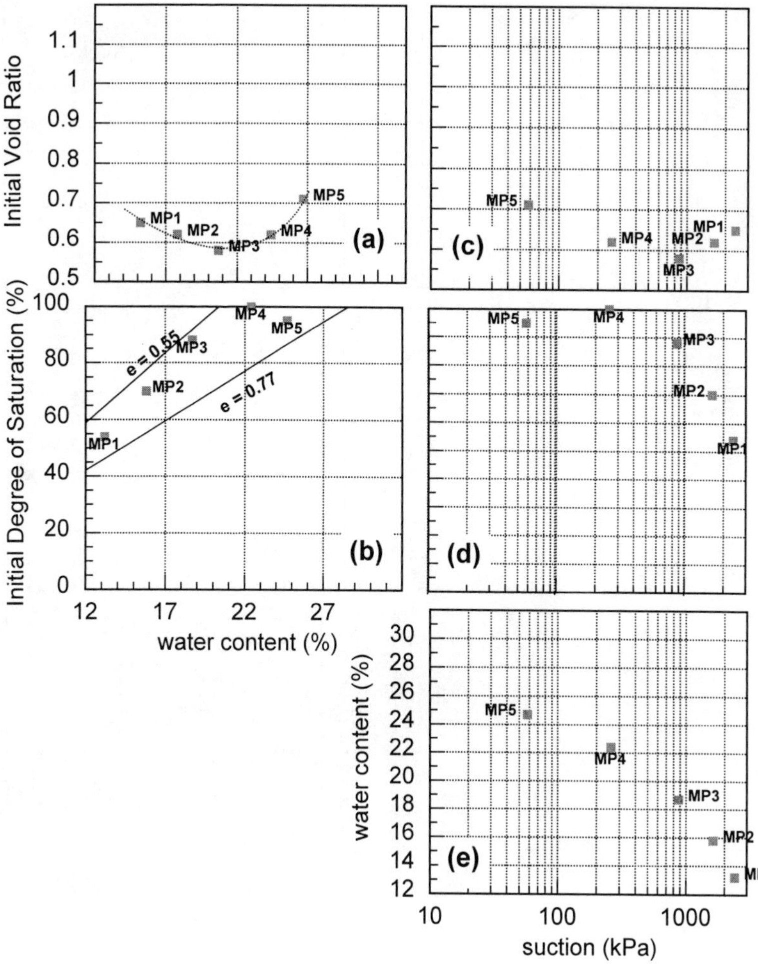

Figure 6: Initial conditions of the samples compacted using the modified proctor.

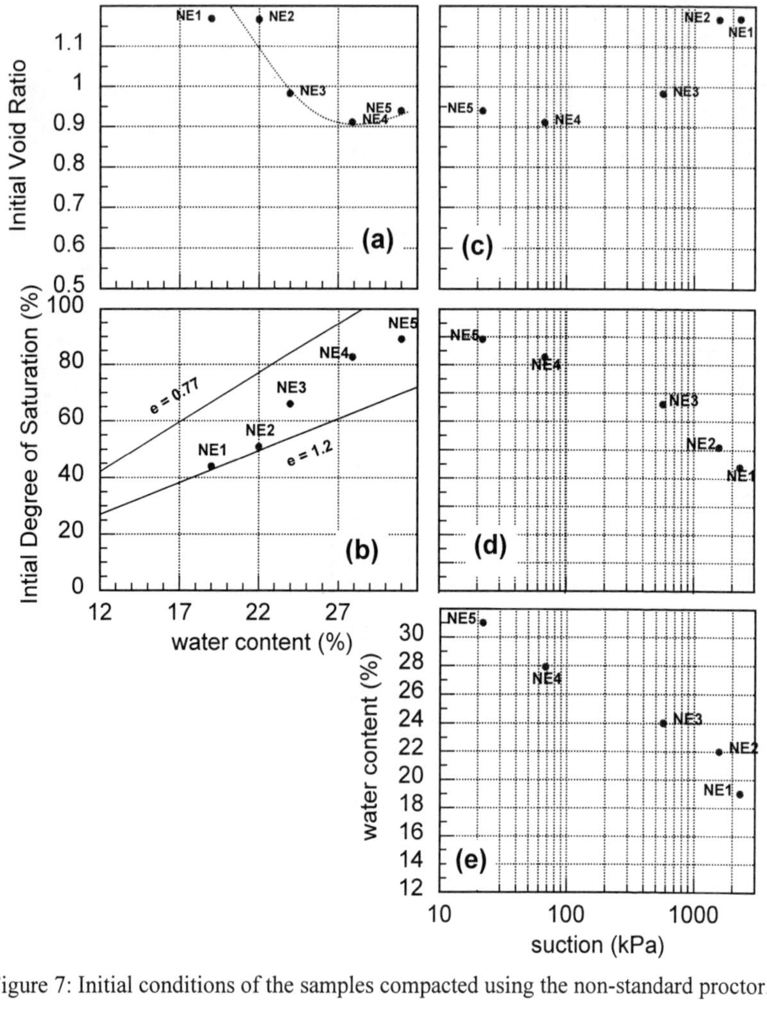

Figure 7: Initial conditions of the samples compacted using the non-standard proctor.

The Soil Water Characteristic Curves

Figures 8, 9 and 10 show the soil water characteristic curves obtained in a drying process for the samples SP, PM and NE, respectively. One can observe on figure 8e and 9e, the different paths followed by the samples according with the initial water content. This behaviour is also related with the initial degree of saturation.

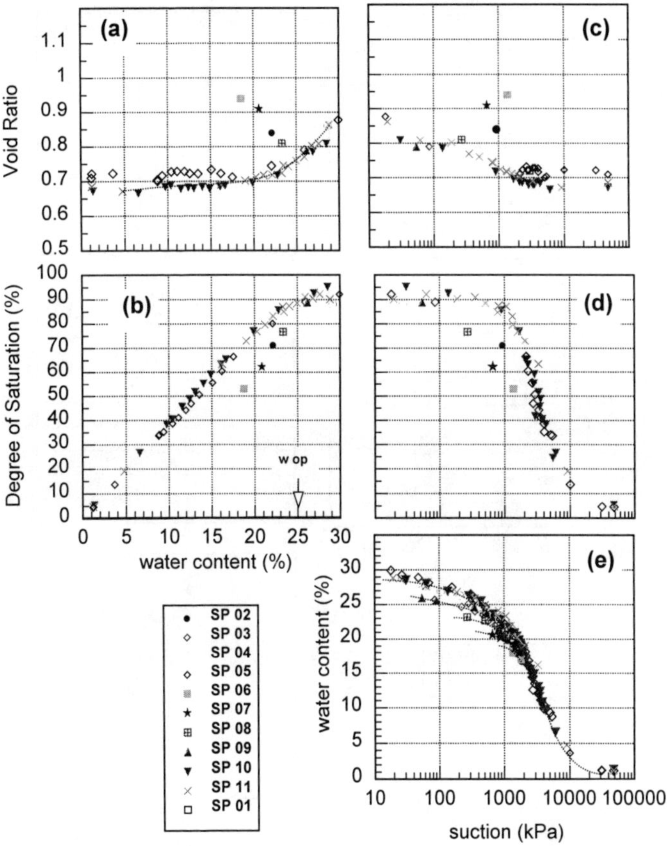

Figure 8: Soil Water Characteristic Curves for the samples compacted using the standard Proctor

Using the results obtained it is possible to identify three types of samples:

1. Samples compacted above the optimum water content (all energies).
2. Samples compacted above the optimum water content but with different void ratio (at a specific energy).
3. Samples compacted at the same void ratio and different water content (at a specific energy).

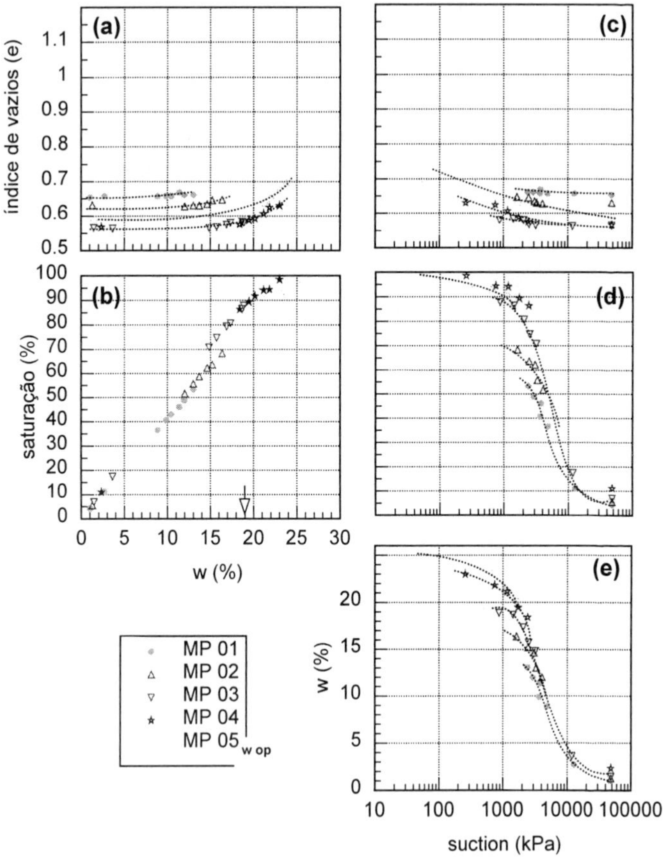

Figure 9: Soil Water Characteristic Curves for the samples compacted using the modified Proctor.

The samples compacted with water content higher than the optimum water content started the drying process at a zone where the amount of water lost (in volume) is equal to the reduction in total soil volume, this is called normal contraction phase (e.g. samples SP 05 and MP 04). These samples presented a volume change bigger than samples compacted at lower water content, although with similar void ratio. Samples starting the drying process on the normal shrinkage phase present high degree of saturation

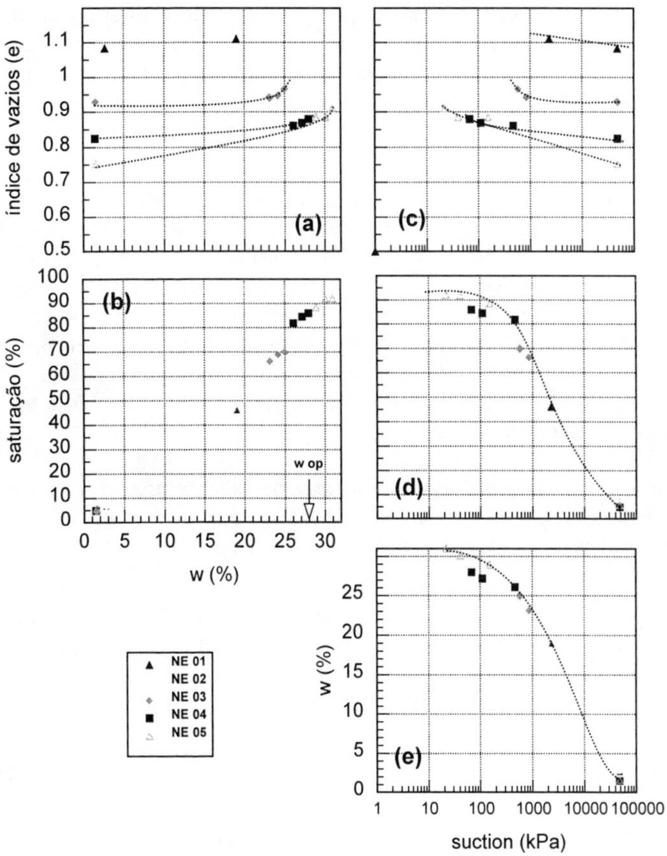

Figure 10: Soil water Characteristic Curves for the samples compacted using the non-standard Proctor.

Samples MP 03 and MP 05 have void ratio of 0,6 and 0,7, respectively. The sample MP 03 is already at a phase where there is little reduction involume due to lost of water, this is called residual shrinkage phase. Sample MP 05 is at the normal phase. The difference observed in the shrinkage process is related to the initial degree of saturation and void ratio.

The samples MP 02 and MP 04 have an initial void ratio of approximately 0,64. For the same value of suction these two samples presented degree of saturation and water content different. Sample MP 02 has a much lower initial degree of saturation (68%) than sample MP 04 (98%). The shrinkage of these samples is affected by the initial degree of saturation and water content. After the general air entry point (GAE) the relation between degree of saturation and water content is practically a straight line for some samples, i.e., void ratio is constant.

Figure 11 resumes the soil water characteristic relations of all samples obtained with the three compaction energies. The higher the energy of compaction the higher is the degree of saturation for the same suction up to the GAE. At some point after the GAE the curves of all energies become practically unique. This phenomenon may be due to the fact that the pore size distribution of the saturated pores is equal. It also should be point out that at very high level of suction the suction measured might be a combination of matrix and total suction, although the osmotic suction in this case seems to be small. It may also be that for higher suctions, the Electro-chemicals effects are stronger than capillary effects.

The curves obtained show, what may be called a pseudo-pre-consolidation effect of the compaction on the storage capacity. For suction values smaller than GAE, the soil behaviour is ruled specially by that effect.

Figure 12 presents the three compaction curves obtained showing the suction for each sample just after compaction. Lines of equal values of suction are also shown. One can observe that there is a dependence of the dry unit weight on the suction after compaction, since the iso-lines of suction are not vertical.

The GAE point occurs for suction values of about 1,000 kPa, 2,000 kPa, and 500kPa respectively for SP, MP and NE samples. The residual water content (θr) is reached for suction value of about 15,000 kPa, with residual degree of saturation (Sr) approximately of 2% for the three curves. After GAE the behaviour of all samples are practically the same. It can be observed that the volume change, up to the GAE, for sample compacted above the optimum water content is 75%, 55% and 35% of the total shrinkage for the energies standard (SP), modified (MP) and non-standard (NE), respectively. In some a great amount of shrinkage is still due to occur after the GAE, down to the zero water content.

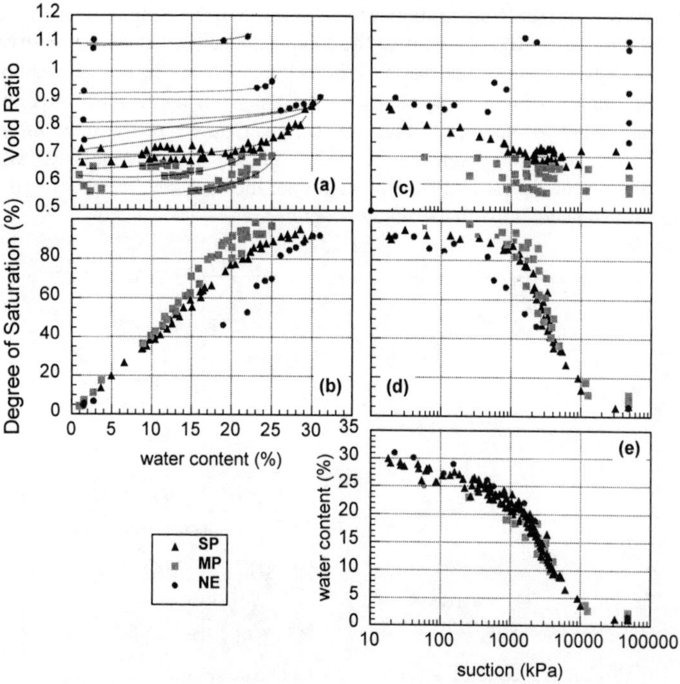

Figure 11: Results of all samples according with energy.

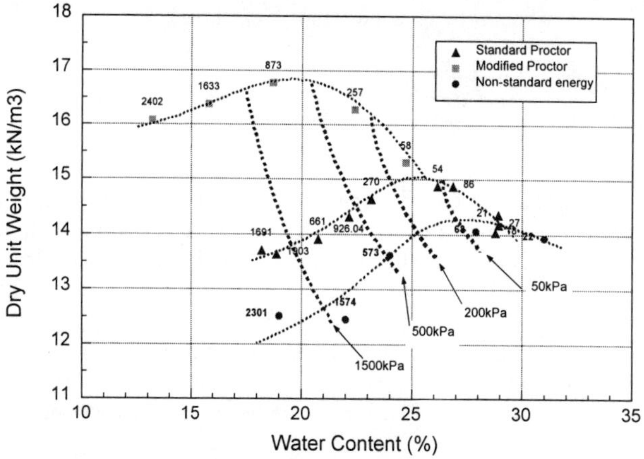

Figure 12: The compaction curves with lines of iso-suction.

Figure 13 presents the drying curves for some of the samples (SP 11, MP 03, MP 05, NE 04 and NE 05) plotted over the compaction curves. It can be observed that during the drying process the samples SP 11 and NE 05 practically follow the compaction curve, rising the dry density until the GAE point. After this value of suction, the gradient of increase of the dry density is reduced. At this point the suction reduces its efficiency of shrinkage. As expected in samples compacted with the modified Proctor the increase in suction could not increase the dry unit weight in the same way it had occurred with the smaller energies.

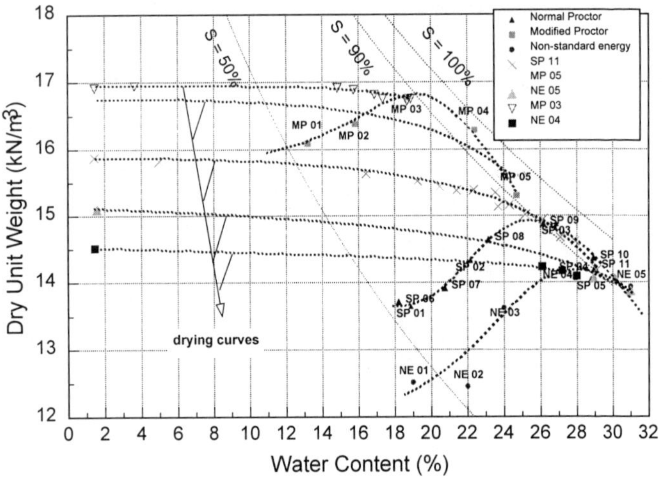

Figure 13: Drying curves of the samples SP11, MP05, NE05, MP03 and NE04, over the compaction curves.

Conclusions

At a particular energy of compaction the initial water content affect the SWCC. During the drying process samples compacted near the optimum water content or below it, always presented smaller suction than samples compacted at wet of optimum

The GAE is not greatly affected by the compaction water content. On the other hand the energy of compaction affects the GAE of the soil. The compaction energy seems to affect the level of suction that is controlled by capillary phenomena.

The soil water characteristic curves embraces not only the relationship between gravimetric or volumetric water content and suction (soil water characteristic curve, SWCC), but also the relation between void ratio and suction. The soil water characteristic curve is affected by the energy of compaction, considering the differences between the energies used in this study. The energy of compaction obviously affects the relation between void ratio and water content and in consequence it affects the quantitative aspect of the relationship between void ratio and suction.

The volume change of the samples after the GAE can be significant, particularly for samples that were compacted using low energy.

References

CHANDLER,R.J.; CRILLY, M.S. and MONTGOMERY-SMITH,G. (1992) - "A low-cost method of assessing clay desiccation for low-rise buildings". - *Proc. of the Institute of Civil Engineering, 92 - n.2 - pp.82-89.*

OLSON, R.E. and LANGFELDER, L.J. (1965) - "Pore pressures in unsaturated soils" - Journal of the Soil Mechanics and Foundations Division - ASCE - July. SM4 - pp.127-150.

MARINHO, F.A.M. and CHANDLER, R.J. (1993) - "Aspects of the behaviour of clays on drying" - Unsaturated Soils - ASCE Geotechnical Special Publication n° 39 - Edited by Houston, S.L. and Wray, W.K. - pp. 77-90.

Acknowledgement

The work presented in this paper is part of a research project sponsored by the São Paulo State agency (FAPESP). The second author is grateful for the sponsorship by the Brazilian Government agency CAPES.

Some Physical Properties of Unsaturated Oil-Contaminated Sand

E.C. Shin[1], and B. M. Das[2]

Abstract

Environmental hazards are created by both on and offshore oil spills which contaminate soil. Oil contamination may also change the physical properties of soil. This paper summarizes the results relating to compaction, hydraulic conductivity, and shear strength of laboratory tests conducted with oil-contaminated sand. Only one type of sand was used for the tests primarily at two relative densities of compaction; however three different oils were used as contaminants to evaluate the effect of their kinematic viscosity on the geotechnical parameters studied. Oil contamination increased the compactibility of sand whereas it reduced the hydraulic conductivity and shear strength.

Introduction

On and offshore oil spills cause many environmental and ecological problems. They may have an adverse effect on the physical and mechanical properties of soils such as compactibility, hydraulic conductivity, and shear strength. A number of studies have been recently published relating to the geotechnical aspects of oil-contaminated soil (Cook et al. 1992; Evgin et al. 1989; Al-Sanad et al. 1995, 1997; Puri et al. 1994; Fan et al. 1994).

In view of the current global emphasis on detailed investigations addressing various aspects of disposal of hazardous wastes, and the problem created by accidental oil spills on and off-shore, a study was initiated to investigate the effect of oil contamination on the properties and behavior of cohesionless soils. This paper presents the results of a

----

[1] Department of Civil Engineering, University of Inchon, South Korea 402-749
[2] College of Engineering and Computer Science, California State University, Sacramento, CA 95819-6023

series of laboratory tests which were conducted as a part of this study, to determine the effect of oil contamination on the compaction, hydraulic conductivity, and shear strength of sand.

Sand and Oil Used for the Tests

Jumoonjin sand, a poorly-graded silica sand in South Korea, was used for the present tests. The physical properties of the sand are given in Table 1. Three different types of oil (Oman crude oil, engine oil, lamp oil) with varying degrees of viscosity were used as contaminants. The kinematic viscosities (ν) for these oils at various temperatures were measured for each oil using the procedure described in ASTM test designation D-445, and these values are shown in Figure 1 along with the standard variation of ν with temperature for water. All laboratory tests reported here were conducted at a temperature of 22° ± 2° C. The specific gravities of the oil (G_o) at 20°C were: Oman crude oil-0.879; engine oil-0.852; lamp oil-0.798.

Table 1. Physical Properties of the Sand

Item	Quantity
Effective size, D_{10}	0.375 mm
Uniformity coefficient, C_u	1.53
Coefficient of curvature, C_z	1.10
Maximum dry unit weight	17.5 kN/m^3
Minimum dry unit weight	13.6 kN/m^3
Specific gravity of soil solids, G_s	2.65
Unified soil classification	SP

Compaction Tests

Standard Proctor compaction tests on the sand were conducted in the laboratory with water and the three types of oil as lubricant. The variations of dry unit weight with the oil and water contents are shown in Figure 2. Note that, in Figure 2, the notations for oil and water content are w_o and w_w, respectively. Based on the results of Figure 2, the variations of maximum dry unit weight [$\gamma_{d(\max)}$] and the degree of saturation (S) at w_{opt} are given in Figure 3. The degree of saturation for the tests conducted with oil as the lubricant can be expressed as

$$S = \frac{w_o G_s}{e G_o} \qquad (1)$$

where e = void ratio

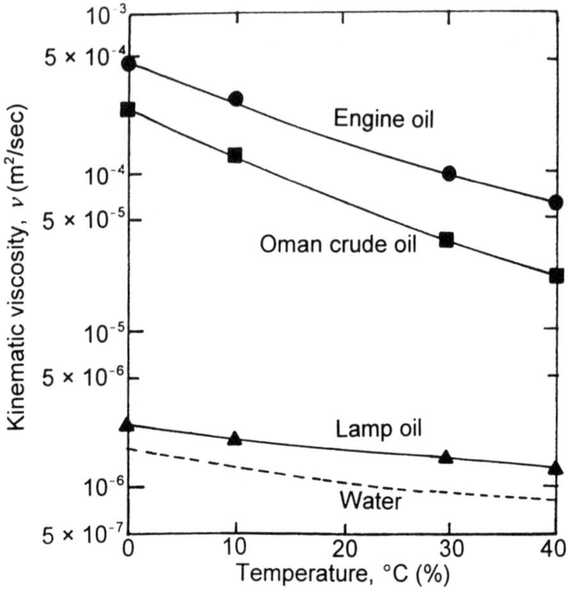

Figure 1 Variation of kinematic viscosity with temperature for three oils and water

From Figure 3 it can be seen that, with increasing kinematic viscosity, the maximum dry unit weight of compaction increases accompanied by a decrease in the degree of saturation. This is most likely due to the fact that oil with a higher viscosity is more effective in reducing the friction between the sand grains, thus helping attain the maximum dry unit weight at a lower degree of saturation.

Hydraulic Conductivity Tests

Laboratory constant head permeability tests were conducted with specimens measuring 100 mm in diameter and 150 mm long. Before conducting the tests, the sand was thoroughly mixed with oil and compacted in the permeameter. The tests were conducted at relative densities of compaction (D_r) of 45% and 75% with oil contents varying from 0% to 6%. Hydraulic conductivities (k) were calculated by collecting water passing through the specimens after a steady state of flow was reached. During the time steady flow was established some oil might have washed away, thereby reducing the oil content at the time of measurement. No verification to that effect was made. The variations of hydraulic conductivity obtained are summarized in Figure 4. Each point shown in this figure is the average of three tests. From Figure 4, the following observations can be made.

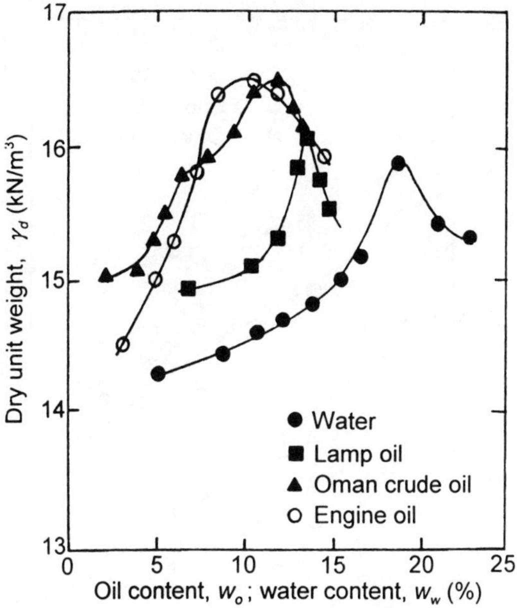

Figure 2 Variation of dry unit weight with oil/water content

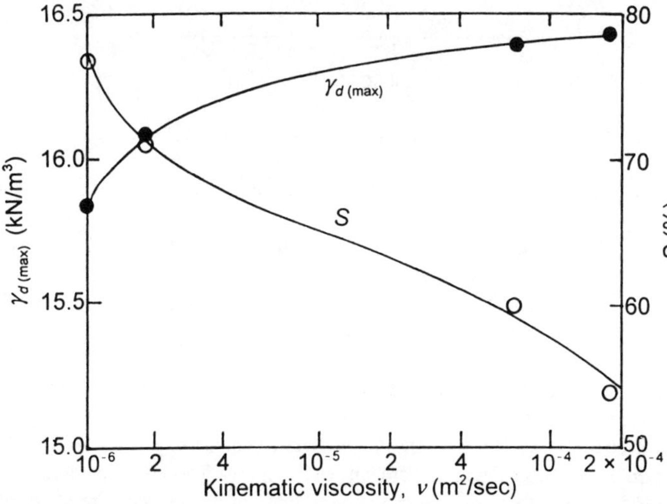

Figure 3. Plot of $\gamma_{d\,(max)}$ and S versus ν

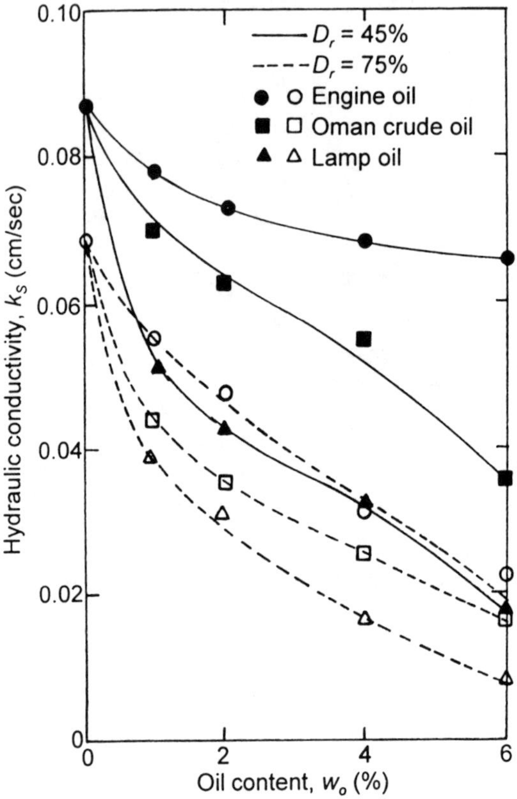

Figure 4 Plot of hydraulic conductivity with oil content

1. For a given oil, the hydraulic conductivity decreases with increasing oil content (w_o).
2. For a given oil content (w_o), the hydraulic conductivity decreases with decreasing kinematic viscosity of oil.
3. For a given oil and oil content, the hydraulic conductivity decreases with increasing relative density of compaction.

In order to evaluate the degree of hydraulic conductivity with the degree of oil saturation and the kinematic viscosity, the variation of $k_s/k_{s=0}$ (k_s = hydraulic conductivity with the degree of oil saturation of S; $k_{s=0}$ = hydraulic conductivity at S = 0) with degree of oil saturation were calculated and are shown in Figures 5 and 6, respectively, for D_r = 45% and 75%. From these plots it can be seen that, for a given sand, relative density of compaction and viscosity of oil, the variation of log ($k_s/k_{s=0}$) with S is approximately linear, or

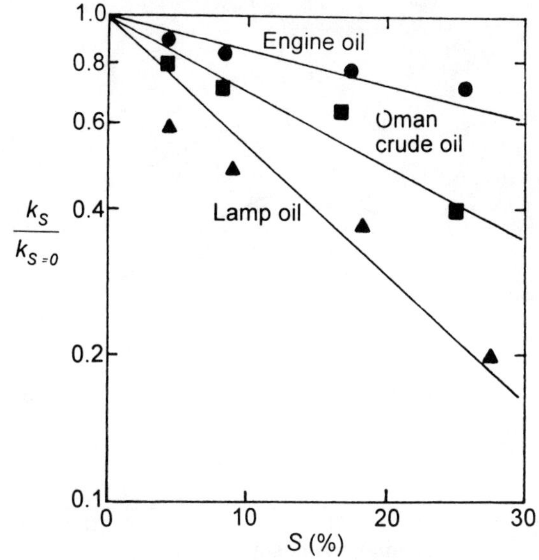

Figure 5 Plot of $k_s/k_{s=0}$ with S (D_r = 45%)

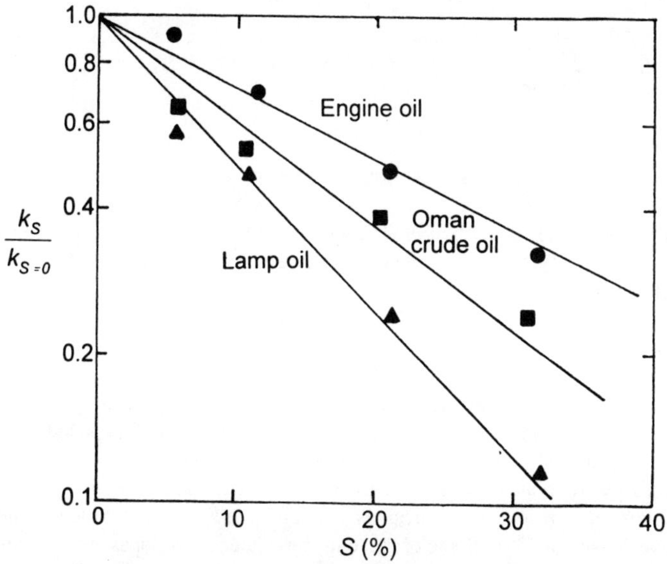

Figure 6 Plot of $k_s/k_{s=0}$ with S (D_r = 75%)

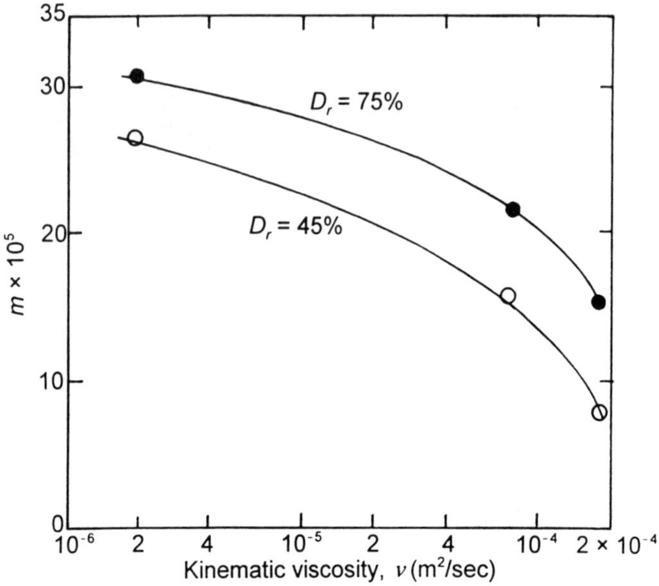

Figure 7 Plot of m vs. ν (D_r = 45% and 75%).

$$\log\left(\frac{k_S}{k_{S=0}}\right) = -mS \tag{2}$$

where m = a constant which is a function of D_r and ν

Figure 7 shows the variation of m with ν for D_r = 45% and 75%. For a given value of kinematic viscosity, the magnitude of m increases with the increase in the relative density of compaction.

Direct Shear Tests

A number of direct shear tests were conducted on dry and oil-contaminated unsaturated sand specimens at relative densities of compaction of 45% and 75%. All specimens measured 64 mm × 74 mm in cross section and had a height of 20 mm. Also, the specimens were tested under three normal stress levels, that is, 31 kN/m^2, 62 kN/m^2, and 87 kN/m^2,. The results of a typical test are shown in Figure 8 (for Oman crude oil at a relative density of 75%). Some of the total stress failure envelopes derived from the test results showed an apparent cohesion intercept (c) of 2 kN/m^2 to 5 kN/m^2. Because of the small specimen dimensions involved and the possible experimental deviations,

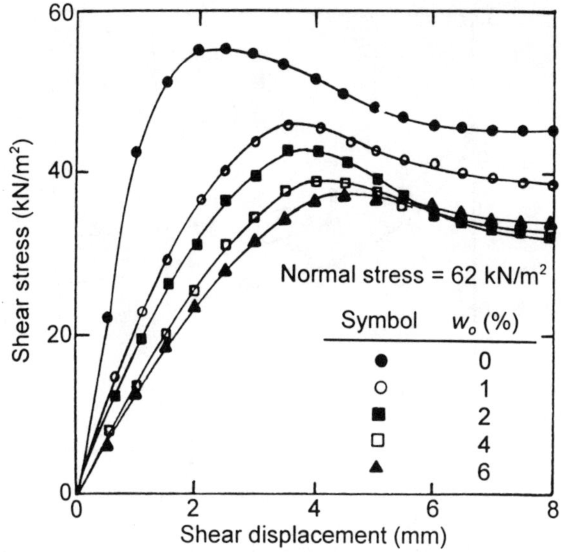

Figure 8 Direct shear test results with Oman crude oil as contaminant ($D_r = 75\%$)

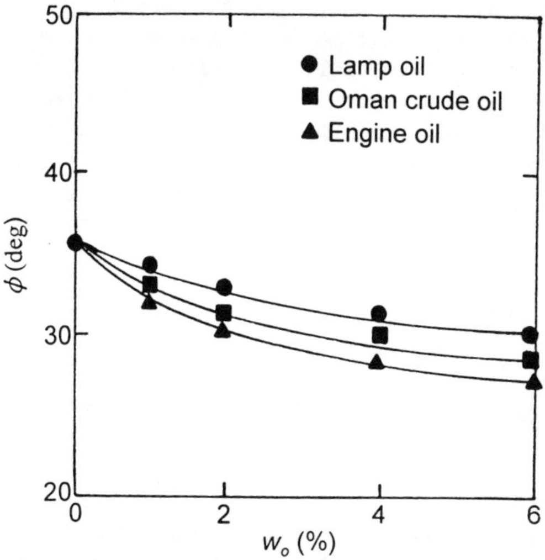

Figure 9 Total stress friction angle vs. oil content ($D_r = 45\%$)

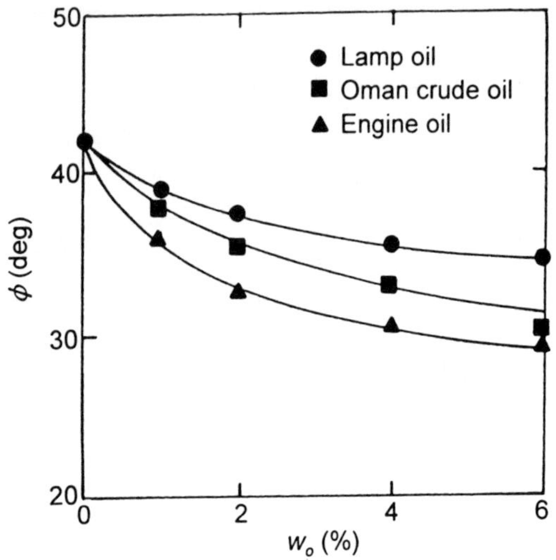

Figure 10 Total stress friction angle vs. oil content (D_r = 75%)

it was not possible to drawn any conclusions about the nature of apparent cohesion. Accordingly, it was decided to assume c to be equal to zero and obtain a total stress friction angle (ϕ) by drawing the failure envelope passing through the origin. This does, at least, provide a sense of the magnitude to which the soil friction angle is affected due to oil contamination. The variations of the total stress friction angle (ϕ) for D_r = 45% and 75% and w_o = 1% to 6% are shown in Figures 9 and 10. From these figures it can be seen that, for a given oil and kinematic viscosity, the magnitude of ϕ decreases with the increase in the oil content. For a given relative compaction of sand and given oil content, the magnitude of ϕ decreases with the increase in kinematic viscosity of oil. The magnitude of decrease of ϕ with oil content can be calculated as

$$\Delta\phi = \frac{\phi_{w_o=0} - \phi_{w_o}}{\phi_{w=0}} \tag{3}$$

where $\Delta\phi$ = decrease in the total stress friction angle

ϕ_{w_o}, $\phi_{w_o=0}$ = total stress friction angle at oil contents w_o and $w_o = 0$, respectively

Based on the present experimental results, the variation of $\Delta\phi$ for various oils, oil contents, and relative densities of compaction were calculated and are shown in Figure 11. It can be seen from Figure 11 that, for a given oil and oil content, the magnitude of $\Delta\phi$ increases with the relative density of the soil. Also for a given oil and relative

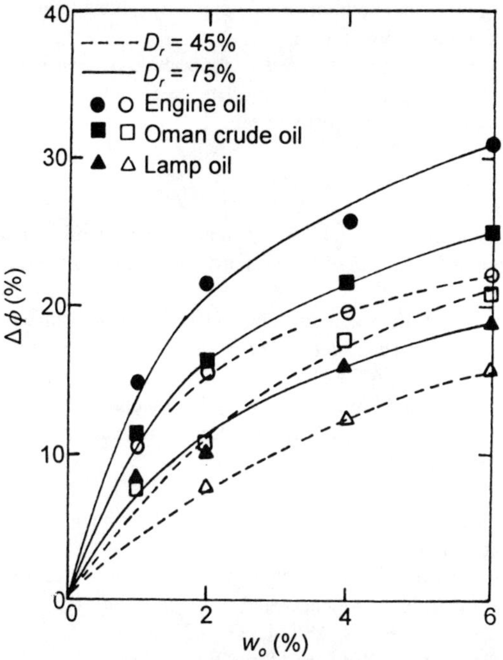

Figure 11 Change in total stress friction angle due to oil contamination

density of compaction, $\Delta\phi$ increases with the increase in oil content. The kinematic viscosity of oil has a great influence of $\Delta\phi$. For a given oil content and relative density, the magnitude of $\Delta\phi$ increases with the increase in kinematic viscosity of the oil. These observations are important because the oil contamination of 2% to 4% can reduce the total stress friction angle by about 15% to 25%. A decrease in ϕ will ultimately affect the stability of shallow foundations, slopes, and other structures and may cause failure

Conclusions

Laboratory test results for compaction, hydraulic conductivity, and shear strength of unsaturated oil-contaminated poorly graded sand were presented. One sand at two relative densities of compaction were tested. Three oils with different kinematic viscosities were used as contaminants. Based on the laboratory test results, the following general conclusions related to poorly graded sand can be drawn:

1. For similar compactive effort, the maximum dry unit weight of compaction of sand increases with the increase in kinematic viscosity of oil. This is accompanied by a decrease in the degree of saturation.

2. Increasing oil content and kinematic viscosity have a tendency to decrease the hydraulic conductivity of a soil.
3. For a given sand, the total stress friction angle decreases with increasing oil content. The magnitude of decrease in the friction angle increases with the increase of kinematic viscosity of oil.

References

Al-Sanad, H.A., and Ismael, N.F. (1997). "Aging effects on oil-contaminated Kuwaiti sand," *Journal of Geotechnical and Geoenvironmental Engineering*, ASCE, 123(3), 290-293.

Al-Sanad, H.A., Eid, W.K., and Ismael, N.F. (1995). "Geotechnical properties of oil-contaminated Kuwaiti sand," *Journal of Geotechnical and Geoenvironmental Engineering*, ASCE, 121(5), 407-412.

Cook, E.E., Puri, V.K., and Shin, E.C. (1992). "Geotechnical properties of crude oil-contaminated sand," *Proceedings*, 2nd International Conference on Offshore and Polar Engineering , Vol. 1, 384-387.

Evgin, E., Amor, F.B., Altaee, A., Lord, S., and Konuk, I. (1989). "Effect of oil spill on soil properties," *Proceedings*, 8th International Conference on Offshore Mechanics and Arctic Engineering, Vol. 1, 715-720.

Fan, C.Y., Krishnamurthy, S., and Chen, C.T. (1994). "A critical review of analytical approaches for petroleum contaminated soil," in *Analysis of Soils Contaminated with Petroleum Constituents*, ASTM STP 1221, 61-64.

Puri, V.K., Das, B.M., Cook, E.C., and Shin, E.C. (1994). "Geotechnical properties of crude oil-contaminated sand," in *Analysis of Soils Contaminated with Petroleum Constituents*, ASTM STP 1221, 75-88.

Characterization of Unsaturated Nonwoven Geotextiles

John C. Stormont[1], M.ASCE and Carl E. Morris[2], M.ASCE

Abstract

Geotextiles are increasingly being used in a wide variety of applications, such as reinforcement, separation and drainage. These geotextiles are often in contact with unsaturated soils; however, the hydrologic behavior of geosynthetic materials in unsaturated environments is most often ignored because the basic characterization of these materials does not exist. Test methods for measuring water retention data and transmissivity under suction have been developed for nonwoven geotextiles. These measurements complement simple contact angle and capillary rise measurements to characterize nonwoven geotextiles under unsaturated conditions. Results indicate that the tested nonwoven geotextiles exhibit properties and behavior similar to coarse, uniform soils. The unsaturated behavior has been found to be hysteretic; that is, the behavior depends on whether the geotextile is being dried or wetted. Penetration of adjacent soil into the geotextile has also been found to affect its behavior. Soil column tests that include nonwoven geotextiles confirm that they will significantly affect unsaturated flow in soils.

Introduction

Geotextiles are increasingly being used in a wide variety of applications, such as reinforcement, separation and drainage to improve the performance of roads, both paved and unpaved; drainage and separation in landfill covers and liners; and reinforcement for slopes, embankments, and retaining walls. These geotextiles are often in contact with unsaturated soils; however, the hydrologic behavior of geosynthetic materials in unsaturated environments is most often ignored because the basic characterization of these materials does not exist.

----

[1]Dept of Civil Engrg., Univ.of New Mexico, Albuquerque, NM, USA, 87131.
[2]Dept of Civil and Mining Engrg., Univ.of Wollongong, Wollongong, NSW 2522, Australia.

Geotextiles vary widely in their fiber composition, pore geometry, and thickness. Consequently, we may expect a wide range of behaviors in unsaturated environments. Most geotextiles designed for permanent functions are fabricated from polymers that are, in general, much less wetting with respect to water than soils are. Thin woven geotextiles with large openings may permit soil-to-soil contact across the geotextile and not significantly affect unsaturated water movement except to reduce the area through which water can move. Other woven geotextiles with small openings may not transmit liquid except under positive pressures. Nonwoven geotextiles are of a sufficient thickness and possess a pore structure that suggests they may behave in a manner consistent with a classic porous medium.

Nonwoven geotextiles have relatively large pore sizes compared to most soils, and consequently their hydraulic behavior may be expected to be similar to a coarse soil. Indeed, nonwoven geotextiles are often used to replace coarse soils in applications that require large saturated hydraulic conductivities. However, materials with large pore sizes will typically possess very low hydraulic conductivities under most unsaturated conditions. A layer of this type of material may impede unsaturated flow, and is often referred to as a capillary barrier or break. Geotextiles have been used as capillary breaks to prevent water from being drawn up beneath a pavement and thus limit frost heave and subsequent thaw weakening (Henry and Holtz, 1997). Geotextiles have also been used in a geocomposite system that forms a capillary break to prevent downward flow and allow water to be drained laterally (Stormont and Stockton, 2000). While capillary breaks may have many positive applications, it is possible that geosynthetics may inadvertently form capillary breaks and have negative consequences. For example, a geosynthetic used beneath a road for reinforcement purposes may serve as a capillary break to downward moving water, causing water to be retained in overlying layers until they become saturated, contributing to the deterioration of the road. A side slope failure of a landfill cover was attributed to a geotextile inadvertently acting as a capillary break (Richardson, 1997).

In this paper, methods and results for the characterization of the unsaturated properties of nonwoven geotextiles are described. Some of these data are fit to functions used to describe the unsaturated properties of soils. Tests of unsaturated flow through soil columns containing nonwoven geotextiles are also described.

Contact angle measurements

The behavior of a geotextile in an unsaturated environment will depend in part upon the attractive forces associated with the water-fiber interface as measured by the contact angle. The lower the contact angle, the easier it is for water to wet the fiber and all other factors being the same, the more water will be held at a given value of suction head. Hydrophillic materials (e.g., glass and most soils) have contact angles near zero; materials with contact angles greater than 90° repel water. The contact angles of polyester and polypropylene fibers, which comprise most nonwoven geotextiles, have been measured (e.g., Berg, 1989; Henry and Patton, 1998). Typical contact angles are in the range of 50° to 80°, which indicate that polyester and polypropylene are only slightly wetting. The results reveal hysteresis in the contact angle, with a higher contact angle measured for wetting compared to drying. Further, polyester has been found to have a somewhat lower contact angle than polypropylene. Hysteresis and product-to-product variability in the contact angle has been attributed to differences in chemical composition of the fibers, surface roughness and the presence of contaminants on the fibers (Berg, 1989). Henry and Holtz (1997) placed soil fines in geotextiles and found that this decreased the contact angle of water on the fibers, apparently because the soil particles cling to the surfaces of the fibers.

Capillary rise measurements

Capillary rise above a water surface provides a measure of the wetting behavior of a porous material. The capillary rise, or capillary depression in the case of a hydrophobic material, is a function of the contact angle of the material as well as the pore structure. Henry and Holtz (1997) and Henry and Patton (1998) report the height of capillary rise for polyester and polypropylene geotextiles ranging from −17 mm to 75 mm, depending on the fiber material and the surface condition.

To identify geotextiles that are more wetting than typical polyester and polypropylene geotextiles, capillary rise measurements have been performed on geotextiles fabricated from alternative materials including nylon and fiberglass. Vertical strips of geotextiles about 50 mm wide were placed with one end immersed in a pan of water. The height of capillary rise was measured both visually and by touch after 48 hours. The water was dyed in order to improve the visual contrast of the wetting front in light colored geotextiles. As shown in Figure 1, the alternative textiles had capillary rise values significantly greater than those of conventional geotextiles. For comparison, the capillary rise in coarse sands is typically in the 100 to 200 mm range.

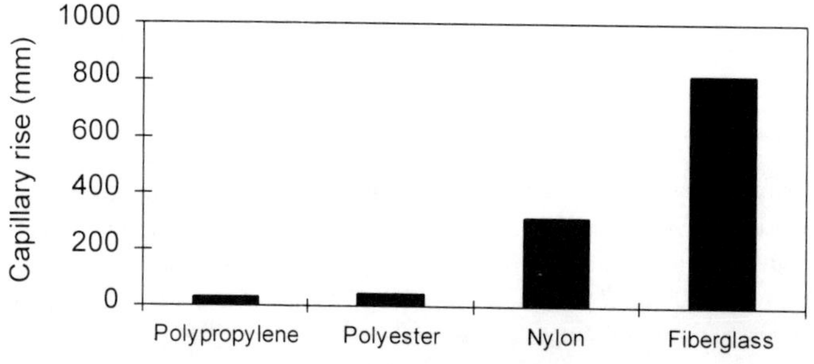

Figure 1 – Capillary rise in various textiles.

Water retention data

Water retention data describe the relationship between water content and negative water pressure (suction head), and is considered a fundamental relationship for a porous medium. Water retention data for nonwoven geotextiles can be obtained by testing specimens in a hanging column apparatus (Stormont et al., 1997). The hanging column is appropriate for suction heads of about 2 m or less, which is the range of interest for nonwoven geotextiles. Geotextile specimens are placed on top of a water-saturated porous plate. The underside of the porous plate is connected by means of a water filled-tubing to a water reservoir. When a geotextile is placed on the porous plate, hydraulic equilibrium is established between the two materials. The suction head in the specimen can be increased by lowering the level of the reservoir below the porous plate. The geotextile specimen either absorbs or exsorbs water to equilibrate with the water in the porous plate as the level of the reservoir changes. The water retention function is determined by systematically raising or lowering the reservoir to change the suction head, allowing the water in the geotextile to equilibrate with the water in the porous plate, and then measuring the specimen's water content. A nominal surcharge of 0.87 kPa was placed on top of the specimens to limit volume change during the tests.

Water retention data are given in Figure 2 for two nonwoven geotextiles described in Table 1. The results are given for the range of 0 mm to 300 mm, as this is the region over which most of the water content changes occur. The specimens were first tested under a wetting path: beginning air-dry, and progressively decreasing the suction head to zero. The specimens were then dried by progressively increasing the suction head.

Table 1 – Nonwoven geotextiles tested with hanging column.

Geotextile	Polymer type	Thickness, mm	Apparent opening size, mm	Mass per unit area, g/m^2	Saturated transmissivity, mm^2/s
A	polyester	1.8	0.04	266	6.8
B	poly-propylene	5.9	0.18	340	39.0

The sharp uptake of water during wetting suggests that these geotextiles have a definable water entry suction head between 10 and 30 mm. The water entry suction head represents the transition of a material from a hydraulically nonconductive to a conductive state. Thus, if water in contact with the geotextile is at a suction head in excess of the geotextile's water entry suction head, water will not flow into the geotextile.

The water retention data exhibit hysteresis. The specimens contained more water during drying than wetting at comparable values of suction head. The water content of many specimens did not decrease during the initial portions of the drying path, indicating that once wetted, some geotextiles may remain relatively wet under small suction heads.

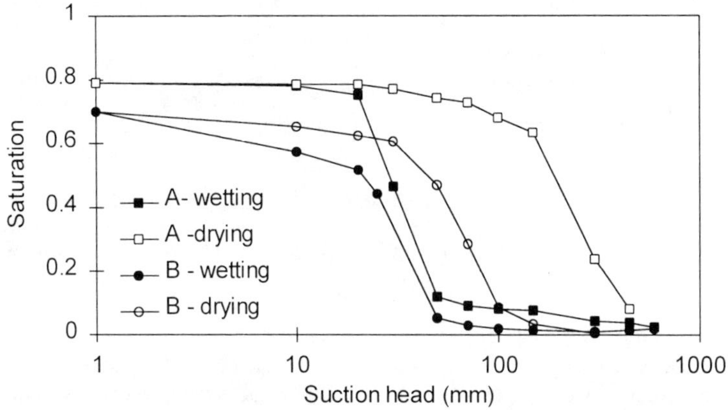

Figure 2 - Water retention data for two nonwoven geotextiles.

Intrusion of soil particles into geotextiles placed in the field may alter their unsaturated hydraulic properties. Consequently, water retention data were obtained for geotextiles with soil intruded into them following the technique of Henry (1998). Tests following a wetting path have been conducted on polyester geotextile specimens (geotextile A from Table 1), with intrusion of sand (particles between 0.15 and 0.075 mm), silt (particles smaller than 0.075 mm but greater than 0.002 mm), and clay particles (less than 0.002 mm). In Figure 3, the water retention data during wetting for soil-intruded specimens along with a clean specimen are given. The data indicate that the intruded soil causes the specimens to initially wet at higher suction heads, that is, have a larger water entry head. The water retention data are not considerably different for the different types of intruded soil.

Geotextiles as barriers to flow in unsaturated soils

Downward flow

Infiltration tests were conducted in soil columns that included nonwoven polypropylene geotextiles. The tests were conducted in 100-mm diameter, 250-mm long plexiglass tubes. The bottom 50 mm of the column was packed with a silty sand (classified as SM) on top of which a nonwoven polypropylene geotextile (geotextile

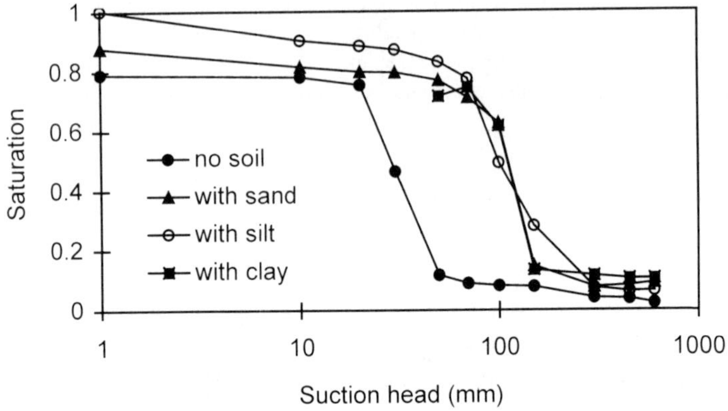

Figure 3 - Water retention data (wetting path) for polyester nonwoven geotextile with intruded soil.

B from Table 1) was placed. The remaining 200 mm of the column was packed with the silty sand. Tensiometers were located just above and below the geotextile. Two additional columns were constructed. One column replaced the lower silty sand layer with a coarse sand (classified as SP), and included the geotextile between the two soils. The final column consisted of the silty sand over the coarse sand without a geotextile. This column constitutes a soil-based capillary barrier due to the contrast of hydraulic properties between the two soils. The soils were packed to achieve a dry density of about 1.6 g/cc, with initial suction heads in the 2 to 3 m range.

Water was added to the top of the column with a peristaltic pump at a constant rate of about $2x10^{-4}$ mm/sec, and the suction heads were monitored. All three columns responded initially in a similar manner: when the wetting front reached the interface at the base of the overlying SM layer, the suction heads decreased to well below 1 m whereas the suction heads below the interface did not substantially change. These results are evidence that downward flow was impeded in each of the columns. Eventually, as continued infiltration decreased the suction heads in the overlying soil, water moved across the interface (termed "breakthrough") and into the underlying soil as evidenced by the decrease in suction heads within the underlying soil. The breakthrough head is taken as the suction head in the soil just above the interface at breakthrough. The breakthrough head should be equivalent to the water entry head as it is coincident with the geotextile becoming conductive.

The measured breakthrough heads are given in Table 2. These results indicate that the breakthrough occurs at a lower suction head in the columns with a geotextile compared to the column without the geotextile, suggesting that this geotextile may serve as a better capillary barrier than the coarse sand. The breakthrough heads from the two tests with a geotextile but different underlying soil type are nearly the same.

This result indicates that breakthrough across the geotextile is independent of the underlying soil. The breakthrough heads in Table 2 are greater than the water entry heads estimated from the water retention data (about 30 to 50 mm), perhaps due to the effects of soil intrusion into the geotextile or condensation of water within the geotextile.

Table 2 – Results of soil column infiltration tests.

Column	Breakthrough head (mm)
SM-GT-SM	150
SM-GT-SP	160
SM-SP	300

Upward flow

Henry (1998) evaluated upward flow in 1.5 m-long, 100-mm diameter cylindrical columns of frost-susceptible soil with water available at the base. In some of the columns, geotextile disks were placed 30 mm above the bottom of the soil column to determine if they would impede upward flow. The water table was located 5 mm below the geotextile. In terms of breakthrough head, water will move upward through the geotextile if its breakthrough head was 5 mm or more, and there will be no water movement if the breakthrough head was less than 5 mm. Following standard procedures for evaluating frost heave, the soil was initially saturated from the bottom upward by raising the water table from the base to top. The water table was then returned to its original position, and the soil was frozen from the top down. The amount of frost heave and the water content in the overlying soil were measured. Three different polypropylene nonwoven geotextiles were evaluated as capillary barriers in these freezing tests. Frost heave was measured in control tests without geotextiles to provide a basis for judging the effectiveness of the geotextiles in impeding upward water flow in response to soil freezing.

Results of these tests indicated that these nonwoven geotextiles functioned well as a capillary barrier in some tests but not in others (Henry, 1998). In the tests where the geotextiles failed to reduce upward flow, the geotextile was apparently wetted during the initial saturation procedure. In the tests where the geotextiles did prevent upward flow into the overlying soil, water did not move across the geotextile during the saturation procedure, which implies the geotextile actually prevented flow under a positive pressure. These results suggest that moistening the geotextiles affects the subsequent hydraulic properties, consistent with the wetting-drying hysteresis observed in the water retention data. To confirm this hypothesis, geotextiles containing soil fines and pre-moistened to a saturation of 30 to 40% were tested in the soil freezing columns. All of these geotextiles were found to be ineffective in impeding upward flow with a water table immediately beneath it, although these results do not indicate how these geotextiles would have functioned with the water table at a greater distance beneath it. Henry (1998) suggested that the geotextiles should be expected to be moist and contain soil fines in the field, and thus, represents the most relevant test condition.

Transmissivity under suction

Measurement of unsaturated flow in nonwoven geotextiles is difficult because the geotextile transitions from being essentially non-conductive to very conductive over a small range of suction. A method has been developed for measuring in-plane transmissivity of geotextiles under suction (Stormont et al., 1998). The unsaturated permeameter consists of a platform that extends above two reservoirs of water. The geotextile lies on the platform, and its ends extend into the reservoirs on both sides. The water level in the reservoirs will be at or below the platform level. Water will rise in the geotextile and result in a suction head in the geotextile equivalent to the distance the platform is above the water level in the reservoir. A total head difference is created along the geotextile by raising one end of the platform and its reservoir. Transmissivities were calculated using the steady-state solution used to calculate transmissivity under positive pressures. The suction head in the geotextile is assumed to remain constant during flow, although it does vary a small amount along its length.

Transmissivities are given in Figure 4 as a function of suction head for geotextiles A and B. These measurements were made under a nominal normal pressure of 0.35 kPa, and with a gradient of 0.25. During initial wetting, the geotextiles were non-conductive until the suction head was reduced to 35 mm for geotextile A and 25 mm for geotextile B. The transmissivity of both specimens increased by about an order of magnitude as they were wetted to near zero suction head. The geotextiles remain transmissive during drying to over 100 mm, well beyond the suction head at which they initially became transmissive. Geotextile A is less transmissive than geotextile B under all values of suction head, consistent with their reported saturated transmissivity values.

The water retention and transmissivity data are consistent. During wetting, the water retention function indicates that the geotextiles do not accept much water until the suction head was reduced to less than 50 mm; the geotextiles had an immeasurably low transmissivity until the suction head was reduced to about 30 mm. The geotextile rapidly takes up water as the suction head is reduced to zero, and coincides with the transmissivity increasing by an order of magnitude. During drying, both the water content and transmissivity of the geotextile remain at greater values compared to the values obtained during wetting. These results suggest that these nonwoven geotextiles will wet and become transmissive under suction, but not until the suction head in the adjacent soil is reduced to about 30 mm.

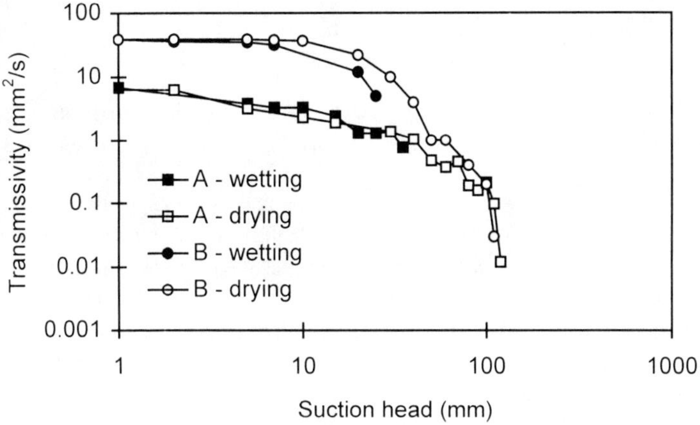

Figure 4 - Transmissivities as a function of suction for Geotextiles A and B.

Fitting parameters for unsaturated hydraulic functions

Continuous functions that describe the water retention and hydraulic conductivity data are commonly applied to soils. One widely used water retention function was developed by van Genuchten (1980). The volumetric water content, θ, and suction head, h, are related by

$$\theta = \theta_r + (\theta_s - \theta_r)\left[1 + (\alpha h)^n\right]^{-m} \tag{1}$$

where the subscripts s and r indicate the saturated and residual values of the water content, α, m and n are fitting parameters, and m = 1-1/n.

By adopting the model for hydraulic conductivity proposed by Mualem (1976), the transmissivity as a function of either water content or suction head can be predicted using the van Genuchten parameters and the saturated transmissivity, T_{sat}. In terms of suction head, the transmissivity is given as

$$T = T_{sat}\,\frac{\{1-(\alpha h)^{mn}[1+(\alpha h)^n]^{-m}\}^2}{[1+(\alpha h)^n]^{m/2}} \tag{2}$$

As is commonly done for soils, water retention data from nonwoven geotextiles can be fit to the van Genuchten's function given by Equation 1. In addition, the unsaturated transmissivity data can be fit to the function given by Equation 2. As an example, the water retention and transmissivity data for Geotextile B were fit to these equations using the RETC software (van Genuchten et al. 1991). The fitting parameters are given in Table 3. The r^2 values for all of the fits were 0.99 or 1.0. Because of the pronounced hysteresis in the water retention data, separate parameters were developed for the wetting and drying paths.

Table 3 – van Genuchten fitting parameters for geotextile B.

test	path	θ_s	θ_r	α (1/mm)	n
water retention	wetting	0.615	0.000	0.0381	6.90
	drying	0.597	0.007	0.0173	5.84
transmissivity	wetting	na	na	0.0378	3.48
	drying	na	na	0.0269	3.25

The transmissivity function (Equation 2) does not explicitly depend on water content, and hence θ_s and θ_r are not given in the table. The fitting parameters for this nonwoven geotextile can be contrasted to those of most soils. The θ_s parameter is considerably larger than that of most soils, and is consistent with the large porosities in these geotextiles. The very low values of θ_r indicate that the geotextiles contain very little water except at small suctions. The values of α, usually assumed to be related to the inverse of the air entry head, are relatively large compared to most soils except very coarse soils. Consistent with the observed rapid change of water content over a small change in suction, the value of n is relatively large.

The good fit to the data suggest that it may be possible to apply functions commonly used for soils to nonwoven geotextiles to extrapolate experimental results as well as facilitate numerical implementation. Ideally, the parameters derived from the water retention and transmissivity data would be the same. One possible cause for the difference in fitting parameters may be that water retention tests involve cross-plane flow and transmissivity measurements involve in-plane flow. Another difference between the two tests was the normal stress applied to the specimens: 0.88 kPa for the water retention test and 0.35 kPa for the transmissivity test.

Conclusions

Test methods for measuring water retention data and transmissivity under suction have been developed for nonwoven geotextiles. These measurements, along with contact angle and capillary rise measurements, serve to characterize the geotextiles behavior in unsaturated environments. The unsaturated behavior of the tested geotextiles were found to be hysteretic; that is, their behavior depends on whether they are being dried or wetted. Penetration of soil into the geotextile has also been found to affect its properties. The nonwoven geotextiles tested exhibit properties and behaviors similar to coarse, uniform soils. Consequently, nonwoven geotextiles can be expected to substantially impede flow in unsaturated soils and often serve as capillary barriers until low values of suction head are reached in the adjacent soil. Tests of flow in soil columns that included geotextiles have confirmed this behavior.

An important consideration for characterizing the unsaturated properties of geotextiles is the condition of the geotextile. In particular, the intrusion of soil fines into the geotextile, as will occur in the field, may alter its unsaturated properties. Testing is underway to evaluate these effects, as well as testing a wide range of woven and nonwoven geotextile specimens. Methods for measuring cross-plane hydraulic conductivity are also under development.

Acknowledgments

Tomas Stockton, Chandradip Ray, Brian Woolsey and Rachelle Ramos assisted with laboratory measurements and development of the fitting parameters.

References

Berg, J.C. (1989). "The use of single-fiber wetting measurements in the assessment of absorbency." In A.F. Turbak and T.L. Vigo (Eds.), *Nonwovens: An advanced tutorial*, Tappi Press, Atlanta, GA, pp.219-239.

Henry, K.S. (1998). "The Use of Geosynthetics to Mitigate Frost Heave in Soils." Ph. D. Dissertation, Civil Engineering Department, University of Washington, Seattle, WA.

Henry, K. S., and Holtz, R. D. (1997). "Capillary Rise of Water in Geotextiles." In S. Knutsson (Ed.), Proceedings, *International Symposium on Ground Freezing and Frost Action in Soils*, Luleå, Sweden, 15–17 April, 227–233.

Henry, K.S. and Patton, S. (1998). "Measurement of the Contact Angle of Water on Geotextile Fibers." *Geotechnical Testing Journal*, *21*(1), 11-17.

Mualem, Y. (1976). "A new model for predicting the hydraulic conductivity of unsaturated porous media." *Water Resources Research*, *12*, 513-522.

Richardson, G. (1997). "Fundamental mistakes in slope design." *Geotechnical Fabrics Report*, *15*(2), 15-17.

Stormont, J.C., Henry, K.S. and Evans, T.M. (1997). "Water Retention Functions of Four Non-woven Polypropylene Geotextiles." *Geosynthetics International*, *4*(6), 661-672.

Stormont, J.C., Evans, T.M., Stockton, T.B., and Ray, C. (1998). "Unsaturated Hydraulic Properties of a Nonwoven Polypropylene Geotextile." *98 Joint Conference on the Environment*, Waste-management Education and Research Consortium, Albuquerque, NM, March 31 - April 1, 213-216.

Stormont, J.C. and Stockton, T.B. (2000). "Preventing Positive Pore Water Pressures with a Geocomposite Capillary Barrier Drain." In *Testing and Performance of Geosynthetics in Subsurface Drainage, ASTM STP 1390*, J.B. Goddard, L.D. Suits and J.S. Baldwin, Eds., American Society for Testing and Materials, West Conshohocken, PA.

van Genuchten, M. Th. (1980). "A closed-form equation for predicting the hydraulic conductivity of unsaturated soils." *Soil Science Society of America Journal, 44*, 892-898.

van Genuchten, M.Th., Leij, F.J., and Yates, S.R. (1991). "The RETC code for quantifying the hydraulic functions of unsaturated soils." *U.S. Environmental Protection Agency report EPA/600/2-91/065*, December.

A Fully Coupled Analysis Procedure for Dynamic Behavior of Unsaturated Soils

Kanthasamy K. Muraleetharan[1], Member ASCE, and Changfu Wei[2]

Abstract

A fully coupled analysis procedure for dynamic analyses of unsaturated porous media is presented. The analysis procedure is based on the finite element implementation of the governing equations derived utilizing the Theory of Mixtures with Interfaces (TMI). TMI explicitly considers the interfaces existing between various bulk phases in a multiphase porous media such as the contractile skin between the pore water and the pore air in an unsaturated soil. The resulting computer code, U_DYSAC2, is used to predict the behavior of an unsaturated kaolin embankment subjected to base shaking and the results are compared with the saturated soil behavior. Unsaturated soil embankment is shown to behave more rigidly than the saturated embankment resulting in lower displacements and higher accelerations. U_DYSAC2 also predicts increase in pore water and pore gas pressures during shaking. The results presented show that it is possible to economically design an unsaturated soil structure for seismic loads using a fully coupled analysis.

Introduction

Unsaturated porous materials are of interest in a number of diverse fields, such as geotechnical, environmental, biomedical, chemical and petroleum engineering, and geophysics and material science. An unsaturated porous material is a mixture consisting of a solid skeleton and several fluids such as water, gas, and oil. In geotechnical engineering, the increasing use of compacted soils in earthquake prone areas, such as in compacted roadway embankments, has created a need for better understanding of the dynamic behavior of these unsaturated soils.

----

[1]Assistant Professor and [2]Graduate Research Assistant, School of Civil Engineering and Environmental Science, University of Oklahoma, 202 W. Boyd, Room 334, Norman, OK 73019

It is also possible to more economically design these geotechnical engineering structures if the true behavior is considered as opposed to assuming the extreme conditions of saturated and dry. The complex nature of the unsaturated soil behavior involving solid skeleton, pore liquid and pore gas has traditionally led the engineers to resort to empirical procedures. However, realistic static and dynamic analysis of multiphase porous media requires a procedure that can rigorously describe the interactions among various phases. Such a procedure is generally called a fully coupled analysis procedure.

A salient feature of unsaturated porous media is the interfaces existing between various bulk phases. These interfaces play a major role in the flow and deformation behavior of multiphase porous media. For instance, many researchers (see Fredlund and Rahardjo 1993) have shown that the contractile skin or the interface between the pore liquid and pore gas plays a major role in the behavior of unsaturated soils. As opposed to its saturated counter part, fully coupled governing equations for unsaturated porous media are poorly developed. Typically, various extensions to Biot's saturated porous media theories (Biot 1962) are used as governing equations when dealing with a multiphase system, for example see Li and Zienkiewicz (1992) and Schrefler and Zhan (1993). In addition, stress-strain behavior of the solid skeleton is described through an effective stress concept. However, from the rigorous mixture theory standpoint, the interactions among various phases in the mixture were considered rather loosely in the above mentioned analyses. In fact, among all the interactions between various phases only the pressure difference between the gas and the liquid (called suction) is considered, and those between the solid phase and the fluid phases were totally excluded. Validity of using a single effective stress for unsaturated soils is also questionable in light of the recent research works showing the effectiveness of two independent stress state variables to describe the stress-strain behavior of the solid skeleton in unsaturated soils (Alonso et al. 1990, Fredlund and Rahardjo 1993). Recently Muraleetharan and Wei (1999a), based on the Theory of Mixtures with Interfaces (TMI), developed a set of governing equations for unsaturated porous media in which the interfacial effects are explicitly considered. The TMI also provides a theoretical basis, based on thermodynamics, for the use of two independent stress state variables.

A fully coupled analysis procedure for dynamic behavior of unsaturated soils based on the finite element implementation of the governing equations developed by Muraleetharan and Wei (1999a) is presented in this paper. A bounding surface elastoplastic model (Muraleetharan and Nedunuri 1998) developed in terms of net stress (total stress minus pore air pressure) and suction is utilized to describe the monotonic and cyclic stress-strain behavior of unsaturated soils. The analysis procedure is implemented into a computer code, U_DYSAC2 (Muraleetharan and Wei 1999b), and preliminary results obtained using this computer code for an unsaturated soil embankment subjected to base shaking are presented. Predicted displacements, accelerations, pore water and pore gas pressures are compared to the corresponding quantities from the saturated embankment analysis.

Governing Equations

Field equations governing the behavior of unsaturated soils have been developed based on the Theory of Mixtures with Interfaces (TMI) (Muraleetharan and Wei 1999a). For completeness, in what follows, these governing equations will be summarized first. Unless explicitly defined, Greek character α , β , or γ denotes all the bulk phases including solid skeleton (s), liquid (ℓ), and gas (g). Combined terms such as $\alpha\beta$ denote the interface between α -phase and β -phase. Tensor quantities are denoted using bold characters.

Mass balance equations are represented by

$$\frac{D^\alpha}{Dt}\left(n^\alpha \rho^\alpha\right) + n^\alpha \rho^\alpha \left(\nabla \cdot \mathbf{v}^\alpha\right) = \sum_{\beta \neq \alpha} \hat{e}^\alpha_{\alpha\beta} \tag{1}$$

where $\frac{D^\alpha}{Dt}$ represents the material derivative with respect to the motion of α -phase, n^α the volume fraction of α -phase (the volume of α divided by the total volume of the mixture) subjected to the so-called saturation condition $\sum_\alpha n^\alpha = 1$, ρ^α denotes the real density of component α , $\mathbf{v}^\alpha$ the velocity, and $\hat{e}^\alpha_{\alpha\beta}$ the exchange rate of mass through interface $\alpha\beta$ into α phase. Neglecting the mass exchange between the solid and the fluids, (1) is subjected to

$$-\hat{e}^\ell_{\ell g} - \hat{e}^g_{\ell g} = 0, \quad \hat{e}^s_{s\alpha} = \hat{e}^\alpha_{s\alpha} = 0 \qquad \alpha = \ell, g \tag{2}$$

Linear momentum balance equations are represented by

$$n^\alpha \rho^\alpha \frac{D^\alpha}{Dt}\left(\mathbf{v}^\alpha\right) - \nabla \cdot \left(n^\alpha \mathbf{t}^\alpha\right) - n^\alpha \rho^\alpha \mathbf{g} = \sum_{\beta \neq \alpha} \hat{\mathbf{T}}^\alpha_{\alpha\beta} \tag{3}$$

where $\hat{\mathbf{T}}^\alpha_{\alpha\beta}$ denotes the exchange rate of linear momentum through interface $\alpha\beta$ into α -phase, $\mathbf{t}^\alpha$ the macroscopic real stress tensor of α -phase, and $\mathbf{g}$ the gravitational acceleration vector. If, among all the interfacial properties, only the interfacial tension on the contractile skin (interface between the liquid and the gas) is considered, the balance of linear momentum of interfaces requires that

$$-\nabla \cdot \left(a^{\ell g} \gamma^{\ell g} \mathbf{I}\right) = -\hat{\mathbf{T}}^\ell_{\ell g} - \hat{\mathbf{T}}^g_{\ell g} \tag{4a}$$

$$0 = -\hat{\mathbf{T}}^s_{sg} - \hat{\mathbf{T}}^g_{sg} \tag{4b}$$

$$0 = -\hat{\mathbf{T}}^s_{s\ell} - \hat{\mathbf{T}}^\ell_{s\ell} \tag{4c}$$

where $a^{\ell g}$ is the surface density of the contractile skin defined by the total area of the contractile skin divided by the control volume; $\gamma^{\ell g}$ can be considered as the macroscopic surface tension. If the isothermal condition is assumed, the balance of energy is automatically satisfied and the associated equations are omitted hereafter.

As far as boundary/initial value problems are concerned, in addition to the balance equations (1)-(4), constitutive relationships must be formulated for the mass exchange terms $\hat{e}^{\alpha}_{\alpha\beta}$, linear momentum exchange terms $\hat{\mathbf{T}}^{\alpha}_{\alpha\beta}$, and stress tensor $\mathbf{t}^{\alpha}$. General forms of these constitutive relationships have been developed based on the TMI (Muraleetharan and Wei 1999a). A direct outcome of the TMI is the flow equation expressed as

$$\rho^{\alpha}\nabla G^{\alpha} - \rho^{\alpha}\mathbf{g} + \rho^{\alpha}\frac{D^{\alpha}\mathbf{v}^{\alpha}}{Dt} = -\frac{1}{n^{\alpha}}\left(\sum_{\beta\neq s}\boldsymbol{\mu}^{\alpha}_{\beta}.\mathbf{v}^{\beta,s} + \sum_{\beta\gamma}\boldsymbol{\mu}^{\alpha}_{\beta\gamma}.\mathbf{v}^{\beta\gamma,s}\right) \qquad (5)$$

$$\alpha = \ell,\, g$$

where G^{α} is the Gibbs' thermodynamical potential of component α and $G^{\alpha} = A^{\alpha} + p^{\alpha}/\rho^{\alpha}$, A^{α} is the free energy. If it is assumed for simplicity that the dependence of A^{α} on the surface density and the change in volume fraction, as well as the coupling flow terms, are neglected, (5) becomes

$$\nabla p^{\alpha} - \rho^{\alpha}\mathbf{g} + \rho^{\alpha}\frac{D^{\alpha}\mathbf{v}^{\alpha}}{Dt} = -\frac{1}{n^{\alpha}}\boldsymbol{\mu}^{\alpha}\cdot\mathbf{v}^{\alpha,s} \qquad \alpha = \ell,\, g \qquad (6)$$

where $\mathbf{v}^{\alpha,s} = \mathbf{v}^{\alpha} - \mathbf{v}^{s}$; $\boldsymbol{\mu}^{\alpha}\,(\alpha = \ell,\, g)$ is a tensor quantity of the second order, and may be given by

$$\boldsymbol{\mu}^{\alpha} = (n^{\alpha})^2(\rho^{\alpha}g)(\mathbf{k}^{\alpha})^{-1}h_{\alpha}^{-1}\left(n^{\ell}\right) \qquad \alpha = \ell,\, g \qquad (7)$$

where $\mathbf{k}^{\alpha}$ is the coefficient of permeability tensor (LT^{-1}), $h_{\alpha}\left(n^{\ell}\right)$ represents the relative permeability (for example see Fredlund and Rahardjo 1993) and is assumed to be a function of volume fraction of liquid n^{ℓ}.

It has been shown that the resultant equation of summing up (4a)-(4c) gives the relationship between the interfacial properties and the pressure difference (Muraleetharan and Wei 1999a). In fact, this formulation represents the well-known soil-water characteristic relationship, which can be assumed as follows,

$$p_c = p^g - p^{\ell} = s_c\left(\varepsilon_v, n^{\ell}\right) \qquad (8a)$$

where ε_v is the volumetric strain of the solid skeleton, and p_c is assumed to be a function of volumetric strain and the volume fraction of liquid (or volumetric water content). For convenience, the inverse form of (8) will be used in the following derivations, i.e.

$$n^{\ell} = v(\varepsilon_v, p_c) \tag{8b}$$

The mass exchange term is a function of the difference of Gibbs' free energy between the liquid and the gas phase (Muraleetharan and Wei 1999a), and, for simplicity, this relationship is assumed to be of the following form

$$\hat{e}^{\ell}_{\ell g} = -\hat{e}^{g}_{\ell g} = c\left(p^{g} - p^{\ell}\right) \tag{9}$$

where c is a material constant. Based on the concepts of the two stress state variables and bounding surface plasticity, an elastoplastic model was recently developed by Muraleetharan and Nedunuri (1998), which may be expressed in the following form

$$\boldsymbol{\sigma}^{N} = \boldsymbol{\sigma} - p^{g}\mathbf{I} = \mathbf{D} : \boldsymbol{\varepsilon} \tag{10}$$

where $\boldsymbol{\varepsilon} = \mathbf{E}$ (Lagrangian strain tensor) for small strains and $\boldsymbol{\sigma}^{N}$ is the so-called net stress tensor and **I** is the second-order unit tensor. This model will used to describe the deformation of the solid skeleton during cyclic and monotonic deformation.

Thus far, the constitutive relationships are yet to be developed for the pressure p^{g} and p^{ℓ}. Assuming only ideal fluids are considered, one has

$$div\ \mathbf{v}\big|_{\alpha} = -\frac{1}{\rho^{\alpha}}\frac{D^{\alpha}\rho^{\alpha}}{Dt} = \frac{1}{K^{\alpha}}\frac{D^{\alpha}p^{\alpha}}{Dt} \tag{11}$$

where $div\ \mathbf{v}\big|_{\alpha}$ denotes the rate of microscopic volumetric strain of α phase, K^{α} the bulk modulus. If the component is incompressible, Eq. (11) is still valid, however, K^{α} can only be considered as a penalty number associated with the incompressibility constraint.

Finite Element Implementation

The governing equations presented in the previous section are solved using the finite element techniques. From now onwards, for convenience, indicial notation is used. It is appropriate to assume that the solid phase (*not the solid skeleton*) is incompressible and the liquid and the gas phases are compressible. Following some manipulations, the governing equations presented in the last section may be changed to the following compact form:

Equilibrium of the mixture $\rho \ddot{u}_i^s - \sigma_{ij,j} - \rho g_i = 0$ (12)

Mass conservation:

liquid plus solid $c_{\ell u}\dot{u}_{i,i}^s + c_{\ell\ell}\dot{p}^\ell + c_{\ell g}\dot{p}^g + U_{i,i}^\ell - \frac{\hat{e}_{\ell g}}{\rho^\ell} = 0$ (13)

gas plus solid $c_{gu}\dot{u}_{i,i}^s + c_{g\ell}\dot{p}^\ell + c_{g\ell}\dot{p}^g + U_{i,i}^g + \frac{\hat{e}_{\ell g}}{\rho^g} = 0$ (14)

where coefficients in (13) and (14) are given as

$$c_{\ell u} = n^\ell + \frac{\partial n^\ell}{\partial \varepsilon_v}, \quad c_{\ell\ell} = \frac{n^\ell}{K^\ell} - \frac{\partial n^\ell}{\partial p_c}, \quad c_{\ell g} = \frac{\partial n^\ell}{\partial p_c} \tag{15a}$$

$$c_{gu} = 1 - n^\ell - \frac{\partial n^\ell}{\partial \varepsilon_v}, \quad c_{g\ell} = c_{\ell g}, \quad c_{\ell g} = \frac{n^g}{K^g} - \frac{\partial n^\ell}{\partial p_c} \tag{15b}$$

and

$$U_i^\alpha = -n^\alpha \left(\mu^{\alpha-1}\right)_{ij}\left(\rho^\alpha \ddot{u}_j^s + p_{,j}^\alpha - \rho^\alpha g_j\right) \qquad \alpha = \ell, g \tag{15c}$$

Boundary conditions are assumed to be

$\sigma_{ij}n_j = \bar{t}_i(t)$	on Γ_σ	$u_i^s = \bar{u}_i(t)$	on Γ_u
$U_i^\ell n_i = \bar{v}_n^\ell$	on $\Gamma_{v\ell}$	$p^\ell = \bar{p}^\ell(t)$	on $\Gamma_{p\ell}$
$U_i^g n_i = \bar{v}_n^g$	on Γ_{vg}	$p^g = \bar{p}^g(t)$	on Γ_{pg}

where $\Gamma_\sigma \cup \Gamma_u = \Gamma$, $\Gamma_\sigma \cap \Gamma_u = \Phi$, and so on.

Displacement of the solid skeleton, the liquid and the gas pressures are used as nodal unknowns. Following standard procedures one may derive the following semi-discretized form of the governing equations (12)-(14):

$$\mathbf{M}_{uu}\ddot{\bar{\mathbf{u}}} + \mathbf{K}_{ug}\bar{\mathbf{p}}^g + \int_\Omega N_{A,j}\sigma_{ij}^N d\Omega = \mathbf{F}_u \tag{16a}$$

$$\mathbf{M}_{\ell u}\ddot{\bar{\mathbf{u}}} + \mathbf{C}_{\ell u}\dot{\bar{\mathbf{u}}} + \mathbf{C}_{\ell\ell}\dot{\bar{\mathbf{p}}}^\ell + \mathbf{C}_{\ell g}\dot{\bar{\mathbf{p}}}^g + \mathbf{K}_{\ell\ell}\bar{\mathbf{p}}^\ell = \mathbf{F}_\ell \tag{16b}$$

$$\mathbf{M}_{gu}\ddot{\bar{\mathbf{u}}} + \mathbf{C}_{gu}\dot{\bar{\mathbf{u}}} + \mathbf{C}_{g\ell}\dot{\bar{\mathbf{p}}}^\ell + \mathbf{C}_{gg}\dot{\bar{\mathbf{p}}}^g + \mathbf{K}_{gg}\bar{\mathbf{p}}^g = \mathbf{F}_g \tag{16c}$$

where the matrices in Eq. (16) are expanded in Appendix I. Eq. (16) can be further reduced into a compact form, i.e.

$$\mathbf{M}\ddot{\mathbf{U}} + \mathbf{C}\dot{\mathbf{U}} + \mathbf{K}_p\mathbf{U} + \mathbf{F}_{\text{int}} = \mathbf{F} \tag{17}$$

where $\mathbf{F}_{int}$ denotes the internal force vector containing the integral term in Eq. (16a). Eq. (17) may be solved using an appropriate time integration scheme. The finite element formulation has been incorporated into a computer code U_DYSAC2 (Muraleetharan and Wei 1999b).

Example Analysis

The finite element mesh of the embankment considered in the example analysis is shown in Fig. 1. The embankment represents a speswhite kaolin centrifuge model tested by Kutter (1982). Kutter's (1982) models were constructed using saturated kaolin and subjected to base shaking. Additional details of the centrifuge model tests and the dynamic analysis of the saturated embankment using the computer code DYSAC2 (Muraleetharan et al. 1997) are presented in Muraleetharan et al. (1994). The exact same mesh shown in Fig.1 was used by Muraleetharan et al. (1994) with a different numbering system for the nodes.

For the unsaturated case, the stress-strain behavior of kaolin was modeled using a bounding surface elastoplastic model (Muraleetharan and Nedunuri 1998). Necessary unsaturated model parameters were obtained from suction controlled triaxial tests on kaolin (Wheeler 1996) and other bounding surface parameters were kept the same as the saturated values used by Muraleetharan et al. (1994). All the model parameters pertaining to unsaturated soil analysis are summarized in Table 1. Other bounding surface model parameters common to both the saturated and unsaturated analyses can be found in Muraleetharan et al. (1994). The soil-water characteristic curve for kaolin was obtained using experimental data given by Aitchison (1961) and the Brooks-Corey relationship (1964). Measured results and the function used in the analysis are shown in Fig. 2. The soil-water characteristic curve was also taken as a function of the volumetric strain of the solid skeleton as indicated in Eq. (8). Permeability coefficients of air and water were also varied according to the Brooks-Corey relationship (1964) starting with the values given in Table 1. An initial degree of saturation (Sr) of 88% (volumetric water content = 0.524, suction = 102 kPa) was considered for the unsaturated analysis. The choice of the initial suction was based on the available model parameters (Wheeler 1996). For comparison purposes, initial stresses used in the saturated analysis were again used in the unsaturated analysis. In the unsaturated analysis, the initial pore water and pore air pressures were set to -102 kPa and 0 kPa (atmospheric), respectively, to reflect the initial suction value of 102 kPa. Mass exchange between air and water was neglected in the present analysis.

The embankment was subjected to the base motion shown in Fig. 3 and displacements (Fig. 3, displacements are positive if the movement is upward or towards the right), accelerations (Fig. 4), and pore fluid pressures and suctions (Fig. 5) were predicted by U_DYSAC2. For comparison purposes, saturated embankment behavior predicted by the previous analysis (Muraleetharan et al. 1994) are also shown in Figs. 3-5. Although the analyses were conducted at the model (centrifuge) scale, all the results are presented in the prototype scale.

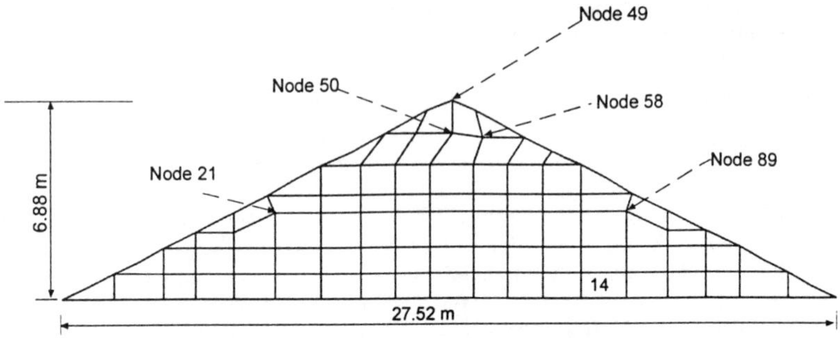

Fig. 1 Finite Element Mesh of the Kaolin Embankment

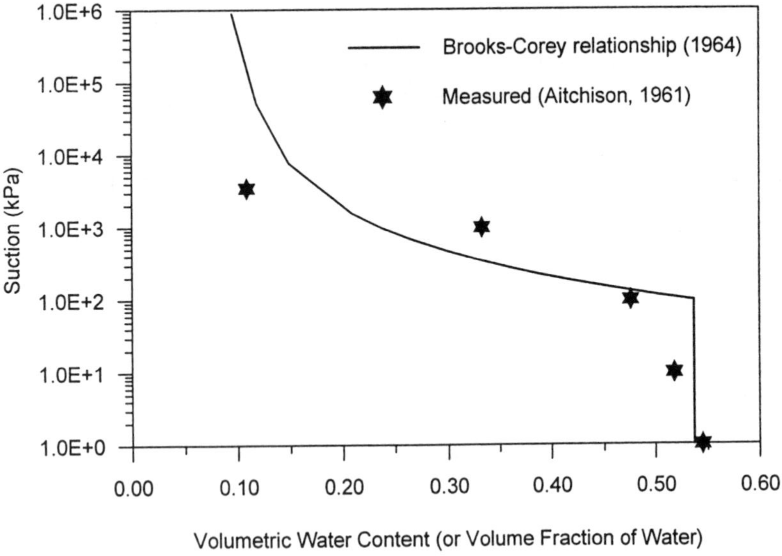

Fig. 2 Soil-Water Characteristic Curve for Kaolin

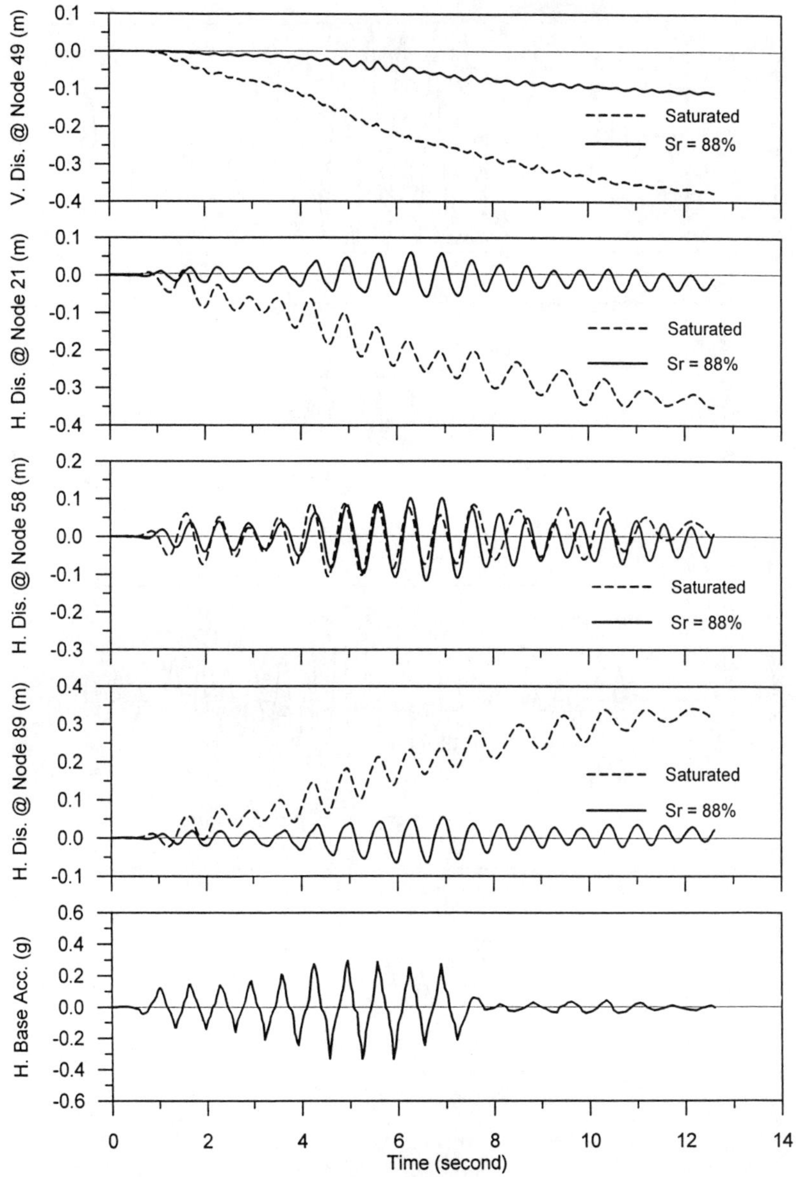

Fig. 3 Saturated and Unsaturated Displacement-Time Histories

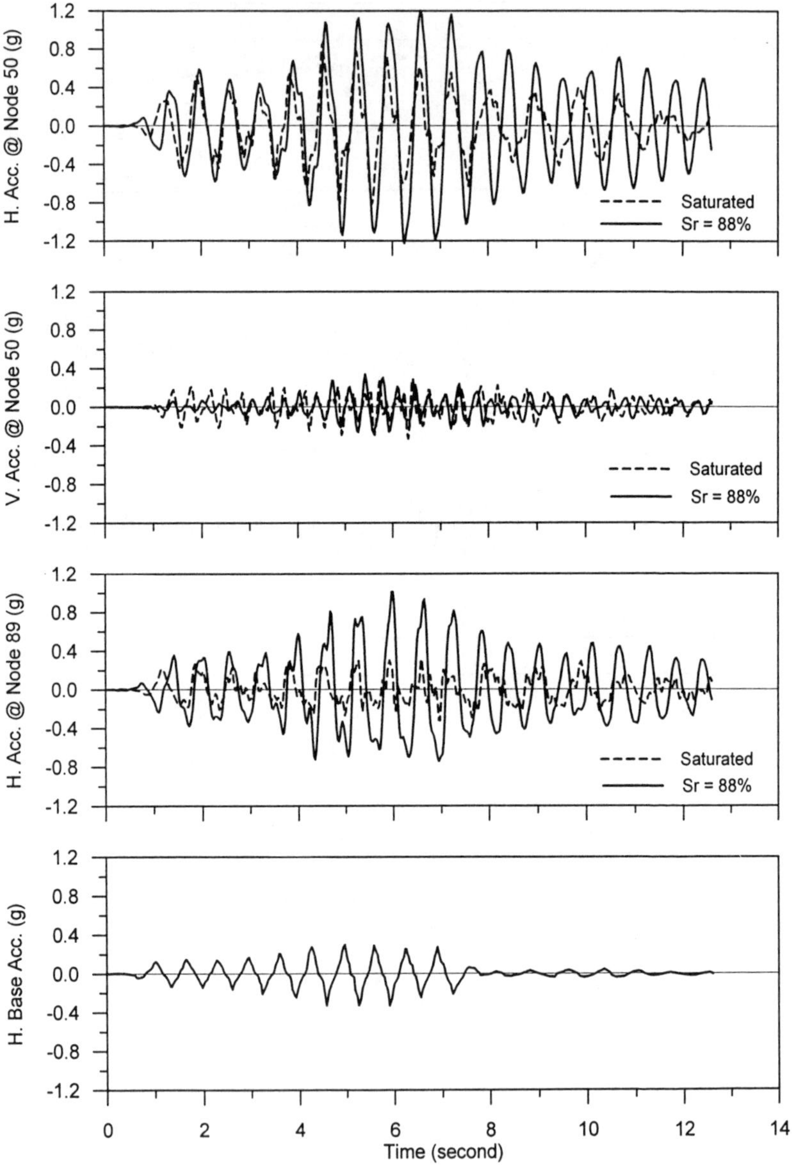

Fig. 4 Saturated and Unsaturated Acceleration-Time Histories

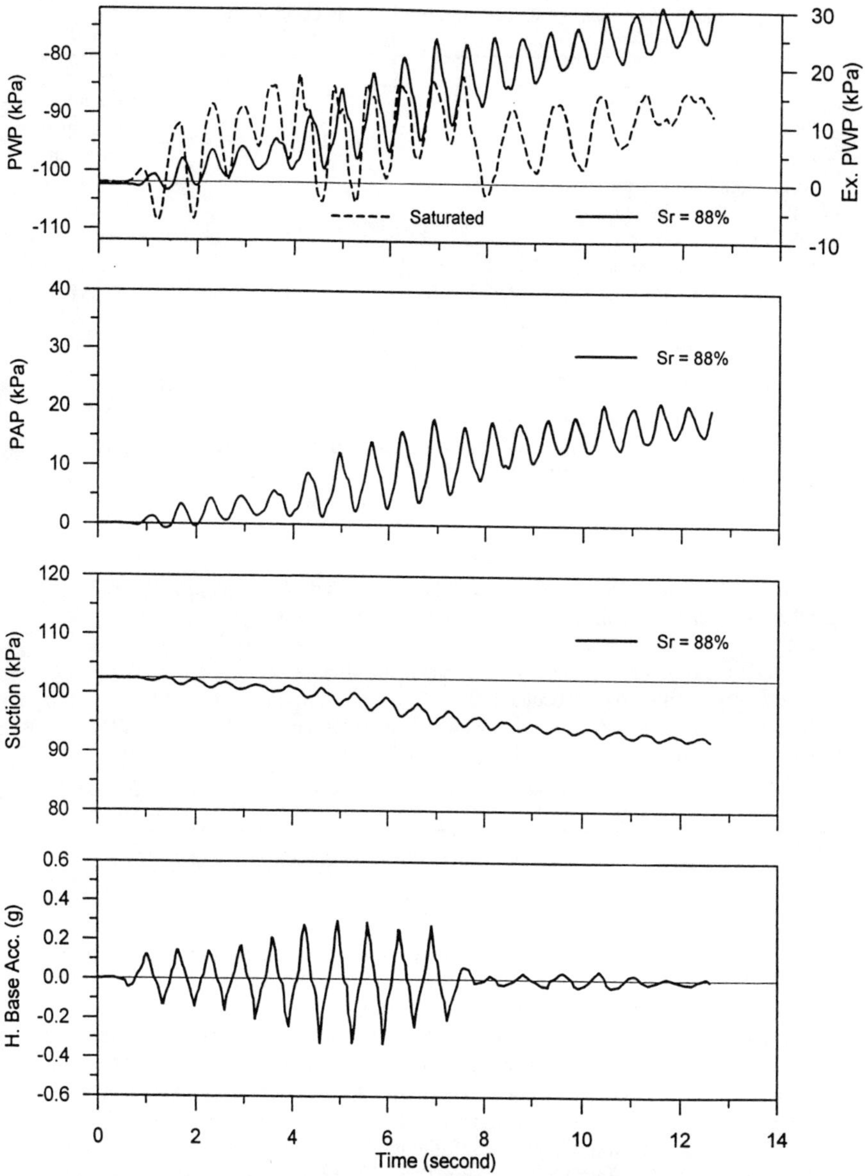

Fig. 5 Saturated and Unsaturated Pore Water Pressure (PWP), Pore Air Pressure (PAP), and Suction Variation in Element #14 with Time

Table 1. Properties of Kaolin and Pore Fluids Used in the Analysis

Property		Value
Density of solid grains		2620 kg/m^3
Density of water		1000 kg/m^3
Density of air		1.22 kg/m^3
Bulk modulus of water		2.2 x 10^6 kPa
Initial Bulk modulus of air		100 kPa
Porosity		0.596
Water coefficient of permeability @ Sr = 1.0		1.7 x 10^{-9} m/s
Air coefficient of permeability @ Sr = 0.0		1.0 x 10^{-5} m/s
Brooks-Corey (1964) Parameters:		
Residual volumetric water content	θ_r	0.1
Pore size distribution index	λ	0.5
Bubbling pressure	p_b	75 kPa
Slope of the isotropic compression line on void ratio-mean net stress plot (Wheeler 1996)	$\lambda(p_c)$	0.128
Parameters describing the change of λ with suction (Alonso et al. 1990)	r	1.570
	β	0.013
Value of specific volume on the isotropic compression line when net mean stress is one atmospheric pressure (Wheeler 1996)	$N(p_c)$	2.122
Slope of the rebound line on void ratio-mean net stress and void ratio-suction plots (Wheeler 1996)	κ	0.02
Slope of the critical state line on deviatoric stress-net mean stress plot on the compression side (Wheeler 1996)	$M_c(p_c)$	0.933
Intersection of the critical state line on the deviatoric stress axis (Wheeler 1996)	$\mu(p_c)$	54.2 kPa

Note: Sr = degree of saturation, p_c = suction and the parameters dependent on suction are indicted with a p_c in the parenthesis.

The results presented in Figs 3-5 are preliminary and further analyses of these results are currently underway. However, number useful conclusions can be made using the predicted results. Based on the comparisons between predicted saturated and unsaturated behavior, it can be concluded that the unsaturated embankment shows a more rigid behavior than the saturated embankment. This rigid behavior is reflected in the lower displacement values for the unsaturated embankment. Larger acceleration values are also predicted for the unsaturated embankment due to an increase in the natural frequency of the embankment towards the predominant frequency of the base motion (0.66 Hz). The increase in strength and stiffness caused by suction is the primary reason for the rigid behavior seen in Figs 3 and 4. The effects of suction will not be considered even in the extreme case of dry soil analysis and predicted displacements will be larger than the unsaturated analysis. Therefore it is possible to economically design an unsaturated soil structure for seismic loads using a fully coupled analysis if moisture conditions are known throughout the life of that structure. Unsaturated embankment also shows lower damping as evidenced by the slower dissipation of the acceleration following the end shaking. Viscosity of the pore fluid contributes to the damping in multiphase porous media. Lower viscosity air will produce less damping than high viscosity water.

Pore fluid pressure-time histories of Element 14 shown in Fig. 5 is typical of the behavior at other locations. Only difference is that in the middle of the embankment, where initial shear stresses are low, less cyclic pore fluid pressures were predicted by DYSAC2 and U_DYSAC2. A detail analysis of the influence of initial shear stresses on the dynamic pore water pressures can be found in Muraleetharan et al. (1994). Although lower peak-to-peak cyclic pore water pressure is predicted in the unsaturated case, residual values (non-cyclic component) are somewhat larger than the saturated case. Considering the larger slope and intercept of the critical state line used for the unsaturated kaolin (Table 1), high degree of saturation, and an almost undrained condition during shaking, predicted pore water pressure behavior is reasonable.

Pore gas pressure also increases substantially during shaking resulting in only a small change in suction. Throughout the embankment suction changes were less than 10 kPa. U_DYSAC2 predicts an increase in degree of saturation (Sr) of Element 14 from 88% to 91.9% at the end of the analysis. Element 14 also undergoes a compressive volumetric strain of 2.82% during the analysis starting with an initial porosity of 0.596. Using these values and assuming undrained conditions, the compressive volumetric strain of air in Element 14 can be approximately calculated as 34%. This compressive volumetric strain corresponds to a pore air pressure increase of 34 kPa (bulk modulus of air is 100 kPa) and is consistent with an increase of about 20 kPa shown in Fig.5. Pore air pressure behavior shown in Fig. 5 includes a partial dissipation of pore air pressures during shaking. Substantial increase in pore air pressure during shaking shown in Fig. 5 is contrary to the standard assumption of atmospheric pore air pressure used in unsaturated soil analysis.

Summary and Conclusions

A fully coupled analysis procedure for dynamic behavior of unsaturated porous media is presented. The governing equations of this analysis procedure are developed based on the Theory of Mixtures with Interfaces (TMI) that considers the effect of the interfaces existing between various bulk phases explicitly. The finite element formulation of the governing equations has been incorporated into a computer code, U_DYSAC2. Predictions made by U_DYSAC2 for an unsaturated embankment subjected to base shaking are presented and compared to the saturated case. Unsaturated embankment behaves relatively rigid and exhibits less damping resulting in larger accelerations and substantially lower displacements. Increase in pore water and pore air pressures are also predicted during shaking. Predicted increase in pore air pressures during shaking makes the standard assumption of atmospheric pore air pressure used in unsaturated soil analysis questionable. The presented results indicate that it is possible to economically design an unsaturated soil structure for seismic loads using a fully coupled analysis if moisture conditions are known throughout the life of that structure. The fully coupled analysis also provides valuable insight into the dynamic behavior of unsaturated soils.

Acknowledgements

This research was supported by a grant (Grant No. CMS 9501718) from the National Science Foundation (NSF). The cognizant NSF program director for the grant is Dr. Clifford J. Astill. Mr. S. Varatharaj's assistance in preparing some figures is also acknowledged.

References

Aitchison, G.D. (1961). "Relationships of moisture stress and effective Stress functions in unsaturated Soils." *Pore Pressure and Suction in Soils*, Butterworths, London, England, 47-52.

Alonso, E.E., Gens, A., and Josa, A. (1990). "A constitutive model for partially saturated soils." *Geotechnique*, 40(3), 405-430.

Biot, M.A. (1962). "Mechanics of deformation and acoustic propagation in porous media." *J. Appl. Phys.*, 33(1), 1482-1498.

Brooks, R.H., and Corey, A.T. (1964). "Hydraulic Properties of Porous Media." *Colorado State University Hydrology Paper No. 3*, Fort Collins, Colorado.

Fredlund, D.G., and Rahardjo, H. (1993). *Soil Mechanics for Unsaturated Soils*, John Wiley & Sons, New York.

Kutter, B.L. (1982). "Centrifugal modeling of the response of clay embankments to earthquakes." *Ph.D. Dissertation*, Cambridge University, Cambridge, England.

Li, X., and Zienkiewicz, O.C. (1992). "Multiphase flow in deforming porous media and finite element solutions." *Computers and Structures*, 45(2), 211-227.

Muraleetharan, K. K., Mish, K.D., and Arulanandan, K. (1994). "A fully coupled non-linear dynamic analysis procedure and its verification using centrifuge test results." *Int. J. Numer. Anal. Mech. Geomech.*, 18, 305-325.

Muraleetharan, K.K., Mish, K.D., Yogachandran, C., and Arulanandan, K. (1997). "User's manual for DYSAC2 (Version 7.0): Dynamic soil analysis code for 2-dimensional problems." *Technical Report*, School of Civil Engineering and Environmental Science, University of Oklahoma, Norman, Oklahoma.

Muraleetharan, K.K., and Nedunuri, P.R. (1998). "A bounding surface elastoplastic constitutive model for monotonic and cyclic behavior of unsaturated soils." *Proceedings, 12th Engineering Mechanics Conference*, ASCE, La Jolla, CA, 1331-1334.

Muraleetharan, K.K., and Wei, C.-F. (1999a). " Dynamic behavior of unsaturated porous media : Governing equations using the theory of mixtures with interfaces (TMI)." *Int. J. Numer. Anal. Meth. Geomech.*, 23, 1579-1608.

Muraleetharan, K.K., and Wei, C.-F. (1999b). "U_DYSAC2: Unsaturated Dynamic Soil Analysis Code for 2-dimensional problems." *Computer Code*, School of Civil Engineering and Environmental Science, University of Oklahoma, Norman, Oklahoma, October.

Schrefler, B. A., and Zhan, X. (1993). "A fully coupled model for water flow and airflow in deformable porous media." *Water Resources Research*, 29, 155-167.

Wheeler, S. J. (1996). " Inclusion of specific water volume within an elasto-plastic model for unsaturated soil." *Can. Geotech. J.*, 33, 42-57.

Appendix I

Element matrices in the semi-discretized finite element equations (Eq. 17) are given below.

$$(\mathbf{M}_{uu})^e = \int_{\Omega^e} \rho N_A^s N_B^s \delta_{ij} d\Omega \qquad (\mathbf{M}_{\ell u})^e = \int_{\Omega^e} \rho^\ell N_{A,i}^\ell \frac{\mu_{ij}^\ell}{n^\ell} N_B^s d\Omega$$

$$(\mathbf{M}_{gu})^e = \int_{\Omega^e} \rho^g N_{A,i}^g \frac{\mu_{ij}^g}{n^g} N_B^s d\Omega \qquad (\mathbf{C}_{\ell u})^e = \int_{\Omega^e} \left(n^\ell + \frac{\partial v}{\partial \varepsilon_v} \right) N_A^\ell N_{B,j}^s d\Omega$$

$$(\mathbf{C}_{\ell\ell})^e = \int_{\Omega^e} \left(\frac{n^\ell}{K_\ell} - \frac{\partial v}{\partial p_c} \right) N_A^\ell N_B^\ell d\Omega \qquad (\mathbf{C}_{\ell g})^e = \int_{\Omega^e} \frac{\partial v}{\partial p_c} N_A^\ell N_B^g d\Omega$$

$$(\mathbf{C}_{gu})^e = \int_{\Omega^e} \left(1 - n^\ell - \frac{\partial v}{\partial \varepsilon_v} \right) N_A^g N_{B,j}^s d\Omega \qquad (\mathbf{C}_{g\ell})^e = (\mathbf{C}_{\ell g})^e$$

$$(\mathbf{C}_{gg})^e = \int_{\Omega^e} \left(\frac{n - n^\ell}{K_\ell} - \frac{\partial v}{\partial p_c} \right) N_A^g N_B^g d\Omega \qquad (\mathbf{K}_{ug})^e = -\int_{\Omega^e} N_{A,i}^s N_B^g d\Omega$$

$$(\mathbf{K}_{\ell\ell})^e = \int_{\Omega^e} N_{A,i}^\ell \frac{\mu_{ij}^\ell}{n^\ell} N_{B,j}^\ell d\Omega \qquad (\mathbf{K}_{gg})^e = \int_{\Omega^e} N_{A,i}^g \frac{\mu_{ij}^g}{n^g} N_{B,j}^g d\Omega$$

$$(\mathbf{F}_u)^e = \int_{\Omega^e} \rho N_A^s b_i d\Omega + \int_{\Gamma_\sigma} N_A \bar{t}_i(t) d\Gamma \qquad (\mathbf{F}_\ell)^e = \int_{\Omega^e} \rho^\ell N_{A,i}^\ell \frac{\mu_{ij}^\ell}{n^\ell} b_j d\Omega - \int_{\Gamma_{vi}} N_A^\ell \bar{v}_n^\ell d\Gamma$$

$$(\mathbf{F}_g)^e = \int_{\Omega^e} \rho^g N_{A,i}^g \frac{\mu_{ij}^g}{n^g} b_j d\Omega - \int_{\Gamma_{vi}} N_A^g \bar{v}_n^g d\Gamma$$

Characterization of Seismic-Compression of Some Compacted Fills

Daniel Whang[1], Michael F. Riemer[2], Jonathan D. Bray[3], Jonathan P. Stewart[4], and Patrick M. Smith[1]

ABSTRACT

Earthquake-induced ground deformations resulting from contractive volumetric strains in unsaturated compacted soil have been observed in numerous earthquakes. This process, termed seismic-compression, is investigated using a robust cyclic simple shear device. Testing is performed on re-compacted specimens using soils retrieved from three sites near Santa Clarita, California which experienced varying levels of ground deformation during the 1994 Northridge earthquake. A key objective of the testing is to evaluate the adequacy of existing design procedures that are based on simple shear testing of clean, uniform sands for soils containing significant fines. A total of seven soils were tested: a clean uniform sand and six different soils from the three field sites possessing varying levels of fines content and plasticity. The soils containing fines were compacted to a range of water contents and dry densities to evaluate the dependence of volume change susceptibility on compaction conditions. Both formation water content and dry density were found to significantly affect the magnitude of volumetric strains induced by cyclic loading. At the same relative compaction and induced shear strain, soils containing fines generally underwent smaller contractive volumetric strains than clean sands. Soil specimens compacted wet of the line-of-optimums generally experienced less seismic-compression than soil specimens compacted on the dry side.

[1] Graduate Student Researcher, University of California at Los Angeles
[2] Assistant Adjunct Professor, University of California at Berkeley
[3] Associate Professor, University of California at Berkeley
[4] Associate Professor, University of California at Los Angeles

INTRODUCTION

Earthquake-induced ground deformations resulting from contractive volumetric strains in unsaturated compacted soil, herein termed seismic-compression, has been recognized as an important design consideration following recent earthquakes, such as the 1994 Northridge and 1995 Kobe events. Ground deformations from seismic-compression can cause severe distress to structures, resulting in significant economic damage. For example, seismic-compression in hillside fills during the Northridge event caused damages on the order of $50,000 to $100,000 per lot at over 2,000 sites, resulting in combined collateral losses of over $100 million (Stewart et al., 1995). Given this unsatisfactory performance, the California Seismic Safety Commission (SSC) recommended in their official Report to the Governor that "seismically induced deformation caused by seismic compaction of fill and underlying alluvium be considered in the design and construction of residential fills" (SSC, 1995).

Previous studies of volumetric strain accumulation in unsaturated soils under cyclic loading have focused primarily on cohesionless coarse-grained soils. The pioneering works of Silver and Seed (1971), Youd (1972), Seed and Silver (1972), and Pyke et al. (1975) used laboratory studies to investigate the volumetric strains induced in dry, clean sands undergoing cyclic loading in level ground situations. They found that the volumetric strains in such sands are influenced by the cyclic shear strain amplitude, the soil's relative density, the number of strain cycles, and multi-directional shaking effects. Current design procedures for dry sands (e.g. Tokimatsu and Seed, 1987) are based on these early studies. However, because these studies were limited to clean sands, they did not systematically investigate the effects of fines content, formation water content, or plasticity index. Pyke (1975) performed limited testing on a clayey sand for back-analysis of settlements at the Jensen Filtration Plant during the 1971 San Fernando Earthquake. His test results indicated that the clayey sand underwent less than a third of the volumetric strain that would have been expected in a sand with relative density D_R = 60% (a reasonable estimate of D_R given the density of the fill). These results suggest that the Tokimatsu and Seed (1987) design procedure should be modified for application to unsaturated compacted fill soils that contain significant fines.

To evaluate the seismic-compression susceptibility of compacted soils with fines, testing was performed on soils retrieved from three sites near Santa Clarita, California that experienced varying levels of ground deformation during the 1994 Northridge earthquake. The testing examines the effects of fines content, formation water content, formation dry density, and plasticity on the seismic-compression potential of these compacted soils. The paper begins by describing the laboratory testing equipment and procedures used. Test results on dry clean sand are then presented to enable comparisons with the results of Silver and Seed (1971). Test results on the soils retrieved from the three sites are presented, and pertinent findings are discussed.

LABORATORY TESTING EQUIPMENT AND PROCEDURES

General

Cyclic simple shear tests were performed to evaluate volumetric strain accumulation under a uniform applied shear strain amplitude. This laboratory testing program included testing of clean sand (Crystal Silica No. 30) and reconstituted bulk specimens from the three field sites (denoted here as Sites A, B, and C). The effect of overburden pressure on volumetric strain has previously been found to be minor (Silver and Seed, 1971). Hence, all tests were performed under the same vertical stress of 101.3 kPa. A loading frequency of 1 Hz was used in all tests.

Test Apparatus

The University of California at Berkeley cyclic simple shear device (UCB-2D) developed by Boulanger et al. (1993) was used to perform all cyclic simple shear (CSS) tests. The main mechanical features of the apparatus are schematically illustrated in Figure 1, and the device's advantages and limitations are discussed in Boulanger et al. (1993, 1995). The UCB-2D apparatus was developed with a stiff vertical table and a rigid base to minimize system compliance associated with potential rocking. Commercially available wire-reinforced membranes were used to laterally confine the cylindrical soil specimens, which were prepared with a diameter of 10.2 cm and a height of 2.3 cm. These membranes minimized lateral expansion of the test specimens, while providing negligible shear stiffness. Since the cross-sectional area of the specimens remain constant, volume changes are evaluated from 3 vertical LVDTs around the specimen which provide continuous measurement of the changes in height throughout the testing process. Whang (2000) found volumetric strain susceptibility of sands tested with wire reinforced membranes to be similar to sands with flexible membranes and K_0 (at rest) cell pressures, suggesting that the use of wire-reinforced membranes should adequately represent the stress state for this study.

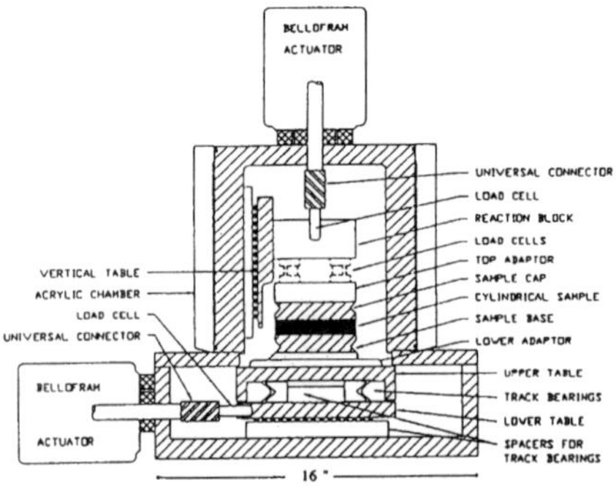

Figure 1. Schematic of the UCB-2D Simple Shear Device (Boulanger et al.1993)

Specimen Preparation

(a) Cohesionless Soils

To optimize the comparison of the present test results with the original work of Silver and Seed (1971), the same sample preparation techniques were used. A screen was placed in the bottom of a mold (which contained a wire-reinforced membrane) and a pre-weighed amount of sand was poured into the mold. The screen was then pulled up through the specimen to ensure uniformity of soil structure. After flattening the specimen and positioning the top cap, the same hand vibrator that was used by Silver and Seed (1972) was employed to vibrate the specimen to a predetermined height that would achieve the appropriate relative density after application of the vertical stress of 101.3 kPa.

(b) Compacted Fills

The specimen preparation methods used for compacted fills were designed to replicate field construction techniques as closely as possible. The water contents and densities were chosen to represent both the in situ soil conditions achieved during construction and those present at the time of the earthquake. The intent was to simulate in the laboratory the in situ soil structure, because previous studies have shown that soil fabric can affect dynamic strengths by up to 45% at the same densities (Mulilis, 1975).

The soil structure of compacted fills containing clay is affected by (Seed and Chan, 1959): (1) the double layer thickness, which in turn is controlled by formation water content, and (2) the shear strains applied to the soil during compaction. Methods of compaction that induce significant shear strain (i.e. kneading compaction) result in a more dispersed arrangement of soil particles than methods of compaction inducing little or no shear strain (i.e. static compaction). Available information suggests that fill at the three field sites was compacted using a sheepsfoot roller. Accordingly, all reconstituted specimens were prepared using kneading compaction to simulate field conditions. A pre-weighed amount of soil was compacted into the mold in two layers using a Harvard Miniature-Compactor. The pressure applied by the Harvard Mini-Compactor was varied through a pressure regulator to achieve the desired density. Lastly, the top of the specimen was leveled using a straight edge to a tolerance of ± 0.01 cm, and holes were filled using the same soil and a rubber mallet.

CRYSTAL SILICA SAND TESTS

All tests on the clean uniform sand were performed on commercially available Crystal Silica No. 30 sand. The maximum and minimum densities of this sand were determined by the Modified Japanese method and dry tipping, respectively, and are $\gamma_{dmax} = 15.87$ kN/m^3 and $\gamma_{dmin} = 12.96$ kN/m^3 and a specific gravity, G_s, of 2.65 was assumed for this material. Although this sand is similar to the Crystal

Silica No. 20 sand used by Silver and Seed (1971), they are not derived from the same parent material.

These tests were performed to compare the results of the widely used Silver and Seed (1971) data from an NGI-type CSS device to the results from the UCB-2D CSS apparatus for two nominal relative densities (D_r = 60% and 80%) over a shear strain range of 0.003 % to 0.6%. No attempt was made to vary the water content since the earlier study on seismic-compression was performed entirely on dry sands. Figures 2 and 3 show test results from the present study compared to the Silver and Seed (1971) results. These figures report a summary of the volumetric strain accumulation at 10 uniform shear strain cycles for all tests. The coefficient of determination values of a straight line fit to the data in log-log space, R^2, were 0.985 and 0.983 in this study and 0.949 and 0.828 in the Silver and Seed (1971) data for D_R = 60% and 80%, respectively.

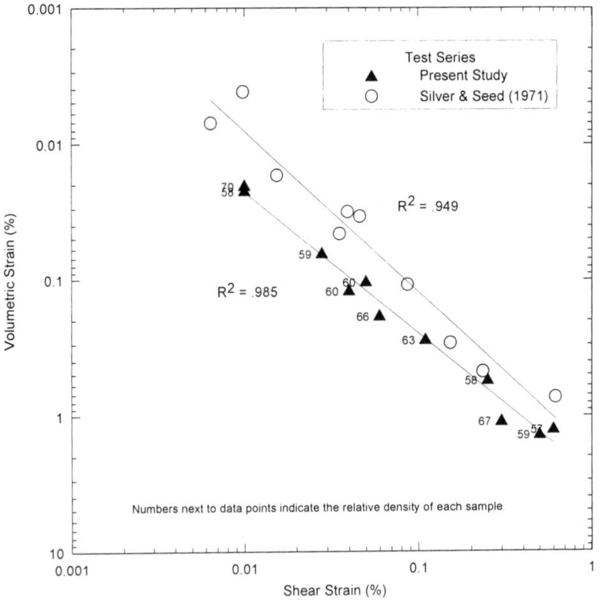

Figure 2. Volumetric Strain of Clean Uniform Sand at D_R = 60% at 10 Cycles

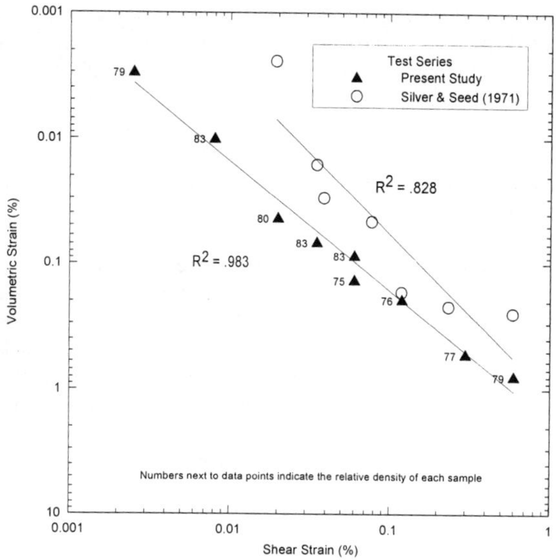

Figure 3. Volumetric Strain of Clean Uniform Sand at D_R = 80% at 10 Cycle

The general trend of the data from the present study is similar to previous findings, with volumetric strains increasing with decreasing relative density and increasing shear strain and number of cycles. However, the measured volumetric strains of the present study are systematically higher than those of Silver and Seed (1971). Two factors may be responsible for the discrepancy between the two data sets. First, the tests were performed on different devices. While Silver and Seed used a cam driven NGI device, all of the present tests were performed on a closed loop computer-automated device with an upgraded, very stiff mechanical design. Second, although the two sands were quite similar, the materials did have slightly different gradations and were not derived from the same parent material.

COMPACTED FILL TESTS

Cyclic simple shear tests were performed on six different reconstituted soils at three shear strain levels (0.12%, 0.4% and 1.2%) for a range of formation water contents and dry densities. The formation water contents and dry densities were chosen to represent the range of in situ conditions at the field sites immediately before the earthquake, which were inferred from pre-earthquake field construction logs and post-earthquake subsurface exploration. Additional tests were performed at water contents both dry and wet of the line-of-optimums to evaluate the effect of formation water content on seismic-compression. Studies investigating the static properties of compacted clay soils (e.g. Seed et al., 1960) have shown that clays compacted dry of optimum have a flocculated soil structure regardless of the method of compaction, whereas clays compacted wet of optimum have increasingly dispersed soil structure with increasing compaction-induced shear strain. Since soil structure is a difficult property to quantify directly, structure is indirectly and qualitatively indexed here in terms of formation water content relative to the line-of-optimums. Fill specimens compacted dry of optimum are expected to have a flocculated structure, and fills compacted wet of optimum are generally assumed to be more dispersed, as the kneading compaction used in this study introduces large shear strains.

Atterberg Limit tests and hydrometer/sieve analyses were performed to characterize each of the tested soils, and the results are summarized in Table 1. The fill specimens are generally classified as silty or clayey sands by the Unified Soil Classification System (either SM or SC), but one soil is a low plasticity sandy silt (ML). The tested soils have fines contents ranging from 37% to 54% and plasticity indices ranging from 7 to 15. Standard and Modified AASHTO Compaction tests (ASTM D 698-91 and D 1557-91, respectively) were performed to evaluate the maximum dry densities and the line-of-optimums for each soil.

Table 1. Summary of Index Properties

Soil	PI	LL	USCS	$\gamma_{d,max}$* (kN/m^3)	Opt. wc* (%)	% Fines
A-1	15	33	CL	21.2	8.5	54
B-1	8	31	SC/SM	20.3	9.5	37
B-2	11	32	SC/ML	19.6	11	45
B-3	7	30	SM	20.0	10	37
C-1	14	35	SC	20.6	8	40
C-2	9	30	SC	20.4	8	48

*Using ASTM D1557-91

The results of a representative cyclic simple shear test are shown in Figure 4. The progressive increase in shear stress amplitude with time during the first 5 cycles to achieve a uniform shear strain amplitude is consistent with expected patterns of behavior. This increase in shear modulus is negligible after 10 cycles. As shown in Figure 4c, the majority of the volumetric strain accumulation occurs within the first few cycles of loading.

All of the test results in this section are summarized by their shear strain amplitude and their volumetric strain accumulation after 15 cycles of loading, corresponding to the approximate number of equivalent uniform stress cycles for a M_W = 7.5 earthquake. In Figures 5 and 6, "RC" denotes relative compaction based on the Modified Proctor standard, and "wc" denotes as-compacted water content. Unlike the results for sands in Figs. 2 and 3, the data for compacted soils in Figs. 5 and 6 are plotted on linear scales for volumetric and shear strain. A number of important trends are apparent from Figs. 5 and 6. First, volumetric strains from seismic-compression are strongly dependent on the number of cycles of shaking and shear strain amplitude, as was found for sands in this study and by

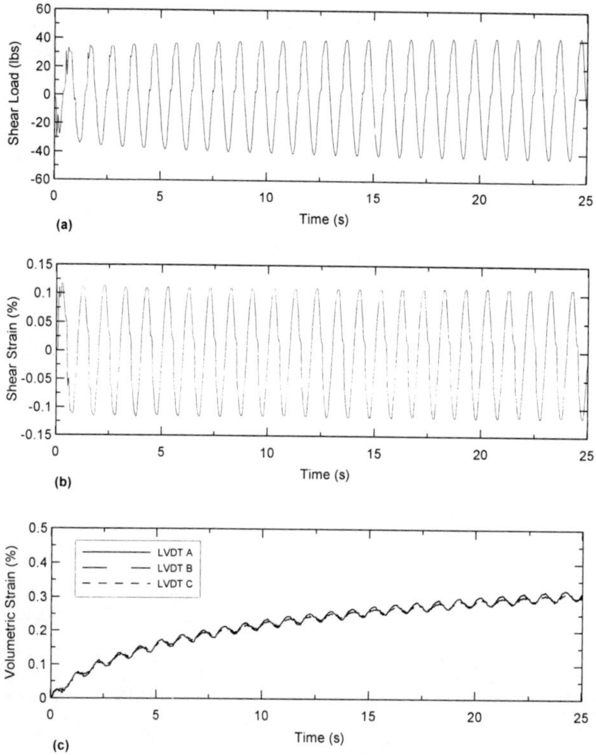

Figure 4. Representative Data for Cyclic Simple Shear Test

Silver and Seed (1971). Second, seismic-compression in the compacted soils varies with formation water content and relative compaction. Settlements decreased with increasing dry density for all soils tested. Test results for Soil A-1 at RC = 80, 82, 84, 88 and 90% at wc = 10% (Figure 5a) clearly show this effect. The effect of formation water content on seismic-compression is less consistent. In specimens from both Sites A and C, increases in formation water contents resulted in reduced volumetric strains at comparable densities. However, volumetric strains in soils from Site B did not necessarily decrease with increasing formation water content. Soil B-2 showed approximately the same settlements when compacted dry and wet of optimum. Two other soils (B-1 and B-3) showed increasing settlement with increasing formation water content. Index testing (sieve analyses, hydrometer tests, Atterberg limits, compaction tests) of Site B soils revealed no particularly unusual characteristics, although soils B-1 and B-3 were found to require relatively little compactive energy to achieve the target densities.

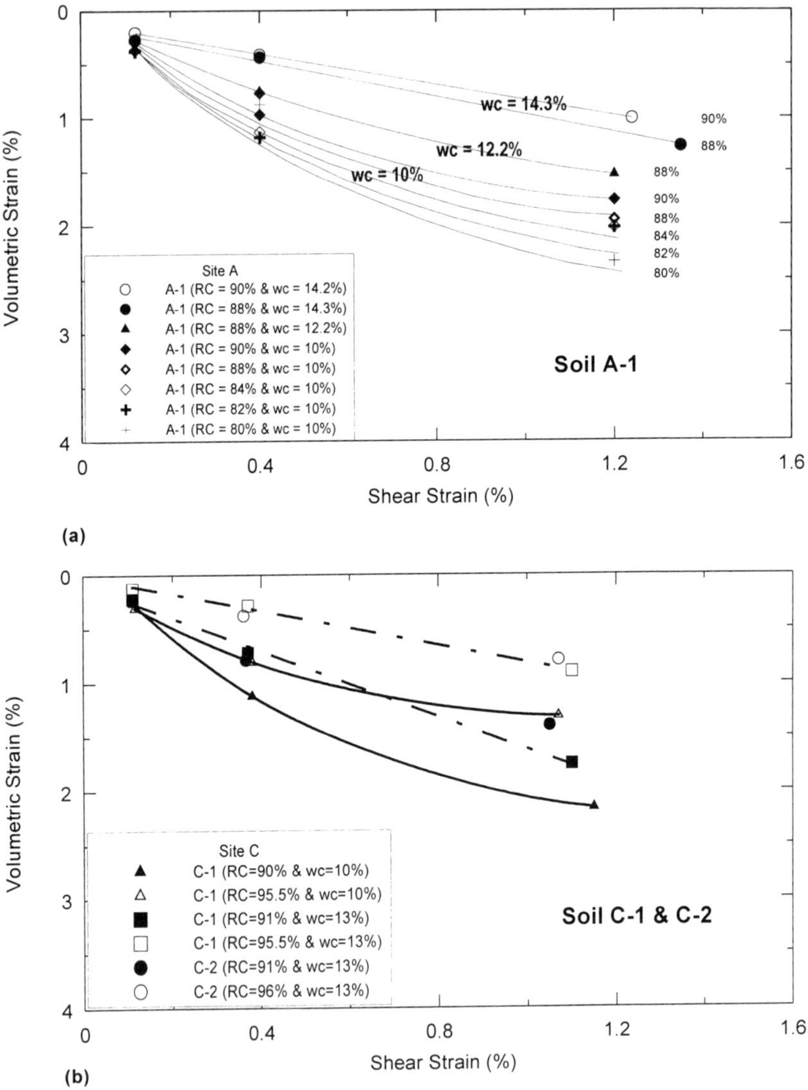

Figure 5. Volumetric Strain vs. Shear Strain Amplitude for Sites A & C

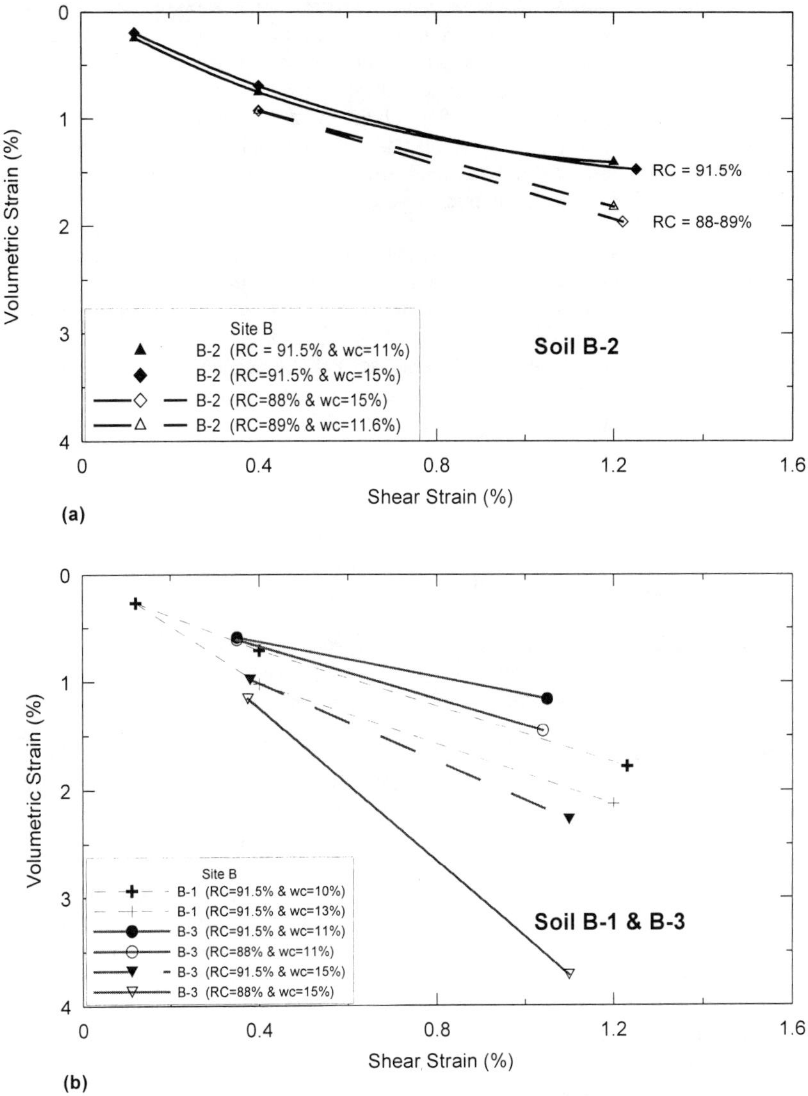

Figure 6. Volumetric Strain vs. Shear Strain Amplitude for Site B

Figure 7 shows the normalized volumetric strain accumulation vs. the number of cycles for all of the tests on compacted fill (mean and mean ± standard deviation). The normalized volumetric strain was computed by normalizing the settlement at each cycle by the settlement at cycle 15. For these reconstituted specimens, approximately 50% of the volumetric strains occurring in 15 cycles took place within the first 3 cycles of loading. This behavior is consistent with the sand behavior reported in Silver and Seed (1971), where the first 10 cycles of loading resulted in as much settlement as the next 40 cycles.

Comparison of results for different soils provides insights into the effects of fines content and plasticity on a soil's susceptibility to seismic-compression. Figures 8 and 9 show a comparison of volumetric strain accumulations at loading cycle 15 between reconstituted fill soils A and C and Crystal Silica No. 30. As relative compaction is used to characterize soils containing significant fines while relative density is used to characterize clean sands, a comparison between the two could only be made with the use of a soil-specific relationship between relative density and relative compaction for Crystal Silica No. 30.

The general relationship between relative compaction and relative density is typically expressed as (Lee and Singh, 1971):

$$D_R\ (\%) = 5\ [RC\ (\%) - 80]; \qquad 80 \leq RC\ (\%) \leq 100 \qquad (1)$$

For Crystal Silica No. 30 sand, the soil specific relationship was found to be:

$$D_R\ (\%) = 546 - 44590/RC\ (\%); \qquad 81.6 \leq RC\ (\%) \leq 100 \qquad (2)$$

Crystal Silica No. 30 has equivalent relative compactions of 91.8% and 95.7% at relative densities of 60% and 80%, respectively. Using this relation, volumetric strain accumulations in the Crystal Silica No. 30 and compacted fills are compared for RC ≈ 90-92 and 95-96% in Figs. 8 and 9 respectively.

For comparable relative densities, seismic-compression in the compacted fill soils is equal to or less than that in the Crystal Silica. Whereas the range of volumetric strains for Crystal Silica No. 30 sand is narrowly banded, the range of volumetric strains for fill soils is much broader. Within this broad range of results for soils from sites A and C, there is a systematic increase in volumetric strain with decreasing water content. In general, specimens compacted wet and dry of optimum can be expected to experience volumetric strains in the lower and upper half of the range of strains for fill soils, respectively.

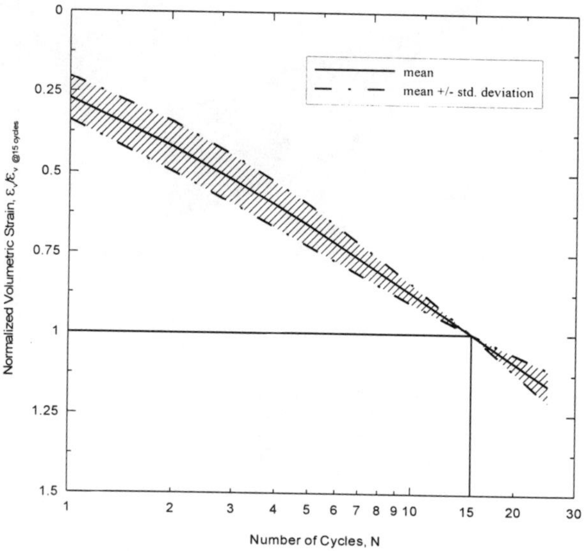

Figure 7. Number of Cycles vs. Normalized Volumetric Strain at 15 Cycles

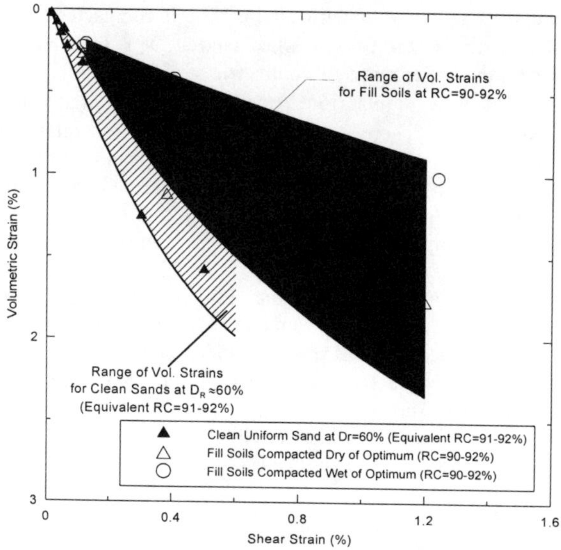

Figure 8. Comparison of Volumetric Strains between Soils A & C and Clean Sands at RC = 92%

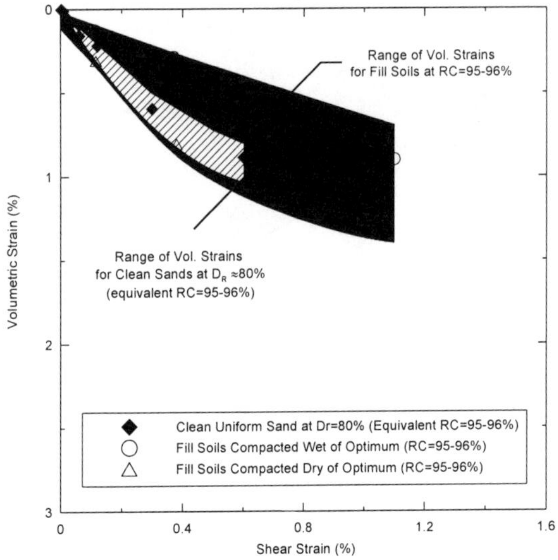

Figure 9. Comparison of Volumetric Strains between Soils A & C and Clean Sands at RC = 96%

The upper range of volumetric strains in soils compacted to RC = 90-92% were less than strains in the Crystal Silica sand at $D_R \approx 60\%$ (equivalent RC $\approx$ 92%). However, the upper range of volumetric strains in Site C soils compacted to RC = 95-96% were similar to Crystal Silica Sand strains at $D_R \approx 80\%$ (equivalent RC $\approx$ 96%). Therefore, it appears that for both relative compaction levels, the seismically induced volumetric strains in clean uniform sands can be considered an upper bound estimate of strains expected in a soil containing significant amount of fines. However, it appears that the influence of fines in decreasing seismic-compression is greater for soils at lower relative densities.

A systematic investigation of the influence of fines content was not performed in this study, and results from one compacted soil cannot be compared with another since indices such as plasticity index and grain size distribution are inconsistent. Therefore, the effect of increasing the fines content from 35% to 50% is not known and should be investigated in future studies. Likewise, the influence of plasticity is not well known, although the above results suggest that non-plastic soils experience more volume change than plastic soils.

CONCLUSIONS

Cyclic simple shear (CSS) testing was performed on seven soils: a clean sand similar to that used by Silver and Seed (1971) and six fill soils from three different field sites. The CSS test results for the clean sand are consistent with those presented by Silver and Seed (1971). Both sets of results indicate that relative density is the critical intrinsic soil property controlling the susceptibility of clean sands to seismic-compression. Results from the reconstituted fill specimens confirmed that dry density is a significant factor affecting susceptibility to volumetric strain accumulation for fine-grained soils. Comparison of the sand and compacted fill test results at a given relative density (or equivalent relative compaction) indicate that seismic-compression can be significantly smaller for fine-grained soils than for clean sands. For engineering purposes, the seismic-compression in clean uniform sands can be considered an upper bound estimate of the strains in soils containing fines.

Another important distinction between the test results for sand and fine-grained soils is the apparent influence of soil structure (as indexed by formation water content) on volume change. Previous studies have shown that static properties of compacted clays such as shear strength, permeability, and wetting-induced volume change are strongly a function of soil structure (e.g. Seed et al., 1960: Brandon et al., 1990; Noorany, 1997). The CSS test results for specimens from Sites A and C suggests that this structure-dependent behavior extends to seismic-compression as well. For these soils, at a given relative compaction, there was consistently less volumetric strain in specimens compacted wet of the line-of-optimums than in specimens compacted dry of the line-of-optimums.

The apparently contradictory results obtained from Site B soils (volume change susceptibility that increases or remains constant with increasing formation water content) suggest that these soils may not follow the simple model relating water content to dispersed vs. flocculated structures. This may have resulted from the relatively low plasticity of these soils (PI of about 9, vs. 15 for A-1 and about 12 for site C). Another possibility is that the shear strains induced during compaction of Site B soils may have been insufficient to develop a dispersed soil structure, even for formation water contents greater than optimum. This hypothesis is supported by the observation that during specimen preparation, these soils required substantially lower levels of compactive energy to achieve the desired densities. For example, approximately a quarter of the pressures used during compaction of other fill soils was needed during preparation of the Site B specimens to RC = 88% and 91.5%. However, it should be noted that the Site B test results are not yet fully understood, and additional studies are required to elucidate the influence of soil structure on the observed behavior.

References

Brandon, T.L., Duncan, J.M., and Gardner, W.S. (1990). "Hydrocompression settlement of deep fills," *J. Geotech. Engrg.*, ASCE, 116(10), 1536-1548.

Boulanger, R.W., Chan, C.K., Seed, H.B., Seed, R.B., and Sousa, J. (1993) "A low-compliance bi-directional cyclic simple shear apparatus," *Geotechnical Testing Journal*, ASTM,16(1), 36-45

Boulanger, R.W., and Seed. R.B. (1995) "Liquefaction of Sand under Bi-directional Monotonic and Cyclic Loading," *J. Geotech Engrg.*, ASCE, 121 (12), 870-878

Lee, K. and Singh, A. (1971) "Compaction of granualar soils," *Proceedings of the 9th Annual Engineering Geology and Soils Engineering Symposium,* April 5-7

Mulilis, John Paul (1975) "The effects of method of sample preparation on the cyclic stress-strain behavior of sands," Ph.D. Dissertation, University of California at Berkeley

Noorany, I. (1997). "Structural fills: design, construction, and performance review," in *Unsaturated Soil Engineering Practice*, Geotech. Special Publication 68, S.L. Houston and D. G. Fredlund (editors), 233-254.

Pyke, R., Seed, H.B., Chan, C.K. (1975) "Settlement of Sands Under Multidirectional Shaking," *J. GeoTech. Engrg. Div.*, ASCE, 101(4), 379-398

Seed, H.B. and Chan, C.K. (1959) "Structure and strength characteristics of compacted clays," ASCE Proceedings Paper 2216, October

Seed, H.B. and Chan, C.K. (1959) "Undrained Strength of Compacted Clays After Soaking," ASCE Proceedings Paper 2293, December

Seed, H.B., Mitchell, J.K., and Chan, C.K. (1960) "The strength of compacted cohesive soils," in Shear Strength of Cohesive Soils, ASCE, Boulder, CO. 877-964

Seed, H.B. and Silver, M.L. (1972) "Settlement of Dry Sands During Earthquakes," *J. Soil Mech. And Foundations Div.*, ASCE, 98 (4), 381-397

Seismic Safety Commission (SCC) (1995), "Northridge earthquake, turning loss to gain: report to governor Pete Wilson in response to governor's executive order W-78-94, " SSC Rpt. No. 94-08, Seismic Safety Commission, State of CA, Sacramento, Nov. 9

Silver, M.L. and Seed, H.B. (1971) "Volume Changes in Sand Due to Cyclic Loading," *J. Soil Mech. And Foundations Div.*, ASCE 97(9), 1171-1182

Stewart, J.P., Bray, J.D., McMahon, D.J., and Kropp, A.L. (1995) "Seismic performance of hillside fills," in Landslides under Static and Dynamic Conditions – Analysis, Monitoring, and Mitigation, Geotech. Special Publication No. 52, ASCE, C.L. Ho and D.K. Keefer (editors), October, 76-95

Tokimatsu, K. and Seed, H.B. (1987) "Evaluation of settlements in sands due to earthquake shaking," *J. Geotech. Engrg.*, ASCE, 113(8), 861-878

Whang, D. (2000) "Seismic-Compression of Compacted Fills," *Ph.D. Thesis*, University of California at Los Angeles

Youd, T.L. (1972) "Compaction of sands by repeated shear straining," *J. Soil Mechanics and Foundations Div.*, ASCE, 98 (7), 709-725

Comparison of Different Procedures to Predict Unsaturated Soil Shear Strength

S.K. Vanapalli [1] and D.G. Fredlund [2]

Abstract: Several procedures have been proposed in the recent years to predict the shear strength of an unsaturated soil. The soil-water characteristic curve has been used as a tool either directly or indirectly in the prediction of the shear strength along with the saturated shear strength parameters in these procedures. This paper provides comparisons between the measured and predicted values of unsaturated shear strength using these procedures for three soils both for limited and large suction ranges. The three soils used in the study for comparisons have different gradation properties, percentages of clay and plasticity index, I_p values. The advantages and limitations associated with predicting the shear strength of unsaturated soils using the procedures is discussed in the paper.

INTRODUCTION

A theoretical framework for unsaturated soil mechanics that parallels saturated soil mechanics is available in terms of stress state variables, namely; net normal stress, $(\sigma_n - u_a)$, and matric suction, $(u_a - u_w)$ where σ_n is the normal stress, u_a is the pore-air pressure an u_w is the pore-water pressure. (Fredlund and Rahardjo, 1993). The framework is based on experimental studies that are costly and time consuming. Several advancements have been made in the prediction of the engineering behavior of unsaturated soils in recent years. The soil-water characteristic curve has been found to be a useful tool in the estimation of engineering properties for unsaturated soils. Examples are the coefficient of permeability and the shear strength functions.

Shear strength forms an important engineering property in the design of numerous geotechnical and geo-environmental structures such as earth dams, retaining walls, pavements, liners, covers, etc. Several procedures have been proposed in the literature during the past five years to predict the shear strength of an unsaturated soil.

[1] Formerly, Civil Engineering Department, University of Saskatchewan, Saskatoon, Canada, S7N 5A9. Present, Civil Engineering Department, Royal Military College, Kingston, Canada, K7K 7B4.
[2] Civil Engineering Department, University of Saskatchewan, Saskatoon, Canada, S7N 5A9.

These procedures use the soil-water characteristic curve as a tool either directly or indirectly along with the saturated shear strength parameters, c' and ϕ', to predict the shear strength function for an unsaturated soil (Vanapalli et al. 1996, Fredlund et al. 1996, Oberg and Sallfors 1997, Khallili and Khabbaz 1998 and Bao et al. 1998).

The philosophy used in each of the prediction procedures proposed by these investigators is different. Comparisons have been provided between predicted and measured values of shear strength for a limited suction range for various soils (i.e., between 0 to 500 kPa).

Escario and Juca (1989) measured the soil-water characteristic curves and the shear strength of three soils prior to the time when any proposals had been made for the shear strength functions. The three soils have different gradation properties, percentages of clay and plasticity indices, I_p. These results are used in this paper to provide comparisons between the predicted and measured shear strength values both for a limited suction range and a large suction range. The study presented in the paper highlights the advantages and limitations associated with the various procedures for predicting the shear strength of unsaturated soils. The simple procedures discussed in this paper are of value in bringing the shear strength theories for unsaturated soils into engineering practice.

EQUATIONS FOR INTERPRETING THE SHEAR STRENGTH OF UNSATURATED SOILS

Bishop (1959) proposed shear strength equation for unsaturated soils by extending Terzaghi's principle of effective stress for saturated soils. Bishop's original equation can be arranged as shown below.

$$\tau = c' + (\sigma_n - u_a)\tan\phi' + (u_a - u_w)[(\chi)(\tan\phi')] \qquad [1]$$

where:

τ	=	shear strength of unsaturated soil,
c'	=	effective cohesion,
ϕ'	=	angle of frictional resistance,
$(\sigma_n - u_a)$	=	net normal stress,
$(u_a - u_w)$	=	matric suction, and
χ	=	a parameter dependent on the degree of saturation

The value of χ was assumed to vary from 1 to 0, which represents the variation from a fully saturated condition to a total dry condition. Several investigators found limitations with respect to the quantification of the parameter χ both theoretically and experimentally.

Fredlund et al. (1978) have proposed a relationship to explain the shear strength of unsaturated soils in terms two independent stress state variables as shown below:

$$\tau = c' + (\sigma_n - u_a)\tan\phi' + (u_a - u_w)\tan\phi^b \qquad [2]$$

The shear strength contribution due to matric suction, ϕ^b, was initially assumed to be linear based on the analysis of limited results published in the literature. Later experimental studies performed over a large range of suction values have shown that the variation of shear strength with respect to soil suction is non-linear (Gan et al. 1988 and Escario and Juca 1989).

Equation [1] can be applied for both the linear and non-linear variation of shear strength with respect to suction.

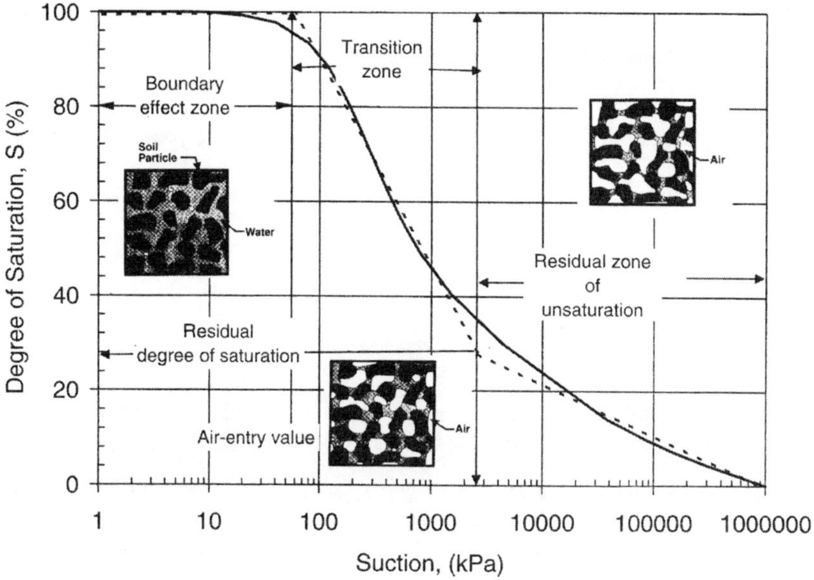

Figure 1. Typical soil-water characteristic curve showing zones of desaturation.

The Relationship between the Soil-Water Characteristic Curve and the Shear Strength of Unsaturated soils

The soil-water characteristic curve defines the relationship between the soil suction and either the degree of saturation, S, or gravimetric water content, w, or the volumetric water content, θ (Figure 1). The soil-water characteristic curve provides a conceptual and interpretative tool by which the behavior of unsaturated soils can be understood. As the soil moves from a saturated state to drier conditions, the distribution of the soil, water, and air phases change as the stress state changes. A typical soil-water characteristic curve with various zones of desaturation are shown in Figure 1.

The wetted area of contact between the soil particles decreases with an increase in the soil suction. There is a relationship between the rate at which shear strength changes in unsaturated conditions to the wetted area of water contact between the soil particles or aggregates. In other words, a relationship exists between the soil-water characteristic curve and the shear strength of unsaturated soils.

Different Procedures for Predicting the Shear Strength of an Unsaturated Soil

Vanapalli et al. (1996) and Fredlund et al. (1996) have proposed a more general, non-linear function for predicting the shear strength of an unsaturated soil using the entire soil-water characteristic curve (i.e., 0 to 1,000,000 kPa) and the saturated shear strength parameters as shown below:

$$\tau = [c' + (\sigma_n - u_a)\tan\phi'] + [(u_a - u_w)\{(\Theta^\kappa)(\tan\phi')\}] \qquad [3]$$

where:

κ = fitting parameter used for obtaining a best-fit between the measured and predicted values, and

Θ = normalized water content, θ_w/θ_s.

The shear strength contribution due to suction constitutes the second part of [Eq. 3], which is:

$$\tau_{us} = [(u_a - u_w)\{(\Theta^\kappa)(\tan\phi')\}] \qquad [4]$$

Equation [3] can also be written in terms of degree of saturation, *S,* or gravimetric water content, *w,* to predict the shear strength yielding similar results.

The entire soil-water characteristic curve data (i.e., 0 to 1,000,000 kPa) is required along with the saturated shear strength parameters in the use of Equation [3]. A best-fit soil-water characteristic curve can be obtained in terms of *a, n,* and *m* parameters using the equation proposed by Fredlund and Xing (1994) which is shown below:

$$\theta_w(\psi) = \theta_s \left[1 - \frac{\ln\left(1 + \frac{\psi}{h_r}\right)}{\ln\left(1 + \frac{10^6}{h_r}\right)}\right] \left[\frac{1}{\ln\left\{\exp(1) + \left(\frac{\psi}{a}\right)^n\right\}}\right]^m \qquad [5]$$

where:

ψ = soil suction,
θ_w = volumetric water content,
θ_s = saturated volumetric water content,
a = suction related to the inflection point on the curve,
n = soil parameter related to slope at the inflection point,
m = soil parameter related to the residual water content, and
h_r = suction related to the volumetric residual water content, θ_r

The shear strength contribution due to suction, $tan\phi^b$ is equal to $tan\phi'$ up to the air-entry value of the soil. In other words, the conventional equation for estimating the shear strength of saturated soils can be used up to the air-entry value for unsaturated soils. The shear strength variation with respect to suction is linear in the *boundary effect* zone. The shear strength contribution due to suction, $tan\phi^b$ is less than $tan\phi'$ in the *transition zone*. Hence, the shear strength variation in this zone is non-linear. The shear strength value gradually starts dropping at high values of suction and reaches saturated shear strength value at 1,000,000 kPa using this equation. Equation [3] is useful to predict the shear strength of unsaturated over the entire range of suction values of 0 to 1,000,000 kPa (i.e., from a fully saturated condition to a total dry condition). Analysis of experimental results presented and shown in the Figures 3 to 8 of this paper using Equation [3] will be referred as **Procedure 1**.

Vanapalli et al. (1996) proposed another equation for predicting the shear strength of unsaturated soils without using the fitting parameter, κ. The equation is given below:

$$\tau = c' + (\sigma_n - u_a)\tan\phi' + (u_a - u_w)\left[\left(\frac{\theta_w - \theta_r}{\theta_s - \theta_r}\right)\tan\phi'\right] \qquad [6]$$

where:

θ_w = volumetric water content,
θ_s = saturated volumetric water content, and
θ_r = residual volumetric water content.

Equation [6] can also be written in terms of degree of saturation, *S,* or gravimetric water content, *w,* to predict the shear strength yielding similar results. To use this equation the residual volumetric water content, θ_r, has to be estimated from the soil-water characteristic curve.

A graphical procedure can be used to define the residual state of saturation (Figure 1). The procedure involves first drawing a tangent line through the inflection point on the straight-line portion of the soil-water characteristic curve. The residual saturation can be defined as the point where the line extending from 1,000,000 kPa along the curve intersects the previous line. A computational technique can also be used to determine the residual suction value. More details about the computational technique are available in Vanapalli et al. (1998). The shear strength of soil may start to decrease beyond the residual state conditions. Analysis of experimental results presented and shown in the figures of this paper using Equation [6] are referred as **Procedure 2**.

Both the Procedures 1 and 2 are consistent with the stress state variable approach satisfying the continuum mechanics concepts. The form of the equation is similar to the Fredlund et al. (1978) equation (i.e., Equation 2).

Oberg and Sallfors (1997) proposed an equation for predicting the shear strength of primarily non-clayey soils such as sands and silts. The proposed equation can be rearranged as follows:

$$\tau = c' + (\sigma_n - u_a)\tan\phi' + (u_a - u_w)[(S)(\tan\phi')] \qquad [7]$$

The χ parameter proposed by Bishop is replaced by the degree of saturation, S, in Eq. [7]. The author's state that the χ -factor proposed by Bishop reflects the fraction of the pore area that is occupied by water (i.e., A_w/A_t) which is approximately equal to the degree of saturation, S. Equation [7] has not been experimentally verified on soils through the measurement of the shear strength and the degree of saturation at the point of failure in the specimens.

Equation [7] suggests that there is a one to one relationship between the degree of saturation, S, and the area of water contact along the shear plane in the soil. Analysis of experimental results presented and shown in the figures of this paper using Equation [7] are referred as **Procedure 3**.

Khallili and Khabbaz (1998) have extended Bishop's equation (1959) (i.e., Equation 1) for predicting the shear strength of an unsaturated soil. An empirical constant has been suggested for the parameter, χ in Equation 1 as given below:

$$\chi = \left\{ \frac{(u_a - u_w)_f}{(u_a - u_w)_b} \right\}^{-0.55} \qquad [8]$$

where:

$(u_a - u_w)_f$ = matric suction in the specimens at failure conditions,

η = a value equal – 0.55 was suggested based on 13 soils data published in the literature

The required parameters for predicting the shear strength of unsaturated soils based on Khallili and Khabbaz (1998) approach is the air-entry value, $(u_a - u_w)_b$ of the soil and the saturated shear strength parameters. Analysis of experimental results presented and shown in the figures of this paper using Khallili and Khabbaz (1998) procedure is referred as **Procedure 4**.

Practicing engineers are most interested in the shear strength behavior in the transition zone (Figure 1). The transition zone lies between the air-entry value and the residual zone of saturation (Vanapalli et al. 1996a). The variation of the soil-water characteristic curve behavior in the transition zone is linear on a semi-logarithmic plot (i.e., variation of degree of saturation, S, or volumetric water content, θ, or gravimetric water content, w versus logarithm of soil suction). Bao et al. (1998) suggest another equation for predicting the unsaturated shear strength taking into account of the linear variation of the soil-water characteristic curve in the transition zone as below:

$$\tau = c' + (\sigma_n - u_a)\tan\phi' + (u_a - u_w)[\xi - \zeta \log(u_a - u_w)]\tan\phi' \qquad [9]$$

where:

$$\xi = \frac{\log(u_a - u_w)}{\log(u_a - u_w)_r - \log(u_a - u_w)_b}$$

$$\zeta = \frac{1}{\log(u_a - u_w)_r - \log(u_a - u_w)_b}$$

$(u_a - u_w)_r$ = soil suction at residual water content conditions

The parameter, ξ, represents the intercept (i.e., on the abscissa) and the parameter, ζ, represents the slope of the linear part of the soil-water characteristic curve respectively. Bao et al. (1998) suggests use of the expression given below for fitting the soil-water characteristic curve data.

$$\frac{(\theta_w - \theta_s)}{(\theta_s - \theta_r)} = \xi - \zeta \log(u_a - u_w) \quad [10]$$

The form and philosophy of the Equation [9] for predicting the shear strength of unsaturated soils is similar to Procedure 2 outlined in this paper.

EXPERIMENTAL DATA USED IN THE ANALYSIS

Escario and Juca (1989) measured the shear strength of three different soils with varying percentages of clay and plasticity indices, I_p, using modified direct shear equipment. Statically compacted specimens were used in the measurement of shear strength under consolidated drained conditions with different net normal stresses. The shear strength for two soils, namely, Madrid gray clay and Red silty clay was measured for a suction range between 0 to 15,000 kPa. The shear strength for Madrid clay sand was measured for a suction range between 0 to 5,000 kPa. The properties of three soils used in their testing program are summarized in Table 1.

Table 1. Summary of soil properties tested by Escario and Juca (1989).

	Madrid gray clay	**Red silty clay**	**Madrid clay sand**
Sand (%)	1	17	87
Silt (%)	27	48	7
Clay (%)	72	35	6
Liquid Limit, w_L (%)	71	33	28
Plasticity Index, I_p (%)	35	13.6	8
Specific Gravity, G_s	2.64	2.66	2.71
Void ratio, e	1.03	0.48	0.38

Escario and Juca (1989) have also measured the soil-water characteristic curves for the three soils for a suction range between 0 to 15,000 kPa. The soil-water characteristic curves were measured with an applied stress of 20 kPa using calibrated springs such that there is a good contact between the specimen and the ceramic stone of the pressure plate apparatus. Figure 2 presents the fitted soil-water characteristic curve data for the entire suction range from 0 to 1,000,000 kPa using Equation 5 (Fredlund and Xing, 1994). It should be noted that the soil-water characteristic curve data in the high suction range were not measured.

The soil-water characteristic curve results suggest that the air-entry and the residual suction values of the soil increases with an increase in the percentage of fines present in the soil. Table 2 summarizes the estimates of air-entry value and the residual suction value based on the construction procedure defined in Figure 1.

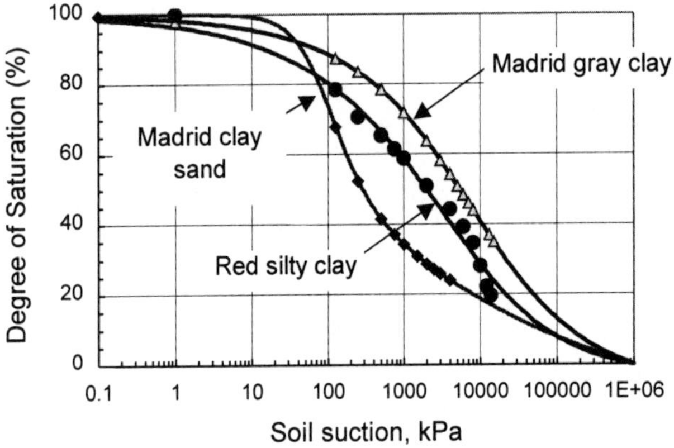

Figure 2. Soil-water characteristic curves for three soils tested by Escario and Juca (1989).

Table 2. Summary of the saturated shear strength parameters and the soil-water characteristic curve data

	Madrid gray clay	**Red silty clay**	**Madrid clay sand**
Air-entry value, $(u_a - u_w)_b$, kPa	110	40	30
Residual suction, ψ_r, kPa	40,000	33,000	12,000
Residual degree of saturation, (%)	23	16	17.6
Fitting parameter, κ (Procedure 1 using Equation 3)	2.8	1.9	1.65
Effective cohesion, c', kPa	30	20	40
Effective angle of internal friction, ϕ' (degrees)	25.3	34	39.5
Net normal stress, $(\sigma - u_a)$, kPa	300	120	120
Saturated shear strength, kPa	173.3	96.3	138.9

The comparisons between the predicted and measured shear strength values are presented using the four different procedures discussed in earlier sections (i.e., Figures 3 to 8). These comparisons are provided both for a limited suction range and as well as a large suction range. The key information required for predicting the shear strength of unsaturated soils using the procedures, along with the saturated shear strength parameters is summarized in Table 2.

ANALYSIS OF RESULTS

Several parameters such as initial compaction water content and the stress state influence the soil-water characteristic curve behavior (Vanapalli et al. 1999). It would appear that the shear strength contribution due to suction can be more accurately estimated using the soil-water characteristic curve that has been determined taking into account the influence of the stress state and the initial water content conditions. The soil-water characteristic curves are conventionally determined in the laboratory only with the application of a nominal normal stress or without the application of any applied vertical stress.

Vanapalli et al. (1998b) studies have shown that the shear strength of Indian Head till specimens can be predicted with a reasonable degree of accuracy using the soil-water characteristic curves that have been measured without considering the influence of stress state. Procedures 1 and 2 presented in this paper were used for analyzing the results. The Atterberg limits of the Indian Head till were as follows: liquid limit, w_L of 35.5% and plasticity index, I_p of 18.7%. The fractions of sand, silt, and clay were 28%, 42% and 30% respectively.

The predicted and measured shear strength values are compared for the three soils in this paper are; namely, Madrid gray clay, Red silty clay and Madrid clay sand. The shear strength of these soils were tested with different net normal stresses (Table 2). The soil-water characteristic curves measured with 20 kPa stress were used in the analyses (Figure 1).

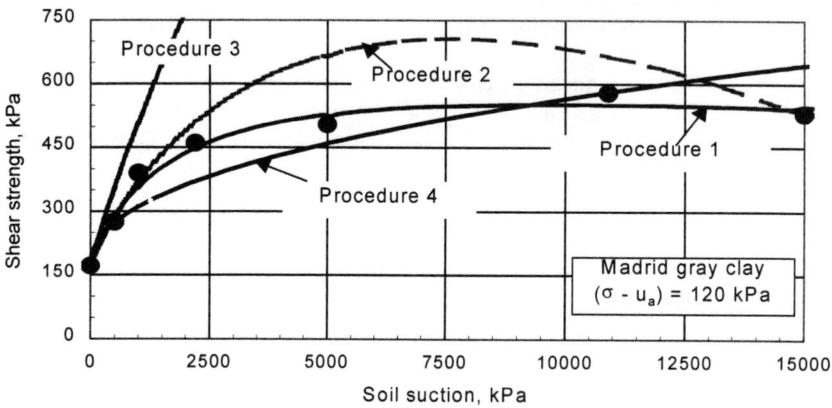

ure 3. Comparison of predicted and measured shear strength values for Madrid gray clay using ferent procedures for a large suction range.

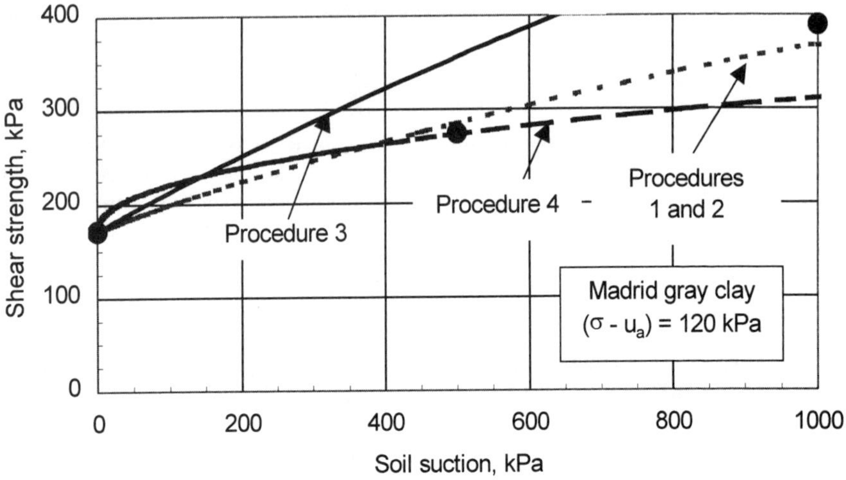

Figure 4. Comparison of predicted and measured shear strength values for Madrid gray clay using different procedures for a large suction range.

Figure 3 shows comparisons between the predicted and measured shear strength values for Madrid gray clay using the four different procedures discussed in the paper for a suction range between 0 to 15,000 kPa. The continuous lines in the figure represent the predicted shear strength values and the symbols are the measured shear strength values. There is a good comparison between the predicted and measured values of the shear strength for the suction range of 0 to 15,000 kPa using Procedure 1 (Figure 3). A fitting parameter value of κ equal to 2.8 was used for obtaining these predictions using Equation [3].

Figure 4 shows comparisons between predicted and measured shear strength values for a limited suction range of 0 to 1, 000 kPa. Procedures 2 and 4 along with Procedure 1 provide good comparisons between the measured and predicted values of shear strength.

Figure 5 shows the comparison between the measured and predicted shear strength values for Red silty clay for the suction range 0 to 15,000 kPa. Procedure 1 provides good comparisons for the suction range of 0 to 15,000 kPa. Procedures 2, 3 and 4 do not provide good comparisons. These observations are consistent with the earlier observations for Madrid gray clay.

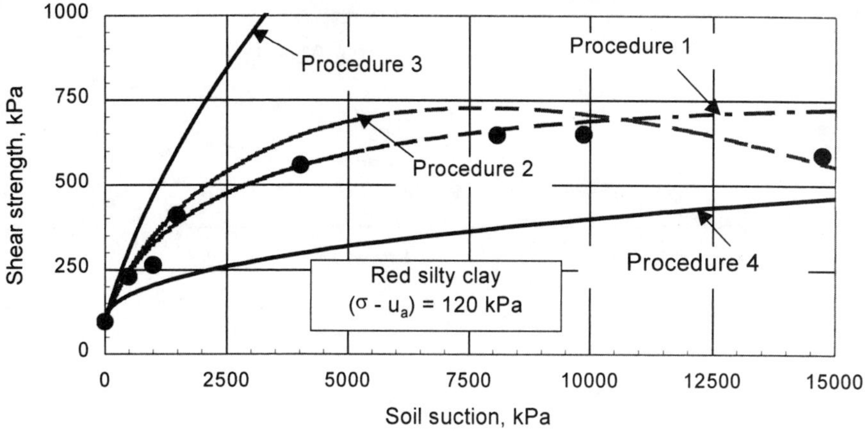

Figure 5. Comparison of predicted and measured shear strength values for Red silty clay using different procedures for a large suction range.

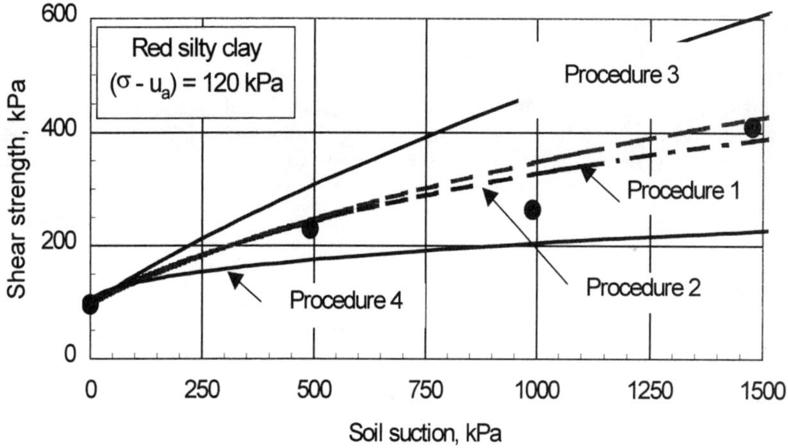

Figure 6. Comparison of predicted and measured shear strength values for Red silty clay using different procedures for a suction range between 0 to 1,500 kPa.

Figure 6 shows the comparison between predicted and measured shear strength values for Red silty clay for a limited suction range between 0 to 1,500 kPa. Procedures 3 and 4 do not provide good comparisons. Procedure 2 provides reasonably good comparisons for a suction range between 0 to 1,500 kPa.

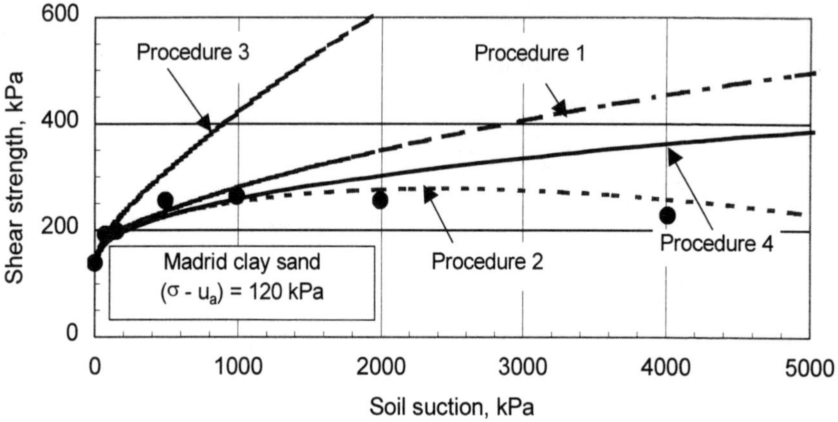

Figure 7. Comparison of predicted and measured shear strength values for Madrid clay sand using different procedures for a large suction range.

Figure 7 provides comparison between the measured and predicted shear strength values for Madrid clay sand for the suction range 0 to 5,000 kPa. Procedure 2 provides good comparison for the suction range of 0 to 5,000 kPa.

Figure 8 shows the comparison between predicted and measured shear strength values for Madrid clay sand for a limited suction range between 0 to 1,000 kPa. Procedure 1, 2, and 4 provide reasonably good comparisons.

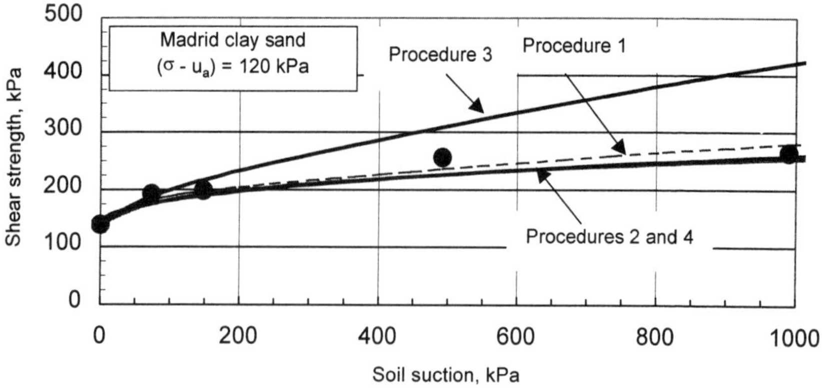

Figure 8. Comparison of predicted and measured shear strength values for Madrid clay sand using different procedures for a suction range between 0 to 1,000 kPa.

Summary and Conclusions

Comparisons between the measured shear strength and predicted shear strength values for three different soils are presented. Four different procedures available in the literature for predicting the shear strength are used both for limited and large suction ranges.

Procedure 1 provides good comparisons between the measured and predicted shear strength values for limited suction ranges between 0 to 1,500 kPa for all the three soils presented in this paper. Unsaturated soil behavior is of practical interest in this suction range. Good comparisons were also observed for two of the three soils for even for a large suction range. This procedure provides better predictions in comparison to the other procedures.

Figure 9 shows the relationship between the fitting parameter, κ, and plasticity index, I_p. Three points of κ versus I_p relationship were obtained from the study presented in this paper (Table 2). The other two points were based on the studies presented by Vanapalli et al. (1996) using Indian Head and Wulfsohn et al. (1996) using a non-plastic soil. A value of κ equal 2.3 provided good correlation between the measured and predicted values of shear strength for Indian Head till with a plasticity index, I_p, of 18.7. A value of κ equal 1 provided good correlation between predicted and measured values of shear strength for the non-plastic soil. The fitting parameter κ may be influenced by other parameters such as soil structure, nature of soil (i.e., slurry consolidated, dynamically compacted or statically compacted, natural) etc. More studies are necessary to establish the uniqueness of this relationship. Correlation such as shown in Figure 9 may be possible with other soil constants.

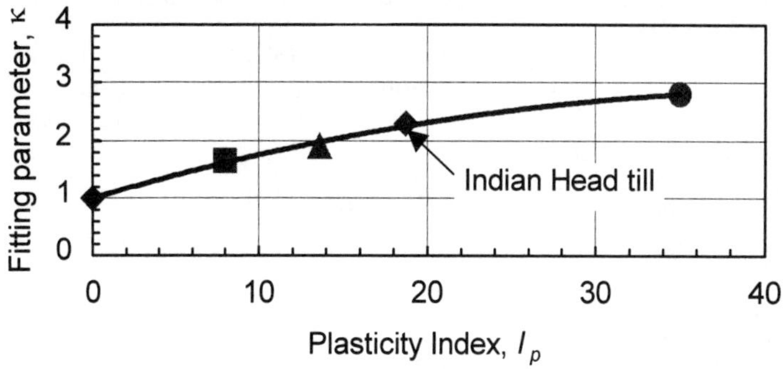

Figure 9. The relationship between the fitting parameter, κ, and plasticity index, I_p

Procedure 2 is provides good comparisons between measured and predicted values for the three soils analyzed in this paper for a limited suction range between 0 to 1,500 kPa. However, the comparisons were not good for predicting the shear strength over a large suction range for two soils. This may be attributed to the possible errors in the estimation of residual suction value from the construction procedure using limited soil-water characteristic curve data.

Procedure 3 does not provide good relations both for limited and as well large suctions ranges for the comparisons undertaken in the study presented in this paper. Figure 8 shows reasonable comparisons for a limited suction range between 0 to 200 kPa for Madrid clay sand.

Procedure 4 provides reasonable estimates for predicting the shear strength of two of the three soils analyzed in the low suction range but does not provide good comparisons for large suction ranges.

The studies presented in this paper show promise of using the procedures predicting the shear strength of an unsaturated soil using the soil-water characteristic curve and the saturated shear strength parameters. More experimental studies are necessary on different types of soils to better understand the shear strength behavior of unsaturated soils and develop better prediction procedures.

REFERENCES

Bao, C.G., Gong, B. and Zhan, L. 1998. Properties of unsaturated soils and slope stability of expansive soil. Keynote Lecture. UNSAT 98, *2nd International Conference on Unsaturated Soils*, Beijing.

Wulfsohn, D., Adams, B.A. and Fredlund, D.G. 1996. Application of unsaturated soil mechanics for agricultural conditions, Canadian Agricultural Engineering, Vol. 38, No. 3, pp. 173-181.

Bishop, A.W. 1959. The principle of effective stress. *Tecknish Ukebland*, 106(39): 859-863.

Escario, V. and Juca. 1989. Shear strength and deformation of partly saturated soils. *Proceedings of the 12th International Conference on Soil Mechanics and Foundation Engineering*, Rio de Janerio, 2: 43-46.

Fredlund, D.G., Morgenstern, N.R., and Widger, R.A. 1978. The shear strength of unsaturated soils. *Canadian Geotechnical Journal*, 15: 313-321.

Fredlund, D.G. and Rahardjo, H. 1993. Soil mechanics for unsaturated soils. John Wiley and Sons Inc., New York.

Fredlund, D.G. and Xing, A. 1994. Equations for the soil-water characteristic curve. *Canadian Geotechnical Journal*. 31: 517-532.

Fredlund, D.G., Xing, A., Fredlund, M.D., and Barbour, S.L.1996. The relationship of the unsaturated soil shear strength to the soil-water characteristic curve. *Canadian Geotechnical Journal*. 33: 440-448.

Gan, J.K.M. and Fredlund, D.G. 1988. Multistage direct shear testing of unsaturated soils. American Society for Testing Materials, *Geotechnical Testing Journal*, 11(2): 132-138.

Khallili, N. and Khabbaz, M.H. 1998. A unique relationship for the determination of the shear strength of unsaturated soils. *Geotechnique*, 48(5); 681-687.

Oberg, A. and Sallfors, G. 1997 Determination of shear strength parameters of unsaturated silts and sands based on the water retention curve, *Geotechnical Testing Journal*, GTJODJ, 20(1): 40-48.

Terzaghi, K. 1943. Theoretical soil mechanics. Wiley Publications, New York.

Vanapalli, S.K., Fredlund D.G., Pufahl, D.E. and Clifton, A.W. 1996. Model for the prediction of shear strength with respect to soil suction. *Canadian Geotechnical Journal*, 33: 379-392.

Vanapalli, S.K., Sillers, W.S., and Fredlund, M.D. 1998a. The meaning and relevance of residual water content to unsaturated soils. 51st Canadian Geotechnical Conference, 1998, Edmonton, pp. 101-108.

Vanapalli, S.K., Pufahl, D.E., and Fredlund, D.G. 1998b. The effect of stress state on the soil-water characteristic curve behavior of a sandy-clay till. 51st Canadian Geotechnical Conference, 1998, Edmonton, pp. 81-86.

Vanapalli, S.K., Fredlund, D.G. and Pufahl, D.E. 1999. The influence of soil structure and stress history on the soil-water characteristics of a compacted till. *Geotechnique, Vol. 49:* 143-159.

EFFECT OF SUCTION ON THE STRENGTH OF UNSATURATED SOIL

Mohd. Raihan Taha[1], Md. Kamal Hossain[2] and Syed Abdul Mofiz[2]

Abstract

Unsaturated soil condition which give rise to negative pore pressure is important in evaluating the stability of geotechnical structures such as slopes, retaining walls, excavations and footings. Its importance is more so in tropical countries, where rainfall and intense chemical weathering have resulted in the formation of such soils. In this study, the shear strength characteristics of granite residual soil were investigated using consolidated drained triaxial tests with special emphasis on the effects of the negative pore pressure, i.e. the matric suction. These tests involved shearing under either a constant net confining pressure and varying matric suctions or under a constant matric suction and varying net confining pressure. The influence of confining pressure and matric suction on shear strength was analyzed and a linear relationship was found to relate matric suction and shear strength.

Introduction

Residual soils cover more than three-quarters of the land area of Peninsular Malaysia. These residual soils are the product of intense weathering, particularly chemical weathering of bedrock in Malaysia's tropical climate conditions. A significant proportion of residual soils is granite residual soil and is widely distributed all over the country. The other kind of residual soil is the sedimentary residual soil derived from the weathering of sedimentary rocks. Many steep slopes in these residual soils often have a deep ground water table above which the soils possess high matric suction. It is well established that the stability of a natural or a cut slope in residual soils depends on the shear strength which is affected by the matric suction. The in-situ matric suction and the shear strength of

[1] Associate Professor, Dept. of Civil and Structural Engineering, Universiti Kebangsaan Malaysia, 43600 UKM Bangi, Selangor D. E., Malaysia.
[2] Research Student, Dept. of Civil and Structural Engineering, Universiti Kebangsaan Malaysia, 43600 UKM Bangi, Selangor D. E., Malaysia.

soils are in turn affected by the climatic condition, particularly rainfall distribution. In addition, other geotechnical structures are also affected by matric suction such as retaining walls, excavation, and foundations. The analysis and design for geotechnical structures affected by matric suction have all been described in detailed by Fredlund and Rahadjo (1993). Thus, it is important to study the shear strength characteristics of the residual soils particularly with respect to matric suction changes. This study attempts to find the shear strength of granite residual soil with respect to changes in soil suction utilizing a modified triaxial test apparatus.

Shear Strength of Unsaturated Soils

Many researchers such as Croney et al. 1952, Bishop 1959, Aitchison 1961, Jennings 1961 and Richards 1966, have developed relationships for effective stress equation for unsaturated soils. However, a more systematic approach was given by Fredlund et al. (1978) who developed a shear strength relationship for unsaturated soils in terms of two independent stress variables:

$$\tau = c' + (u_a - u_w) \tan\phi^b + (\sigma - u_a) \tan\phi' \tag{1}$$

where: c' = effective cohesion, σ = total stress, u_a = pore-air pressure, u_w = pore water pressure, ϕ' = effective angle of friction, ϕ^b = angle indicating the rate of increase in shear strength with respect to matric suction, $(u_a - u_w)$. The two-dimensional graphical procedure to evaluate of shear strength parameters from unsaturated triaxial tests was discussed by Fredlund and Rahardjo (1993).

Soil Properties

The soil used in this study was obtained from a residual granite soil formation in Cheras, just 8 km south of Kuala Lumpur, the capital city of Malaysia. The block sampling technique was adopted for soil sampling in which thirty 300 mm cube samples were hand cut, wax around the sides, place in wooden boxes and stored in the laboratory until required for testing. The mean dry density from these field tests was 13.6 kN/m^3 with natural moisture content of about 31%. The overall mean dry density obtained from laboratory test was found to have a value of 13.52 kN/m^3. The specific gravity of this soil (2.55 to 2.61) shows a change from the specific gravity of fresh granites (2.65 to 2.68). Based on Unified Soil Classification System (UCS), the soil was grouped as "clay with high plasticity" (CH). It was also classified "A-7-6" according to the AASHO classification system. Optimum moisture content and maximum dry density were found to be 23% and 1497.5 kg/m^3, respectively from compaction tests.

Testing Program

The testing program consisted of a series of consolidated drained triaxial tests on unsaturated specimens are shown in Table 1. These tests were performed under either constant net confining pressure (σ_3 - u_a) and varying matric suction, (u_a - u_w) or under a constant matric suction and varying net confining pressures. The tests reported in this paper were carried out under high effective confining pressure (600 -1000 kPa) and high matric suction (50 - 400 kPa). A strain rate of 0.00166 %/min was employed that ensured no pore pressure change required in drained test.

Table 1: Test program for CD triaxial test on unsaturated specimens

Test Type	Test no.	σ_3 (kPa)	u_a (kPa)	u_w (kPa)	(σ_3-u_a) (kPa)	(u_a - u_w) (kPa)	Strain Rate (%/min)	Remarks
CD	1	1000	400	200	600	200	0.00166	Constant (u_a - u_w)
	2	1200	400	200	800	200		
	3	1400	400	200	1000	200		
	4	800	200	150	600	50		Constant (σ_3 - u_a)
	5	900	300	200	600	100		
	6	1000	400	200	600	200		
	7	1200	600	200	600	400		

Testing Equipment

The computer based automated testing and control systems described by Menzies (1989) was modified for testing of unsaturated soil specimens. The changes from saturated triaxial testing system to unsaturated triaxial testing system consist of both software and hardware elements. Major modification of the triaxial testing system comprised the installation of high air entry disk on pedestal to separate the pore air and pore-water supplies and a digital pressure-volume controller for measuring and control of pore air pressure and pore air volume change. Fig. 1 shows the full layout of the computerized triaxial testing system for unsaturated soil. In this set-up, Bishop-Wesley triaxial cell set was modified to carry out triaxial test on suction induced soil specimens. The layout of the modified triaxial cell is shown elsewhere (Hossain, 1999). Key features of the modified cell base included placement of a 15 bar high air entry disc, and an additional port below the disk to flush out diffused air during testing. An additional circular collar was placed between base and triaxial cell cylinder which provides 12 access ports for connection of air/hydraulic connector or transducer.

The computer controlled system and the versatile cell was equally adapted to carry out classical as well as advanced tests. A microprocessor collects the data from transducer automatically at prescribed intervals. The data were transmitted by the controlling microprocessor for recording, processing and production of results, which could be displayed on the screen or tabulated or plotted by a plotter.

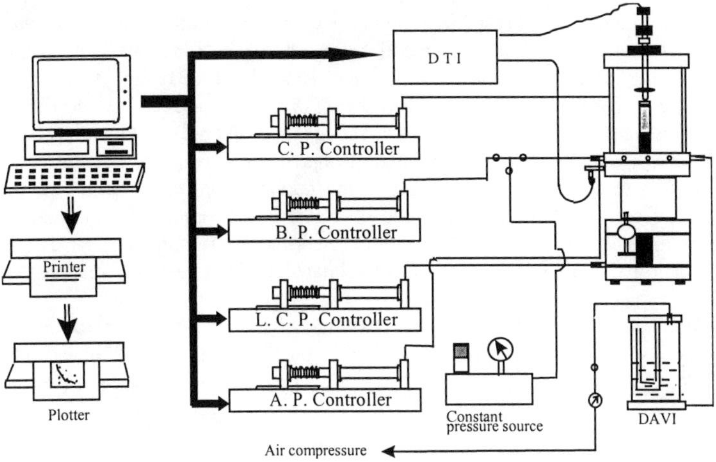

Fig. 1. The layout of computerized triaxial testing system for unsaturated soil.

Testing Procedure

The triaxial specimens (50 mm diameter and 100 mm high) were prepared by trimming from block samples. The specimens were set up between a saturated high air entry disk at its bottom and a porous disk at the top. A rubber membrane was placed over the sample using a membrane stretcher. The rubber membrane was coated with silicon grease, to reduce diffusion of air, and a second rubber membrane was placed around it. O-rings were placed over the membranes on the bottom pedestal and upper cap. Saturation of test specimens was achieved by continuously increasing the cell pressure and back pressure. An unsaturated soil sample was prepared from a saturated sample by applying a desaturation and consolidation process. Air pressure was applied at the top of the sample and maintained at a higher value than the back pressure. Water was driven out from the soil sample against the back water pressure, through the high air entry disk and into the back pressure controller, until equilibrium was established under the applied pressures. A total of 4 to 5 days was required to complete this consolidation and desaturation. The specimens were sheared under a constant strain rate 0.00166 %/min and the air and water pressure valves were remained open. The tests involved shearing under either a constant matric suction and varying net confining pressures or under a constant net confining pressure and varying matric suctions. To carry out the tests under a constant matric suction, the net matric suction, $(u_a - u_w)$, was kept constant at 200 kPa and the net confining pressure, $(\sigma_3 - u_a)$, was varied i.e. 600, 800 and 1000 kPa.

Results

The deviator stress vs. axial strain curves obtained from these tests are shown in Fig. 2. A strain-hardening type of stress-strain curves was obtained for most of these tests. The deviator stress vs. strain curve does not exhibit a peak point, even at large strains. The limiting strain failure criterion (i.e. 18%) was selected to represent the failure condition (Fredlund and Rahardjo 1993). Fig. 2 illustrates that under a constant matric suction, shear stress increase as net confining pressure increases. The water volume change and overall specimen volume change during compression are presented in Figs. 3 and 4, respectively. Although pore-air was forced into the specimen during compression, total volume of the specimen decreases. In other words, the specimen volume reduced during shear. The amount of contraction was low at higher confining pressure (Fig. 4). Blight (1967) reported similar behavior of consolidated drained tests on unsaturated silts.

The extended Mohr-Coulomb failure envelope at a specific matric suction is presented in Fig. 5 as a horizontal projection on the τ vs. (σ - u_a) plane. The failure envelope has an ordinate intercept of 97 kPa and a slope angle of 26.5°. From the triaxial test on saturated specimens of this residual granite soil, the effective angle of internal friction, ϕ' and effective cohesion, c' were found to have values of 26.5° and 40 kPa, respectively (Taha et al. 1999). Ting and Ooi

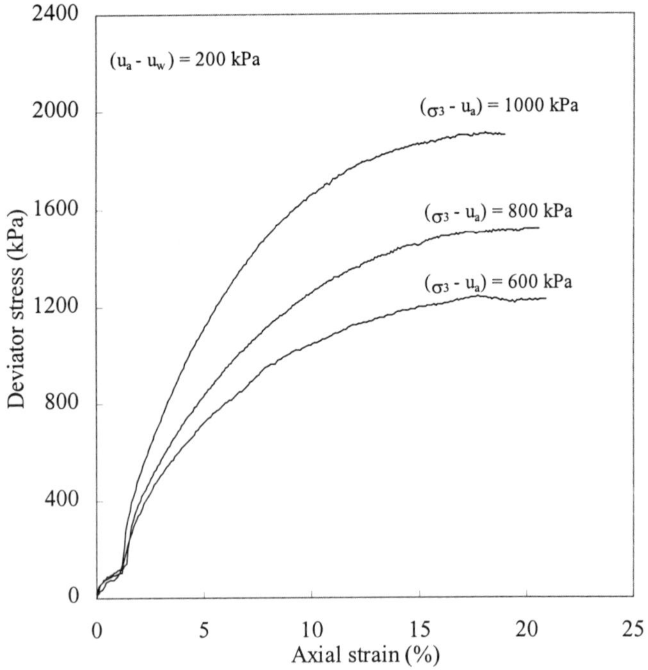

Fig. 2. Deviator stress vs. axial strain curves from CD tests under a constant matric suction of 200 kPa.

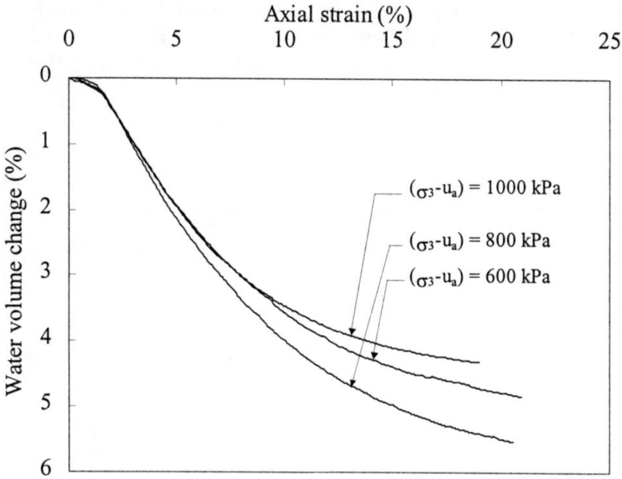

Fig. 3. Water volume change vs. axial strain curves from CD tests under a constant matric suction of 200 kPa.

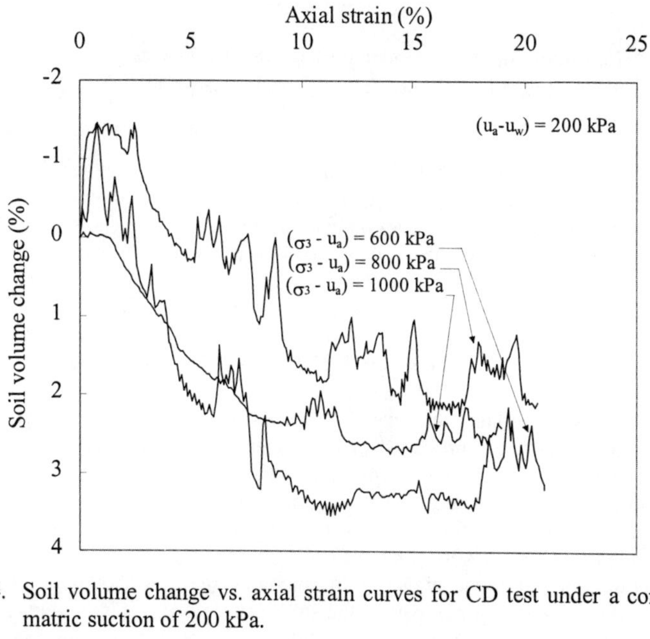

Fig. 4. Soil volume change vs. axial strain curves for CD test under a constant matric suction of 200 kPa.

(1976) also mentioned a similar angle of internal friction but less cohesion intercept. This difference is possibly due to difference in factors such as sampling location, depth, and the heterogenic nature of the residual soil. It was found from the tests that the unsaturated specimens have value of ϕ' which is equal to that of saturated specimens. This finding led to the conclusion that the value of ϕ' for saturated CD tests can be used in equation (1) to predict the drained shear strength of the unsaturated specimens. However, Salman (1995) did not find similar ϕ' values for fine granular soil at low suction values.

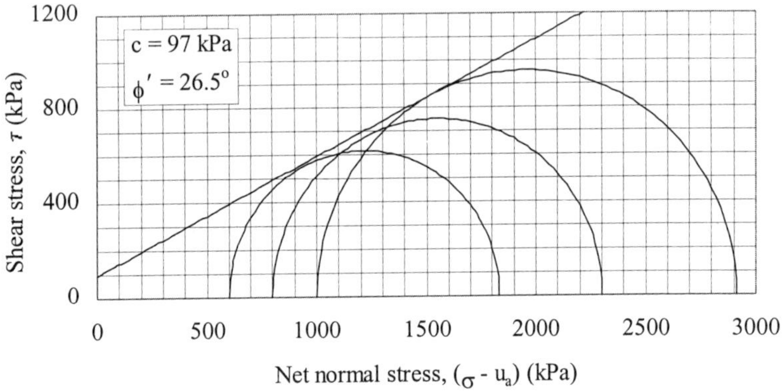

Fig. 5. Two-dimensional presentation of failure envelope projected on the τ vs. $(\sigma - u_a)$ plane for CD test under a constant matric suction of 200 kPa.

In the other approach, the net confining pressure, $(\sigma_3 - u_a)$, was kept at 600 kPa and the matric suction, $(u_a - u_w)$, was varied at 50, 100, 200 and 400 kPa. The deviator stress versus strain curves obtained from the tests are shown in Fig. 6. A strain-hardening type of deviator stress-strain curves was also obtained in most of these tests. The deviator stress vs. strain curve again does not exhibit peaks, even at large strains. Fig. 6 illustrates that at a constant net confining pressure shear strength increase as matric suction increases. Such phenomenon was also obtained by Blight (1967), Mashhour et al. (1995) and Gan and Fredlund (1995).

The water volume and overall specimen volume (air and water) change during compression are presented in Figs. 7 and 8, respectively. Similar shear stress development and water volume change behavior were observed in both test procedure involving increasing net confining pressures and increasing matric suctions. The volume of the specimens initially increased or dilate then started to decrease (about 3% - 4% axial strain) and continue to contract until failure. The amount of contraction was higher at low suction (Fig. 8). It should be noted that due to effect of end restrain, part of the specimens (usually middle part) reaches the ultimate strength and has higher contraction than the rest portion of the specimen. Thus, the measured specimen volume change using the volume change unit (back pressure controller) is the average volume change to the specimen.

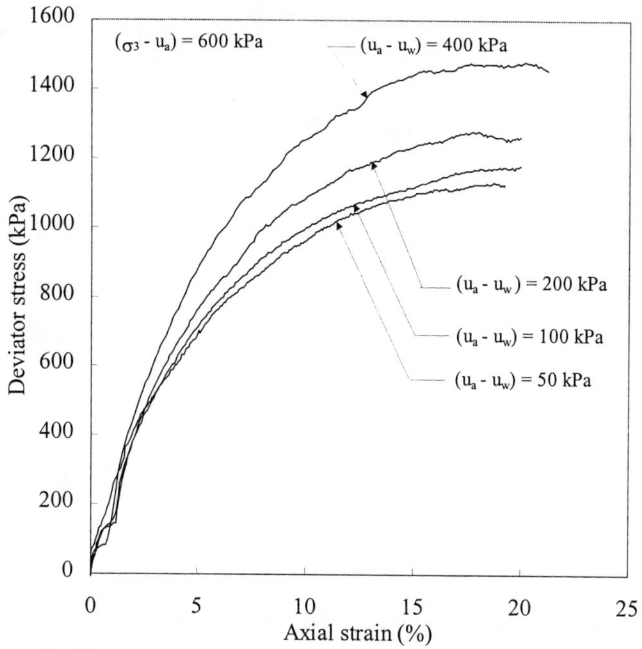

Fig. 6. Deviator stress vs. strain curves from CD tests at a constant net confining pressure of 600 kPa.

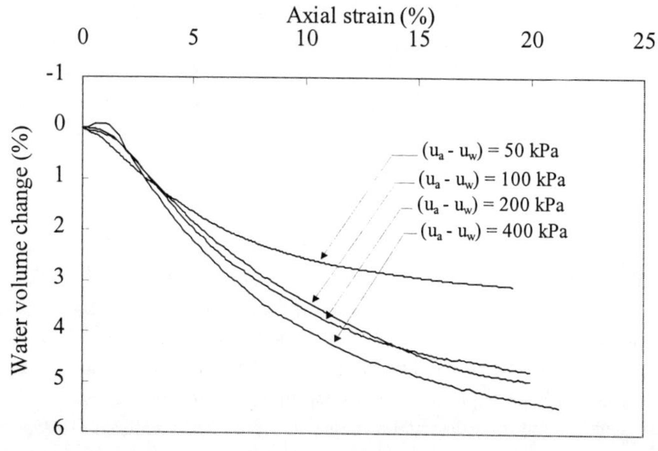

Fig. 7. Water volume change vs. axial strain curves from CD tests at a constant net confining pressure of 600 kPa.

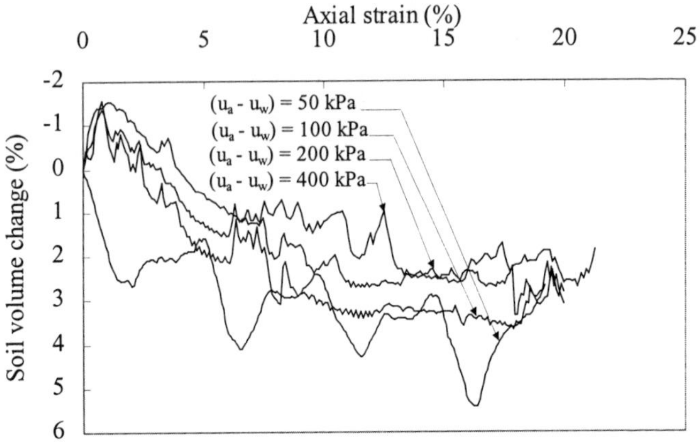

Fig. 8. Specimen volume change vs. axial strain curves from CD tests at a constant net confining pressure of 600 kPa.

A typical two dimensional projections of the failure envelope on the shear stress, τ vs. $(\sigma - u_a)$ plane are presented in Fig. 9(a). The intersections between the failure envelope and the ordinate are plotted in Fig. 9(b). For a constant net confining pressure, the shear strength increased with increasing matric suctions, as illustrated in Fig. 9(a). For a planar failure envelope, the internal friction angle, ϕ', remains essentially constant under saturated and unsaturated conditions. The effect of matric suction is clearly shown by the ϕ^b angle in Fig. 9(b) which show a linear relationship between matric suction and shear strength. The ϕ^b angle which describes the rate of change in shear strength due to a change in matric suction was found to have a value of about 17.8^o. Rohm and Vilar (1995), however, obtained a hyperbolic relationship between shear strength and matric suction for lateratized sandy soil from Brazil.

Ho and Fredlund (1982) reported a comparable $\phi^b = 15.3^o$ for undisturbed decomposed granite from Hong Kong in multistage CD triaxial test. The results from the CD test showed that the angle ϕ^b is less than the internal friction angle, ϕ', for the residual granite soil. This indicates that the shear strength of unsaturated soil is more affected due to change in effective confining pressure than the change in matric suction. Fredlund (1985) measured the ϕ^b angles for various soils and the experimental results showed that the ϕ^b is always smaller than or equal to the internal friction angle, ϕ'. Gan and Fredlund (1995) showed that ϕ^b is equal to ϕ' at low matric suction when the soil is close to saturation.

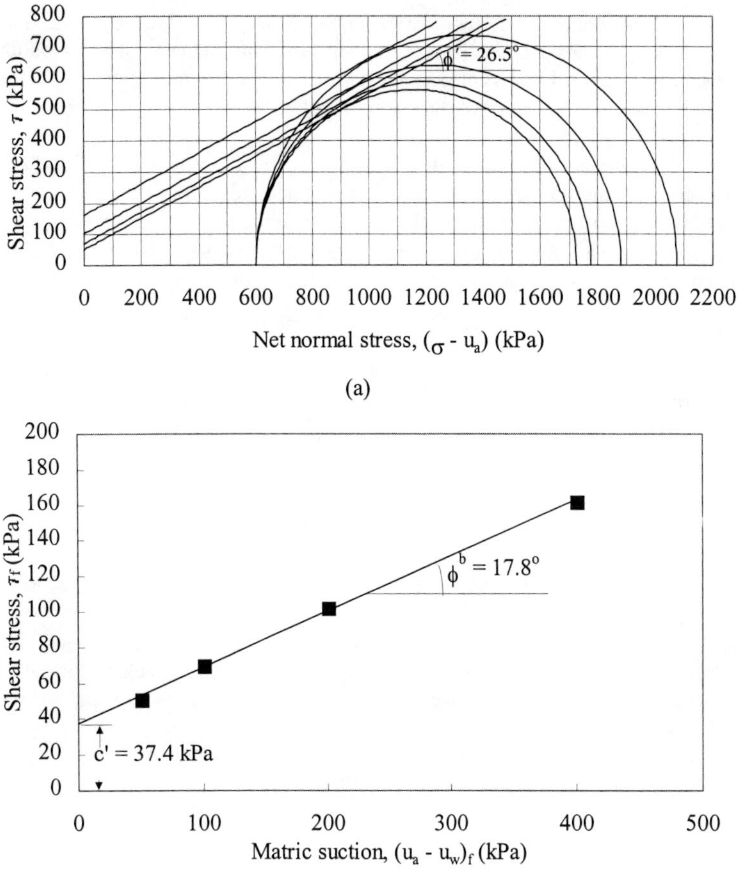

(b)

Fig. 9. Two-dimensional presentation of failure envelop for residual granite soil specimen. (a) Failure envelope projected onto the τ vs. $(\sigma - u_a)$ plane; (b) intersection line between the failure envelope and the τ vs. $(u_a - u_w)$ plane.

The c′ obtain from saturated CD test can be either greater or less than the c′ obtained from unsaturated CD test. The value of c′ obtained from unsaturated triaxial tests may differ with the c′ value obtained from triaxial tests due to linear interpretation of failure envelope, in the case of curve failure envelope (Fredlund and Rahardjo, 1993). Similar phenomenon was reported by Ho and Fredlund (1982) from Satija's (1978) test data, and Wong (1984) from Ho's (1981) test data.

The unsaturated CD test results plotted in terms of stress state variables (Figs. 5 and 9) shows almost linear relationship. This indicates that the failure surface can be considered as planer. Therefore, Mohr-Coulomb criterion (i.e., Eq. 1) can be used to determine the shear strength of an unsaturated soil.

Soil parameters c′, ϕ', and ϕ^b can be used for geotechnical design. Some commercial computer programs have incorporated this for a better or a more reliable design especially in tropical soils. In the laboratory, the conventional soil parameters c′ and ϕ' can be readily obtained. However, to obtain ϕ^b, the conventional triaxial apparatus will require some modification.

Conclusions

The following conclusions can be drawn from the results of the experimental program:

(a) The effective friction angle from the CD test on saturated and unsaturated specimens are similar and thus the ϕ' from saturated CD tests can be used to estimate the shear strength of unsaturated test saturated.
(b) The soil suction affects the shear strength. An increase in soil suction increases the shear strength of the soil.
(c) The average rate of increase in shear strength with respect to matric suction ϕ^b was 17.8^o for matric suction up to 400 kPa.
(d) A linear relation between shear strength vs. matric suction was found. This finding indicates the suitability of using a planar failure envelope for natural residual granite soil at the magnitude of matric suction frequently encountered in practice.
(e) The shear strength of unsaturated soil is more affected due to change in effective confining pressure than the change in matric suction.

References

Aitchison, G.D. (1961). "Relationship of moisture and effective stress functions in unsaturated soil," *in Pore Pressure and Suction in Soils*, Butterworths, pp. 47-52.

Bishop, A.W. (1959). "The principle of effective stress," *Teknisk Ukeblad*, vol. 106, pp. 859-863.

Blight, G. E. (1967). "Effective stress evaluation for unsaturated soils," *ASCE J. Soil Mech. Found. Eng. Div.*, vol. 93, SM2, pp. 125-148.

Croney, D. (1952). "The movement and distribution of water in soils," *Geotechnique*, vol. 3, pp. 1-16,

Fredlund, D. G. (1985). "Theory formulation and application for volume change and shear strength problems in unsaturated soils," *Proc. 11th Int. Conf. Soil Mech. Found. Eng.*, San Francisco.

Fredlund, D. G., Morgenstern, N. R. and Widger, R. A. (1978). "The shear strength of unsaturated soils," *Can. Geotech J.*, Vol. 15, No. 3, pp. 313-322.

Fredlund, D. G. and Rahardjo, H. (1993). "Soil mechanics for unsaturated soils," John Wiley & Sons Inc., New York.

Gan, K. J. M. and Fredlund, D. G. (1995). "Shear strength behavior of two saprolitic soils," Proc. 1st Int. Conf. Unsaturated Soils, vol. 1, Balkema, Rotterdam, pp. 71-76.

Hossain, M. K. (1999). "Shear strength characteristics of decomposed granite residual soils in Malaysia," Ph.D. Thesis (unpublished), Universiti Kebangsaan Malaysia, 224 pp.

Ho, D. Y. F. (1981). "The shear strength of unsaturated Hong Kong soils," M. Sc. Thesis, University of Saskatchewan, Saskatoon, Saskachewan, Canada.

Ho, Y. F. and Fredlund, D. G. (1982). "A multi-stage triaxial test for unsaturated soils," ASTM Geotech. Testing J., Vol. 5, pp. 18-25.

Jennings, J.E. (1961). "A revised effective stress law for use in the prediction of the behavior of unsaturated soils", *in Pore Pressure and Suction in Soils*, Butterworths, pp. 26-30.

Mashhour, M. M., Ibrahim, M. I. & El-Eemam, M. M. (1995). "Variation of unsaturated soil shear strength parameters with suction," *Proc. 1st Int. Conf. Unsaturated Soils*, vol. 3, pp. 1487-1493.

Menzies, B. K. (1989). "A computer controlled hydraulic triaxial testing system," *ASTM STP977*. Advanced triaxial testing of soil and rock. ASTM, Philadelphia.

Richards, B.G. (1966). "The significance of moisture flow and equilibria in unsaturated soils in relation to the design of engineering structures built on shallow foundations in Australia," *Symp.on Permeability on Capillary*, ASTM.

Rohm, S.A., and Vilar, O.M. (1995). "Shear strength of an unsaturated sandy soil," *Proc. 1st Int. Conf. Unsaturated Soils*, Vol. 1, Balkema, Rotterdam, pp. 189-193.

Salman, T.H. (1995). "Triaxial behavior of partially saturated granular soils," *PhD Thesis*, University of Sheffield.

Satija, B. S. (1978). "Shear behavior of partially saturated soils," *Ph. D. thesis*, Indian Institute of Technology, Delhi, 327 pp.

Taha, M. R., Hossain, M. K., and Mofiz, S. A. (1999). "Effect of Matric Suction on the Shear Strength of Unsaturated Granite Residual Soil," *Journal of the Institution of Engineers*, Malaysia (Accepted for publication).

Ting, W. H. and Ooi, T. A. (1976). "Behavior of a Malaysian Residual Granite Soil as a Sand-Silt-Clay Composite Soil," *Geotechnical Engineering*, Vol. 7, pp. 67-79.

Wong, C. K. (1984). "The shear strength of unsaturated soils," *Ph. D. thesis*, University of Glasgow. 278 pp.

The Influence of Fabric on the Shear Behaviour of Unsaturated Compacted Soils

David Geoffrey Toll[1]

Abstract

Fabric plays a vital role in determining the engineering behaviour of compacted soils, particularly under unsaturated conditions. Clayey materials compacted dry of optimum moisture content develop an aggregated or 'packet' fabric. The presence of aggregations causes the soil to behave in a coarser fashion than would be justified by the grading. This means that the friction angle ϕ^a (representing the strength component due to net stress) depends on the fabric of the soil. Dry of optimum moisure content, $\phi^a > \phi'$ (the angle of friction in saturated conditions). Similarly, the packet fabric also affects the angle ϕ^b (representing the strength component due to matric suction). Dry of optimum moisture content, ϕ^b becomes significantly lower than ϕ' and at low water contents drops to zero. The normal practice for determining the shear strength properties of unsaturated soils (by multi-stage tests in which the suction is increased from an initial low suction state) will result in the packet fabric being destroyed in the first stage of shearing. Therefore, to properly study the shear strength behaviour of unsaturated compacted soils, tests should be performed on the soil in an 'as compacted' condition.

Introduction

The term fabric is used to mean the geometrical arrangement of particles within a soil. The effect of fabric on the behaviour of unsaturated soils can be of great significance, particularly for compacted soils (Toll, 1990). Croney *et al* (1958) pointed out that compacted clays do not exist as a uniform mass of clay particles, but that aggregations of clay particles form, separated by comparatively large air voids. Microscopic observations of the aggregates in clays (Barden and Sides, 1970; Collins, 1983; Zein, 1985) and in a silt (Delage *et al*, 1996) have demonstrated this to be the case.

---

[1] School of Civil and Structural Engineering, Nanyang Technological University, Singapore 639798 (on leave from School of Engineering, University of Durham, Durham, DH1 3LE, UK)

Brackley (1973; 1975) proposed a model of unsaturated clay fabric in which the clay particles grouped together in 'packets'. He suggested that the voids within the aggregations were saturated, while the inter-packet voids were filled with air. The soil within the aggregations or 'packets' is held together by suction, and the packets act as individual particles. Zein (1985) showed that although compacted materials are highly aggregated on the dry side of optimum moisture content, aggregation does not exist on the wet side.

The influence of this packet fabric on the shear strength of unsaturated soils is considered in this paper. A conceptual picture of fabric in an unsaturated compacted soil is first presented. The influence of the fabric on the angles of friction (relating both to the net stress and the matric suction) are then presented for a lateritic gravel from Kenya. The implications of these results are considered in terms of testing procedures for unsaturated compacted soils.

Compaction and Fabric

Compaction tests were performed on a lateritic gravel from Kenya (Kiunyu gravel) in CBR moulds using three levels of compactive effort. The properties of Kiunyu gravel are given in Table 1. The compactive efforts were equivalent to British Standard (BS) light (Proctor) compaction (2.5kg rammer falling through 300mm, 62 blows/layer, 3 layers of soil); BS heavy (Modified Proctor) compaction (4.5kg rammer falling through 450mm, 62 blows/layer, 5 layers of soil) and an intermediate level (4.5kg rammer falling through 450mm, 30 blows/layer, 5 layers of soil) (British Standard 1377, 1990). The compaction curves are given in Figure 1.

Table 1. Properties of Kiunyu gravel

Gravel fraction	55 %
Sand fraction	30 %
Silt fraction	7 %
Clay fraction	8 %
Liquid Limit of fines*	56-66 %
Plastic Limit of fines*	27-31 %
Plasticity Index of fines	29-35 %
Specific gravity	3.2

* not dried before testing

Lines of constant degree of saturation (S_r) are shown in Figure 1. It can be seen that the degree of saturation at optimum moisture content (OMC) for BS light compaction, intermediate and BS heavy compaction falls around 95%, 90% and 85% respectively. This is in agreement with the general pattern shown by a large number of compaction tests by K.B. Woods (Road Research Laboratory, 1952: p. 158) where optimum moisture content occurred between 4-8% air voids (S_r = 80-90%).

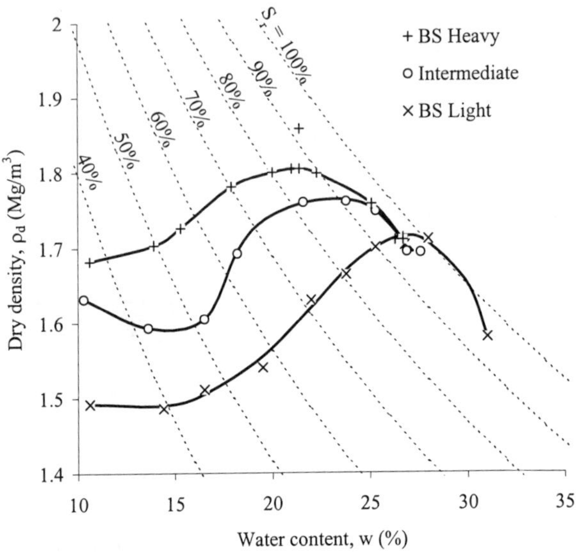

Figure 1. Compaction curves for Kiunyu gravel

Zein (1985) identified the degree of aggregation (the percentage of material present as aggregations) of a clayey soil for two levels of compaction (BS light and BS heavy). He noted that at OMC for BS light compaction (and also wet of optimum) there were no aggregations and matrix material dominated. However, the degree of aggregation increased as the water content reduced below OMC for BS light compaction; at water contents below 70% of OMC the material was completely aggregated with no matrix material. For BS heavy compaction the material showed a degree of aggregation of 70% at OMC, with no aggregations wet of optimum.

Brackley's (1973; 1975) model of unsaturated clay fabric is based on the assumption that the inter-packet voids are filled with air while the intra-packet voids are saturated. In such circumstances, the number and size of air voids (and hence the degree of saturation) is closely linked to the degree of aggregation.

The link between degree of saturation and degree of aggregation is supported by Zein's observations, and the observations of degree of saturation for different levels of compaction. Figure 1 shows that at OMC for BS light compaction the degree of saturation is around 95%; Zein's observations indicate no aggregation for this condition. The same would be true for BS heavy compaction, wet of optimum, where the degree of saturation approaches 95%. However, at a degree of saturation of 85% (OMC for BS heavy compaction) Zein's observations suggest a degree of aggregation of 70%. The degree of aggregation rises to 100% as the degree of saturation drops below 50%.

It might be thought strange that aggregation would be present at OMC at the higher level of compaction (BS heavy) whereas no aggregation is present at OMC under BS light compaction. This can be explained by an examination of the suctions present during compaction. Figure 2 shows suctions measured using a CBR apparatus with suction measurement described by Grace *et al* (1985) and Toll (1988). The apparatus used the axis-translation technique and had high air entry pore water pressure probes installed at three levels in the CBR mould.

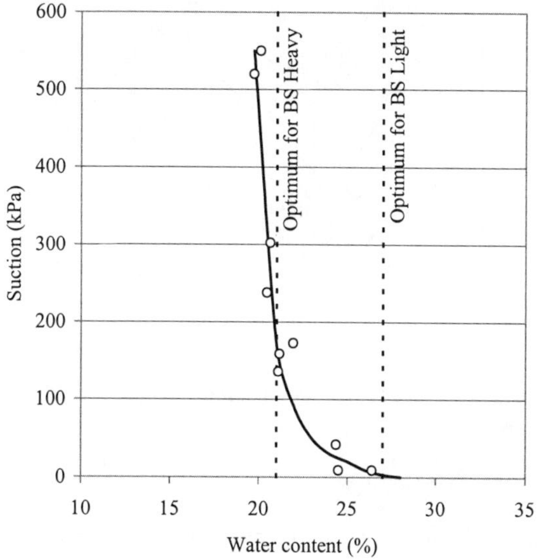

Figure 2. Suctions measured in compacted samples

It can be seen from Figure 2 that at the optimum moisture content for BS light compaction, the matric suctions are very small (<10 kPa). Therefore, any aggregations present would be easily remoulded and broken down, even by a light compactive effort. However, at OMC for BS heavy compaction, suctions are of the order of 150 kPa. Clearly a suction of this magnitude will be able to impart significant strength to the aggregations and restrict them from being remoulded. Although a high density is obtained, it is achieved by packing the aggregations together more closely, rather than remoulding and breaking them down completely.

The preceding discussion leads to the conceptual picture of compacted soil fabric illustrated in Figure 3. For soils compacted to degrees of saturation of 90% and over, the material would be expected to be non-aggregated. As the degree of saturation drops, the amount of aggregation increases rapidly and reaches a fully aggregated condition for degrees of saturation below 50%.

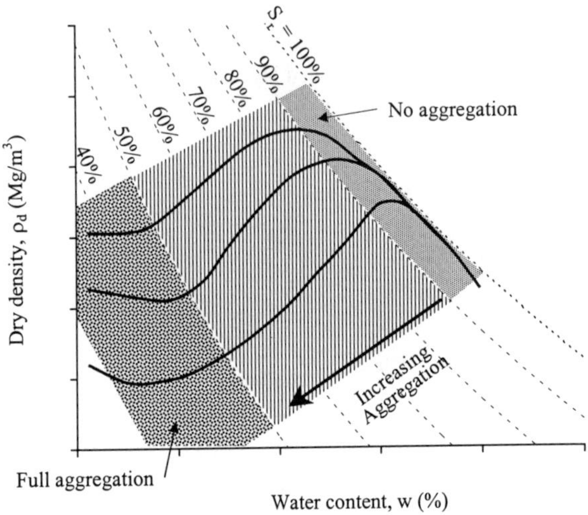

Figure 3. A conceptual picture of compacted fabric in clayey soils

The Shear Strength of Unsaturated Soils

Fredlund *et al* (1978) suggested the following equation for the shear strength of unsaturated soils:

$$\tau = c'' + (\sigma - u_a) \tan \phi^a + (u_a - u_w) \tan \phi^b \qquad [1]$$

where τ is shear strength
σ is the total stress
u_a is the pore air pressure
u_w is the pore water pressure
c'' is the cohesion when the two stress variables $(\sigma - u_a)$ and $(u_a - u_w)$ are zero
ϕ^a is the friction angle with respect to net stress $(\sigma - u_a)$
ϕ^b is the friction angle with respect to matric suction $(u_a - u_w)$

Fredlund *et al* went on to suggest that, when the matric suction is zero, the $(\sigma - u_a)$ plane will have the same friction angle as the $(\sigma - u_w)$ plane. Therefore, they suggested that ϕ^a is the same as ϕ' (the friction angle with respect to effective stress $(\sigma - u_w)$). Fredlund and Rahardjo (1993: p. 238) suggest that this friction angle "appears to be essentially equal to the effective angle of internal friction obtained from shear strength tests on saturated soil specimens". Fredlund *et al* (1978) also suggested that c" is the same as c' (the effective cohesion).

Therefore, making these two assumptions, equation [1] becomes the following:

$$\tau = c' + (\sigma - u_a) \tan \phi' + (u_a - u_w) \tan \phi^b \qquad [2]$$

and it is this equation that is commonly quoted (eg Fredlund and Rahardjo, 1993).

It has also been shown that the failure envelope in relation to matric suction is non-linear (Escario and Saez, 1986; Fredlund *et al*, 1987). This means that ϕ^b is not a constant, but reduces with increasing suction.

Toll (1990) considered the behaviour of unsaturated soils from a critical state standpoint. He suggested the following relationship for soils sheared to a critical state under unsaturated conditions:

$$q = M_a (p - u_a) + M_w (u_a - u_w) \qquad [3]$$

where q is the deviator stress $(\sigma_1 - \sigma_3)$
p is the mean total stress $(\sigma_1 + \sigma_2 + \sigma_3)/3$
M_a is the critical state stress ratio with respect to net mean stress $(p - u_a)$
M_w is the critical state stress ratio with respect to matric suction $(u_a - u_w)$

For comparison with this approach, equation [1] can be rewritten for critical state conditions as:

$$\tau_c = (\sigma - u_a) \tan (\phi^a)_c + (u_a - u_w) \tan (\phi^b)_c \qquad [4]$$

where $(\phi^a)_c$ is the critical state angle of friction with respect to $(\sigma - u_a)$
$(\phi^b)_c$ is the critical state angle of friction with respect to $(u_a - u_w)$

and taking the cohesion intercept to be zero for critical state conditions (Atkinson, 1993: p. 116).

Angles of Friction for Kiunyu Gravel

Toll (1988) presented data from triaxial tests on an unsaturated compacted lateritic gravel from Kenya. The properties of the material (Kiunyu gravel) are summarised in Table 1. Samples were compacted at a range of water contents (17-28%) to achieve dry densities between 1.44 and 1.85 Mg/m^3. The tests were constant water content triaxial tests performed on samples "as compacted" with no change in suction prior to shearing. Matric suctions were measured during the tests using the axis-translation technique; a high air entry pore water pressure probe provided an additional measurement of suction at the mid-height of the specimen.

The results of these tests were originally analysed by Toll (1990) in terms of the critical state stress ratios (M_a and M_w). However, the data has been reinterpreted within the extended Mohr-Coulomb criteria proposed by Fredlund *et al* (1978). Equation 4 has been used for critical state conditions. The values of ϕ^a and ϕ^b are presented in Figure 4.

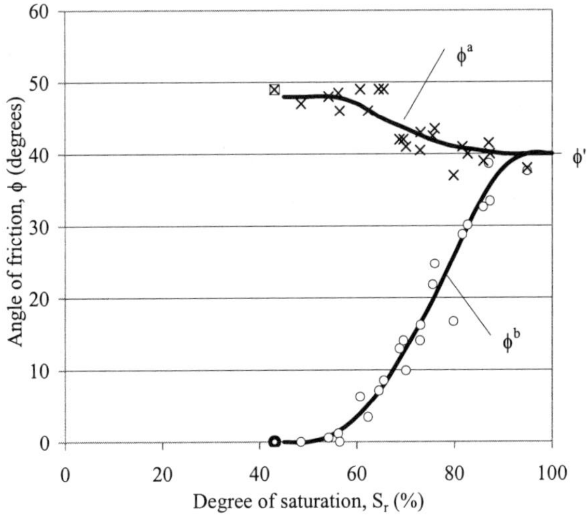

Figure 4. Critical state angles of friction for Kiunyu gravel

It can be seen from Figure 4 that the friction angle ϕ^a (representing the strength component due to net stress (σ-u_a)) is not a constant. Figure 4 suggests that ϕ^a depends on the degree of saturation of the soil. This can be explained by the fact that the different degrees of saturation represent different soil fabrics, as suggested by the conceptual model outlined in Figure 2.

The presence of a packet fabric in unsaturated compacted soils causes the soil to behave in a coarser fashion than would be justified by the grading. The packets are held together by suction and act as individual larger particles. The larger effective particle size produces more frictional behaviour and hence a higher angle of friction. This means that for samples compacted dry of OMC ($S_r < 90\%$) ϕ^a is higher than in saturated conditions (ϕ'). However, at OMC, or wet of optimum, ϕ^a is equal to ϕ'.

Figure 4 also suggests that the angle ϕ^b (representing the strength component due to suction (u_a-u_w)) is also affected by the packet fabric. For samples compacted dry of OMC the water phase retreats into the packets. Therefore the contribution to strength from suction reduces. ϕ^b becomes significantly lower than ϕ' and at low water contents drops to zero. However, at optimum moisture content, or wet of optimum, ϕ^b is equal to ϕ'.

It is also possible that the friction angles could be affected by void ratio, although it would not be expected that critical state angles would show such an effect. The test results were determined from 'end of test' conditions when no significant changes in deviator stress or suction were observed (Toll, 1988). However, in some cases, changes in volume were still observed, and therefore the effect of void ratio should be considered. Figure 5 shows a plot of the angles of friction plotted against void ratio at the end of each test.

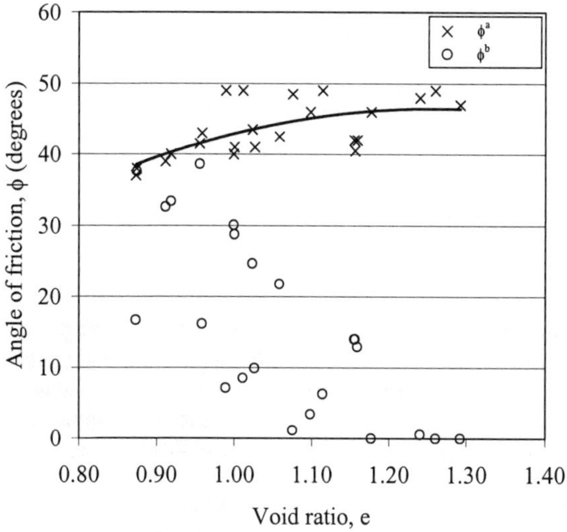

Figure 5. Variation of critical state angles of friction with void ratio

Figure 5 does show a relationship between void ratio and ϕ^a. However the trend is for ϕ^a to increase with increased void ratio (i.e. lower density) whereas a decrease in angle of friction would be expected for an increase in void ratio. Therefore it is clear that the variation in ϕ^a is due to the effects of fabric (as represented by degree of saturation) rather than void ratio per se. The fact that a trend with void ratio exists is due to the fact that there is close relationship between degree of saturation and void ratio, as would be expected for samples compacted on the dry side of optimum. Samples compacted further away from optimum have a lower degree of saturation and achieve a lower density.

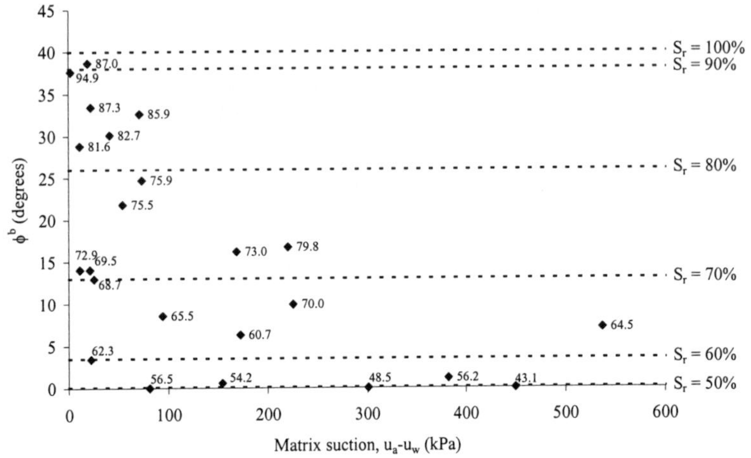

Figure 6. Variation in ϕ^b with matric suction. Values of degree of saturation (%) are given alongside the data points.

The data for ϕ^b shown in Figure 4 is also show plotted against the matric suction at critical state conditions in Figure 6. The values of the corresponding degree of saturation is given alongside each data point. It can be seen that similar values of ϕ^b are obtained for a wide range of suction values. For example, values of ϕ^b of 1° or less were measured in five samples with matric suctions ranging from 80 kPa to 450 kPa. The common element in all these tests is that the degree of saturation is around 50%. Another example is two tests that show a similar value of ϕ^b of 7-8° even though the suction values are 94 kPa and 537 kPa. Again the common element is a degree of saturation of around 65%. Therefore it can be seen that it is the degree of saturation (as a representation of fabric) that is the controlling factor for ϕ^b rather than the matric suction values.

Bands for degree of saturation have been drawn in Figure 6 which show a clear grouping of data points by degree of saturation values. These bands are consistent with the relationship shown by Figure 4.

Discussion

The results presented here are for a clayey gravel with a clay fraction of less than 10% (Table 1). However, it is clear that the role of fabric is still important in this type of material. Fabric studies of compacted Kiunyu gravel by Toll *et al* (1987) showed that the coarser grains (gravel and sand size) were surrounded by a matrix of silt and clay. At microfabric level it could be seen that coarse silt particles were surrounded by finer silt particles and clay. The finer material was therefore very well distributed throughout the soil fabric. This means that a predominantly granular material with only a small amount of clay can be subject to similar fabric effects as those observed in more clayey materials.

It is clear from Figure 4 that the assumption that ϕ^a is equal to ϕ' will not be correct for soils that exhibit significant fabric differences between the unsaturated and saturated states. The packet fabric in compacted soils is maintained by suction, and provides the intra-packet strength that resists packet breakdown during shear. However, the same material, compacted under the same conditions but then subjected to saturation, will demonstrate a loss of the intra-packet strength and the packets will be easily broken down during shear. Therefore the observation that ϕ^a "appears to be essentially equal to the effective angle of internal friction obtained from shear strength tests on saturated soil specimens" (Fredlund and Rahardjo, 1993: p. 238) must be viewed with caution.

It is normal for the shear strength properties of unsaturated soils to be measured in multi-stage tests in which the suction is increased from an initially saturated (or low suction) state (Ho and Fredlund, 1982; Gan and Fredlund, 1988; Fredlund and Rahardjo, 1993). If this procedure is followed for a compacted soil, the packet fabric is likely to be destroyed in the first stage, when the soil is sheared under low suction. Therefore, the true behaviour of the compacted soil, at its compacted water content, will never be properly observed. It is for this reason that many of the results published seem to suggest that $\phi^a = \phi'$. Therefore, for a proper understanding of the behaviour of compacted soils (or other soils with a distinct fabric supported by suction) the soil should be tested at its initial state before altering the suction to other levels.

The variation in ϕ^b shown in Figure 4 should not be confused with the non-linearity of the failure envelope in the u_a-u_w plane as shown by Escario and Saez (1986) and Fredlund *et al* (1987). This latter non-linearity is a result of changes in degree of saturation (or volumetric water content) as the matric suction is changed (Fredlund *et al*, 1995). The variation shown in Figure 4, although expressed in terms of degree of saturation, is largely due to fabric differences. A further change in ϕ^b would be expected if the suction was varied after compaction.

It is interesting to speculate what changes in ϕ^b would occur in a compacted soil if the suction were altered from its initial state. If the suction were increased by drying, then the packet fabric would be sustained and the only factor producing a change (reduction) in ϕ^b would be the reducing degree of saturation (volumetric water content). However, if the suction were to be reduced by wetting, this could induce a collapse of the initial fabric (either due to the wetting process itself, or in subsequent shearing). In this case ϕ^b could increase quite considerably, and by a greater amount that would be expected based purely on the volumetric water content changes.

Conclusions

Clayey materials compacted dry of optimum moisture content develop an aggregated or 'packet' fabric. These effects can be observed even in a predominantly granular material that has less than 10% clay. It is suggested that the degree of saturation of a compacted soil can give an indication of the amount of aggregation present.

The presence of aggregations causes the soil to behave in a coarser fashion than would be justified by the grading. This means that the friction angle ϕ^a (representing the strength component due to net stress) depends on the fabric of the soil. Dry of optimum moisture content, $\phi^a > \phi'$ (the angle of friction in saturated conditions).

The packet fabric also affects the angle ϕ^b (representing the strength component due to matric suction). Dry of optimum moisture content, ϕ^b becomes significantly lower than ϕ' and at low water contents drops to zero. The water phase withdraws into the packets and suction makes no contribution to the overall strength of the soil.

The normal practice for determining the shear strength properties of unsaturated soils has been to perform multi-stage tests in which the suction is increased from an initially saturated (or low suction) state. If this procedure is followed for a compacted soil, the packet fabric is likely to be destroyed in the first stage of shearing, when the soil is sheared under low suction. It is for this reason that many of the results published seem to suggest that $\phi^a = \phi'$. Therefore, to properly study the shear strength behaviour of unsaturated compacted soils, tests should be performed on the soil in its 'as compacted' condition.

References

Atkinson, J.H. (1993) *The Mechanics of Soils and Foundations,* London: McGraw Hill.

Barden, L. and Sides, G.R. (1970) *Engineering Behaviour and Structure of Compacted Clay,* Proc. ASCE, Vol. 96, SM4, pp 1171-1200.

Brackley, I.J.A. (1973) *Swell Pressure and Free Swell in a Compacted Clay*, Proc. 3rd Int. Conf. Expansive soils, Haifa, Vol. 1, pp 169-176, Jerusalem: Academic Press.

Brackley, I.J.A. (1975). *A Model of Unsaturated Clay Structure and its Application to Swell Behaviour,* Proc. 6th African Conf. Soil Mechanics and Foundation Engineering, Durban, Vol. 1, pp. 65-70, Rotterdam: Balkema.

British Standard 1377 (1990) *British Standard Methods of Test for Soils for Civil Engineering Purposes,* London: British Standards Institution.

Collins, K. (1983) *Scanning Electron Microscopy of Engineering Soils,* Geoderma, Vol. 30, pp. 243-252.

Croney, D., Coleman, J.D. and Black, W.P.M. (1958). *Movement and Distribution of Water in Relation to Highway Design and Performance,* Highway Research Board Special Report No. 40, Washington: Highway Research Board.

Delage, P., Audiguier, M., Cui, Y-J, and Howat, M.D. (1996) *Microstructure of a Compacted Silt,* Canadian Geotechnical Journal, Vol. 33, pp 150-158.

Escario, V. and Saez, J. (1986) *The Shear Strength of Partly Saturated Soils,* Geotechnique, Vol. 36, No. 3, pp. 453-456.

Fredlund, D.G., Morgenstern, N.R. and Widger, R.A. (1978). *The Shear Strength of Unsaturated Soils,* Canadian Geotechical Journal, 15, pp 313-321.

Fredlund, D.G., Rahardjo, H. and Gan, J.K.M. (1987). *Non-Linearity of Strength Envelope for Unsaturated Soils,* Proc. 6th Int. Conf. Expansive Soils, New Delhi, pp 49-54, Rotterdam: Balkema.

Fredlund, D.G. and Rahardjo, H. (1993) *Soil Mechanics for Unsaturated Soils*, New York: Wiley.

Fredlund, D.G., Xing, A., Fredlund, M.D. and Barbour, S.L. (1995) *The Relationship of the Unsaturated Soil Shear Strength to the Soil-Water Characteristic Curve,* Canadian Geotechnical Journal, Vol. 32, pp. 440-448.

Gan, J.K.M. and Fredlund, D.G. (1988) *Multistage Direct Shear Testing of Unsaturated Soils,* ASTM Geotechnical Testing Journal, Vol. 11, No. 2, pp. 132-138.

Grace, H., Hight, D.W. and Toll, D.G. (1985) *Some Recent Work on the Use of Locally Occurring Materials for Bases of Bituminous Surfaced Low Volume Roads,* Proc. 1st Int. Conf. Geomechanics of Tropical, Lateritic and Saprolitic Soils, Brasilia, Brasil, Vol. 2, pp 295-307.

Ho, D.Y.F. and Fredlund, D.G. (1982) *Increase in Strength due to Suction for Two Hong Kong Soils,* Proc. ASCE Specialty Conf. Engineering and Construction in Tropical and Residual Soils, Hawaii, pp. 263-295.

Road Research Laboratory (1952) *Soil Mechanics for Road Engineers,* London: Her Majesty's Stationery Office.

Toll, D.G. (1988). *The Behaviour of Unsaturated Compacted Naturally Occurring Gravel,* PhD thesis, University of London.

Toll, D.G. (1990). *A Framework for Unsaturated Soil Behaviour,* Geotechnique, Vol. 40, No. 1, pp 31-44.

Toll, D.G., Hight, D.W. and Shaw, H.F. (1987) *The Role of Soil Fabric in Determining the Engineering Behaviour of Compacted Lateritic and Quartzitic Gravels*, Zeitschrift für Geomorphologie, Suppl. Bd. 64, pp 133-144.

Zein, A.K.M. (1985) *Swelling Characteristics and Microfabric of Compacted Black Cotton Soil*, PhD thesis, University of Strathclyde, Glasgow.

SHEAR BEHAVIOR OF COMPACTED SILTY LOESS

John J. Bowders[1], J. Erik Loehr[2], and Jacob W. Owen[3]

ABSTRACT

Loess, a uniform-sized, wind-blown, cohesive soil can be problematic in geotechnical engineering applications. Natural deposits have high strength and exhibit stable vertical slopes; however, dramatic collapse can occur when loess becomes saturated. In remolded, compacted applications, the behavior of loess is less well known. Regionally (central USA), embankments of compacted loess can exhibit deformation and slope failure after years of service. The initial phase of a larger effort to characterize the strength behavior of compacted loess is reported herein. This phase is limited to assessing the short-term, undrained shear strength of compacted loess. Undrained strengths of compacted loess were measured to be up to 500 to 650 kPa. Higher compactive effort, dry of optimum water content, maximizes the undrained strength. Modified Proctor energy increased undrained shear strength by up to 2.5 to 3.0 times higher than loess compacted at the same water contents using standard energy. Loess compacted at or greater than optimum moisture content, as determined in the standard Proctor test, had similar undrained strength (100 to 200 kPa) regardless of the compactive effort. In order to maximize short term strength of compacted loess, use the greatest compactive effort and compact the soil well dry of optimum moisture content.

INTRODUCTION

Loess is a uniform-sized, cohesive wind-blown sediment found widespread across the central United States (Figure 1). Particle sizes typically range between 0.01 to 0.05 mm. The cohesion results from small amounts of clay or calcareous constituents. In natural deposits, loess maintains high vertical hydraulic conductivity due to the presence of extensive root holes and structure features of the loess. Loess is well-known for its ability to maintain vertical slopes for extended periods of time and also for its sometimes sudden collapse when inundated. The engineering properties and behavior of natural deposits of loess are well documented although deposits can have significant variability due to wind deposition (MSHD 1978). However, the engineering behavior of remolded, compacted loess is under reported.

[1]Associate Prof, Civil & Environmental Engrg, University of Missouri, Columbia, MO 65211-2200
Ph 573.882.8351, Fx 573.882.4784, bowders@missouri.edu
[2]Assistant Prof. Civil & Environmental Engrg, University of Missouri, Columbia, MO 65211-2200
Ph 573.882.6380, Fx 573.882.4784, eloehr@missouri.edu
[3]Undergraduate Research Assistant, Civil & Environmental Engrg, University of Missouri, Columbia, MO 65211-2200

Figure 1 – General location of loess deposits in the United States (Gibbs and Holland 1960)

In Missouri, a substantial portion of the land surface of the state is covered with loess. Development and construction often requires the use of loess in a remolded, compacted state for controlled fills. Embankments whose constituents include recompacted loess have shown significant deformation and in a few cases, slope failure (Figure 2). These embankments have often been in service for up to 30 years before deformation was noted. There is concern that over the long-term, the strength of compacted loess may gradually deteriorate.

(a) (b)

Figure 2 – (a) Typical loess deposit in Missouri exhibiting stable vertical slope. (b) Slope failure in compacted loess embankment.

In this paper, we report on phase 1 of a project to document the shear behavior of compacted loess. The overall objective of the project is to better understand the stability processes in embankments of compacted loess based on complete characterization of the strength parameters of compacted loess. Phase one, reported herein, involves determining the undrained shear strength of samples of compacted loess. Subsequent phases of the project will focus on drained strength behavior.

MATERIALS AND METHODS

Soil samples were taken from a site near Easley in southwestern Boone County, Missouri approximately 18 km southwest of Columbia Missouri (Figure 3). The site is a steep Peoria loess deposit located on the east side of the Missouri River (Krusekopf and Scrivner 1962). The area is severely eroded, revealing vertical faces of the loess deposit. Disturbed samples were taken using a hand shovel from the vertical face and at the base of the face and stored in sixteen 30-L buckets. Care was taken to avoid organic material while sampling the soil. Soil samples were stored in the laboratory in 120 liter plastic containers. The containers used were not air tight as it was not critical in this testing application.

Tests performed to characterize the soil and measure the undrained shear strength of compacted loess are summarized in Table 1. All tests generally followed ASTM standards with the exception of the compaction tests, which were modified to permit compaction of specimens for triaxial testing. The energy levels for compaction tests were recalculated based on the 7.4-cm diameter by 14-cm high mold as shown in Table 1.

Table 1. Summary of laboratory tests performed to characterize and document the behavior of compacted loess.

Test Performed	Procedure	No. of Samples
Standard Proctor	ASTM D698 (627 kN*m/m^3) (13100 ft*lbs/ft^3)	10
Modified Proctor	ASTM D1557 (2,728 kN*m/m^3) (57,100 ft*lbs/ft^3)	10
Reduced Energy (24 kN, 15 blows/ layer, 3 layers)	(338kN*m/m^3) (7100ft*lbs/ft^3)	10
Grain Size Analysis	ASTM D422	1
Atterberg Limits	ASTM D4318	1
Specific Gravity	ASTM D854	1
Unconsolidated-Undrained Triaxial Test (Q-test)	ASTM D2850	30

Specimens for triaxial testing were compacted in 7.4 cm diameter by 14 cm tall molds at five different molding water contents using three different compaction energies, corresponding to compactive efforts consistent with the Modified and Standard proctors and a "reduced" Standard proctor compactive effort. Specimens were prepared at moisture contents of 12%, 14%, 16%, 18%, and 20% to permit evaluation of specimens compacted well below and above the optimum moisture contents for all three compactive efforts. The desired moisture contents were obtained by weighing out 8000g samples, placing these samples in a mixer, and adding water to reach the desired moisture contents. After thorough mixing, samples were sealed in plastic bags, placed in 30 liter buckets with sealed lids, and then stored for a period of at least 24 hours in an environmental room (21 degrees C, 100% relative humidity) to ensure complete hydration of the soil. Following hydration, specimens were compacted in the triaxial molds, extruded, weighed and prepared for triaxial testing. The actual moisture content of each specimen was measured from the excess soil remaining after compaction of the specimen. A total of thirty unconsolidated-undrained (Q-type) triaxial tests were performed at confining pressures of 70 kPa and 140 kPa. Specimens prepared at identical molding water content but different compactive efforts were tested at both confining pressures.

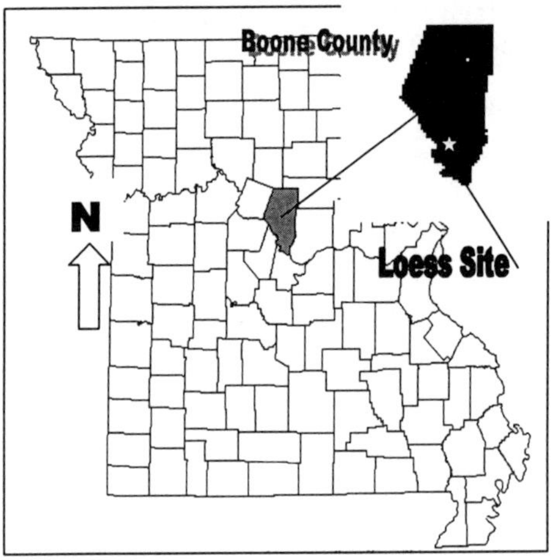

Figure 3 – Site location of Easley loess samples, southern Boone County, Missouri, USA.

RESULTS AND DISCUSSION

The index properties of the Easley loess are given in Table 2. The natural moisture content at the time of sampling ranged from 2% to 6% (dry weight basis). The liquid limit was 35 and the plasticity index was 15, qualifying the soil as a low plasticity silt (ML) according to the Unified Soil Classification System. All of the loess was finer than 0.074 mm (No. 200 sieve) and it contained a clay size (<0.002 mm) fraction of 22% by weight (Figure 4).

The loess was compacted using three different compactive efforts – standard Proctor (627 kN*m/m^3), modified Proctor (2,728 kN*m/m^3) and reduced energy (338kN*m/m^3). The optimum moisture content and maximum dry unit weights for the specimens compacted using standard Proctor energy were 16% and 17.3 kN/m^3, respectively. The optimum moisture content and maximum dry unit weight for the other energies are given in Table 3. As shown in Figure 5, the maximum dry unit weight increased with increasing compactive effort and the optimum water content decreased as typical of most soils.

The degrees of saturation for the specimens tested at 70 kPa confining stress are given in Tables 4 and 5. The degree of saturation increased with increased molding water content. Above 18% molding water content, specimens compacted at all three energies exhibited degrees of saturation of 90% or higher. The specimens compacted with the highest energy maintained the highest degrees of saturation at all water molding water contents, and the degrees of saturation were above 80% for all specimens at the highest energy.

Table 2 – Index properties for Easley loess

Property	Value
In Situ Moisture Content	2%-6%
Specific Gravity	2.68
Atterberg Limits	LL=35, PL=20,PI=15
USCS Classification	ML

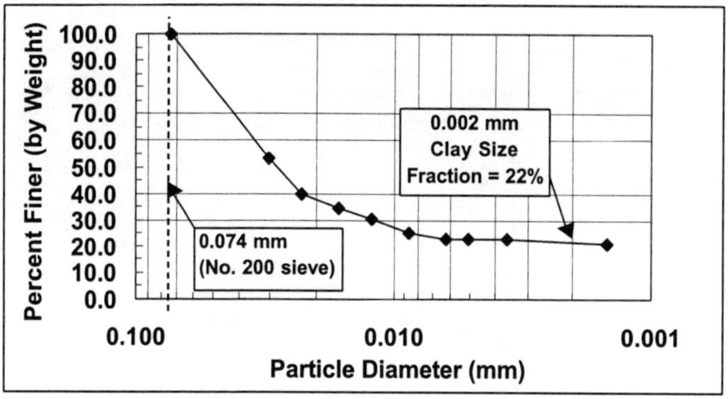

Figure 4 – Grain size distribution of Easley loess.

Typical shear stress-strain curves for the compacted loess during unconsolidated-undrained shearing are shown in Figure 6. The results shown in Figure 6 are for 3 specimens all compacted at a molding water content of 16 percent and tested under a total confining stress of

Table 3 – Optimum moisture contents and maximum dry densities of loess compacted using a range of energies.

Effort		w% opt	γ_{dmax}	
	kN-m/m^3	%	kN/m^3	(pcf)
Reduced	338	18	17.0	108
Standard	927	16	17.3	110
Modified	2728	14	18.8	120

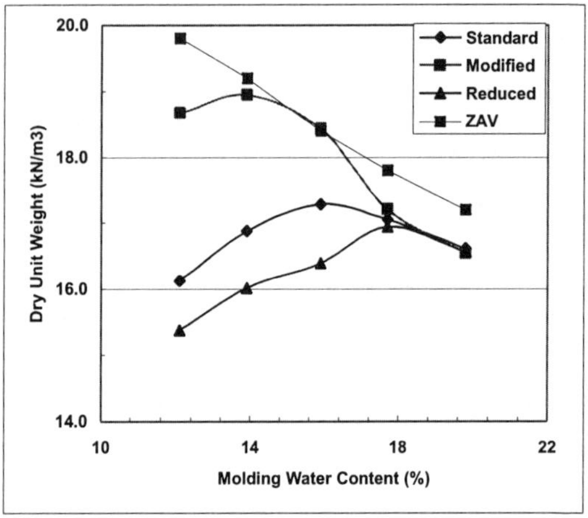

Figure 5 – Dry unit weight versus molding water content for compacted Easley loess (specimens used in the undrained triaxial tests with confining stress of 70 kPa).

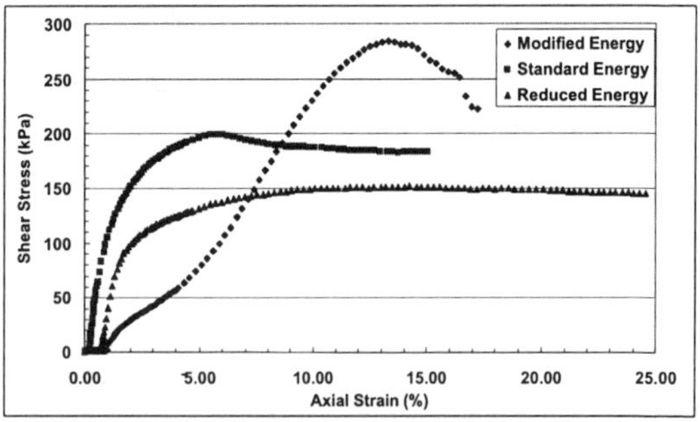

Figure 6 – Typical stress – strain response for compacted Easley loess during unconsolidated-undrained triaxial tests. Results shown are for specimens compacted at a molding water content of 16% and tested under a 70 kPa confining stress.

70 kPa. For all of the testing, specimens compacted wet of optimum moisture content exhibited a stress-strain behavior similar to that shown for the modified energy specimen in Figure 6, i.e., the shear resistance increased as the strain or deformation increased. Specimens compacted wet of optimum exhibited dramatic increases in shear resistance.

The results of the unconsolidated-undrained triaxial tests (Q-tests) performed at confining stresses of 70 kPa and 140 kPa are presented in Tables 4 and 5, respectively. The undrained shear stength obtained from the Q-tests are plotted as a function of molding water content in Figures 7 and 8 for the tests at 70kPa and 140kPa, respectively. These figures show the undrained strength of the compacted loess is higher with increased compactive effort. The greatest strength (550 to 650 kPa) being obtained for specimens compacted at the highest compactive effort (2,728 kN*m/m^3) and the lowest molding water content (12%). The ratio of the strengths (τ_{peak} Ratio, Tables 4 and 5) of modified or reduced energy to standard energy specimens is plotted in Figures 9 and 10. The strength was up to 2.5 to 3.0 times higher than specimens compacted at the same water content but using standard energy (627 kN*m/m^3).

The strength of the specimens compacted using modified energy decreased with increasing molding water content. The specimens compacted at standard and reduced energy maintained a relatively constant strength for specimens compacted dry of the optimum moisture content for standard energy (16%). Strength decreased for all specimens as the molding water content increased beyond optimum moisture content (16%).

Comparing the strengths for all of the specimens tested at a confining stress of 140 kPa to those tested using 70 kPa shows that the strength at higher confining pressure was slightly greater for all specimens. The specimens were not backpressure saturated prior to shearing, thus the degrees of saturation varied as shown in Tables 4 and 5. The greatest difference in strength (Figs 7 and 8) was for specimens compacted dry of optimum (w = 16%). This corresponds to the greatest difference in degrees of saturation (Fig 6). Consider, for specimens compacted at the same energy and same degree of saturation, application of higher total confining stresses results in higher effective stresses in the specimens and accordingly, increased shear strength. This is nicely shown by comparing the strength results shown in Figures 7 and 8.

A ratio (τ_{peak} Ratio) of the strength at modified or reduced energy to the strength at standard energy was calculated for each specimen compacted at the same molding water content. The results for the specimens tested at 70 kPa and 140 kPa are shown in Figure 9. Undrained shear strengths were 2.5 to 3.0 times greater for specimens compacted very dry of optimum using modified compactive effort. Specimens compacted dry of optimum, using reduced energy exhibited strengths that were about 75 percent of that of the standard energy specimens. All specimens, regardless of compactive effort, exhibited similar strengths when compacted wet of optimum, as illustrated in Figure 9. The dry densities (Fig. 5) of all the specimens compacted wet of optimum are similar and it stands to follow that their strengths should be similar as well.

A ratio of undrained strength measured at a total confining stress of 140 kPa to that at 70 kPa versus degree of saturation is shown in Figure 10. The ratio decreases, i.e., strengths are closer, as the degree of saturation of the specimen increases.

Table 4 – Results of samples prepared for Q-tests using σ_3 = 70 kPa

Sample No.	Compactive Effort	Molding w%	Undrained Shear Strength (kPa)	γ_{dmax} kN/m^3	Sr (%)	τ_{peak} Ratio (--)
11	Standard	12.1	174	16.1	52	1
21	Standard	13.9	201	16.9	67	1
31	Standard	15.9	200	17.3	82	1
41	Standard	17.7	158	17.1	88	1
51	Standard	19.8	105	16.6	91	1
M11	Modified	12.1	542	18.7	80	3.1
M21	Modified	13.9	467	18.9	97	2.3
M31	Modified	15.9	284	18.4	100	1.4
M41	Modified	17.7	138	17.2	90	0.9
M51	Modified	19.8	94	16.5	90	0.9
R11	Reduced	12.1	138	15.4	46	0.8
R21	Reduced	13.9	149	16.0	58	0.7
R31	Reduced	15.9	149	16.4	71	0.7
R41	Reduced	17.7	149	16.9	86	0.9
R51	Reduced	19.8	98	16.6	91	0.9

τ_{peak} Ratio is the ratio of peak strength for modified (or reduced) compactive effort to that for standard compactive effort.

Table 5 – Results of samples prepared for Q-tests using σ_3 = 140 kPa

Sample No.	Compactive Effort	Molding w%	Undrained Shear Strength (KPa)	γ_{dmax} kN/m^3	Sr (%)	τ_{peak} Ratio (--)
12	Standard	12.1	259	16.3	53	1
22	Standard	13.9	266	17.0	68	1
32	Standard	15.9	289	17.5	85	1
42	Standard	17.7	218	17.2	90	1
52	Standard	19.8	135	16.7	92	1
M12	Modified	12.1	625	18.8	82	2.4
M22	Modified	13.9	606	18.9	95	2.3
M32	Modified	15.9	365	18.4	100	1.3
M42	Modified	17.7	197	17.1	89	0.9
M52	Modified	19.8	122	16.5	90	0.9
R12	Reduced	12.1	218	15.5	47	0.8
R22	Reduced	13.9	202	16.0	58	0.8
R32	Reduced	15.9	213	16.4	71	0.7
R42	Reduced	17.7	214	17.0	87	1.0
R52	Reduced	19.8	117	16.5	90	0.9

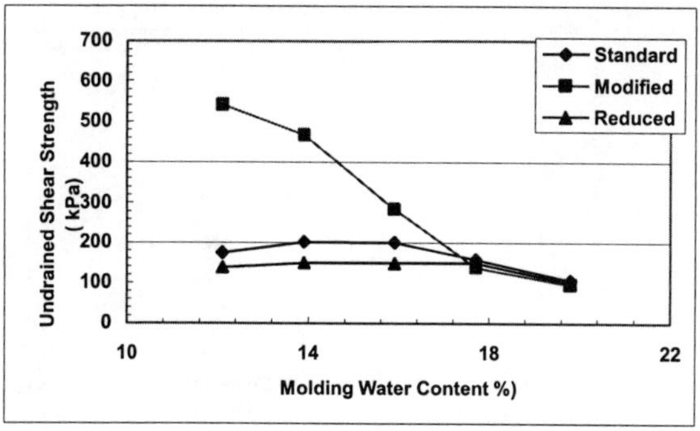

Figure 7 – Undrained shear strength for compacted specimens of Easley loess at a confining pressure of 70 kPa.

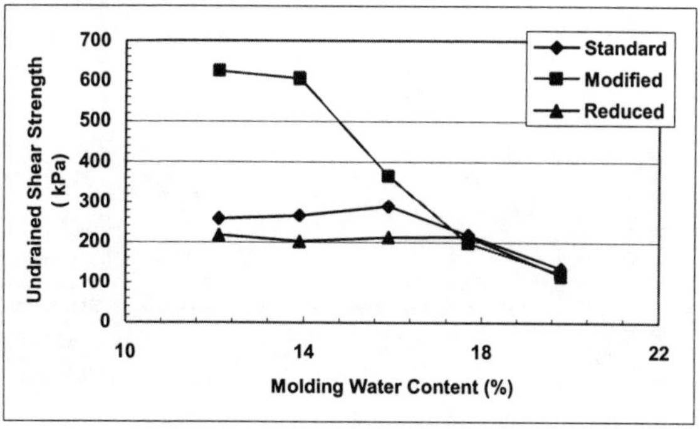

Figure 8 – Undrained shear strength for compacted specimens of Easley loess at a confining pressure of 140 kPa.

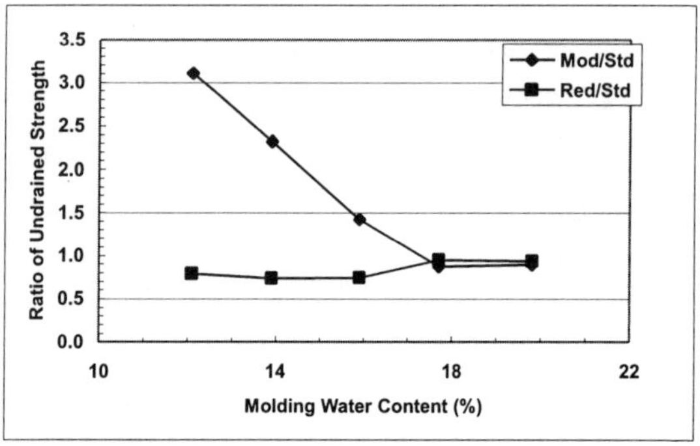

Figure 9 – Ratio of the peak undrained shear strength obtained for modified and reduced energy compaction relative to the undrained strength at standard compaction energy for specimens tested under a 70 kPa confining stress.

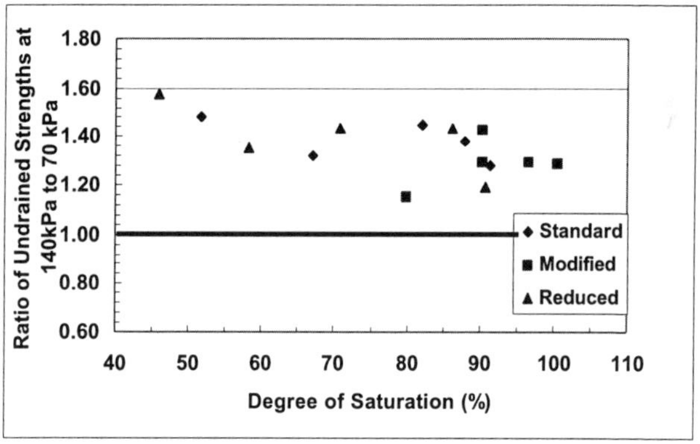

Figure 10 - Ratio of the peak undrained shear stress for specimens compacted at the same molding water content but tested using different confining stresses (140 kPa or 70 kPa) versus the initial degree of saturation of the specimen.

PRACTICAL IMPLICATIONS

Several points for practical application are evident from the results.

- For this particular soil, compacting wet of optimum yielded similar undrained shear strengths regardless of compactive effort used. This is because the range in compactive efforts considered resulted in nearly the same dry density at the higher water contents.

- When compacted wet of optimum, this loess exhibited similar strengths regardless of compactive effort.
- When compacted wet of optimum, the loess exhibited a strain-hardening behavior, i.e., shear resistance increased with deformation over considerably large strains.
- The laboratory results indicate that in order to maximize short term, undrained strength of compacted loess, the loess should be compacted with maximum energy, dry of optimum moisture content; however, in practice quite the opposite has been found to be practicable (Koenig and Potter, 2000):
    - Loess is typically compacted to 85 to 95 percent of standard Proctor density and at a moisture content at or slightly above optimum. This is done for several reasons: it is difficult to compact loess when it is dry of optimum, compacting to higher energies and drier than optimum tends to result in a brittle material which cracks quite readily upon differential settlement, and in the case of impoundments, any cracks are initiation points for piping failures.
    - The Natural Resources Conservation Service (formerly the Soil Conservation Service) has built many successful embankments up to 10 m high at slopes from 2.5H:1V to 3H:1V using compacted loess.

CONCLUSIONS

Loess, a uniform-sized, wind-blown, cohesive soil can be problematic in geotechnical engineering applications. Natural deposits have high strength and exhibit stable vertical slopes; however, the material shows dramatic collapse when it becomes saturated. In remolded, compacted applications, the behavior of loess is less well known. The initial phase of a larger effort to characterize the strength behavior of compacted loess has been reported herein. This phase is limited to assessing the short-term, undrained strength of compacted loess. The following conclusions are presented based on this phase of the project:

- Undrained strength of compacted loess can be high. For the loess used in this study, the maximum strengths were 500 to 650 kPa.
- High compactive effort, dry of optimum water content, maximizes the undrained strength.
- Modified Proctor energy increased the undrained shear strength by a factor of 2.5 to 3.0 over the strength of loess compacted at the same water contents using standard energy.
- Loess compacted at molding water contents equal to or greater than optimum moisture content has similar undrained strength (100 to 200 kPa) regardless of the compactive effort.
- Although laboratory results indicate undrained strength is maximized at higher compaction energies and lower molding water contents, successful field practice is to compact at 85 to 90 percent of standard Proctor energy at water contents at or above optimum.

The next phase of this project is to investigate the long-term strength behavior of compacted loess. Consolidated-undrained triaxial tests with porewater pressure measurements are underway.

ACKNOWLEDGMENTS

The authors are greatly appreciative of Mr. Chee-Keong Boey, former undergraduate research assistant at UMC, for making the extensive preparations for conducting the testing. In addition, the assistance of Professor David Hammer (Soil & Atmospheric Sciences, UMC) and Mr. Dennis Potter (Natural Resources Conservation Service) in obtaining the loess and providing their endless knowledge of soil formation and deposition is appreciated. Finally, we thoughtfully recognize Dr. D. Todd Mooney for his technical review and input on the manuscript.

REFERENCES

Gibbs HJ and Holland WY (1960) *Petrographic and Engineering Properties of Loess*, Engineering Monograph No 28, US Bureau of Reclamation, Denver, pp37.

Koenig R and Potter C (2000) Natural Resources Conservation Service, USDA, Missouri State Office, Columbia, MO, Personal Communication.

Krusekopf HH and Scrivner CL (1962) *Soil Survey of Boone County, Missouri*, Series 1951, No. 12, Washington, U.S. Govt. Print Off..

MSHD (1978) “Development of Design Criteria for Cut Slopes in Loess,” Missouri Cooperative Highway Research Program, *Final Report, No 74-1*, Missouri State Highway Department, pp70.

Design Strength Parameters of a Slope on Unsaturated Gneissic Residual Soil

By Roberto Q. Coutinho, M. ASCE[1]
João B. Souza Neto[2]
Flávio Q. Costa[3]

Abstract

This paper presents basic engineering properties of a gneissic residual soil from a natural slope in which a greater landslide occurred. Emphasis is given to the design strength parameters obtained from laboratory tests. A stability analysis is done taking into consideration unsaturated condition. The landslide causal factors are also presented and discussed. The study results have shown that considering a low value of initial soil matric suction in the field the strength parameters mobilized in the first time failure are characterized by the laboratory results (peak or deformation near to the critical state condition).

Introduction

The State of Pernambuco presents some roads with slope instability problems that periodically require repair and a great cost to the state government. In this context, the Geotechnical Group of the Federal University of Pernambuco established a joint research project with the Pernambuco State Road Department. This project aimed at making geological-geotechnical study of the materials (residual soil/rock), slope stability evaluation, and a proper understanding of the failure mechanism. The site studied is a natural non-homogeneous and unsaturated slope located in the road PE-89, Machados, Pernambuco, Brazil, where a significant landslide occurred years ago (Figure 1). This paper will present and discuss the basic engineering properties with emphasis on design strength parameters of a gneissic residual soil together with the stability analysis results considering the unsaturated condition and the landslide causal factors.

---

[1]Associated Professor, Federal University of Pernambuco, Dept. of Civil Engineering, Recife – Brazil
[2]Doctorate Student, COPPE/UFRJ – Dept. of Civil Engineering / UFPE, Recife – Brazil
[3]Civil Engineering M.Sc., Federal University of Pernambuco, Recife - Brazil

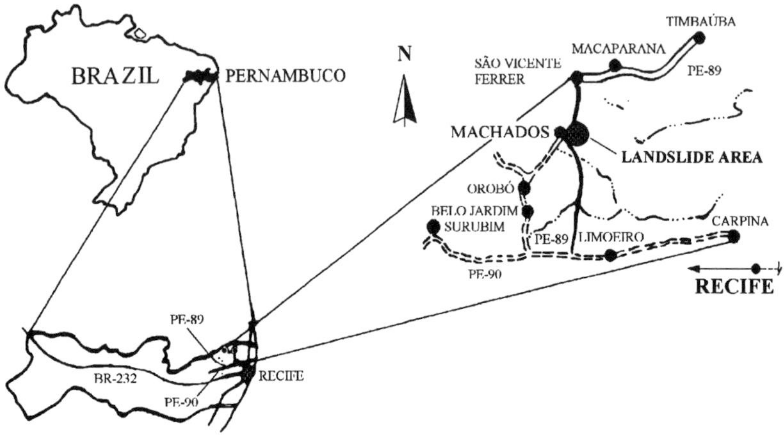

Figure 1. Location of the landslide / site investigated

Site Characterization of The Investigated Area

The natural slope investigated is formed of a gneissic residual soil / weathered rock presenting an elevation of 350 m above sea level and an average declivity of 35°. The landslide presents a ruptured surface having a length of 170 m and an average width of approximately 80 m (Figure 2). The region has a hot and humid climate with rainfall averaging 1200 to 1600 mm per year having a higher intensity from April to august period (wet season). In the year the landslide occurred the rainfall average per month during the wet period was in the order of 200 mm.

The field investigation performed in the landslide was carried out by using topographic profile of cross sections, engineering - geologic mapping, subsurface exploration soil / rock – SPT / RQD, vertical trenches, soil moisture content profiles, field instrumentation - piezometer, and sampling in soils and in rocks. Figure 2 shows the locations of the activities in the field exploration program. The information presented in this paper is from Areas 1 and 3. More details can be found in Coutinho et al. (1997) and Coutinho et al. (1998).

In the subsurface exploration five boreholes investigation tests were performed, in the investigated area, being four in soil and rock (SPT / Core boring - RQD) in the contour of the landslide area, and one (core boring) in the exposed rock, inside of the landslide area (Figure 2 – Area 1 to Area 6). The procedures were in accord of the Brazilian standards NBR 6484 Dec/1980 and ABGE (1991). Hand augers (dry method) were used to drill the boreholes into the residual soils. Water table was not observed in the borehole investigation.

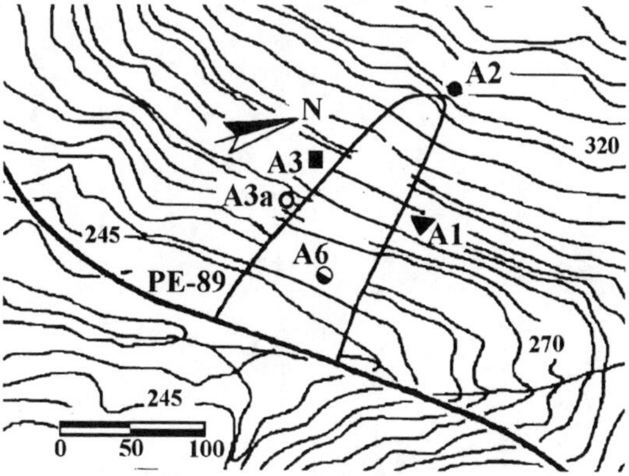

Figure 2. Landslide 1 – Investigated areas (A1 to A6)

Figure 3 presents results of a complete borehole performed in the Area 1 together with some physical and mineralogical properties of the materials. In general, the mature residual soil (horizon B) is composed of porous sandy clay having an average thickness of 2 meters. The young residual soil (horizon C/Cr) is a silty sandy with mica, but still preserving some characteristics of parent rock (foliation of around 30° with horizontal), and thickness that can reach 15 meters. The rock (biotite gneiss) presents a high degree of fracturing with low RQD values in some depths. Similar geotechnical profile was observed in the boreholes performed (Area 2 and Area 3). The results suggest that the failure surface correspond to depth limit of the standard penetration test (the approximated definition of the contact soil-rock).

The moisture content profiles were periodically determined in Areas 1, 2 and 3a, from hand auger samples, to observe the moisture content variation during the dry and wet seasons (Figure 4). The results show the moisture content values between 10% and 24% in general depth decreasing. In the profiles located on the slope the most significant variations occurred at a depth of approximately two meters (mature residual soil), and at a depth of approximately four meters in the profile located on the top of the landslide area.

The hydraulic conductivity of the soils was determined through a field test using the Guelph permeameter. The Guelph permeameter is an in-hole constant-head permeameter, employing the Mariotte Principle, and yields essentially a "point" measurement (Soilmoisture, 1991). For any soil suction (as measured in cm of water), the unsaturated hydraulic conductivity can be predicted by the following equation: $K = K_{fs} [e^{(\alpha)(\psi)}]$, where K_{fs} is the saturated hydraulic conductivity, α is a constant, which is dependent on the porous properties of soil, $\alpha = K_{fs} / \phi_m$; and ψ is soil water suction in cm of water. The results of the tests (Figure 5) show that the mature residual soil, a sandy clay presents a higher saturated hydraulic conductivity (K_{fs}=1.1 ± 0.48 x 10^{-5}m/s) than the young residual soil, a silty sand (K_{fs} = 0,41 ± 0,15

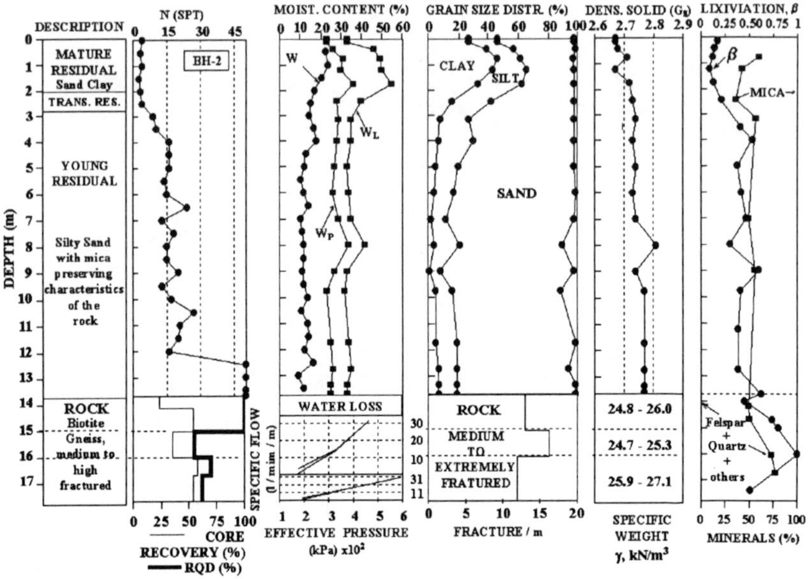

Figure 3. Result of a borehole (SPT/RQD) and characteristic properties – Area 1.

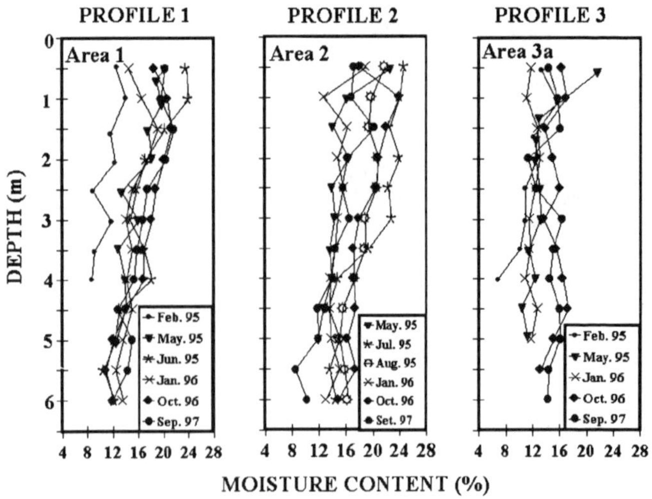

Figure 4. Moisture content profiles – Areas 1, 2 and 3a

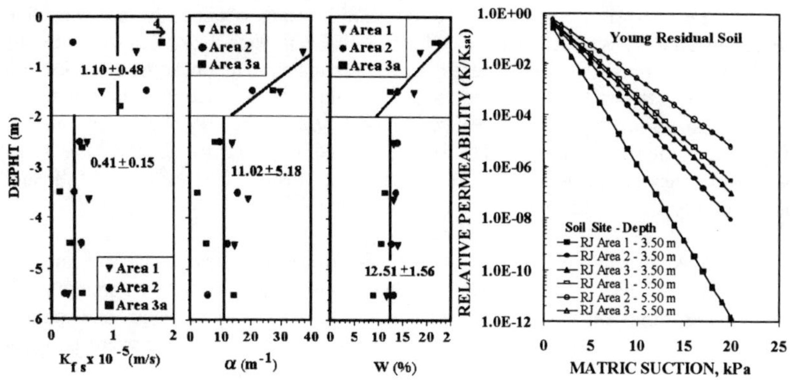

Figure 5. Results of permeability test (Guelph permeameter)

x 10^{-5} m/s). This behavior is due to the granular and porous structure of the mature residual soil. The variation with suction of the relative hydraulic conductivity to the young residual soil (K_r=K/K_{fs}) indicate a strong reduction in the K_r values to small values of suction in this soil influencing the flow behavior of the slope.

Geotechnical Properties

An intense laboratory investigation was done to evaluate geotechnical properties of the residual soils using disturbed and undisturbed – block samples. A summary is presented in this paper with emphasis on the strength parameters. More details can be found in Coutinho et al. (1998) and Souza Neto (1998).

The results of the characterization test in the mature residual soil showed clay fraction around 40% and a plasticity index in the interval of 12 to 20%, being it classified as CL/MH in the Unificated Classification. The young residual soil presented around 70% of sand fraction, plasticity index range of 6 to 16 %, being classified as SM/SC with a significant presence of mica. The plasticity index in the young residual soil is high considering the clay fraction in the order of 4% and the kaolinite as the predominant clay mineral. This fact may be related to the high mica content (43 to 51 %) in this material (Figure 3).

Due to limitation in the traditional geotechnical classification applied to the tropical residual soils, new methods have been proposed. Considering the Vargas (1992) method the samples of the young residual soil were basically classified in the SM/ML_m group (silty sand with mica) and this classification showed a strong relationship with the geotechnical and mineralogical characteristics of these materials. Under the MCT classification (Nogami & Villibor, 1985) the majority of these samples (young residual soil) were classified between groups NS' (non-lateritic silty soils) and NA' (non-lateritic sandy soils). These results are also in accordance with the geotechnical behavior of the soil.

The soil-water characteristic curve (SWCC) was obtained by pressure plate test using four or five specimens from block samples to each curve. Figure 6 presents the results obtained from the different soils investigated. The observed behavior is in accordance with the texture of the soils. The values of the suction in the field were determined during June-August/95 period (wet season) by the filter paper method from the young residual soil. The results ranged from 10 to 80 kPa. The initial suction of the specimens used in the pressure plate tests in laboratory was estimated in the range of 20-70 kPa similar to the field values.

Conventional direct shear tests were performed by using square specimens of 101.6 mm side and 25 mm of height to obtain drained shear strength parameters. The tests were done on submerged specimens prepared perpendicular and parallel directions to the foliation, and on natural moisture content specimens prepared parallel direction to the foliation plane. In this stage of the research the anisotropy investigation was done only in submerged specimens, because the equipment used in this laboratory tests program was a conventional direct shear tests that does not have a controlled suction system. Table 1 presents a summary of the initial condition of the specimens used in the tests. These results probably are similar to the field values due to type of sampling (block samples) and the short time of storage in the wet room. Table 2 and Figure 7 present a summary of the strength parameters results obtained in the residual soils from two investigated areas.

The young residual soils showed the effective friction angle (ϕ') values in the range of 29.3° to 31.4° with an average of 30° while the effective cohesion parameter (c') values were in the following ranges: c'=2.8 - 7.0 kPa to the young residual soil from Area 1; and c'=0.0 – 38.3 kPa to the young residual soil from Area 3. The higher c' values from tests with specimens in the natural moisture content, and the lower values to the tests with submerged specimens prepared parallel to foliation, as expected. The strength parameters of submerged specimens prepared parallel to the foliation plane are similar to both areas investigated. However, the other conditions of tests showed some difference in the results (an apparent higher influence of the suction and anisotropy in the young residual soil from Area 3).

It was also considered another condition to the shear strength parameters to include the possibility to have some horizontal deformation before general failure occur (see Stability Evaluation). In this case the shear strength parameters were obtained considering horizontal deformation higher than 10 mm in the direct shear tests. This condition probably is near to the critical state (see Figure 8). The results for the strength parameters were: c'=0 kPa (both Areas) and ϕ'=25° (Area 1) and ϕ'=30° (Area 3).

Sandroni (1977) (from Lacerda & Almeida 1995) studied the gneissic residual soils and found that the mineralogy of the sand fraction, including the mica and feldspar proportions, correlates well with the soil strength which tends to be higher as the feldspar content increases and the mica content decreases (Figure 9).

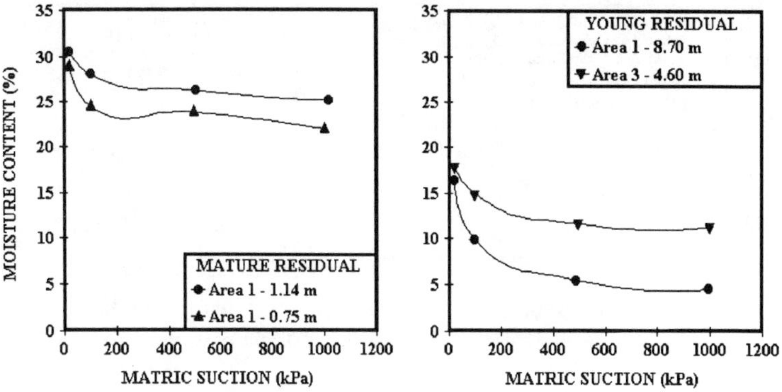

* Note: Initial condition in the same order of the Table 1

Figure 6. Soil-water characteristic curve (SWCC)

Table 1 – Initial condition of the specimens – conventional direct shear.

Site	Depth (m)	Foliation	Test Condition	W_o (%)	e_o	γ_d (kN/m^3)	S_o (%)
Area 1 (MR)	0.75	---------	Submerged	16.3-20.7	0.85-1.04	16.3-17.4	53.0-60.0
		--------	Nat. M. C	17.5-19.8	0.85-1.05	16.1-17.8	52.1-61.1
Area 1 (YR)	8.70	Parallel	Submerged	15.3-16.9	0.83-0.89	16.7-17.4	45.9-53.1
		Perpend.	Submerged	13.1-14.4	0.74-0.82	17.3-17.7	48.7-52.9
		Parallel	Nat. M. C	14.7-16.5	0.87-0.94	15.7-16.9	46.8-48.6
Area 3 (MR)	1.90	---------	Submerged	23.1-25.9	1.19-1.33	14.4-15.3	47.9-55.5
		---------	Nat. M. C.	22.3-13.1	1.26-1.48	13.7-15.0	42.3-50.6
Area 3 (YR)	5.60	Parallel	Submerged	13.8-17.1	0.83-0.96	16.7-170.8	48.6-57.6
	4.60	Perpend.	Submerged	17.3-18.9	1.02-1.13	15.2-16.2	43.4-50.9
		Parallel	Nat. M. C.	15.2-16.0	1.06-1.15	14.9-15.7	36.6-42.2

* Note: G_s in the order of 2.74 to 2.83 to mature residual soil and 2.75 to 2.81 to young residual soil

Table 2 – Strength parameters – conventional direct shear

Site	Soil	Depth (m)	Foliation	Test Condition	c' (kPa)	ϕ' (°)
Area 1	Mature Residual	0.75	---------	Submerged	10.8	31.9
			--------	Nat. M. C	71.4	41.6
	Young Residual	8.70	Parallel	Submerged	2.8	29.3
			Perpendicular	Submerged	5.3	31.4
			Parallel	Nat. M. C	7.0	29.5
Area 3	Mature Residual	1.90	---------	Submerged	0.0	35.5
			---------	Nat. M. C.	15.8	37.8
	Young Residual	5.60	Parallel	Submerged	0.0	30.5
		4.60	Perpendicular	Submerged	10.2	30.3
			Parallel	Nat. M. C.	38.3	31.2

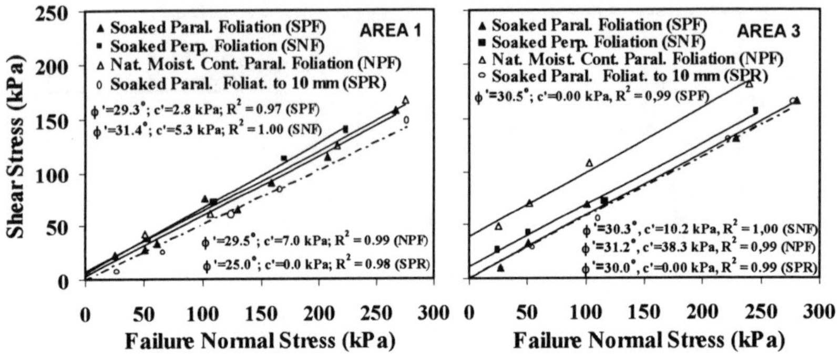

Figure 7. Results of conventional direct shear tests in the young residual soil

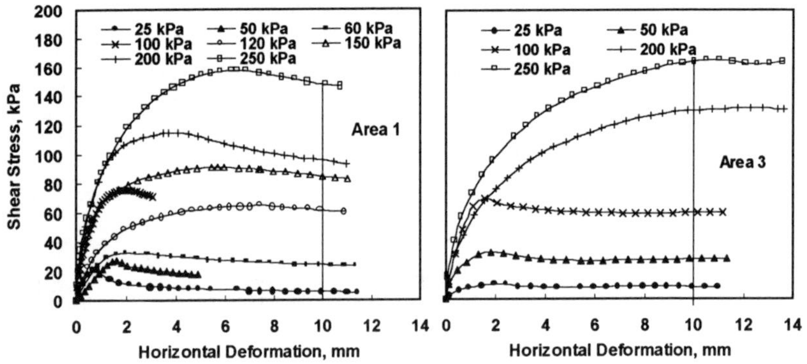

Figure 8. Curve deformation vs. shear stress to young residual soil – direct shear tests in submerged specimens prepared in the foliation parallel direction.

Mineralogy study in the sand fraction of the young residual soil (see Figure 3) from Area 1 showed 40 to 50% of mica in the soil. Similar results are probably to occur in the other areas investigated. The results (c', ϕ') obtained in this study in the young residual soil from Area 1 and Area 3 are in agreement with Sandroni (1977) investigation (Figure 9).

The submerged specimens prepared perpendicular to foliation showed higher values of c', indicating some influence of the anisotropy in this soils (Figure 10). Considering the in situ total stress, the increase in the shear strength is 11% in Area 1 and 17% in Area 3, although results of the literature suggest low or no influence of the anisotropy. The parallel directions to the foliation plane to the slope direction (adversely oriented mass discontinuities) plus the high mica content are potentially predisposition factors.

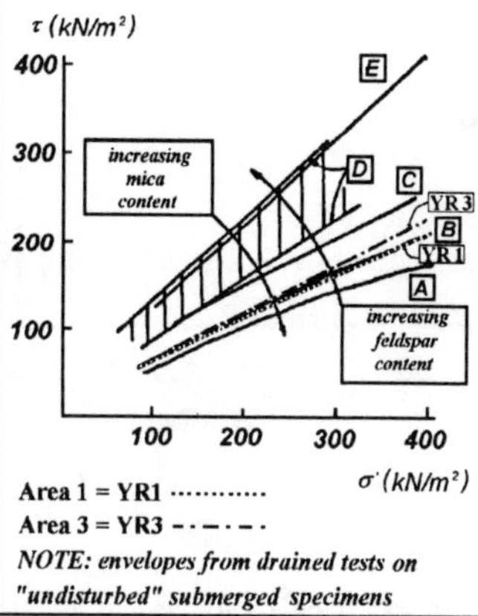

AUTHOR SITE	GRAIN SIZE (%)			VOIDS RATIO	STRENGTH ENVELOPE	COARSE GRAINS MINERALS
	SAND	SILT	CLAY			
CAMPOS (1974), Catingueiro	92	5	3	0.76	CURVED A	Q: 30 - 60 F ~ 0 M: 40 - 70
VARGAS (1951), Vila Anastácio	40	44	16	1.0	CURVED B	High mica content some feldspar
SERAPHIM (1974), Estrada de Furnas	75	19	6	0.95	$c' = 40\ kN/m^2$ $\phi' = 29.5°$ C	Q: 48 - 60 F: 11 - 23 M: 22 - 28
SANDRONI (1973), Rio	50 to 80	15 to 30	5 to 10	0.30 0.90	$c' = 17$ to $35\ kN/m^2$ $\phi' = 34°$ to $43°$ D	Low mica content
SERAPHIM (1974), Estr. de Furnas	83	13	4	0.70	$c' = 25\ kN/m^2$ $\phi' = 43°$ E	Q: 48 - 58 F: 30 - 42 M: 6 - 14

NOTE: Q = QUARTZ F: FELDSPAR M: MICA

Figure 9. Influence of the mineralogy in the soil strength (After Sandroni, 1977).

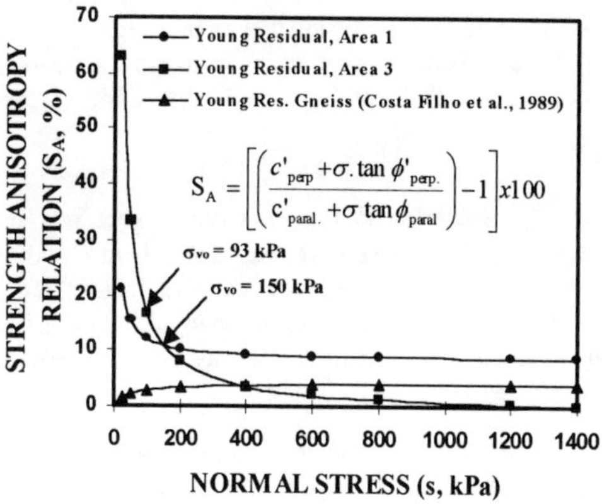

Figure 10. Anisotropy of the young residual soil - Submerged direct shear tests.

More recently direct shear tests with controlled suction were performed in the young residual soil from Area 1 for valuation of the influence of the suction in the strength parameters. The time allowed for equilibrium to each sample was in general 8 days and the strain rate was 0.0003 mm/sec. The results obtained using suction value of 25 kPa are presented in the Figure 11 and show strength parameters c'=42.3 kPa and ϕ'=24.1°. This values can be inconsistent with the results from the conventional tests, because to the difference in the initial condition of the specimens (lower water content and higher density). This fact is probably due to a higher time of storage in the wet room.

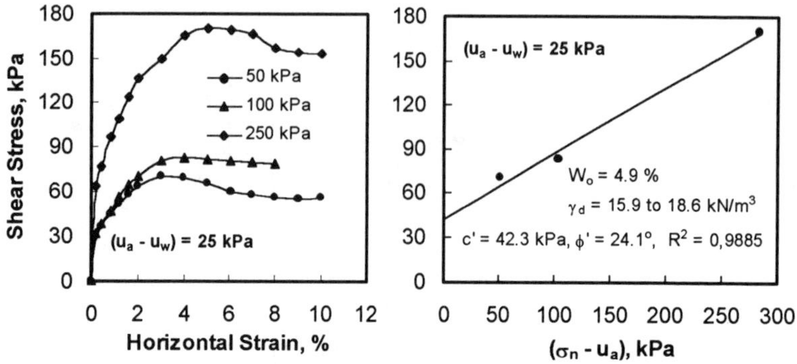

Figure 11. Results of direct shear tests with controlled suction – young residual soil (Area 1)

The study of the collapsibility (Coutinho et al., 1997) established that the soils can be classified as a non-collapsible soil ($\Delta e/(e_o+1) < 2\%$) if they meet natural condition and $\sigma_v=\sigma_{vo}$ indicating that this phenomena cannot be considered as a predisposition factor to the landslide occurrence. In the same way oedometer laboratory tests results suggest that the residual soils investigated can be classified as a non-expansive soil. The maximum swelling deformation after to be submerged under a small stress of 1 kPa was 0.5% of the initial height of the specimen.

Stability Evaluation

The slope stability evaluation in this research project was done as a two-dimensional unsaturated stability analysis. It was divided into two steps: 1) slope flow analysis for the rainfall history just before the failure; 2) slope stability evaluation using a limited equilibrium analysis (Sarma, 1979), considering the pore pressure distribution obtained from the flow analysis. More information about this analysis is presented in Costa (1996).

Two conditions were considered in the analysis: the cross section before the road construction (slope natural condition) and the cross section after the construction (failure condition). For the failure condition, two hypotheses were considered for the landslide mechanism. In the first case, the landslide occurred in only one step, which means one complete cross section for all slope (cross section 2 - Figure 12a). In the second, the landslide occurred in two steps: one small landslide close to the foot of the slope, and the other one giving continuity to the top of the slope (cross sections 10-8 - Figure 12a). Figure 12b presents the typical soil profile / cross section for the case of landslide in only one step (section 2).

Flow Analysis

In the flow analysis due to the unknown initial suction in the slope during the failure period, a parametric study was done considering a constant value for the initial suction in the entire slopes in the range of 4 to 60 kPa. These values of suction are basically in the range observed in the field by filter paper method during the wet season. The geotechnical parameters (hydraulic relationship between moisture content and suction, etc) used in the analysis were those described before, taking into consideration the saturated hydraulic conductivity of the saprolite (Ks – layer below the young residual soil – see Figure 12b) higher or equal to the hydraulic conductivity of the young residual soil (Ks ≥ Kyr; Ks > 5 x 10^{-5} m/s or Ks = 5 x 10^{-5} m/s, respectively).

The results of the analysis showed basically two types of behavior: i) for initial suction less or equal to 10 kPa, the rain infiltration reached the top of the rock (adopted as the impervious surface), saturating this part of the slope and causing positive pore pressure along this impervious surface; ii) for initial suction higher than 10 kPa, the rain infiltration did not saturate the region of the impervious surface, keeping the entire slope in an unsaturated condition with negative pore pressure.

Stability Analysis

The drained shear strength parameters adopted in the stability analysis were a function of the results obtained from conventional direct shear tests of the mature and young residual soils. Figure 13a presents the hypotheses considered in the study to the drained shear strength in the saturated condition for all soils adopting c'=20 kPa and ϕ'=45° to the saprolite. The range of values used for c' (MR and YR soils) is inside of those observed in the direct shear drained test (Figure 7) considering a low value for suction. The saprolite (or very weathered rock) is the layer below the residual soil (Figure 12b) just after the depth limit of the SPT (the approximate definition of the contact soil – rock). The values adopted for the strength parameters were considering it a stronger material than the young residual soil above. The "ϕ^b parameter related to the matric suction (Fredlund et al., 1978) was adopted equal to ϕ', in function of the small values of suction observed in the field during the wet season. The research of the critical sliding surface was done considering a non-circular surface varying inside a region that included the observed failure surface.

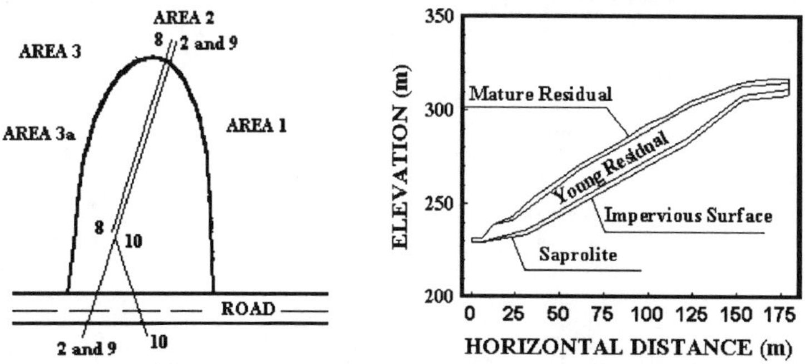

Figure 12. a) Cross-sections-stability analysis, b) Typical soil profile-cross section 2.

Figure 13b shows the results of the critical safety factor for the landslide case in only one step (cross section 2) for the different hypotheses for the cohesion and suction investigated and considering Ks > Kyr. The behavior of the FS shows a continuous reduction with the reduction of the initial suction considered in the analysis. The reduction rate in the FS increases for suction values equal or smaller to 10 kPa (Figure 13b), probably due to the positive pore pressure occurrence in the slope during the rain infiltration. For the case presented here, it can be also observed that in the conditions of c'=0 to 5 kPa (hypotheses a and b), the FS is very close to 1.0 for initial suction in the order of 7 to 10 kPa. In the same conditions as above the FS results for the two-step landslide case (cross sections 10-8 – Figure 12a) were very similar to the one presented, pointing out the possibility of occurrence of both landslide mechanisms.

The critical sliding surface obtained in all conditions considered in the stability analysis presented a similar behavior. All of them were closed to the failure surface observed in the field. The slope stability evaluation for the cross section 9 before the road (Figure 12a) showed that for initial suction ≤ 10 kPa, the natural condition FS was in the order of 15% higher than the FS after the construction of the road (Figure 13b). Such result characterizes the possibility of low natural FS occurrence and a significant landslide hazard in the area. This result is an indication that the road cut was an aggravating or a triggering factor in the landslide.

Back analysis

A back-analysis was done considering apparent cohesion $c^* = c' + (u_a - u_w) \tan\phi^b$ to find which strength parameters can explain the failure and to evaluate the strength parameters from laboratory tests. Table 3 presents a summary of the results obtained to the section 2 (Figure 12a). The results show that strength parameters

Hypotheses for cohesion	Cohesion -c' (kPa)		
	MR	YR	SP
(a)	0,00	2.88	20.00
(b)	5.00	5.00	20.00
(c)	10.00	10.00	20.00
(d)	15.00	15.00	20.00
MR = Mature Residual,	$\phi' = 30°$		
YR = Young Residual,	$\phi' = 30°$		
SP = Saprolite,	$\phi' = 45°$		

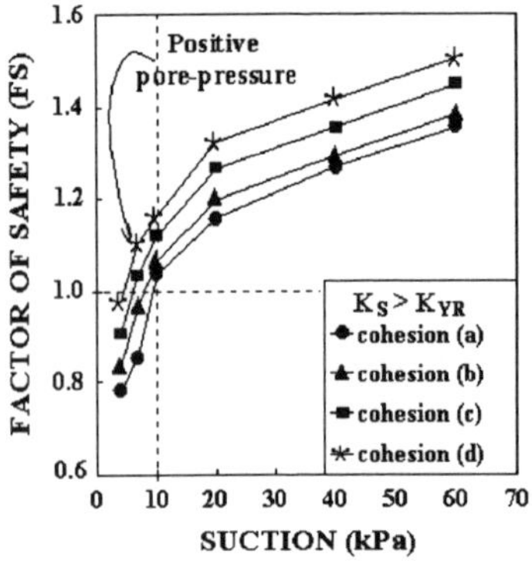

a) b)

Figure 13. a) Strength parameters considered; b) FS-Suction Relationship/section 2 obtained from back-analysis were similar to those determined from laboratory tests (peak or near to the critical state), indicating that results from laboratory tests can explain the failure of the slope considering low value of initial suction.

Table 3. Summary of the results of back-analysis to the section 2

Friction angle, ϕ' (^{o})							
28.5		29.0		29.5		30.0	
c* (kPa)	FS	c* (kPa)	FS	c* (kPa)	FS	c* (kPa)	FS
5	1.019	3	1.007	1	1.003	1	1.024
7	1.046	5	1.039	3	1.024	3	1.044
--------	---------	--------	--------	5	1.053	5	1.073

Commentaries

Leroueil et al. (1996) proposed a geotechnical characterization of slope movements considering four distinct stages: a) the pre-failure stage, when the soil is essentially overconsolidated and intact; b) the onset of failure characterized by the formation of a continuous shear band or surface through the entire soil mass; 3) the post-failure which includes movement of soil mass involved in the landslide, from just after failure and until it essentially stops; and 4) the reactivation stage, when the soil mass slides along one or several pre-existing shear surface; this reactivation can be occasional or continuous with seasonal variations of the rate of movement. Each of the four stages involves mechanical phenomena, controlling laws and parameters, which are very different.

In first time failure, local failure is initiated when stress state reaches the peak strength envelope corresponding to the age or strain of the slope. After the peak, stress conditions progressively move towards the critical state line, and part of the shear stress is transferred for the neighboring soil elements, which could then reach local failure. This phenomenon of progressive failure can extend into a continuous shear surface trough the entire soil mass and then trigger a landslide. The average strength parameters at failure can be somewhere between the peak and critical state values. For intact clays, the shear strength parameters can be characterized by a friction angle corresponding to critical state, and by an empirical cohesion. In the case of ductile materials, the strength mobilized in situ is close to that obtained in the laboratory tests on specimens of appropriate size so as to include typical features of the materials. In unsaturated soils their strength increases with the matric suction in the soil. Thus, an important point is the permeability or the rainwater infiltration of the unsaturated soil. The authors consider that the failure is the result of a complex process, and the selection of representative in situ parameters is generally difficult.

In the present study, the stability analysis using the peak strength parameters obtained in laboratory tests can explain the failure of the slope. The initial results also show that in order to obtain FS $\cong$ 1.0 it is necessary a low or very low value of suction. This hypothesis is expected to be possible from in situ suction measurement observed in this research and considering the real time condition during an intense or prolonged rain. Considering the strength parameters for horizontal displacement higher than 10 mm (c' = 0 and $\phi' = 25^o - 30^o$; see Figure 7), that means strength parameters close to critical state, the necessary value of the initial suction to obtain FS ≅ 1.0 is a little higher, which it is also possible to occur.

Popescu (1994) and Leroueil et al (1996) presented a method for reporting and investigating landslide causal factors considering ground conditions, geomorphologic processes, physical processes, and man-made processes. In this context causal factors can be classified as: 1) preparatory causal or in situ predisposition factors which make the slope susceptible to movement; 2) triggering or aggravating causal factors which initiate movement; and 3) revealing factors which give an evident evolution of slope. In the present study the probable predisposition factors are the direction of the foliation plane parallel to the direction of the slope (adversely oriented mass discontinuities) and the high mica content in the young residual soil. The excavation of the slope to construct the road is an aggravating or a triggering factor. The intense rainfall or the prolonged high precipitation is also a probable triggering factor through of the reduction in the suction and the possibly development of positive pore pressure.

Conclusion

The basic engineering properties and the strength parameters results from laboratory tests has been in accordance with what is described in literature. It is also observed a small anisotropic behavior in the submerged shear strength. The slope stability analysis shows that it is possible to have a low factor of safety (1.15-1.20) for the natural condition of the slope during the wet season, characterizing a significant landslide hazard in the area. The results of stability analysis indicate that the laboratory strength parameters (peak or close to critical state) may explain the failure with a very low or low value of suction. The probable predisposition factors are the high mica content in the young residual soil and the direction of foliation plane parallel to the direction of the slope (adversely oriented mass discontinuities). For the natural condition, the soils are not collapsible or expansible suggesting that these phenomena cannot be considered a predisposition factor. The aggravating or triggering factors are the excavation of the slope to construct the road and the intense rainfall or prolonged high precipitation.

Acknowledgements

The authors are grateful for all support received from the MCT/FINEP, CNPq-Brazilian Research Council and the Road Department of the State of Pernambuco. Professors Willy A. Lacerda (COPPE / UFRJ), Tácio M.P. de Campos (DEC/PUC-RJ) and Edmilson Lima (DEGEO/UFPE) for their helpful contributions to the project as consultants.

References

Costa, F. Q. (1996) Problemas de Estabilidade de Taludes em Solo Residual de Gnaisse – Análise do Escorregamento do Espinhaço da Gata, PE-89 / Pernambuco. *Tese de MSc.*, Eng. Civil/UFPE, 113p.

Costa Filho, L.M., Döbereiner, L., Campos, T.M.P. & Vargas Jr, E. (1989) Fabric and Engineering Properties of Saprolites and Laterites. *Proceeding of the XII International Conference son Soil Mechanics and Foundation Engineering.* Moscow, vol.4, 2463-2476.

Coutinho, R. Q., Costa, F. Q. & Souza Neto, J. B. (1997) Geotechnical characterization and Slope Stability Evaluation of a Slope in Residual Soil from Pernambuco, Brazil. *2nd Pan-American Symposium on Landslide.* Rio de Janeiro/Brazil, vol. 2, 287-298.

Coutinho, R. Q., Souza Neto, J. B., Barros, M. L. S., Lima & E. S., Carvalho, H. A. (1998) Geotechnical characterization of a young residual soil/gneissic rock of a slope in Pernambuco, Brazil. *2nd International Symposium on The Geotechnics of Hard Soils and Soft Rocks*, vol.1, Naples, Italy, 115-126.

Lacerda, W. A.& Almeida, M. S. S. (1995) State-of-the-Art-Lecture: Engineering Properties of Regional Soils: Residual Soils and Soft Clays. *IX PCSFE.* Guadalajara / Mexico.

Leroueil, S., Vaunat, J., Picarelli, L., Locat, J., Lee, H. & Faure, R. (1996). Geotechnical Characterization of Slope Movements. *Proc. of the Seventh International Symposium on Landslides.* Trondheim, 53-74.

Leroueil, S., & Locat, J. (1998). Slope Movements - Geotechnical Characterization, Risko Assessment and Mitigation. *XI Danube European Conference on Soil Mechanic and Geotechnics Engineering.* Porec, Croacia.

Nogami & Villibor (1985). Additional considerations about a new geotechinical classification for Tropical Soils. *First International Conference on Geomechanics in Tropical Lateritic and Saprolitic Soils.* Brasilia, Brazil, vol.4, 165-174.

Popescu, M. E. (1994). A suggested method for reporting landslide causes. *BULLETIN of the International Association of Engineering Geology.* LCPC, Paris. n° 50, 71-74.

Sandroni, S. S. (1977). Micromorphology and the interpretation of a landslide in gneissic residual soil. *Int. Symp. on the Geotechnics of Structurally Complex formations.* Ass. Geotecnica Italiana. Capri, vol.1, 423-431.

Sarma, S.K. (1979). Stability analysis of embankments and slopes. *Journal of the Soil mechanics and Foundation Division.* ASCE, vol.105. n°. GT12, 1511-1524.

SOILMOISTURE (1991). Guelph Permeameter 2800. Operating Instrctions. Soilmoisture Equipment Corp. Santa Barbara, USA.

Souza Neto, J. B. (1998). Características geotécnicas do solo residual de gnaisse da encosta Espinhaço da Gata, PE-89, Machados - PE. *Tese de MSc.*, Eng. Civil/UFPE, 250p.

Vargas, M. (1992). Identification and classification of tropical soils. *Proc. US/Brazil Geotechnical Workshop on Applicability of Classical Soil Mechanics Principles to Structured Soils.* Belo Horizonte, Brazil, vol.1, 200-205.

Unconfined Compressive Strength of a Silty Soil and Kaolin below the Residual State

T.Nishimura[1]
D.G.Fredlund[2]

Abstract

Arid regions are widely distributed around the world. Soil water fluxes are related to the actual evaporation that is closely related to ground surface temperature and relative humidity. A soil can be dried such that its water content is below residual conditions. Of particular interest in this paper is the shear strength of very dry soils. This paper reports results on the relationship between total suction and shear strength beyond the residual water content state. This study involved performing unconfined compression tests for compacted, unsaturated soils and measuring the soil-water characteristic curves. A silty soil and kaolin were used in this test program. Two types of compacted, unsaturated soils were subjected to total suctions as high as 124,000 kPa and the shear strengths were measured. The soil-water characteristic curves were measured over a wide range of suctions. The residual state for the silty soil is clearly defined. The residual state for the kaolin is difficult to define. The shear strength in the residual state appears to increase slightly. This paper concludes that the failure envelope is essentially horizontal in the residual states.

1. *Introduction*

The fundamental theories for unsaturated behavior have become widely accepted in the geotechnical field. Unsaturated soil regions are generally located near ground surface. These regions are classified with in a saturated/unsaturated soil profile as following; dry soil with a discontinuous water phase, two phase zone

[1]Associate Professor of Civil Engineering, Department of Civil Engineering, Ashikaga Institute of Technology, 268 Omae, Ashikaga Tochigi, Japan.
[2]Professor of Civil Engineering, Department of Civil Engineering, University of Saskatchewan, 57 Campus Drive, Saskatoon, Saskatchewan, Canada.

with a continuous water phase, capillary fringe zone with discontinuous air phase, and saturated zone. The interface between the soil and the atmosphere produces a moisture flux boundary at ground surface. The assessment of appropriate boundary conditions at the ground surface is important. The geotechnical engineers have to solve practical problems involving unsaturated soils with highly negative pore-water pressures. The very high suction is produced by evaporation near ground surface. It is necessary to study the relationship between shear strength of an unsaturated soil and the evaporation process. Performance of a waste soil barrier systems is closely related to the change in shear strength of soils subjected to high suctions. Experimental relationship between shear strength and suction undergoing desiccation is needed in order to predict unsaturated soil property functions below residual state.

2. *Purpose of This Study*

This study involved performing unconfined compression tests for compacted, unsaturated soils subjected to high suctions. A silty soil and kaolin as fine-grain soils were used in this test program. Compacted unsaturated soil was desiccated using a relative humidity equilibrium technique until its water content was below residual water content. This study provides the relationship between total suction and shear strength of very dry compacted unsaturated soil. The results obtained from the unconfined compressive tests suggested that the shear strength has a relationship with the soil-water characteristic curve. This study concluded that the failure envelopes are essentially horizontal in the residual state of unsaturation.

3. *Literature*

3.1 *Evaporation at Ground Surface*

Recent studies provide design criteria and performance modeling for the estimation of evaporation rate for a composite soil cover system constructed on an acid generating waste rock dump in New Brunswick (Yanful, Bell and Woyshner, 1993). To prevent the movement of water into and through waste disposal facilities, compacted earth barriers intended to minimize the migration of contaminants from the facilities have became an essential component of the geotechnical design (Daniel, 1983). Silvestri, Soulie, Lafleur, Sarlis and Benkkouche (1990) mentioned that there are extensive settlement problems in lightweight structures related with soil volume changes beneath shallow foundations caused by precipitation and evapotranspiration balances. Engineered soil covers cannot be designed without evaluating the evaporation at the soil surface. Morris, Graham and Williams (1992) reviewed the occurrence and morphology of cracks in dry-climate regions of Australia and Canada. They mentioned that the settlement and cracks initially appear on the surface of the soil and propagate downward forming a cracked soil column with a desiccated surface crust. Reclamation operations for soft cohesive waste-disposal sites require the development of a desiccated surface crust with an adequate strength in order to support workers and equipment. A theory for modeling the consolidation and desiccation process of soft fine-grained soils after deposition was developed by Abu-Hejleh and Znidarcic (1995). Konrad and Ayad (1997) proposed a framework based on the theory of linear elastic fracture mechanics, showing the phenomenon of crack propagation for a cohesive soil under going desiccation.

Wilson, Fredlund and Barbour (1997) presented a theoretical approach in which a Dalton-type mass transfer equation was used to predict the evaporative fluxes from non vegetated soil surfaces. It is important to realize that some of the water loss comes directly from the soil due to evaporation, and some comes indirectly from the soil through the plants (i.e., transpiration) (Silvestri, Soulie, Lafleur, Sarlis and Benkkouche, 1990). Direct evaporation of water from a soil surface is subjected to purely physical influences (i.e., temperature, relative humidity, wind and duration of daylight), whereas transpiration is basically a biological process, partly dependent on plant physiology and soil conditions. Barton (1979) suggested soil evaporation may be estimated on the basis of the humidity and the water content of the near surface soil. Hammel, Papendick and Campbell (1981) described actual evaporation from soil as a function of soil-water content and included the water flow processes in the soil with combined moisture and temperature gradients. Granger (1989) stated that evaporation from unsaturated soil surface is a function of the actual vapor pressure at the soil surface.

3.2 *Residual Stage of Unsaturation*

Unsaturated soil properties are closely related to the relationship between the amount of water in soils and soil suction. The relationship between the water content and soil suction is known as the soil-water characteristic curve. Its role is essential for unsaturated soil mechanics. The soil-water characteristic curve is studied adequately in order to develop predictive numerical models for unsaturated soil behavior. The amount of water in soil with increase suction is generally quantified in terms of volumetric water content, θ, gravimetric water content, w, or degree of saturation, S. The entire suction range of the soil-water characteristic curve is required primarily to investigate evaporation process near ground surface.

An initially saturated soil specimen begins to desaturate as a result of an increase in suction in the soil. The desaturation stages, generally, are classified along the desorption branch of a soil-water characteristics curve; namely, the boundary effect stage, the transition stage (i.e., with primary and secondary transition portions) and the residual stage of desaturation (White, Duke and Sunada, 1970, Vanapalli, Fredlund, Pufahl and Clifton, 1996). White, Duke and Sunada (1970) have provided the original concepts for the different stages during the desaturation process. Vanapalli, Fredlund, Pufahl and Clifton (1996) developed different saturation stages on the soil-water characteristic curve in order to predict the shear strength in terms of soil suction.

In the boundary effect stage all soil pores are filled with water. The soil behavior can be explained using saturated soil mechanics in the boundary effect stage. The first important point on the soil-water characteristic curve is the air-entry value, $(u_a - u_w)_b$. This value of suction identifies the point at which air enters the largest pores of the soil. When suction is larger than the air-entry value, the soil starts to desaturate in the transition stage. The amount water in the soil decreases significantly with increasing suction in the transition stage. Eventually large increases in suction lead to a relatively small changes in water content in the residual stage of unsaturation.

A coarse-grained soil such as a gravel or sand shows a tendency to desaturate at a fast rate with increasing suction. Since the water storage ability depends on grain size distribution, the rate of desaturation decreases with an increase in fine content. The residual suction, $(u_a - u_w)_r$, can be expected to be higher for fine-grained soils. The suction corresponding to the residual water content approximates the wilting point of many plants. Fredlund (1998) mentioned that water in the liquid phase drains from most of the soil pores when the suction is less than 1500 kPa. He suggested that water content corresponding to a suction of about 1500 kPa may be used as the residual water content. van Genuchten (1980) indicated that a wilting point of 1500 kPa corresponds to the residual water content for practical field problems. The pore-water at suctions greater than the residual suction may move through the pores in the vapor phase.

When the wetted contact area has reduced significantly, the volume change of an unsaturated soil becomes negligible. The suction in the residual stage makes a small contribution to an increase in shear strength compared with the suction in the transition stage. Vanapalli, Fredlund and Pufahl (1999) concluded that soil-water characteristics are not significantly influenced either by the soil structure aggregation or the stress history in the high suction ranges (i.e., 20,000-1,000,000 kPa). Fredlund (1998) commented that the shear strength of an unsaturated soil remains essentially constant beyond residual state.

3.3 *Shear Strength of Unsaturated Soil*

The relationship between soil suction and the shear strength of unsaturated soil is important for many geotechnical engineers. The concept of stress state variables to describe the unsaturated soil behavior were introduced by Fredlund and Morgenstern (1977). Fredlund, Morgenstern and Widger (1978) proposed a shear strength equation for unsaturated soils using two independent stress state variables, the net normal stress, $(\sigma - u_a)$, and the matric suction, $(u_a - u_w)$.

$$\tau = c' + (\sigma - u_a)\tan\phi' + (u_a - u_w)_b \tan\phi^b \quad (1)$$

where, τ = shear strength, c' = effective cohesion intercept, σ = total normal vertical stress, ϕ' = effective angle of internal friction with respect to net normal stress, u_a = pore-air pressure, u_w = pore-water pressure, ϕ^b = angle of internal

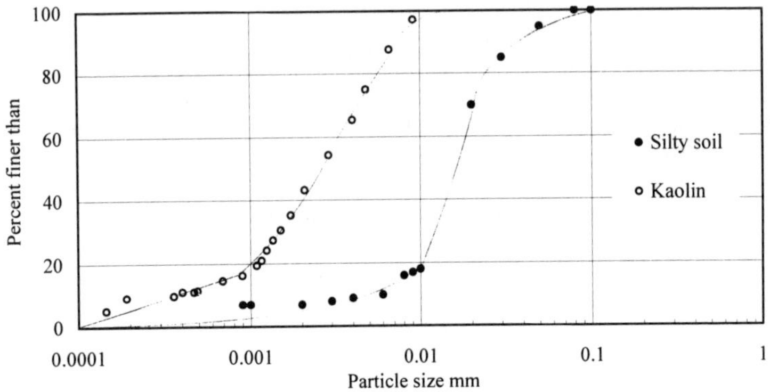

Fig. 1. Grain size distribution for a silty soil and a kaolin

friction with respect matric suction.

The unsaturated soil shear strength equation proposed by Fredlund, Morgenstern and Widger (1978) describes a planar surface tangential to the Mohr-Coulomb circles at failure. Fredlund, Rahardjo and Gan (1987) showed non-linear failure envelopes using a modified direct shear apparatus. The slope of the shear strength versus matric suction envelope can also be negative (Donald, 1956). Escario and Saez (1987) conducted direct shear tests on three compacted unsaturated soils by controlling suction up to 15,000 kPa. They found that the relationship between shear strength versus suction was non-linear. Escario and Saez (1987) have observed an increase in shear strength up to 11,000 kPa for the grey clay. However, for the suction range from 11,000 kPa to 150,000 kPa, a slow drop in shear strength was observed. Escario and Juca (1989) reported that direct shear strength of a clay sand increases up to a suction of 1,000 kPa. Beyond 1,000 kPa, the slope of the shear strength versus soil suction, ϕ^b, appears to approach a negative value.

Mahalinga-Lyer and Williams (1985) carried out unconsolidated undrained triaxial tests on two lateritic soils. The results show that the slope, ϕ^b, of the failure surface with respect to the matric suction decreases sharply in the suction range 0-1,000 kPa and then decreases slightly in the suction range 1,000-8,000 kPa for the lateritic soil developed on basalt. For the lateritic soil developed on sandstone, ϕ^b, decreases sharply in the suction range 0-200 kPa and then decreases slightly in the range from 200 kPa to 1,800 kPa. They showed that the failure envelope with respect to the matric suction was non-linear for lateritic soils, and mentioned that suction has no effect on the shear strength when the degree of saturation was lower than 53-55 %.

4. *Test Procedure*

A silty soil and kaolin were used in this test program. The silty soil is non-plastic soil and the kaolin is a fine-grain cohesive soil. Grain size distributions for both soils are shown in Fig.1. The two fine – grain soils were compacted statically. Each soil specimen had a height of 100 mm and a diameter of 50 mm. The initial water contents of the silty soil and the kaolin were 9.6 % and 31.7 %, respectively. All soil specimens were placed directly into a temperature and relative humidity controlled chamber in order to apply a high total suction. The chamber used in this test program could control the relative humidity in a range from 20 % to 90 % at a temperature of 30 °c. The thermodynamic relationship between soil suction and the partial pressure of the pore-water vapor can be written as follows:

$$\psi = -\frac{RT}{v_{w0} w_v} \ln\left(\frac{\overline{u_v}}{\overline{u_{v0}}}\right) \tag{2}$$

where, ψ = soil suction or total suction (kPa), R = universal (molar) gas constant [i.e., 8.31432 $J/(molK)$], T = absolute temperature [i.e., $T = (273.16 + t^\circ)(K)$], t° = temperature $(^\circ c)$, v_{w0} = specific volume of water or the inverse of the density of water [i.e., $1/\rho_w$ (m^3/kg), ρ_w = density of water, w_v = molecular mass of water vapor (i.e., 18.016 $kg/kmol$), $\overline{u_v}$ = partial pressure of pore-water vapor (kPa), $\overline{u_{v0}}$ = saturation pressure of water vapor over a flat surface of pure water at the same temperature (kPa).

The term, u_v/u_{v0} is called relative humidity, RH(%). If a temperature of 20 °c is used as the reference, the above equation (2) can be written to give a fixed relationship between total suction (in kPa) and relative vapor pressure (i.e., relative humidity).

$$\psi = -135022 \ln\left(\frac{\overline{u_v}}{\overline{u_{v0}}}\right) \tag{3}$$

Figure 2 shows a plot of relative humidity versus total suction. Fredlund and Xing (1994) mentioned that there is a maximum total suction value corresponding to a zero relative humidity in any porous medium. The total suction corresponding to zero relative humidity or zero water content is approximately 1,000,000 kPa. This value is supported experimentally for a variety soils (Croney and Coleman, 1961).

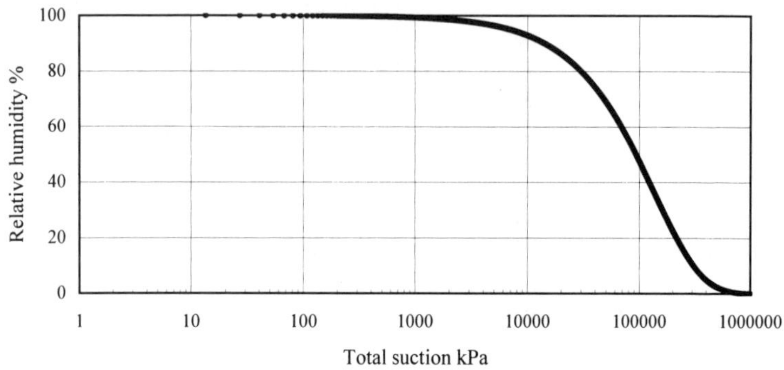

Fig. 2. Relationship between relative humidity and total suction

The test program selected relative humidity of 80 %, 70 %, 60 %, 50 % and 40 %. Each soil specimen was subjected to the relative humidities for a month. For example, a relative humidity of 80 % and 40 % corresponds to a soil suction of 30,000 kPa and 124,000 kPa, respectively. Desaturation of a soil started as a result of evaporation in the relative humidity chamber. After the soil specimen had reached equilibrium with the selected relative humidity, an unconfined compression test was conducted at a rate of axial strain of 0.5 mm per minute.

At the end of the unconfined compression test, the water content of the soil specimen was measured in order to evaluate the soil-water characteristic curve. The soil-water characteristics curve for the two soil were measured using a pressure plate apparatus, glass desiccator containing saturated salt solutions (i.e., vapor equilibrium technique) and relative humidity technique. The above mentioned three methods were used to develop the soil-water characteristics curve over the entire soil suction. The air pressure in the pressure plate apparatus was increased to a maximum of 490 kPa. The vapor equilibrium technique (relative humidity technique) was used to impose suction larger than 490 kPa. Five different salt solutions were used in the glass desiccator: lithium chloride ($LiCl \cdot H_2O$), magnesium chloride ($MgCl_2 \cdot 6H_2O$), magnesium nitrate ($Mg(NO_3)_2 \cdot 6H_2$), sodium chloride (NaCl) and potassium sulphate (KSO_4). The above five salt solutions controlled the atmospheric humidity in the glass desiccator (Yong, 1967).

5. *Experimental Results and Interpretation*

5.1 *Soil-Water Characteristic Curve*

The soil-water characteristic curve can be described as a measure of the

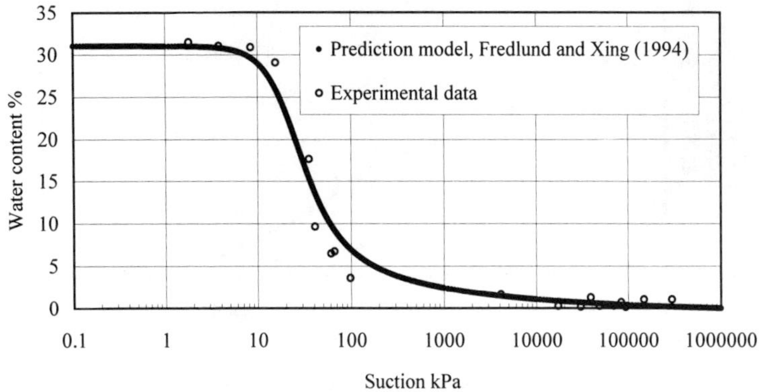

Fig. 3. Soil-water characteristic curve for a silty soil

water-holding capacity (i.e. storage capacity) of the soil, when subjected to various values of suction. The distribution of the soil, water and air phases changes as the stress state changes (e.g. shear strength, volume change). Changes in these phases (soil, water and air) take on different forms and influence the engineering behavior of unsaturated soils.

The relationship between amount of soil water and soil suction was evaluated using a pressure plate apparatus, salt solution desiccator and a relative humidity chamber. The soil-water characteristic curves for a silty soil and a kaolin were tested from a saturated condition to a totally dry condition (i.e. for soil suctions ranging from 0 to 1000,000 kPa) and are shown in Figs. 3 and 4. The soil-water characteristic curve for the kaolin is different from the soil-water characteristic curve of the silty soil. The key features of the soil-water characteristic curve are the air-entry value, the residual water content and the slope of the straight line portion of the desorption branch. The air-entry value for the two soils can be obtained by extending the constant slope portion of the soil-water characteristic curve to intersect the suction axis at saturated condition. Evaluated air-entry value for the silty soil and the kaolin are 10 kPa and 100 kPa, respectively. When matric suction exceeds the air-entry value, desaturation starts in the liquid phases. The soil dries rapidly with increasing suction. The slope of branch in the transition stage for the silty soil is larger than that of the kaolin. The rate of desaturation is higher for the silty soil.

The definition of the residual water content is not easy, as the shape of the soil-water characteristic curve varies depending on soil properties. An empirical, graphical procedure has been suggested for the residual state (Vanapalli, Fredlund and Pufahl, 1999). The residual state of saturation can be defined as the point where the line extending from 1000,000 kPa along the curve intersects the previous tangent line. The silty soil shows a residual water content of 2.5 % and a residual

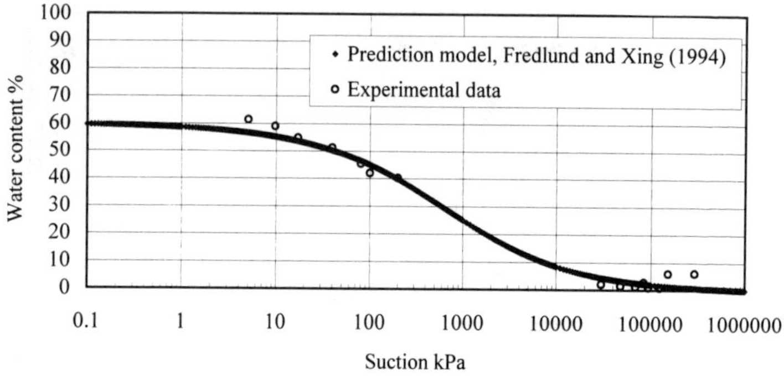

Fig. 4. Soil-water characteristic curve for a kaolin

suction of 200 kPa using above mentioned graphical procedure. On the soil-water characteristic curve of the kaolin there is not a clear point, where the residual suction can be found. A decrease in the water content continues with soil suction. In other words, the soil-water characteristic curve of kaolin has no distinct breaking point in the high suction range.

5.2 *Unconfined Compressive Strength and Failure Envelope*

Unconfined compressive test results for the silty unsaturated soil and the kaolin unsaturated soil subjected to a high suction are summarized in Tables 1 and 2. The initial relative humidity selected in this test program is less than 80 %. Soil suction corresponding to this relative humidity is larger than 30,000 kPa. An unconfined compressive test was performed on a specimen with the initial water content in order to compared the results with the compressive strength of the unsaturated soil subjected to higher suctions. Total suction in Tables 1 and 2 is computed using a relative humidity versus total suction theoretical Eq.2.

The water content of the soil specimen decreases with increasing total suction or decreasing relative humidity. The decrease of water content with total suction, however, is not significant. Desaturated silty soil and kaolin are in the residual state of saturation. The decreasing in the axial strain at failure was noticed during compression testing. A decrease in the axial strain at failure is predictable due to a decreasing wetted area of contact between soil particle due to the high suction. Below the residual stage water content, when deviator stress reached to peak, there is similarity in the axial strain regardless of soil suction value.

The relationship between unconfined compressive strength and total suction is presented in Figs. 5 and 6. The shear strength of the desaturated silty

Table 1 Summary of unconfined compression test for a silty soil

Relative humidity %	Total suction kPa	Unconfined compressive strength kPa	Water content %	Axial strain at failure %
80	30000	32.4	0.304	0.14
70	48000	38.8	0.269	0.19
60	69000	58.3	0.188	0.15
50	94000	59.2	0.158	0.32
40	124000	39.8	0.105	0.29
Initial condition	41	28.8	9.6	0.56

In case of "Initial condition", suction means matric suction.

Table 2 Summary of unconfined compression test for a kaolin

Relative humidity %	Total suction kPa	Unconfined compressive strength kPa	Water content %	Axial strain at failure %
80	30000	551	2.16	1.34
70	48000	763	1.56	1.4
60	69000	608	1.21	1.32
50	94000	986	0.95	1.73
40	124000	1000	0.88	1.43
Initial condition	900	394	31.7	3.31

In case of "Initial condition", suction means matric suction.

soil and kaolin increases slightly with suctions in the suction range from 30,000 kPa to 124,000 kPa. In the residual stage of saturation, there is a slight increase in shear strength of soil related to essentially no change in water content (i.e., no change in wetted area of contact) with increasing of soil suction. The angle of friction, ϕ^b, with respect to suction are computed from unconfined compressive strength versus suction relationship shown in Figs 5 and 6. The angle of friction, ϕ^b, for the silty soil and kaolin are 0.02 degrees and 0.3 degrees, respectively. The failure envelope shows an essentially horizontal line in residual stage.

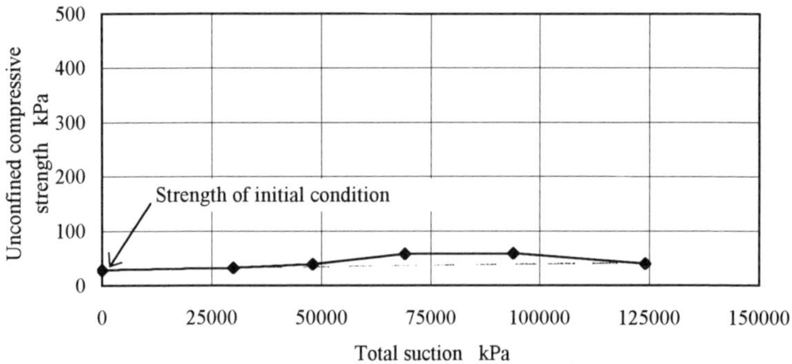

Fig. 5. Relationship between unconfined compressive strength and total suction for a silty soil

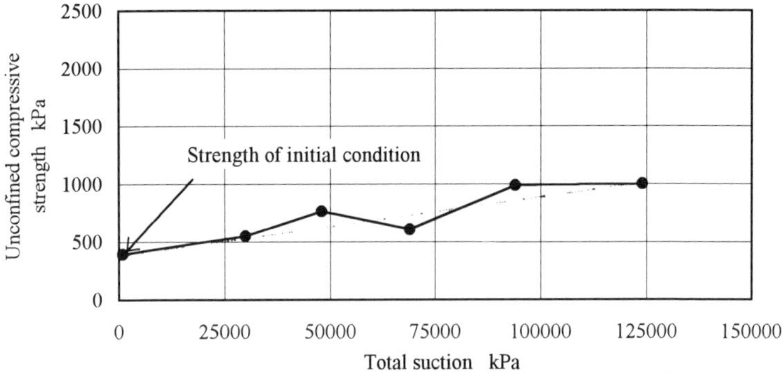

Fig. 6. Relationship between unconfined compressive strength and total suction for a kaolin

6. *Conclusions*

The soil-water characteristic curve and the unconfined compressive strength under the residual state of unsaturation were obtained for a silty soil and a kaolin during this study. The soil-water characteristic curve was obtained using a pressure plate apparatus, vapor pressure equilibrium technique and relative humidity equilibrium technique. The shear strength of an unsaturated soil subjected to high suction increases slightly with increasing suctions. An essentially horizontal envelope was obtained in the residual state of saturation. The soil-water characteristic curve in entire suction range (i.e. 0-1000,000 kPa) is essential to the prediction of unsaturated soil shear strength.

References

Abu-Hejleh, A.N. and Znidarcic, D. (1995), Desiccation theory for soft cohesive soils. *Journal of Geotechnical Engineering, American Society of Civil Engineers, Geotechnical Engineering Division*, Vol. 121, No. 6, 493-502.

Barton, I.J. (1979), A parameterization of the evaporation from nonsaturated surfaces, *Journal of Applied Meterology*, Vol. 18, 43-47.

Croney, D. and Coleman, J.D. (1961), Pore pressure and suction in soils, In Proceedings of the Conference on Pore Pressure and Suction in Soils, Butterworths, London, 31-37.

Daniel, D.E. (1983), Shallow land burial of low-level radioactive waste, *Journal of Geotechnical Engineering, American Society of Civil Engineers,* Vol. 109(1), 40-55.

Donald, I.B. (1956), Shear strength measurements in unsaturated non-cohesive soils with negative pore pressures, *Proceedings of 2nd Australia and NewZeland Conference on Soil Mechanics and Foundation Engineering, Christchurch, NewZealand*, 200-205.

Escario, V. and Saez, J. (1987), Shear strength of partly saturated soils versus suction, *Proceedings of 6th International Conference on Expansive Soils, New Delhi, India*, Vol. 2, 602-604.

Escario, V. and Juca, J.F.T. (1989), Strength and deformation of partly saturated soils, *Proceedings of the Twelfth International Conference on Soil Mechanics and Foundation Engineering, RIO DE JANEIRO*, Vol. 1, 43-46.

Fredlund, D.G., and Morgenstern, N.R. (1977), Stress state variables for unsaturated soils, *Journal of the Geotechnical Engineering Division, ASCE*, Vol. 103(GT5), 447-466.

Fredlund, D.G., Morgenstern, N.R. and Widger, R.A. (1978), The shear strength of unsaturated soils, *Canadian Geotechnical Journal*, Vol. 15, 313-321.

Fredlund, D.G., Rahardjo, H. and Gan, J-K.M. (1987), Non-linearity of strength envelope for unsaturated soils, *Proceedings 6th International Conference on Expansive Soils, New Delhi, India*, Vol. 1, 49-54.

Fredlund, D.G. and Xing, A. (1994), Equations for the soil-water characteristic curve, *Canadian Geotechnical Journal*, Vol. 31, 521-532.

Fredlund, D.G. (1998), Bringing unsaturated soil mechanics into engineering practice, *Proceedings of the Second International Conference on UNSATURATED SOILS, Beijing China*, Vol. 2, 1-36.

Granger, R.J. (1989), Evaporation from natural non-saturated surfaces, *Journal of Hydrology*, Vol. 111, 21-29.

Hammel, J.E., Papendick, R.I. and Campbell, G.S. (1981), Fallow tillage effects on evaporation and seedzone water content in a dry summer climate, *Soil Science Society of American Journal*, Vol. 45, 1016-1022.

Konrad, J-M. and Ayad, R. (1997), An idealized framework for the analysis of cohesive soils undergoing desiccation, *Canadian Geotechnical Journal*, Vol. 34, 477-488.

Mahalinga-Iyer, U. and Williams, D.J. (1985), Unsaturated strength behaviour of compacted lateritic soils, *Geotechnique*, Vol. 45, No. 2, 317-320.

Morris, P.H., Graham, J. and Williams, D.J. (1992), Cracking in drying soils, *Canadian Geotechnical Journal*, Vol. 29, 263-277.

Silvestri, V., Soulie, M., Lafleur, J., Sarkis, G. and Bekkouche, N. (1990), Foundation problems in Champlain clays during droughts. I:Rainfall deficits in Montreal (1930-1988), *Canadian Geotechnical Journal*, Vol. 27, 285-293.

Vanapalli, SK., Fredlund, D.G., Pufahl, D.E. and Clifton, A.W. (1996), Model for the prediction of shear strength with respect to soil suction, *Canadian Geotechnical Journal*, Vol. 33, 379-392

Vanapalli, S.K., Fredlund, D.G., and Pufahl, D.E. (1999), The influence of soil structure and stress history on the soil-water characteristics of a compacted till, *Geotechnique*, Vol. 49, No. 2, 143-159.

van Genuchten. (1980), A closed-form equation for predicting the hydraulic conductivity of unsaturated soils, *Soil Science Society of American Journal*, Vol. 44, 892-898.

White, N.F., Duke, H.R. and Sunada, D.K. (1970), Physics of desaturation in porous materials, *Journal of the Irrigation and drainage division, Proceeding of the American Society of Civil Engineers*, No. IR2, 165-191.

Wilson, G.W., Fredlund, D.G. and Barbour, S.L. (1997), The effect of soil suction on evaporative fluxes from soil surfaces, *Canadian Geotechnical Journal*, Vol. 34, 145-155.

Yanful, E.K., Bell, A.V. and Woyshner, M.R. (1993), Design of a composite soil cover for an experimental waste rock pile near Newcastle, New Brunswick, Canada, *Canadian Geotechnical Journal*, Vol. 30, 578-587.

Yong, J.F. (1967), Humidity control in the laboratory using salt solutions-A review, *Journal of Applied Chemistry*, Vol. 17, 241-245.

Shear Strength and Swelling Stabilization of Unsaturated clayey soil Using Pozzolanic Material

Dr. Mousa F. Attom, Ph.D.
Assistant Professor of Civil Engineering
Jordan University of Science and Technology
Irbid - Jordan

Dr. Salah A. Taqieddin , Ph.D.
Associate Professor of Civil Engineering
Jordan University of Science and Technology
Irbid - Jordan

Tamara Mubeideen
Graduate Student of Civil Engineering
Jordan University of Science and Technology
Irbid - Jordan

ABSTRACT

This paper presents the effect of using pozzolanic material as a new stabilizing agent on shear strength and swelling properties of unsaturated clayey soil. Three clayey soils were selected from the northern part of Jordan and used for this research. Different percentages of pozzolanic material by dry weight of the soils namely; 2.5%, 5.0%, 7.5% and 10.0% were added to the selected soils. Atterberg's limits tests, pr
octor density test, direct shear test, unconfined Compression test, swelling pressure test and swell potential test were conducted on specimens treated with pozzolan. It was found that the increase in pozzolanic material will decrease the plasticity index and increase the unconfined compressive strength of the soils use. Furthermore, the addition of 10.0% of pozzolanic material by dry weight of the soil could eliminate both the swelling pressure and the swell potential of the used soils.

INTRODUCTION

Soil stabilization refers to the improvement of soil engineering and physical properties such as increasing its strength and reducing or eliminating its expansion. Geotechnical engineers used many materials to stabilize the soils. The increases of soil bearing resistance by the addition of cement and fly ash to the soil was studied by Rizkallah (1989). Basma (1991) showed that the lime can be used as soil stabilizing agent against volume change and compressibility. The subgrade of natural soil at Gatewick, Stansted, Heathrow and Glaskow airports was stabilized by using both cement and lime Heath (1992). The same material was used by Kennedy et al. (1987) to improve the engineering properties of the soil. A study was conducted by Al-Homoud et al. (1995) to use the bitumen as soil stabilizing agent to control the swelling and collapsibility of the soil. Nishida (1992) investigated the use of waste rock powder as soil stabilizing agent.

Geologically, Tel Ermah area in northern part of Jordan containss of millions of tons of pozzolanic material. Because the pozzolanic material is cheap to obtain, the government of Jordan started to make use of this material in the industries such as the cement industry. Nothing found in the literature about using the pozzolanic material as soil stabilizing agent. Since the chemical composition of the pozzolana consists of oxides of calcium, magnesium, alumium, iron, sodium, potasum and titanium in addition to silica (Khateeb, 1989). The addition of this composition to the clayey soil may improve the soil engineering and physical properties.

OBJECTIVE

The objective of this paper is to study the use of Jordanian pozzolanic material as stabilizing agent to increase soil strength and prevent or eliminate soil swelling. To achieve the objective of this research, three types of soils of different engineering physical properties from northern part of Jordan were selected and brought to soil testing laboratory at Jordan University of Science and Technology. These samples were tested in accordance with ASTM Standard Methods for Testing and Materials.

EXPERIMENTAL PROGRAM AND TEST RESULTS

A full testing program was performed on the three selected clayey soils. A study by Tuncer et al. (1990) indicated that the soil in northern part of Jordan has medium to high expansive properties.
The following tests : Atterberg's limits (ASTM D-4138), standard proctor density (ASTM D-1557), direct shear (ASTM D-3080), unconfined compression (ASTM D-2166), swelling pressure (ASTM D-2435), and swell potential (ASTM D-2435) were conducted on samples treated with 0.0%, 2.5%, 5.0%, 7.5% and 10.0% of pozzolanic material. For the direct shear, unconfined compression, remolded samples with pozzolanic material at 2.5%, 5.0%, 7.5% and 10% by dry weight of the soil and prepared at optimum water content and 90% percent of the maximum dry density. In case of swelling tests the remolded samples were prepared at 95% relative compaction 15% initial water content and then mixed the four different percentages of the pozzolanic material. All remolded samples were wrapped carefully with a plastic cover and kept in the laboratory room for 72 hours before testing.

Natural Physical Properties of the Soils Used

The primary physical properties such as gradation, Atterberg,s limits, maximum dry density, optimum water content and specific gravity of the three soils were determined under its natural conditions in accordance with the American Society for Testing and Materials (ASTM) standard specification and then classified according to Unified Soil Classification System (USCS).

Table 1 shows the physical properties of the soil used in this study.
After the engineering properties of the soil were found, the three types of soils are then mixed with pozzolan at four percentages and prepared for the prementioned tests. The following is a brief description of the conducted tests.

Standard Proctor Compaction Test

A representative air dried sample from each type of soil was obtained and passed through U.S. No. 4 sieve and the moisture content was determined. Five batches from each soil were obtained and mixed with the four different pozzolanic percentages. Different amount of water was added to each batch so that the water content was about 3 to 4 percent between the different batches and mixed thoroughly and placed in a covered container for 24 hours for curing. The standard compaction test was then performed.

Strength Tests : Direct Shear Test and Unconfined Compression Test.

In the direct shear test the three types of soil were mixed with four percentages of pozzolan (2.5%, 5.0%, 7.5%, 10.0%) and remolded at 95% relative compaction at the optimum water content. The direct shear test standard procedure ASTM D-3080 was conducted on the treated samples after 48 hours from remolding.
In the unconfined compression test the soil is passed through U.S. No. 40 sieve. Each type of soil was mixed with pozzolan at 2.5%, 5.0% and 7.5% and 10.0% by dry weight of the soil. Test specimens were remolded in the standard unconfined compression mould at optimum moisture content and 95 % relative compaction according to Standard Proctor Compaction test and wrapped with a plastic cover to prevent any moisture content losses. The standard unconfined compression test was conducted on treated specimens after 48 hours from remolding time at a strain rate of 2 percent per minute until feasible failure or 15% strain whatever took place first.

Swelling Test - Constant Volume Test

The swelling pressure in this method is defined as the pressure that prevent the soil sample from swelling when water is added. Zero swell test was conducted on remolded samples mixed with pozzolan at the four percentages at the maximum dry density and 15% initial water content. The prepared specimens were placed in the standard one dimensional odometer device under seating load of 6.9 kPa (1 psi) and water is added. As the sample starts in swelling, a convenient load increment was added to the sample to prevent the swelling. The final pressure that prevent the sample from swelling is defined as the swelling pressure of the soil.

Swell Potential Test

The sample in this test is prepared at the maximum dry density and optimum water content. The swell potential is calculated as the ratio of the maximum swell height of the sample due wetting to the initial height of the sample upon placed in the odometer cell under an applied pressure of 7 kN/m². Enough amount from each type of soil was mixed thoroughly with the different percentages of pozzolan and remolded in the one dimensional odometer ring. The treated specimens were then placed in the one dimensional odometer device and water is added. Readings of swelling deformation due to addition of water were taken at 0.25, 0.5, 1, 2, 4, 8, 15, 30, 60, 120 and 1440 minutes.

ANALYSIS AND DISCUSSION OF THE RESULTS

Plastic Index and Maximum Dry Density

The effect of pozzolanic material on the plastic limits and the maximum dry density of used soils are shown in Figures 1 and 2 respectively. It is obvious from Figure 1 that the increase of pozzolan in the soils will reduce the plastin limit. This reduction will be higher for the sample which has more plasticity. The reason behind that is the addition of none-plastic material to a plastic material will result in a decrease in the plasticity of the soil. Figure 2 depicts the effect of pozzolan on the maximum dry density of the soil. It is clear that the increase in percentage of pozzolan somewhat increases the maximum dry density of the soil especially for the soil of higher percentages of clay. This is because the clayey soil with pozzolan form a heavier weight for the same volume resulting in increase in the maximum dry density.

Srength Properties: Unconfined Compression Strength, Angle of Internal Friction and Cohesion

Figure 3 shows the effect of the addition of pozzolan on the unconfined compressive strength of the soils. The three types of tested soil behave in the same trend in increasing the unconfined compressive strength with the

addition of pozzolan. This increase is due to the reaction between the pozzolan composition and soil minerals within the 72 hours curing period which produce a cementatiuos bonds that result in increasing the unconfined compressive strength. This cementatoius bond is more apparent in the soil 3 which has more clay content than in soil 1.
The effect of pozzolan on angle of internal friction and on cohesion are shown in figure 4 and 5, respectively. Figure 4 shows that the addition of pozzolan up tp 5% by dry weight of the soil will increase the angle of internal friction. Further increase in the pozzolan will reduce the angle of internal friction. This trend is more obvious in the soil of higher percentages of clay content. Figure 5 indicates that the addition of pozzolan up to 5% by dry weight of the soil will reduce the cohesion. Further increase in the amount of pozzolan will increase the cohesion of the soils resulting in increasing overall shear strength of the soils..

Swell Pressure and Swell Potential

Figure 6 and 7 depicts the effect of addition pozzolanic material on soil swelling pressure and swell potential of the soil, respectively. It is clear that the addition of pozzolan will reduce both the swelling pressure and consequently the swell potential of the three types of soils. This reduction is due to the bonding produced from the reaction between the pozzolanic silica and aluminium oxides and hydrated lime in the clayey soil which produce cementatious calcium silicates and calcium aluminates results in a significant reduction in both swelling pressure and swell potential. Moreever, it was found that the addition of 10% of pozzolan by dry weight of the soil will almost eliminate the both the swelling pressure and the swell potential for the three types of soils.

CONCLUSION

The potential of using pozzolanic material as a new stabilizing agent was studied on some soil properties. The following conclusions are reached:
1) The addition pozzolan decreases the plastic index and slightly increases the maximum dry density of the soil. This result promise for the use of pozzolan as a new stabilizing agent.
2) An increase in amount of pozzolan in the soil will increase the unconfined compressive strength.
3) The addition of pozzolan up to 5% by dry weight of the soil gives the highest vale of angle of internal

friction and lowest value of cohesion
4) The swelling pressure and swell potential can be eliminated by the addition of 10% of pozzolan by dry weight to the clayey soil. 5) Highly plastic soil is a good candidate for soil iprovement by using pozzolanic material

References

1) Al-Homoud, Azem S., Khedaywi, Taisir and Al-Ajlouni, Abdallah M. (1995): " Studies on the control of swell and collapse potential of selected Jordanian soils using asphalt stabilization," Japanese Society of Soil Mechanics and Foundation Engineering, Vol. 35, No. 2, Japan.
2) Basma, A. A. and Tuncer, E. R. (1991): " Effect of lime on the volume change and compressibility of expansssive clays," Transportation Research Board, Washington, D.C., TRR No. 1295, pp. 54-61, U.S.A.
3) Heath, D. C. (1992): "Application of lime and cement soil stabilization at BAA airports," Proceedings of the Institutions of Civil Engineers, Transport Vol. 95, No. 1, P 11-49.
4) Kennedy, T. W., Smith, R., Holgreen, R. J. and Tahmoressi, M. (1987): " An evaluation of lime and cement stabilization," Transportation Research Board, Washington, D. C., TRR No. 1119, pp. 11-25, U.S.A.
5) Khateeb, F. (1989):"Studies of volcanic tuff in Jordan," Ministry of Energy and Mineral resources; Natural Resources Authorties Economic Geology diversion, Amman, Jordan, 53 pp.
6) Nishida, Kazuhoko and Kubio, Yorimichi (1992): " Utilization of waste rock powder for stabilization of soil," Journal of Society of Materials Science, Vol. 40, No. 459, p 1532-1537, Japan.
7) Rizkallah, V. (1989): " Stabilization of difficult soils in developing countries," Proceeding of the International Conference on Soil Mechanics and Foundation Engineering, Vol. 1, p 301-304, Netherland.
8) Tuncer, E. R., Basma, A. A. and Taquiddin, S. (1990): " Geotechnical properties of some selected Irbid soils," Jordan University of Science and Technology, Report No. 14/87, Jordan.

Table 1. Physical properties of the soil used

Properties		Unit	SOIL1	SOIL2	SOIL3
Grain Size	Sand (2mm-75μm) Silt(75μm-2μm) Clay (<2μm)	% % %	6 15 79	12 36 52	25 39 36
Consistancy Limits	Liquid Limit Plastic Limit Plastic Index	% % %	81 31 50	64 34 30	50 24 26
Compaction	$\gamma_{d\ max}$ W_p	g/cm³ %	1.46 34	1.5 25	1.61 22
Activity		-	0.63	0.57	0.72
Specific Gravity		-	2.7	2.6	2.6
Unified Soil Classification[a]			CH	MH	CL

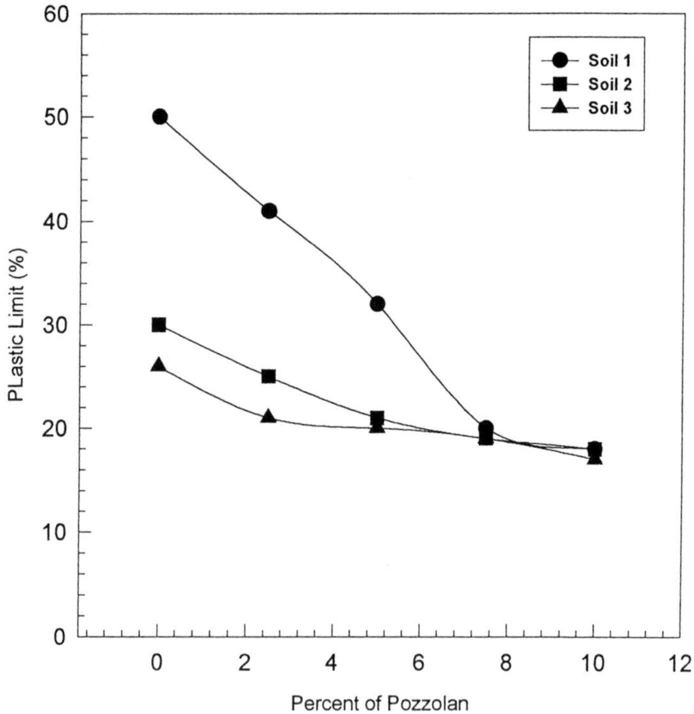

Figure 1. The effect of Pozzolan on plastic index of the soil.

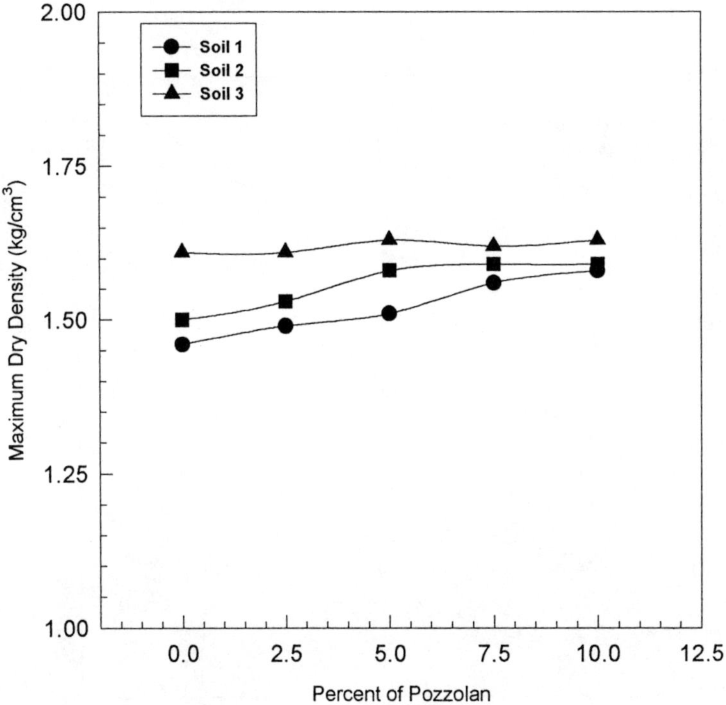

Figure 2. The effect of Pozzolan on maximum dry density of the soil.

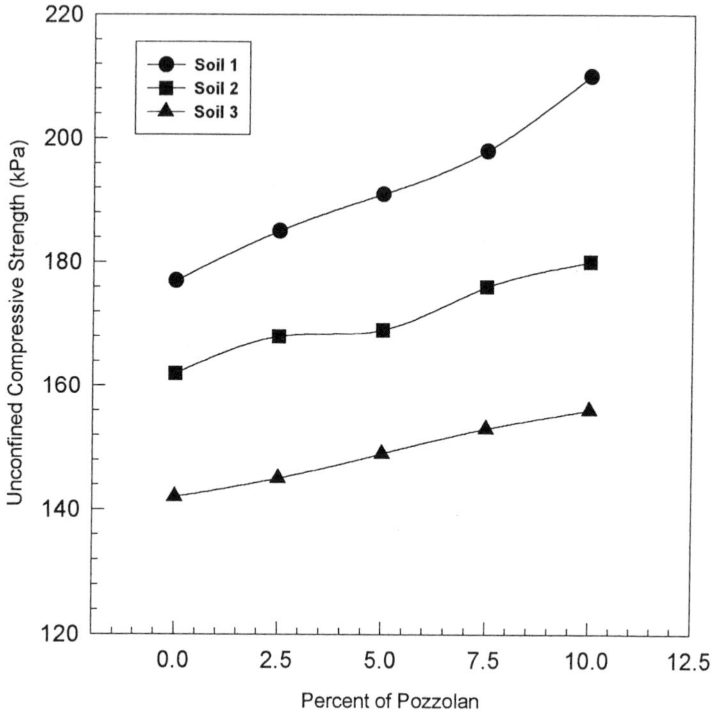

Figure 3. The effect of Pozzolan on unconfined compressive strength of the soil.

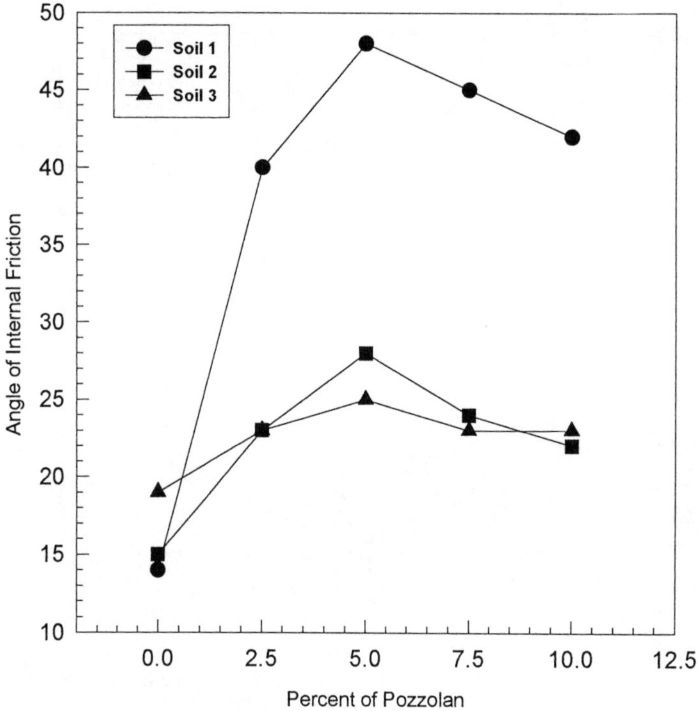

Figure 4. The effect of pozzolan on angle of internal friction of the soil.

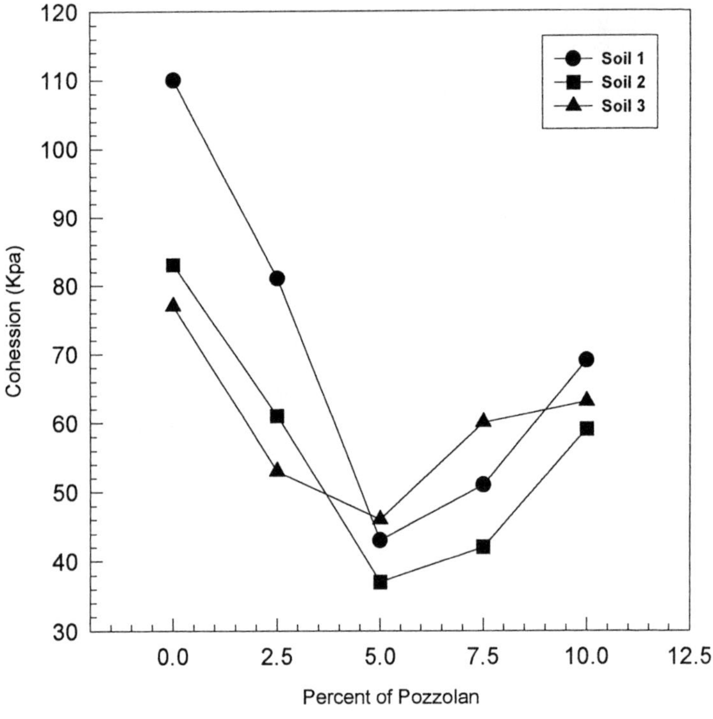

Figure 5. The effect of pozzolan on cohession of the soil.

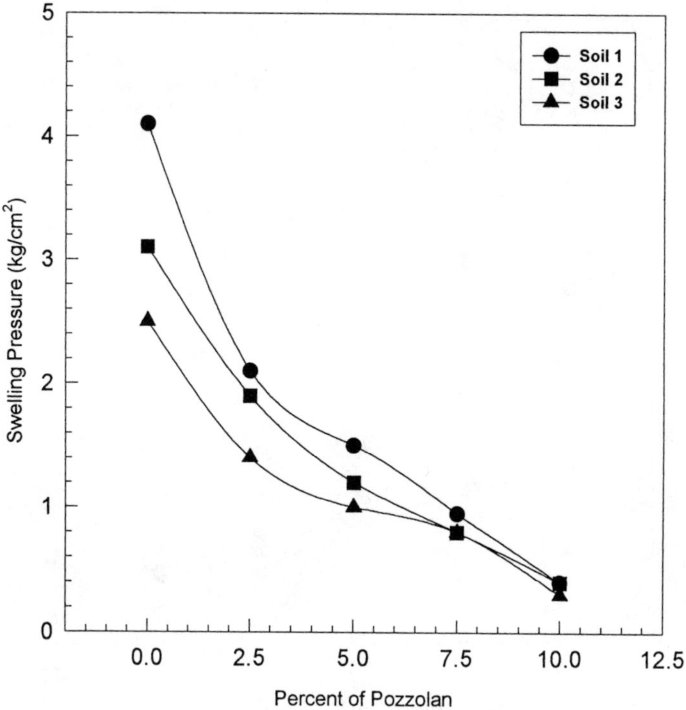

Figure 6. The effect of pozzolan on swelling pressure of the soil.

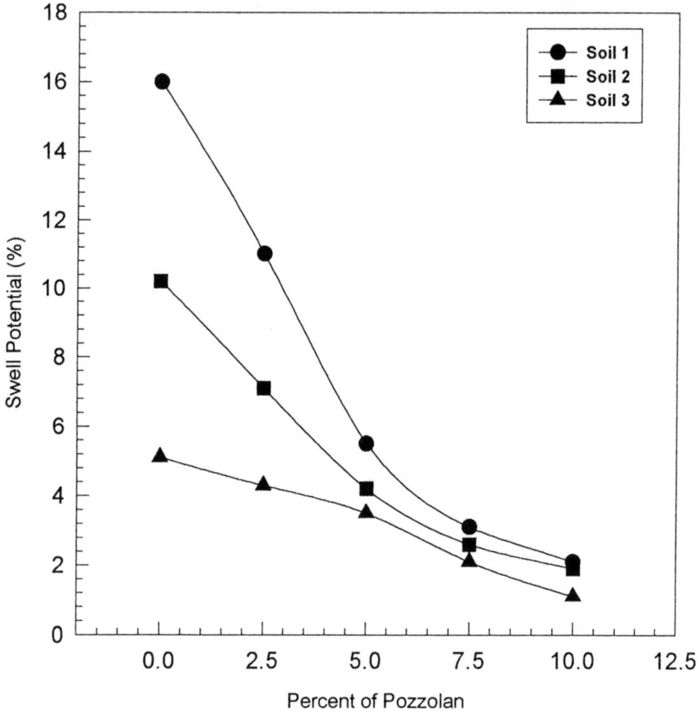

Figure 7. The effect of pozzolan on swell potential of the soil.

STABILISATION OF TROPICAL BLACK CLAY WITH CEMENT AND PULVERISED COAL BOTTOM ASH ADMIXTURE

Kolawole 'Juwonlo Osinubi[1], Member, ASCE

ABSTRACT

Pulverised coal bottom ash (PCBA) obtained as residue from the firing of coal was tested on cement stabilised tropical black clay classified as A-7-6 or CL in the AASHTO and Unified Soil Classification Systems, respectively. PCBA at stepped concentrations (5,10,15, 20,25 and 30%) were admixed with soil - cement mixtures containing 2,4,6 and 8% cement by weight of dry soil, respectively, in order to establish the soil improving potentials of the pulverized coal bottom ash. Unconfined compressive strength (UCS) and California bearing ratio (CBR) tests on the admixture stabilised tropical black clay specimens (i.e. soil - cement - PCBA) yielded peak values at 5% PCBA concentration, in all cases considered, which were markedly higher than values obtained for the natural soil and soil - PCBA combination.

Keywords: California bearing ratio (CBR), compaction, unconfined compressive strength (UCS), durability, laboratory studies.

1.0 INTRODUCTION

Expansive soils are in the group of problematic soils encountered by geotechnical engineers. They are confined to the semi-arid regions of the tropical and temperate climate zones and are abundant where the annual evaporation exceeds the precipitation and can be found anywhere in the world (Chen,1975).

[1] Snr. Lect., M.NSE, M. G-I (ASCE), Regd. Engr.(COREN). Dept. of Civil Engrg., Ahmadu Bello University, Zaria, Nigeria.

Expansive soils in some parts of the world are named `Black Cotton Soil'. These tropical black clays range from light grey to dark grey and black in colour. The name has been given because of their black colour and great suitability for growing cotton. Thus the terms tropical black clay and black cotton soil can be used interchangeably. This group of soil is characterised by the presence of montmorillonite in the mineralogy which is capable capable of large volume changes from the dry to the saturated state. When wet they swell and exert high swelling pressures.

Deposits of black cotton soil in the field show a general pattern of cracks during the dry season of the year. Cracks measuring 70 mm wide and over 1 m deep have been observed (Adeniji,1991) and may extend up to 3 m or more in case of high deposits. These soils are poor materials to employ for highway or airfield construction because they contain high percentages of plastic clay. In areas where they occur, usually there are no suitable natural gravel or aggregates and most deposits cover singnificantly large areas that avoiding them is not possible.

An appreciable improvement in the soil properties of lime modified as well as lime and cement stabilized black cotton soil has been recorded (Ola,1983; Balogun,1991a,b; Osinubi,1995). These improvements though adequate to meet the immediate requirements of the specification for the use of the soil in roads and highway construction, but are costly in execution. It is therefore neccessary to add other cheaper additives to achieve the same objectives.

Pulverised fuel ash has been mainly used as a supplement or replacement for lime and cement in soils showing poor pozzolanic properties. Past and recent research has established the potential of fly ash for use in a variety of construction applications such as fills and pavements. Nicholson et al. (1994) reported that lime - fly ash admixture stabilization has shown tremendous potential as an economic method to update the geotechnical properties of tropical Hawaiian soil. An investigation of the effect of fly ash on soft clay reported by Indraratna et al. (1995) showed that after two weeks of curing, lime and fly ash treated Bangkok clay attained compressive strengths two to three times greater than those of the natural clay.

Inspite of the large amount of past and recent researches carried out on fly ash, the use of PCBA for soil stabilization has barely taken off the the ground and its stabilizing potential is undetermined. The aim of this study is to evaluate the use of PCBA as an admixture in cement stabilization of tropical black clay.

2.0 DESCRIPTION OF MATERIALS USED IN THE STUDY

Soil: The soil sample used for this study was collected by the method of disturbed sampling at Njime command, New Marte (latitude 12° North and longitude 13° 45' East) in Borno state, Nigeria. The soil is dark grey in colour and this is associated with the calcium rich basic igneous parent rocks from which it is formed. X-ray diffraction carried out on soils from this area by Ola (1983) shows that it contains a mixture of kaolinite and montmorillonite with the latter being predominant. The results of tests for the identification of the soil and the determination of its properties are presented in Table 1. The soil is classified under the A-7-6 subgroup of the AASHTO Soil Classification System (Standard,1986) or CL in the Unified Soil Classification System (Annual,1992). The particle size distribution of the natural soil is shown in Fig. 1.

Cement: See Table 2 for the standard oxide composition.

Pulverised Coal Bottom Ash: The PCBA was collected at the disposal sites of the Nigerian Railway Corporation in Jos and its oxide composition is summarised in Table 2.

3.0 METHODOLOGY

Laboratory tests were performed on the soil sample collected in accordance with BS 1377 (1990) for the natural soil and BS 1924 (1990) for the stabilized soil. California bearing ratio (CBR) tests were conducted in accordance with the Nigerian General Specification (1970) which stipulates that specimens be cured in the dry for six days, then soaked for 24 hours before testing.

Tests were first conducted on recompacted soil specimens without cement and admixture followed by additional tests in which cement and PCBA were added in different amounts in order to evaluate the changes in the engineering properties of the soil. Sixteen mix sets of the soil and cement with admixture were prepared. These mixes consist of pulverised natural soil treated with 2,4,6 and 8% cement admixed with 5, 10, 15 and 20% PCBA by weight of dry soil. Predetermined amount of water obtained from moisture - density relationships was later added to the dry mixture and thoroughly mixed before compaction at the energy of the standard Proctor.

The tests performed on the soil with and without admixures include index tests used to classify the natural soil as well as particle size analysis, compaction

TABLE 1: TEST RESUTS OF NATURAL SOIL

PROPERTY	QUANTITY
Percentage passing BS sieve No. 200	60.0
Liquid limit, %	62.3
Plastic limit,%	24.2
Plasticity index, %	38.1
Linear shrinkage, %	18.6
Group Index	24
AASHTO Classification	A-7-6
USCS Classification	CL
Maximum Dry Density, Mg/m^3	1.71
Optimum Moisture Content, %	18.0
Unconfined Compressive Strength, kN/m^2	300
California bearing ratio, %	8
pH	7.2
Free swell,%	76.25
Swelling pressure at OMC, kN/m^2	127.0
Specific gravity	2.40
Colour	Dark grey

TABLE 2: OXIDE COMPOSITION OF PULVERISED COAL BOTTOM ASH

OXIDE	PULVERIZED COAL BOTTOM ASH, %	ORDINARY PORTLAND CEMENT*, %
CaO	2.24	63
SiO_2	26.4	20
Al_2O_3	9.35	6
Fe_2O_3	2.25	3
MgO	0.60	0.86
$Na_2O + K_2O$	0.35	1
Mn_2O_3	0.50	-
SO_3	0.97	2
TiO_2	0.35	-
Loss on ignition	4.45	2

* After Czernin, 1962.

tests, unconfined compression tests and California bearing ratio tests. Compaction was carried out at the energy of the standard Proctor only because this is easily achieved in the field. In order to determine strength properties, specimens were tested as either uncured or cured for 7, 14 and 28 days in the humidity room at a temperature of 25 ± 2°C and relative humidity of 100%.

4.0 ADMIXTURE STABILISATION REACTION

The underlying causative factor for improved properties of soil due to stabilising and admixing is the chemical reaction between stabiliser/admixture and soil particles. This is usually supplemented by compaction and curing.

For soil - cement mixtures, hydration of cement (primary reaction) and the reaction between the free lime product of hydration of cement and the silica and alumina of the clay fraction of the soil as well as the unused components of cement (secondary reaction) were identified by Kedzi (1979). The primary reaction results in the formation of calcium silicate hydrates (CSH), which are cementing substances and are responsible for the initial high strength development while the secondary reaction is pozzolanic in nature and is responsible for the time dependent gain in strength.

Table 2 indicates that except for the percentage of calcium oxide which is very low other chemical components of PCBA nearly match the composition of ordinary Portland cement. Consequently, a form of reaction that can be expected for soil - cement - pulverised coal bottom ash could be a combined action of cement and pulverised coal bottom ash a single stabiliser to obtain very rapid initial strength development.

Reaction between cement and PCBA has the resultant effect of improved workability and reduction in swell, shrinkage and plasticity of the treated soil. However, long term reaction results in gain in strength as mentioned earlier while additional bonding is provided by compaction as this enhances the close packing between particles (mechanical bonding). Soaking and curing, respectively, provide hydraulic and thermal bonding effects.

5.0 RESULTS AND DISCUSSIONS

Evaluation of Untreated Black Cotton Soil

Particle size analysis shows that more than 60% of soil fraction passes the BS sieve No. 200. Also x-ray diffraction on soils from the same area shows that the predominant clay mineral in black cotton soil is montmorillonite (Ola, 1983). The geotechnical properties of the natural soil are summarised in Table 1, while its particle size distribution curve is shown in Fig.1.

The index properties obtained show that the soil belongs to the A-7-6 subgroup of the AASHTO Soil Classification System (AASHTO: Standard, 1986) or CL in the Unified Soil Classification System (ASTM: Annual, 1992). This type of soil does not meet the requirements for subgrade of road pavements.

Control Test

This test was conducted for soil with 5% optimum PCBA concentration only. The compaction test results obtained gave the maximum dry density (MDD) to be 1.59 Mg/m^3 at an optimum moisture content (OMC) of 24%. The maximum UCS of 500 kN/m^2 at 5% PCBA was obtained after 28 days curing and this shows an increase above the 300 kN/m^2 for the natural soil.

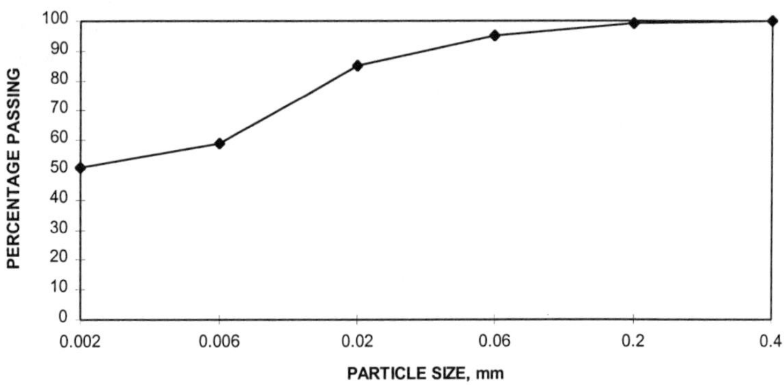

Fig.1: Particle size distribution of natural soil

Stabiliser and Admixture Tests

Moisture-density relationship

The effect of cement/PCBA on MDD and OMC on black cotton soil are shown in Figs. 2 and 3, respectively. Generally, there is a decrease in MDD with increasing concentrations of PCBA. This might not be unconnected with the effect of particle size distribution and specific gravities of soil, cement and pulverised coal bottom ash, which are 2.40, 3.15 and 1.01, respectively. When the black cotton soil with specific gravity of 2.40 is mixed with cement and PCBA, the tendency was for the mixture to decrease in dry density due to the lower specific gravity of PCBA. Increases in OMC recorded were probably due to the larger quantities of water required for the hydration of cement and pulverised coal bottom ash to take place.

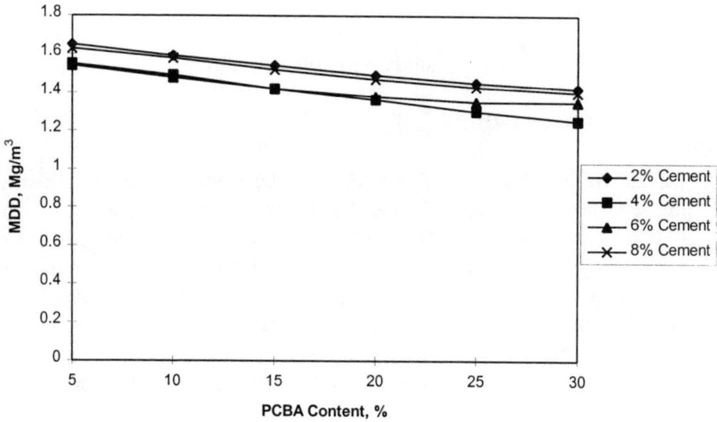

Fig.2: Variation of MDD with pulverised coal bottom ash content for soil-cement-PCBA mixtures.

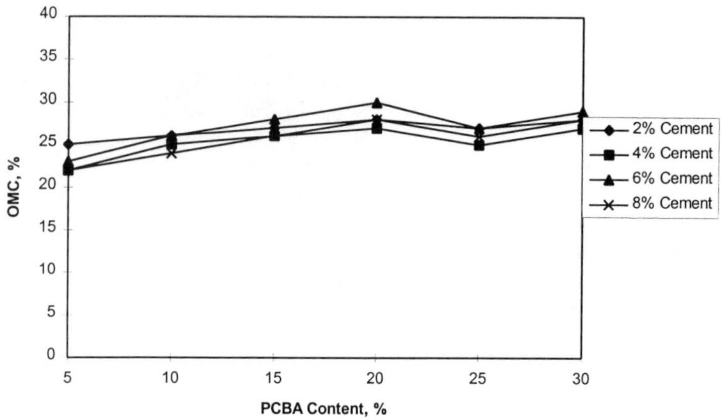

Fig.3: Variation of OMC with pulverised coal bottom ash content for soil-cement - PCBA mixtures.

Strength Characteristics

California bearing ratio: The relationship between the soaked CBR values of the specimens prepared and cement stabilised black cotton soil admixed with PCBA is shown in Fig. 4. The results recorded indicate that the CBR values of the soil - cement mixtures decreased with increased PCBA concentrations. An average peak CBR value of 130% was obtained for specimens with 5% PCBA and 8% cement by weight of dry soil. If CBR values of 10%, 30% and 80% (standard Proctor compaction) are adopted as criteria to be met for subgrade, sub-base and base courses, respectively, then the specimens with the peak value can be conveniently used as a base material.

Unconfined compressive strength: Variation of UCS with PCBA concentrations from 5 - 30% for soil - cement - PCBA mixtures when cured for 7 days; cured for 7 days and soaked for another 7 days; cured for 14 days and 28 days durations are shown in Figs. 5, 6, 7 and 8, respectively. An average peak 7 day UCS of 900 kN/m^2 was recorded for specimens containing 5%

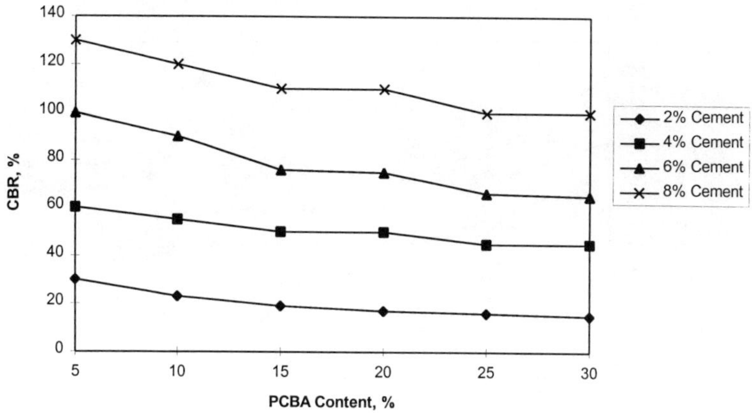

Fig.4: Variation of soaked CBR with pulverised coal bottom ash content for soil-cement-PCBA mixtures.

PCBA and 8% cement by weight of dry soil. This is about 53% of the 1700 kN/m^2 recommended by TRRL (1977) for base courses. However, if the UCS of 687 - 1373 kN/m^2 recommended by Ingles and Metcalf (1972) for soil-cement mixtures used as sub-base of lightly trafficked roads is taken into consideration then the material prepared can be used for sub-base. A factor which could have affected the strength gain of the soil - cement - PCBA mixtures is the maximum dry densities of the different mixtures that decreased with increase in PCBA contents due to the low specific gravity of the admixture as shown in Fig. 2..

Immersion in water, which was used to evaluate durability by the determination of loss of strength, of specimens cured for 7 days and soaked for another 7 days (see Fig. 6) as compared with specimens cured for 14 days (see Fig. 7) showed that the peak strength at 14 days was equal to the unsoaked values for mixtures containing 5% PCBA and cement contents of 6% and 8%, respectively. Consequently, the durability is acceptable since there were no losses in strength for the mixtures mentioned. However, it should be noted that BS 1924 (1990) does not specify any value for resistance of loss in strength measured by the immersion test. It was also observed that soil - cement - PCBA mixtures with PCBA concentrations greater than 20% and cement

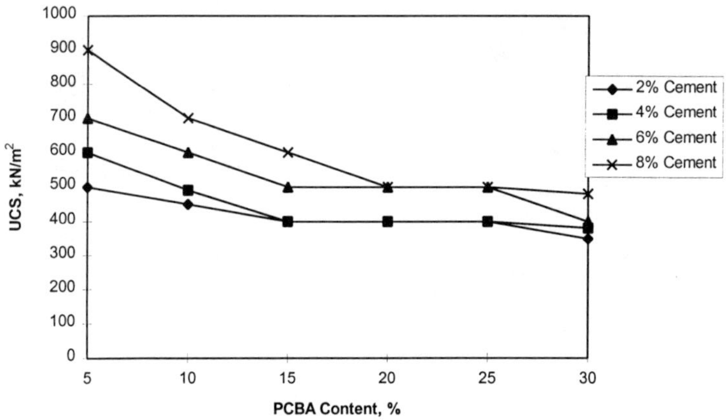

Fig 5: Variation of UCS (7 days curing) with pulverised coal bottom ash content for soil-cement-PCBA mixtures.

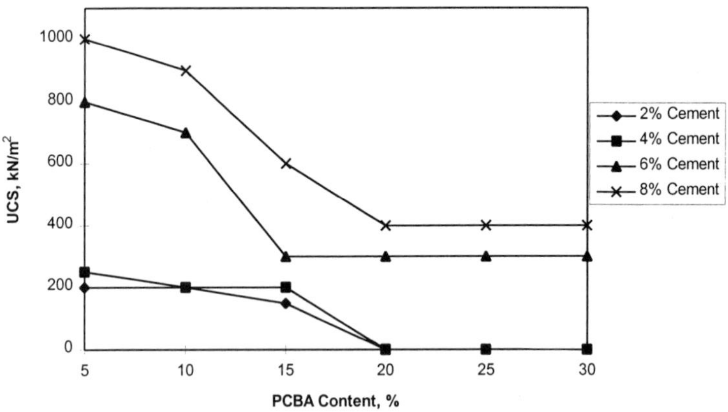

Fig. 6: Variation of UCS (7days cured + 7 days soaked) with pulverised coal bottom ash content for soil-cement-PCBA mixtures.

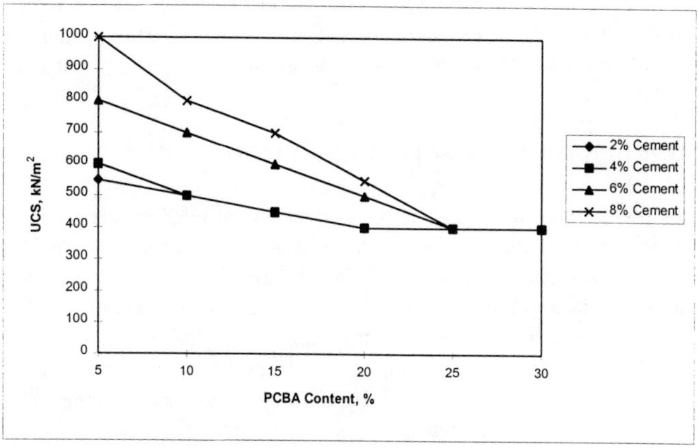

Fig.7: Variation of UCS (14 days curing) with pulverised coal bottom ash content for soil-cement-PCBA mixtures.

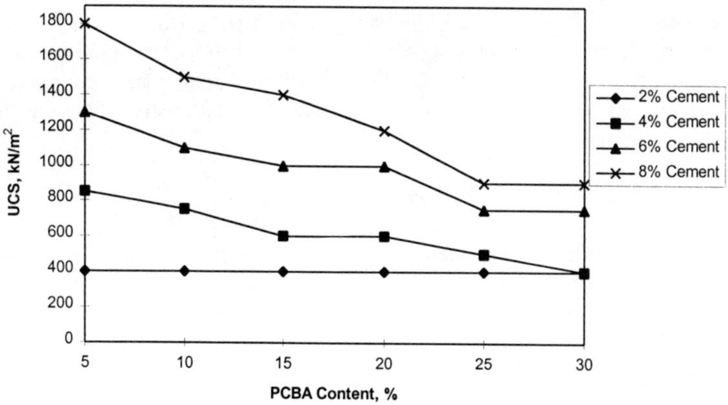

Fig.8: Variation of UCS (28 days curing) with pulverised coal bottom ash content for soil-cement-PCBA mixtures.

contents less than 4% of dry weight of soil totally lost their strength at the end of 7 days soaking after the mandatory 7 days curing period as shown in Fig. 6. Fig. 8 shows that an average peak strength value of 1800 kN/m^2 was also obtained at 5% PCBA and 8% cement content after 28 days curing. Generally, the UCS decreased with increasing PCBA concentrations for all durations and types of curing.

6.0 CONCLUSION AND RECOMMENDATION

Results of experiments on the soil improvement potential of pulverised coal bottom ash on cement stabilised black cotton soil have been presented. The use of 5% by weight of dry soil pulverised coal bottom ash admixed with 8% cement stabilised black cotton soil (A-7-6 or CL in the AASHTO and Unified Soil Classification Systems, respectively) increased the CBR value and UCS from 9% and 300 kN/m^2 for the untreated soil to 130% and 900 kN/m^2, respectively. However, the durability of the UCS specimens is assumed to be acceptable since no loss in strength was recorded using the immersion test prescribed in BS 1924 (1990).

Generally, the strength property enhancement were found to be markedly above the minimum requirement for good subgrade quality in accordance with the Nigerian General Specification (1970) and the treated soil can be used as sub-base of lightly trafficked roads inspite of the high CBR value of the soil - cement - PCBA mixture. The study also highlighted the potential of economically utilising a low cost pozzolanic industrial by-product as an admixture in the cement stabilisation of deficient black cotton soil. Consequently, the cost of improving the strength characteristics of black cotton soils can be reduced.

7.0 ACKNOWLEDGMENT

Mr. A. Toro conducted the test for this study in the Civil Engineering Department Soil Mechanics Research Laboratory, Ahmadu Bello University Zaria, Nigeria. His support is greatly appreciated.

8.0 REFERENCES

AASHTO (1986) Standard Specifications for Transportation Materials and Methods of Sampling and Testing. 14th Ed., Am. Assoc. of State Hwy and Transp. Officials (AASHTO), Washington, D. C.

Adeniji, F.A. (1991)."Recharge function of vertisolic vadose zone in sub-sahelian Chad Basin". Proc. 1st Inter. Conf. on Arid Zone Hydrology and Water Resources, Maiduguri, pp.331-348.

ASTM (1992). Annual book of ASTM standards. Vol. 04.08, American Society for Testing and Materials, Philadelphia.

Balogun, L.A. (1991a). "Foundation investigation and design techniques for residential buildings on expansive black cotton soils". The Nigerian Engineer, Vol. 20, No. 2, pp.34-41.

Balogun, L.A. (1991b)."Effect of sand and salt additives on some geotechnical properties of lime-stabilized black cotton soil". The Nigerian Engineer, Vol. 26, No. 4, pp.15-24.

BS 1377, (1990). Methods of Testing Soils for Civil Engineering Purposes. British Standards Institute, London.

BS 1924, (1990). Methods of Test for Stabilized Soils. British Standards Institute, London.

Chen, H.F. (1975). Foundations on Expansive Soil. Elsevier Scientific Publishers, Amsterdam.

Czernin, w.,(1962). Cement Chemistry and Physics for Civil Engineers, Crosby Lockwood, London.

Indraratna, B., Balasubramaniam, A.A. and Khan, M.J. (1995). "Effect of fly ash with lime and cement on the behaviour of soft clay". Quart. J. of Engrg. Geol., Vol. 28, pp.131-142.

Ingles, O.G. and Metcalf, J.B. (1972). Soil Stabilization -Principles and Practices. Butterworth, Sydney.

Kedzi, A.(1979). Stabilized Earth Roads. Elsevier, Amsterdam.

Nicholson, P.G., Kashyap, V. and Fuji, C.F. (1994)."Lime and fly ash admixture improvement of tropical Hawaiian soils". Transportation Research Record No. 1440, National Academy Press, Washington D.C., pp.71-78.

Nigerian General Specifications, (1970). Roads and Bridgeworks. Federal Ministry of Works, Lagos, Nigeria.

Ola, S.A.,(1978). "Geotechnical properties and behaviour of some stabilized Nigerian lateritic soils". Quart. J. Engrg. Geol., London, Vol. 11, pp.145 -160.

Ola, S.A., (1983). "The geotechnical properties of the black cotton soils of north-eastern Nigeria". In: Tropical Soils of Nigeria in Engineering Practice, S.A. Ola (Editor), Balkema, Rotterdam, pp.85-101.

Osinubi, K.J.,(1995). "Lime modification of black cotton soil". Spectrum Journal, Kaduna, Vol. 2, Nos. 1 and 2, pp.112-122.

TRRL (1977). A Guide to the Structural Design of Bitumen-surfaced Roads in Tropical and Subtropical Countries. Road Note 31, Transport and Road Research Laboratory, H.M.S.O., London.

Simulating Deformation of an Excess Spoil Fill

Richard A. Rohlf,[1] M. ASCE, Xiangwu Zeng,[2] A.M. ASCE, and Larry G. Wells[3]

Abstract

Deformation of an excess spoil fill subjected to infiltration and groundwater seepage is simulated with a matric-stress critical state model. The model includes groundwater and elastoplastic soil constitutive relationships and is formulated in terms of fluid head and net stress plus matric stress, where matric stress is the stress generated by matric suction. Results of the simulation are presented in terms of the pore-water pressure distribution, phreatic surface, shear strain distribution, and settlement at selected points on the fill.

INTRODUCTION

The analysis of unsaturated soils is complicated by the existence of negative pore water pressure which can significantly affect mechanical behavior. One of the characteristics of unsaturated soil behavior that has proved difficult to analyze is the phenomenon of collapse. Collapse occurs when a soil under a constant confining stress experiences a decrease in volume when moving from an unsaturated to a saturated state.

Two factors are required for collapse to occur: (1) a loose soil structure, and (2) a water content less than saturation. Some soils at their natural water content can support a heavy load with only a small amount of consolidation, but when wetted they undergo a considerable reduction in volume. Structures built on collapsible soils can suffer severe damage if these soils become saturated (Dudley, 1970).

The disposal of excess spoil generated by coal mining operations in the Appalachian Coal Field is often accomplished by placing this material in valley fills. Regulations governing the construction of excess spoil fills contain special provisions

[1]Envir. Engr. Consultant, Kentucky Dept. for Surface Mining Reclamation and Enforcement, Frankfort, KY 40601

[2]Assoc. Prof., Dept. of Civil Engrg., Univ. of Kentucky , Lexington, KY 40506

[3]Prof., Dept. of Biosys. and Agric. Engrg., Univ. of Kentucky, Lexington, KY 40506

for "durable rock fills" which allow fills to be constructed by end dumping in a single lift. These fills are dependent on gravity segregation of the material to form an underdrain as the fill is constructed. Over the last several years, concern has increased over the amount of fine material occurring in some excess spoil fills (Robinson and Ventura, 1983). Fine material decreases the overall permeability of the fill and can block or hinder the development of an adequate underdrain. In addition, end dumped fills that contain excess fine material are initially loose, unsaturated, and are subject to unsaturated soil collapse as a result of wetting from rainfall infiltration and groundwater seepage. The amount and timing of collapse may be important in determining the stability of these fills. On the one hand, the reduction in volume which occurs during collapse may induce strain hardening and thus improve overall stability. On the other hand, if a fill is near saturation a rapid reduction in volume could produce increased pore pressure and lead to mud slides and failure.

Due to the limited availability of flat land for development in the Appalachian Coal Field, reclaimed mine sites located near existing communities are attractive sites for economic development projects. These projects can utilize both reclaimed mine benches and the tops of fills for constructed facilities. Understanding the long-term deformation behavior of mine spoil, and especially the potential for collapse, is crucial for foundation design and for ensuring the stability of embankments and fill slopes.

Environmental groups have recently become concerned about the loss of streams and ecosystems which occur with the construction of excess spoil fills. A multi-Federal agency task force is considering fill design requirements which would minimize stream loss and the fill "footprint". These requirements would result in the disposal of more spoil in the mine backfill and create higher fills placed on steeper slopes thus increasing the risk of stability problems and producing a conflict between environmental and geotechnical interests.

This paper describes the development of a model for simulating shear and volume behavior in unsaturated soil including unsaturated soil collapse. The model is used to simulate deformation of an excess spoil fill subjected to infiltration and groundwater seepage.

BACKGROUND

Stress State Variables

Summaries describing the development of stress state variables and constitutive relationships for unsaturated soils can be found in Fredlund and Rahardjo (1993), Alonso et al. (1990), Toll (1990), and Lloret and Alonso (1980). Initial research focused on the development of an effective stress equation similar to that used for saturated soils. Bishop(1959) proposed the relationship

$$\sigma' = \sigma - u_a + \chi(u_a - u_w) \tag{1}$$

where σ' is effective normal stress, σ is total normal stress, u_a is pore-air pressure, u_w is pore-water pressure, χ is a parameter initially assumed to be a function of degree of saturation, and $(u_a - u_w)$ is matric suction. Equation 1 predicts a decrease in effective stress under a reduction in matric suction and thus indicates soil expansion under wetting rather than contraction or collapse.

Primarily due to difficulties in defining the parameter χ for volume change behavior, the search for an effective stress relationship was gradually abandoned in favor of independent stress state variables. Fredlund and Morgenstern (1977) formally proposed the use of net stress $(\sigma_{ij} - u_a\delta_{ij})$ and matric suction $(u_a - u_w)\delta_{ij}$, where σ_{ij} is the total stress tensor and δ_{ij} is the Kronecker delta.

In contrast to the work of Fredlund and Morgenstern (1977), Karube (1988) described matric suction as an internal stress component which can be included in constitutive relationships without explicit addition as an independent stress state variable. For soils in which apparent cohesion was assumed to be a function of matric suction only, Karube proposed that the ultimate strength envelope be defined by

$$q_f = M[p_a + f(\mu)] \tag{2}$$

where q_f is the deviator stress at failure, $(\sigma_1 - \sigma_3)$; p_a is mean net stress, $(\sigma_1 + 2\sigma_3)/3 - u_a$; $f(\mu)$ is a function which converts matric suction, μ, into mean stress; M is the slope of the failure line; and σ_1 and σ_3 are the major and minor principle stresses, respectively. The stress produced by matric suction, $f(\mu)$, was determined from the intersection of the uncorrected ultimate strength line with the p_a-axis.

Review of Collapse Models

Lloret and Alonso (1980) coupled the one-dimensional air and water continuity equations with state surfaces for void ratio and degree of saturation to develop a model for volume change behavior including swelling and collapse. The state surfaces were functions of the independent stress variables $(\sigma - u_a)$ and $(u_a - u_w)$ and were based on the experimental work of Matyas and Radhakrishna (1968). Collapse is determined directly from the void ratio state surface.

Alonso, et al. (1990) developed a constitutive model which addresses some of the basic observed behaviors of unsaturated soil including (1) compression stiffness variation with changes in matric suction; (2) variation in amount of collapse as a function of stiffness and hence matric suction; and (3) irreversible response under stress and suction reversals. The model was formulated using the independent stress state variables, $(\sigma - u_a)$ and $(u_a - u_w)$, and incorporated separate volume change relationships for mean stress and matric suction. The yield surface was also formulated as a function of both mean stress and matric suction. At saturation, the model reduces to the modified Cam clay yield function. One of the fundamental aspects of the model is the loading-collapse (LC) yield curve expressed by the relationship

$$\frac{p_o}{p^c} = \left(\frac{p_o^*}{p^c}\right)^{\left[\frac{\lambda(0)-\kappa}{\lambda(\mu)-\kappa}\right]} \tag{3}$$

where p_o is the preconsolidation stress for unsaturated conditions; p_o^* is the preconsolidation stress for saturated conditions and serves as a hardening parameter; p^c is a reference stress, $\lambda(0)$ and $\lambda(\mu)$ are the virgin compression stiffness parameters for saturated and unsaturated conditions, respectively; and κ is the elastic recompression stiffness parameter applicable to both saturated and unsaturated conditions. Equation 3 defines a set of yield, p_o, values in p_a-μ space for each matric suction. The LC yield curve controls both yielding in response to an increase in mean net stress and collapse which occurs with a reduction in matric suction.

Wheeler and Sivakumar (1995) proposed an elasto-plastic critical state framework for unsaturated soils based on data from controlled suction triaxial tests. The framework was formulated in terms of the state variables of mean stress, deviator stress, matric suction, and specific volume. Within this four dimensional space a normal compression line, critical state line, and state boundary surface were defined. An associated flow rule and yield surface equivalent to the modified Cam clay model were assumed. The model employs an LC yield curve with evolution of the yield curve controlled by Equation 3. Thus, the model is similar to that proposed by Alonso et al. (1990).

Thomas and He (1995) coupled the governing differential equations for heat, moisture, and air transfer with a non-linear elastic stress-strain relationship to develop a model which is capable of predicting unsaturated soil deformation including heave and collapse due to wetting. Collapse is determined with a state function which expresses void ratio as a function of mean net stress and matric suction.

Bolzon et al. (1996) extended the general plasticity model developed by Pastor et al. (1990) for saturated soil to address unsaturated soil behavior including stiffness induced by matric suction and collapse. Bishop's relationship [Equation 1] was substituted for effective mean stress, and the hardening parameter for the yield surface included matric suction as an independent stress parameter. The evolution of the yield function is controlled by a relationship which is formally identical to the LC yield curve [Equation 3] developed by Alonso et al. (1990) except that Bishop's stress is substituted for mean net stress. The strain hardening modulus is a function of Bishop's stress and matric suction, which allows the model to compute a maximum collapse at a critical mean net stress.

MATRIC-STRESS CRITICAL STATE MODEL

The University of Kentucky Department of Biosystems and Agricultural

Engineering has been developing a model for predicting shear and volume behavior of unsaturated soil including collapse and a maximum collapse at a critical pressure. Essential elements of the model are described in the following sections with additional information available in Rohlf and Wells (1999).

Definition of Matric Stress

Rohlf et al. (1997) used the intergranular stress tensor developed by Matyas and Radhakrishna (1968) to define matric stress as the stress generated by matric suction. By applying equilibrium conditions along a curved surface normal to solid contact points in an unsaturated soil matrix, Matyas and Radhakrishna (1968) showed that the intergranular stress could be expressed as

$$\overline{\sigma_{ij}} = \left(p_a + A_w\mu + \int T dx\right)\delta_{ij} + S_{ij} \tag{4}$$

where $\overline{\sigma_{ij}}$ is the intergranular stress or equivalent stress transmitted by solid contacts; A_w is a measure of pore space saturation; T is surface tension; x is the perimeter of the air-water meniscus; and S_{ij} is deviatoric stress. Since $\int T dx$ is a function of water content and thus μ, Equation 4 can be written

$$\overline{\sigma_{ij}} = [p_a + f(\mu)]\delta_{ij} + S_{ij} \tag{5}$$

where

$$f(\mu) = A_w\mu + \int T dx \tag{6}$$

can be considered as a matric stress term. The matric-stress critical state model is developed under the assumption that the state of a soil element can be described by the intergranular stress tensor, $[\sigma_{ij} - u_a\delta_{ij} + f(\mu)\delta_{ij}]$; void ratio, e; water content, w; and soil fabric. Matric stress is treated as an applied confining stress.

Ultimate Strength and Energy Methods for Determining Matric Stress

Although matric suction can be measured directly, adequate methods for the direct measurement of matric stress are not yet available, requiring that matric stress be determined by indirect methods. The following describes methodologies for determining matric stress from ultimate strength and energy relationships.

An ultimate strength procedure for determining matric stress is developed under the assumptions that (1) soil has no true cohesion; (2) apparent cohesion is due to matric suction only; and (3) slope of the failure line, M, is independent of degree of saturation such that $q_f / [p_a + f(\mu)]_f = M$ for all $S > S_t$, where S is degree of saturation, S_t is a lower threshold degree of saturation, and $[p_a + f(\mu)]_f$ is mean net stress plus matric stress at failure. The first and second assumptions eliminate from consideration cemented soils and soils which exhibit cohesion due to electrochemical forces between soil particles. The third assumption eliminates the requirement for separate friction parameters for mean net stress and matric stress.

As indicated in Figure 1, plotting the deviator stress versus mean net stress, p_a, yields a failure line with slope, M_a, and cohesion intercept, q_c

$$q_f = M_a p_a + q_c \tag{7}$$

Plotting the deviator stress versus mean net stress plus matric stress, $p_a + f(\mu)$, yields a failure line with slope, M, and cohesion intercept of zero

$$q_f = M[p_a + f(\mu)] \tag{8}$$

Equations 7 and 8 can be equated and solved to give matric stress at failure

$$f(\mu) = \frac{p_a(M_a - M) + q_c}{M} \tag{9}$$

Additional details on the development of the ultimate strength matric stress formulation can be found in Rohlf et al. (1997).

An energy matric stress relationship can be developed under the same assumptions as the ultimate strength procedure plus an additional assumption that soil behavior is described by an energy relationship such as the modified Cam clay model.

By following Karube's suggestion and replacing p_a with $p_a + f(\mu)$, the modified Cam clay work equation can be written (Roscoe and Burland, 1968)

$$\eta^2 + 2D\eta - M^2 = 0 \tag{10}$$

where $\eta = q/[p_a + f(\mu)]$. Solving for η gives

$$\eta = \frac{-2D + (4D^2 + 4M^2)^{1/2}}{2} \tag{11}$$

With η determined, matric stress can be calculated from

$$f(\mu) = \frac{q - \eta p}{\eta} \tag{12}$$

Equations 11 and 12 allow matric stress to be calculated at any point along the stress path where q, p_a, D, and M are known.

Groundwater Governing Equations

The flow of groundwater in a saturated or unsaturated porous medium can described by

$$\frac{\partial v_i}{\partial x_i} = (n\frac{\partial S}{\partial h} + nS\gamma_w\beta)\frac{\partial h}{\partial t} + S\frac{\partial \varepsilon_v}{\partial t} \tag{13}$$

where v_i is specific discharge; n is porosity; S is degree of saturation; γ_w is fluid unit weight; β is fluid compressibility; ε_v is volumetric strain; h is head, $h = z + u_w/\gamma_w = z + \psi$, z is elevation head, and ψ is pressure head.

The soil-water characteristic curve provides a relationship between matric suction and degree of saturation for the groundwater constitutive relationship. To provide a smooth transition in matric stress between unsaturated and saturated conditions, the soil-water characteristic function can be formulated as a function of matric stress

$$(u_a - u_w) = [1 + g(1 - S)]f(\mu) \tag{14}$$

where g is a constant. At saturation, pore-air pressure is zero and $f(\mu) = -u_w$. Differentiation of Equation 14 gives the storage term in Equation 13

$$\frac{\partial S}{\partial h} = \gamma_w \frac{1}{gf(\mu) - [1 + g(1 - S)]\frac{\partial f(\mu)}{\partial S}} \tag{15}$$

Soil Governing Equations

The equilibrium equation for a differential soil element is given by

$$\frac{\partial \sigma_{ij}}{\partial x_j} - F_i = 0 \tag{16}$$

with $\overline{\sigma}_{ij} = \sigma_{ij} - u_a \delta_{ij} + f(\mu)\delta_{ij}$ for unsaturated conditions, and $\sigma'_{ij} = \sigma_{ij} - u_w \delta_{ij}$ for saturated conditions, where σ_{ij} is total stress, F_i is body force, and σ'_{ij} is effective stress. Body forces incorporated in the formulation include saturated unit weight for saturated conditions, and unit weight of solids, unit weight of fluid, and fluid drag for unsaturated conditions.

Ultimate Strength and Volume Change Relationships

As suggested by Karube (1988), the ultimate strength line in the matric-stress critical state model is given by Equation 2 (Figure 1).

Volume change relationships for unsaturated constant water content hydrostatic compression tests are presented in Figure 2. Volume behavior considering mean net stress alone can be described by

$$e = e_o - \lambda_a \operatorname{Ln}(p_a / p_{ao}) \tag{17}$$

where λ_a is the slope of the normal compression line for mean net stress and e_o is the void ratio at p_{ao}. Volume behavior considering mean net stress and matric stress can be described by following Karube's (1988) suggestion and replacing p_a with $p_a + f(\mu)$

$$e = e_o - \lambda \operatorname{Ln}\left[\frac{p_a + f(\mu)}{p_{ao} + f(\mu)_o}\right] \tag{18}$$

where λ is the slope of the normal compression line and e_o is the void ratio at $p_{ao} + f(\mu)_o$. Differentiating Equations 17 and 18 and equating differentials gives

$$\lambda = \lambda_a \frac{p_a + f(\mu)}{p_a} \frac{1}{1 + \frac{\partial f(\mu)}{\partial p_a}} \tag{19}$$

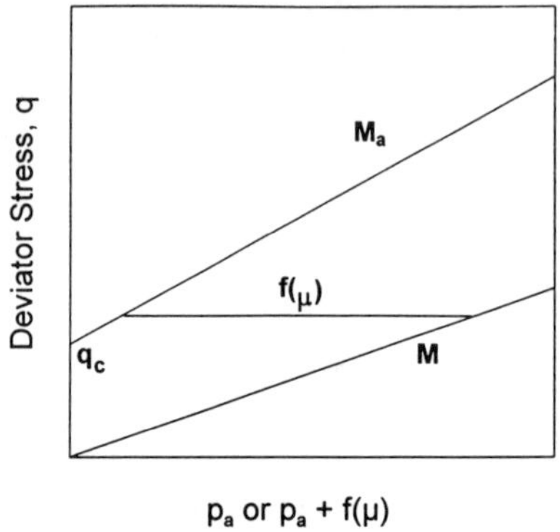

Figure 1. Matric Stress Schematic

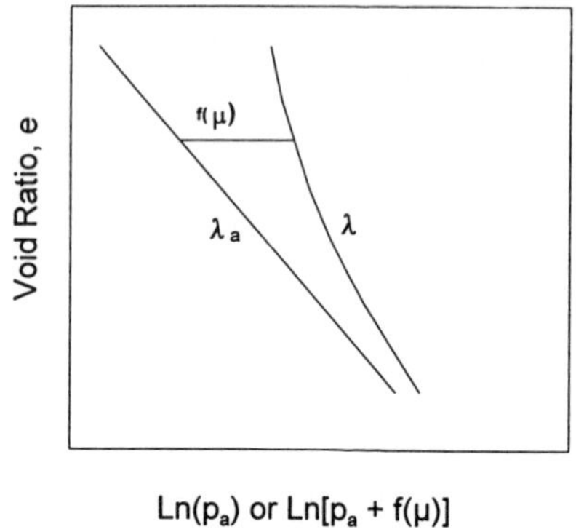

Figure 2. Volume Behavior Schematic

which corrects λ_a for the effects of mean net stress and matric stress.

State Functions for Matric Stress and Slope of the Normal Compression Line

Analysis of matric stress values determined from ten sets of triaxial tests conducted at low confining pressures of 0 - 7 kPa resulted in a matric stress state function of the form (Rohlf et al., 1997)

$$f(\mu) = ae^{b} \exp[c(1 - S)^{2}], \quad S \prec S^{*} \tag{20a}$$

$$f(\mu) = f(e, S^{*}) \frac{(1-S)}{(1-S^{*})}, \quad S \geq S^{*} \tag{20b}$$

where a, b, and c are coefficients and S^* a degree of saturation which defines the breakpoint between the two matric stress relationships. The coefficients can be determined from three triaxial tests conducted at different void ratios and degrees of saturation or by regression if more than three values are available. At saturation, Equation 20a reduces to $f(\mu) = ae^{b}$ which is considered to be the air entry pressure.

Since Atterberg limits have been used for years to describe the consistency of remolded soils, slope of the normal compression line, λ_a, was assumed to be a function of water content. Analysis of hydrostatic compression tests showed that the slope of the normal compression line could be represented by the relationship

$$\lambda_a = \exp(d + fw) \tag{21}$$

where w is water content, and d and f are coefficients with d dependent on the initial void ratio. The coefficients d and f can be determined from two hydrostatic compression tests conducted at the same initial void ratio but different water contents. If hydrostatic compression tests are conducted at different initial void ratios as well as different water contents, then additional points would be required to account for the variation of λ_a with initial void ratio (Rohlf and Wells, 1999). A relationship similar to Equation 21 can also be developed for slope of the recompression line κ_a.

Collapse Yield Curve

The collapse yield curve is given by Equations 18 - 21 and is illustrated in Figure 3 for a range in water content and mean net stress. A soil elementary volume under hydrostatic loading at constant water content, w, follows a single normal compression line in $p_a + f(\mu)$ space with varying slope, λ. However, if water content increases at constant load, p_a, the soil elementary volume moves to a new normal compression line with an accompanying reduction in void ratio. Lines of constant mean net stress, p_a, are shown in Figure 3. The reduction in void ratio occurs because the new normal compression line has a greater slope than the previous normal compression line.

In the matric-stress critical state model, the normal compression line and collapse yield curve are the same. With an increase in water content at constant load, the normal compression line moves to the left producing an unstable stress state which lies outside the yield surface. Collapse occurs as the stress state follows the yield surface.

Finite Element Implementation

The previously described matric-stress critical state model is implemented in a two-dimensional saturated-unsaturated groundwater flow and elastoplastic stress-strain finite element program (Rohlf and Wells, 1999). The model includes groundwater and soil constitutive relationships and is formulated in terms of fluid head and net stress plus

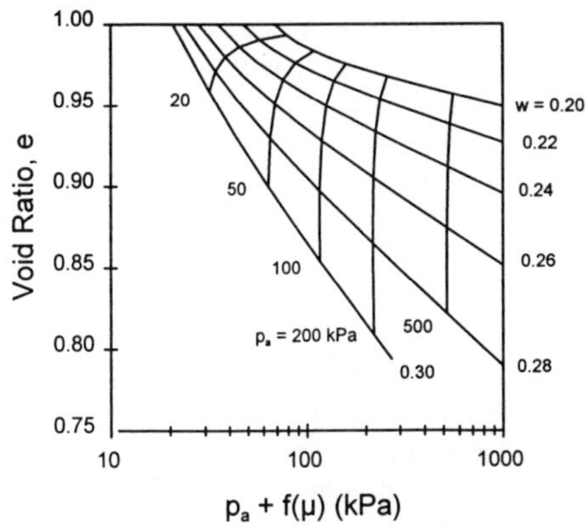

Figure 3. Collapse Yield Curve

matric stress. Matric stress is included in both shear and volume relationships using the state function described by Equation 20. The soil constitutive relationship uses the modified Cam clay yield function.

A Galerkin and finite difference solution is used for the groundwater equation, and a virtual displacement solution is used for the soil equation. The solution is uncoupled and loops over both the groundwater and soil equations until convergence is achieved. Since slope of the normal compression line, λ_a, is a function of water content, solving for unsaturated soil collapse requires integration of the soil constitutive relationships from initial conditions for each incremental change in water content.

DESCRIPTION OF THE STUDY SITE

An end dumped excess spoil fill associated with a multiple seam contour and mountaintop removal coal mining operation located in eastern Kentucky has been selected for study. The completed fill is approximately 370 m long, 150 m wide, and 30 m deep at the valley centerline. Construction of the fill began in 1995 and reclamation was completed in the fall of 1998. Groundwater is flowing from near the toe and a couple of points approximately one-third of the fill height up from the toe. Possible sources of water include infiltration from rainfall and groundwater flow from a coal seam. A two-dimensional limit equilibrium stability analysis using average soil strength parameters ($c = 9.6$ kPa, $\varphi = 30°$, $\gamma = 19.6$ kN/M^3) developed by Huang (1978) and an estimated phreatic surface based on the upper seepage points gives a factor of safety of 1.16 using the simplified Bishop method.

Continued inflow from infiltration and groundwater sources could further increase water content and the phreatic surface, leading to significant deformation or failure. Approximately forty surveying monuments were installed across the face of the fill in December 1999 to monitor for both slope failure and unsaturated soil collapse. If movement is detected, it is anticipated that piezometers and an inclinometer will be installed to monitor water levels and determine movement along shear zones.

DEFORMATION SIMULATION

Model Input Parameters

Input parameters required for the soil constitutive relationship include coefficients for the matric stress, normal compression, and recompression state functions; friction angle, φ; specific gravity of solids, G_s; Poisson's ratio, v; and initial void ratio, e_o. Input parameters for the groundwater constitutive relationship include the coefficient for the soil-water characteristic curve; saturated coefficient of permeability, k_s; fluid unit weight, γ_w; fluid compressibility, β; and unsaturated permeability versus matric suction relationship. A linear function was used to express the relationship between unsaturated water permeability and matric suction, $k_w = k_s - r\mu/\gamma_w$, where r is a constant. The finite element mesh contained 79 linear triangular and quadrilateral elements.

Collection and testing of soil samples from the project fill have not yet been conducted and estimated parameters were used for this analysis. Parameters were estimated from soil strength testing conducted on mine spoil by Huang (1978), testing of a silt loam conducted by Rohlf et al. (1997), and comparison of model simulations with oedometer and isotropic collapse experiments conducted by Tadepalli and Fredlund (1991), and Wheeler and Sivakumar (1995). Parameters based on testing of silt loam were adjusted to account for the larger particle sizes exhibited by mine spoil. Parameters used in the analysis of the excess spoil fill are presented in Table 1.

Loading, Boundary and Initial Conditions

The model uses gravity loading to simulate construction of a fill over time. An initial soil solids specific gravity greater than zero must be used to maintain numerical stability in the solution. It is convenient to express the loading used in the analysis as a percentage of full loading using $(G_s - G_{si}) / G_s \times 100$ where G_{si} is the initial solids specific gravity. The use of fractional loading results in under prediction of strain for self loading and collapse but does not similarly affect stress calculations. In the subsequent analyses, the fill was assumed to be constructed essentially instantaneously (1 day) with $G_{si} = 0.35G_s$ and a loading of 65%.

The analysis assumes that the fill is subjected to a combination of long-term average rainfall infiltration and groundwater inflow from a coal seam. Rainfall infiltration was simulated by decreasing matric suction at the surface from an assumed initial condition of 45.1 kPa to 8.8 kPa over a period of 9 days. The surface matric suction was then maintained at 8.8 kPa. The initial matric suction of 45.1 kPa produced a degree of saturation of 0.44. The final boundary matric suction produced a degree of saturation of approximately 0.78. The coal seam was assumed to have a pore pressure head, ψ, of 3 m. Pore-air pressure was assumed to be atmospheric throughout the simulation.

Results of Simulation and Discussion

Deformation of the excess spoil fill subject to the previously described input parameters, and loading, boundary and initial conditions was simulated with the matric-stress critical state model. The finite element analysis terminated at 64 days due to numerical instability caused by reaching ultimate strength in elements located near the toe of the fill. Reaching ultimate strength in elements near the toe does not necessarily indicate initiation of a progressive failure condition which would lead to failure of the entire fill. The phreatic surface at 64 days is presented in Figure 4 and pore-water pressure head is presented in Figure 5. Figures 4 and 5 show a shallow zone of positive pore pressure on the bench and slope with a somewhat deeper zone of positive pore

Table 1. Model Input Parameters

Parameter	Description	Value
a	Matric-stress state function, Eq.(20)	4.01 kPa
b	Matric-stress state function, Eq.(20)	-2.25
c	Matric-stress state function, Eq.(20)	5.08
S^*	Breakpoint degree of saturation, Eq.(20)	0.9
d_λ	Normal compression state function, Eq.(21)	-7.9
f_λ	Normal compression state function, Eq.(21)	17.0
d_κ	Recompression state function, Eq.(21)	-13.3
f_κ	Recompression state function, Eq.(21)	29.4
φ	Friction angle	30°
G_s	Solids specific gravity	2.68
ν	Poisson's ratio	0.3
e_o	Initial void ratio	1.0
g	Soil-water characteristic curve, Eq.(14)	2.33
r	Unsaturated permeability - matric suction	9.6×10^{-3} /day
k_s	Saturated water permeability	0.1 m/day
γ_w	Unit weight of water	9.81 kN/m^3
β	Water compressibility	5.0×10^{-7} /kPa

pressure near the toe of the fill. Figure 5 shows remaining zones of negative pore-water pressure head in the center of the fill with $\psi < -4$ m.

Deformation in the x-direction (horizontal) versus deformation in the z-direction (vertical) is displayed in Figure 6, and z-deformation versus time is displayed in Figure 7 for three finite element nodes located on the surface of the fill. The positions of the finite element nodes are shown in Figure 4. The z-deformation versus time graph contains three distinct zones: (1) self loading during simulated construction of the fill (1 day); (2) collapse during application of the matric suction boundary condition from 45.1 kPa to 8.8 kPa (days 2 - 10); and (3) collapse at a constant matric suction boundary condition of 8.8 kPa (days 11 - 64). Examination of Figure 7 indicates that approximately 66 % of the z-deformation occurred as a result of collapse. Additional collapse would occur as zones of negative pore-water pressure move closer to the boundary condition (Figure 5).

Shear strain incurred during self loading and collapse is displayed in Figure 8. Comparison of shear strain values at the end of loading (1 day) with shear strain values at 64 days shows that approximately 64% of the shear strain occurred as a result of collapse.

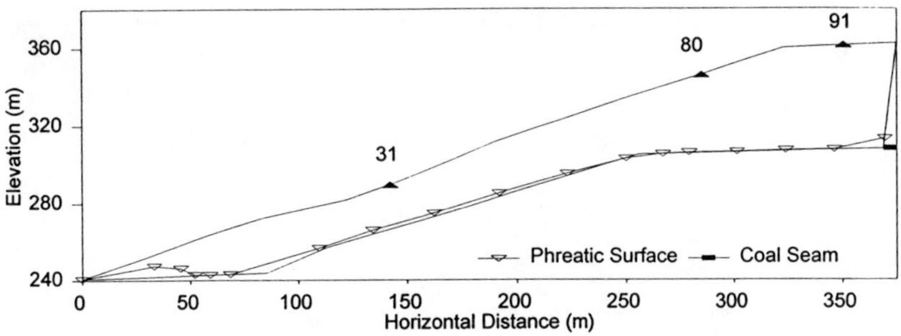

Figure 4. Phreatic Surface at 64 Days

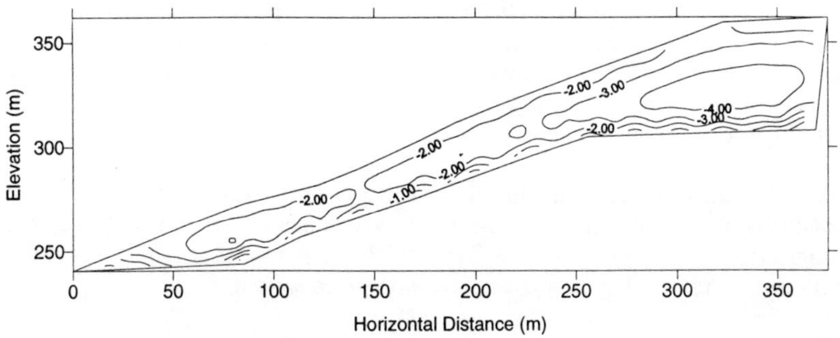

Figure 5. Pore Pressure Head at 64 Days (m)

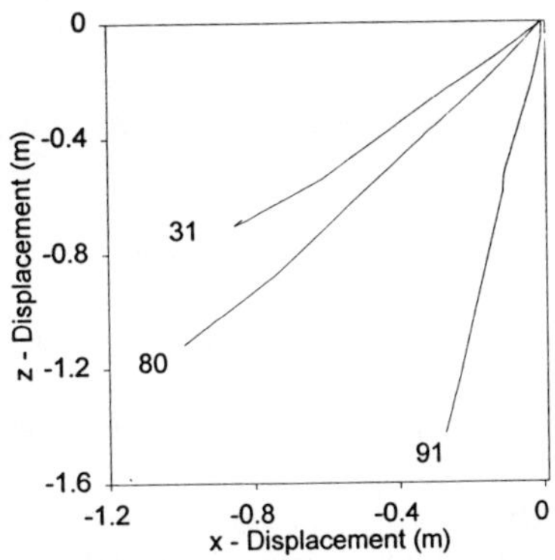

Figure 6. x Versus z-Displacement

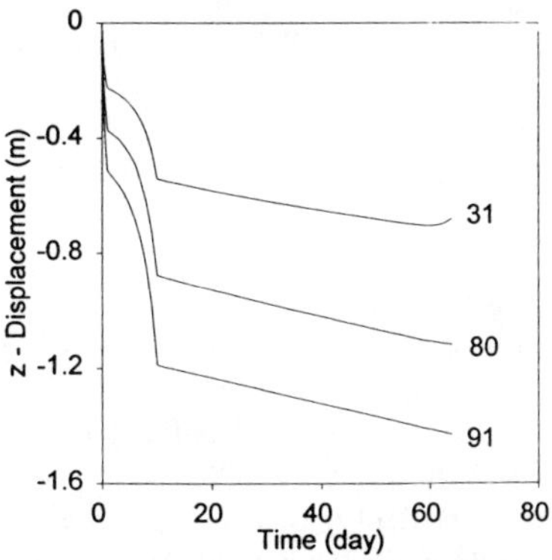

Figure 7. z-Displacement Versus Time

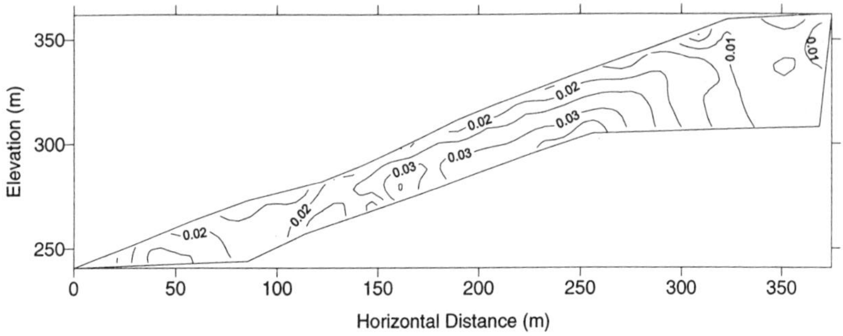

Figure 8. Shear Strain at 64 Days

The above described simulation represents the response of the excess spoil fill to a hypothetical long-term infiltration and groundwater inflow event. Under actual conditions, considerable collapse would take place as the fill is wetted by precipitation during construction. Collapse which occurs during construction would be indistinguishable from settlement incurred due to self loading.

The model demonstrates that future collapse will be dependent on the intensity and duration of storm events and amount of groundwater inflow. Future large storm events may be capable of producing surface deformation and shear strain that can be measured with surveying techniques and a slope inclinometer.

REFERENCES

Alonso, E.E., Gens, A., and Josa, A. (1990). "A constitutive model for partially saturated soils." *Geotechnique* 40(3), 405-430.

Bishop, A.W. (1959). "The principle of effective stress." *Teknisk Ukeblad* 106(39), 859-863.

Bolzon, G., Schrefler, B.A., and Zienkiewicz, O.C. (1996). "Elastoplastic soil constitutive laws generalized to partially saturated states." *Geotechnique* 46(2), 279-289.

Dudley, J.H. (1970). "Review of collapsing soils." *J. Soil Mech. Found. Div.*, ASCE, 96(SM3), 925-947.

Fredlund, D.G., and Morgenstern, N.R. (1977). "Stress state variables for unsaturated soils." *J. Geotech. Engrg.*, ASCE, 103 (GT5), 447-466.

Fredlund, D.G. and Rahardjo, H. (1993). *Soil mechanics for unsaturated soils*. John Wiley, New York, NY.

Huang, H. H. (1978). *Stability of spoil banks and hollow fills created by surface mining*, Institute for Mining and Minerals Research, University of Kentucky, Lexington, KY.

Karube, D. (1988). "New concept of effective stress in unsaturated soil and its proving test." *Advanced triaxial testing of soil and rock*, ASTM STP 977, Robert T.

Donaghe, Ronald C. Chaney, and Marshall L. Silver, Eds., ASTM, 539-552.

Lloret, A. and Alonso, E.E. (1980). "Consolidation of unsaturated soils including swelling and collapse behavior." *Geotechnique* 30(4), 449-477.

Matyas, E.L. and Radhakrishna, H.S. (1968). "Volume Change Characteristics of Partially Saturated Soils." *Geotechnique* 18(4), 432-448.

Pastor, M., Zienkiewicz, O.C., and Chan, A.H.C. (1990). "Generalized plasticity and the modelling of soil behaviour." *Int. J. Numer. Anal. Methods Geomech.* 14, 151-190.

Robinson, M. K. and Ventura, J. D. (1983). "Disposal of Excess Spoil: durable rock fills." In *1983 Symposium on surface mining hydrology, sedimentology and reclamation*, D. H. Graves (ed.), University of Kentucky, Lexington, KY, 179-187.

Rohlf, R. A., Barfield, B.J., and Felton, G.K. (1997). "Ultimate strength matric stress relationship." *J. Geotech. and Geoenv. Engrg.*, ASCE, 123(10), 938-947.

Rohlf, R. A. and Wells L.G. (1999). "Matric-stress critical state model for soils." In *Fourth international conference on constitutive laws for engineering materials*. R.C. Picu and E. Krempl (eds), Rensselaer Polytechnic Institute, Troy, NY, 363-366.

Roscoe, K.H. and Burland, J.B. (1968). "On the generalized stress-strain behaviour of 'wet' clay", in *Engineering Plasticity*, J. Heyman and F.A. Leckie (eds), Cambridge University Press, Cambridge.

Tadepalli, R. and Fredlund, D.G. (1991). "The collapse behavior of a compacted soil during inundation." *Can. Geotech. J.* 28, 477-488.

Thomas, H.R. and He, Y. (1995). "Analysis of coupled heat, moisture and air transfer in a deformable unsaturated soil." *Geotechnique* 45(4), 677-689.

Toll, D.G. (1990). "A framework for unsaturated soil behaviour." *Geotechnique* 40(1), 31-44.

Wheeler, S.J. and Sivakumar, V. (1995). "An elasto-plastic critical state framework for unsaturated soil." *Geotechnique* 45(1), 35-53.

Stress-Strain Behavior of and Hyperbolic Parameters for Structured/Cemented Silts

Timothy D. Stark[1], M., ASCE, Robert M. Ebeling[2], M., ASCE, and Kenneth R. Daly[3], A.M., ASCE

Abstract

Isotropically consolidated-drained triaxial compression tests were performed on undisturbed structured/cemented silt to gain a better understanding of the shear behavior of and hyperbolic stress-strain parameters for this material. The results show that the hyperbolic stress-strain parameters for the structured/cemented silt are anisotropic and differ significantly from the hyperbolic parameters developed for silt reconstituted at the field total unit weight and water. As the confining pressure in a triaxial compression test approaches the preconsolidation pressure, the effects of the structure/cementation are reduced due to bond breakage, and the structured/cemented silt exhibits a stress-strain behavior similar to that of reconstituted silt.

Introduction

The hyperbolic stress-strain model has been shown to be valid for modeling the nonlinear stress-strain behavior of soils prior to failure in soil-structure interaction analyses (Duncan and Clough 1971; Ebeling and Mosher 1996). Ten parameters (modulus number, K, modulus exponent, n, bulk modulus number, K_b, bulk modulus exponent, m, cohesion intercept, c, angle of internal friction at a confining pressure equal to atmospheric pressure, ϕ_0, change in friction angle with increasing pressue, $\Delta\phi$, failure ratio, R_f, unload/reload modulus number, K_{ur}, and unload/reload modulus exponent, n_{ur}) are employed in the hyperbolic stress-strain relationships developed by Duncan et al. (1978) for virgin loading and unloading/reloading conditions. Duncan et al. (1978 and 1980) used the results of isotropically consolidated-drained triaxial compression tests on compacted granular and clayey soils to provide guidance when selecting hyperbolic stress-strain parameters for nonlinear finite-element analyses.

[1]Professor of Civil Engineering, University of Illinois at Urbana-Champaign, 205 N. Mathews Ave., Urbana, IL 61801, 217/333-7394, t-stark1@uiuc.edu

[2]Research Civil Engineer, U.S. Army Engineer Waterways Experiment Station, 3909 Halls Ferry Road, Vicksburg, MS, 39180-6199

[3]Senior Staff Engineer, GeoSyntec Consultants, Huntington Beach, CA 92648

Stark et al. (1991 and 1994) expanded this guidance to freshly deposited, normally consolidated silts and clayey silts. The present study provides this guidance for naturally occurring structured/cemented silts. This was accomplished by conducting oedometer and isotropically consolidated-drained triaxial compression tests. Structured/cemented and reconstituted specimens were used to evaluate the importance of the structure/cementation on the stress-strain behavior and anisotropy of silts. Also, to investigate the effect of inundation on the drained stress/strain behavior of silts, triaxial compression tests were conducted on structured/cemented and reconstituted specimens at the natural degree of saturation and after laboratory saturation. Finally, to estimate the unload/reload modulus and the effect of unloading/reloading on the degradation of structure/cementation, unload/reload triaxial compression tests were conducted.

Mississippi Loess

The Mississippi loess belt is approximately 110 to 193 km wide, extending eastward from the bluffs along the Mississippi River. Generally, the loess is less than 3 m thick except at the bluffs where it is up to 30 m thick. Mississippi loess contains mainly silt and clay size particles. Scanning electron microscope photographs reveal that the silt particles are subangular to subrounded (Fig. 1). The cementing agents in the loess are predominantly carbonates, iron salts, and clays in various combinations (Krinitzsky and Turnbull 1967). Fig. 1 also presents a close up view of the cementation between two silt particles.

Fig. 1. Scanning electron microscope photograph of cementation joining two silt particles (magnification equals 500 times)

Two undisturbed block samples were hand excavated from a nearly vertical 10-m-high loess bluff at the Waterways Experiment Station in Vicksburg,

Mississippi. One of the block samples was used to obtain specimens trimmed in the field orientation and the other block was used to obtain specimens trimmed 90 degrees from the field orientation. The field orientation is referred to as the vertical orientation, and the horizontal orientation corresponds to 90 degrees from the field orientation. Hydrometer analyses revealed that the clay-size fraction, the material finer than 0.002 mm, of the light-brown loess is approximately 10 to 12 percent. Approximately 2 to 3 percent of the loess is fine sand, shells, and organic particles that do not pass the U.S. Standard Sieve No. 200. The loess classifies as a low plasticity silt according to the Unified Soil Classification System and the liquid and plastic limits, natural water content, initial void ratio, degree of saturation, total unit weight, and specific gravity of solids are 30, nonplastic, 19.9%, 0.793, 68.3%, 17.9 kN/m^3, and 2.71, respectively,

Oedometer Testing

Oedometer tests were conducted in accordance with ASTM Standard Test method D2435-94 ("Standard" 1994) to investigate the effect of structure/cementation and inundation on the compressibility of both undisturbed and reconstituted silt specimens and to estimate the preconsolidation pressure. The structured/cemented silt was obtained by trimming a specimen directly from an undisturbed block sample into a rigid oedometer ring in the vertical orientation. Test results show that the partially saturated structured/cemented specimen, trimmed in the field orientation, yielded a preconsolidation pressure, according to Casagrande's construction, of approximately 958 kPa. The corresponding compression and recompression indices, obtained from the void ratio-applied vertical stress relationships, are approximately 0.26 and 0.017, respectively. The oedometer test results were plotted in terms of applied vertical stress and not effective stress because the specimens were partially saturated and capillary pressures were probably present.

An oedometer test was also conducted on a partially saturated reconstituted specimen, obtained by compacting remolded silt in a rigid oedometer ring. The silt was compacted directly into a rigid oedometer ring at the natural water content of 19.9 percent and total unit weight of 17.9 kN/m^3. As expected, the test results show that the reconstituted specimen is more compressible than the structured/cemented specimen (Stark et al. 1996). The reconstituted preconsolidation pressure is approximately 383 kPa, which is less than the structured/cemented value of 958 kPa by a factor of about 2.5. The corresponding compression and recompression indices obtained from the void ratio-applied stress relationships for the reconstituted specimens are estimated to be 0.16 and 0.026, respectively. The compression index is greater for the structured/cemented silt than the reconstituted silt probably because of the collapse of the structure/cementation at stresses that exceed the preconsolidation pressure. The values of the recompression index are less for the structured/cemented silt than the reconstituted silt because of the stiffer behavior at stresses less than the preconsolidation pressure.

Two oedometer tests were conducted to investigate the effect of soaking, or distilled water inundation, on the stiffness and compressibility of the structured/cemented silt trimmed in the vertical orientation. The specimens were inundated by filling the chamber surrounding the specimen container with distilled water, and the test was performed according to ASTM Standard Test Method D2435-94 ("Standard" 1994). However, the specimens were soaked at an applied vertical stress of 115 kPa and 1,101 kPa. An applied vertical stress of 115 kPa corresponds to the recompression range of the silt. An applied vertical stress of 1,101 kPa exceeds the preconsolidation pressure of approximately 958 kPa, and thus corresponds to the virgin compression range.

A negligible difference between the compressibility of the partially saturated and inundated specimens of the structured/cemented silt was observed in the recompression range. Therefore, inundation in the recompression range does not significantly increase compressibility or decrease the stiffness of the structured/cemented silt. However, inundation in the virgin compression range resulted in an increase in axial strain of approximately 1 percent at an applied vertical stress equal to 1,101 kPa. After inundation and an increase in applied vertical stress, the silt exhibited a stress-strain behavior similar to that of the partially saturated specimen.

In summary, inundation of structured/cemented silt in the recompression range did not significantly change the compressibility or stiffness. Therefore it may be concluded that inundation using distilled water does not damage or dissolve the natural structure/cementation of this loess. However, soaking in the virgin compression range may cause an increase in axial strain or a decrease in void ratio. This should be considered when constructing on or inundating structured/cemented silts.

TRIAXIAL COMPRESSION TESTS

Preparation of Structured/Cemented and Remolded Triaxial Specimens

The structured/cemented triaxial specimens were trimmed using a trimming lathe, a fine wire saw, and a 0.3m long razor blade. One set of specimens was trimmed from a block sample oriented in the field orientation and the other set was trimmed from a block sample oriented 90 degrees from the field orientation. The reconstituted triaxial test specimens were fabricated using a stainless steel mold. The remolded silt was compacted directly into the stainless steel mold at the natural water content of 19.9 percent and total unit weight of 17.9 kN/m^3.

Stress-Strain Behavior of and Hyperbolic Parameters for Partially Saturated Silt

A series of isotropically consolidated-drained (ICD) triaxial compression tests were conducted on partially saturated, structured/cemented silt specimens. The specimens were trimmed and tested at the natural water content of approximately 19 percent. Therefore, water was not introduced to the specimens before, during, or after the tests. These tests were conducted at an axial displacement rate of 0.2 mm/minute, which corresponds to an axial strain rate of 1.7 percent/minute. Since the material was partially saturated and the drainage valve to the specimen was open during the consolidation and shear phases of the tests, this axial displacement rate was deemed suitable. However, no water entered or exited the specimens during the tests. Shearing commenced after the specimens came to equilibrium under the applied confining stress (σ_3). The specimens were partially saturated and capillary stresses were probably present at the time of shearing. As a result, the confining pressure is not referred to as an effective stress or a consolidation stress. In addition, the resulting shear strength and hyperbolic parameters are not referred to as effective stress parameters. Since the natural or in situ degree of saturation is approximately 59 percent, no volumetric strain measurements were made during these tests and thus no hyperbolic volume change parameters were derived.

The test results illustrate the effect of structure/cementation and confining pressure on the drained stress-strain behavior of structured/cemented silts. Figs. 2.a through 2.d present the stress difference-axial strain relationships from the four ICD triaxial compression tests on partially saturated, structured/cemented silt (field orientation). The stress difference is defined as the major principal stress (σ_1) minus σ_3. The confining pressures for the four ICD triaxial tests shown in Fig. 2 range from 47.9 to 552 kPa. The resulting Mohr-Coulomb stress cohesion and angle of internal friction are 46 kPa and 28 degrees, respectively (see Table 1).

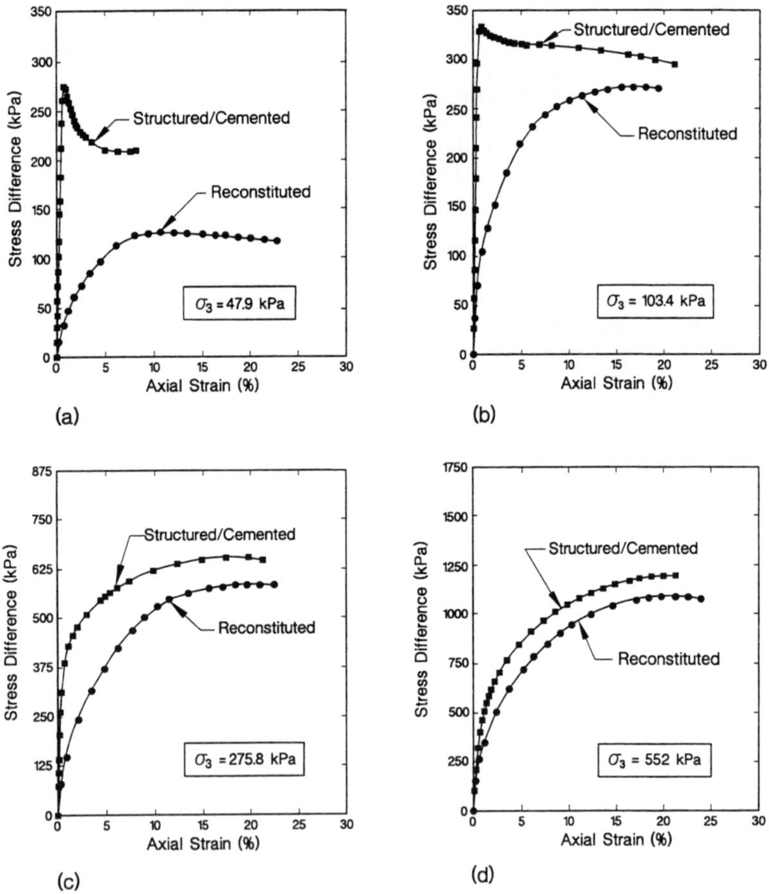

Fig. 2. ICD triaxial compression tests on partially saturated structured/cemented and reconstituted silt tested at field orientation and reconstituted silt at a confining pressure of a) 47.9 kPa, b) 103.4 kPa, c) 275.8 kPa, and d) 552 kPa

Tests on reconstituted specimens were conducted at the same confining pressures as the tests on the structured/cemented silt specimens. These ICD triaxial compression test results are superimposed in Figs. 2.a through 2.d for comparison purposes. At confining pressures of 47.9 kPa and 103.4 kPa, the stress-strain relationship of the structured/cemented silt exhibits a pronounced peak strength and a post-peak strength loss. As the confining pressure increases to 275.8 kPa, the difference in stiffness and maximum stress difference decreases. At a confining pressure of 552 kPa (Fig. 2.d), the structured/cemented and reconstituted silt specimens exhibit similar stiffness and shear strength characteristics. The structured/cemented silt test results in Fig. 2.d no longer exhibit the post-peak behavior attributed to the structure/cementation as shown in Fig. 2.a. Since a confining pressure of 552 kPa is still less than the preconsolidation pressure of approximately 958 kPa, differences in the relationships shown in Figure 2.d are expected. Tests results in Fig. 2 show that the confining pressure can overcome (and possibly "break") the structure/cementation of the undisturbed silt resulting in a stress-strain behavior that approaches the reconstituted silt behavior as the confining pressure approaches the preconsolidation pressure. The cohesion and friction angle are 0 kPa and 30 degrees, respectively, for the reconstituted silt. Therefore, the structure/cementation results in a higher value of cohesion (46 kPa) and a similar value of friction angle (28 degrees) as the reconstituted specimens.

Table 1 - Mohr-Coulomb Shear Strength and Hyperbolic Stress-Strain Parameters for Partially Saturated Structured/Cemented and Reconstituted Silt

Type of Specimen	Average Initial Total Unit Weight (kN/m^3)	Average Initial Water Content (%)	Cohesion (kPa)	Friction Angle (degrees)	Modulus Number (K)	Modulus Exponent (n)	Failure Ratio (R_f)
Structured/ Cemented (Vertical Orientation)	16.7	17.2	46	28	1200	-0.4	0.85
Reconstituted	17.9	19.9	0	30	80	0.8	0.85
Structured/ Cemented (Horizontal Orientation)	17.9	19.9	12	31	305	-0.1	0.95

Table 1 presents the hyperbolic stress-strain parameters obtained from the ICD triaxial compression tests on the partially saturated, structured/cemented and reconstituted silts. The hyperbolic stress-strain parameters for the structured/cemented and reconstituted silt specimens were obtained using the previously reported Mohr-Coulomb shear strength parameters and the best geometric agreement between measured and hyperbolic stress-strain relationships. The geometric agreement was emphasized at axial strains of less than 5 percent to provide a reasonable estimate of the tangent modulus at small strains. The geometric agreement is emphasized at axial strains less than 5 percent because the hyperbolic stress-strain model does not simulate strain softening behavior. The initial estimate of the hyperbolic stress-strain parameters was obtained using the procedure recommended by Duncan et al. (1980) in which the stress difference at 70 and 95 percent of the maximum stress difference are used to estimate the tangent modulus. Figs. 3.a (applied confining stress of 47.9 kPa) and 3.b (applied confining stress of 552 kPa) present typical comparisons of the actual and hyperbolic stress-strain relations for the structured/cemented silts trimmed in the field orientation. It can be seen from Fig. 3a that the best geometric agreement between the actual and hyperbolic relationships occurs at an axial strain less than about 2 percent. Afterwards the structured/cemented silt undergoes a strain softening behavior and the hyperbolic model over predicts the actual stress-strain behavior. However, at an applied confining pressure of 552 kPa, the geometric agreement is better because the structured/cemented silt does not exhibit a strain softening behavior. Table 1 shows that the modulus exponent, n, is negative for the partially saturated, structured/cemented silt. Typically, the modulus exponent is positive, which reflects an increase in stiffness or tangent modulus with increasing effective confining pressure. However, the removal or possible breakage of the structure/cementation with increasing confining pressure causes a decrease in tangent modulus. This behavior is unique to structured/cemented soils and should be incorporated into design considerations.

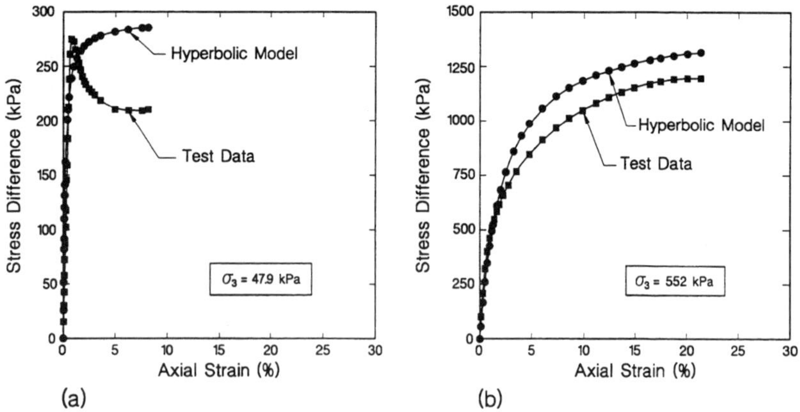

Fig. 3. Comparison of measured and hyperbolic stress-strain relationships of partially saturated structured/cemented silt tested at a confining pressure of a) 47.9 kPa and b) 552 kPa

Figs. 4.a and 4.b present typical comparisons of the actual and hyperbolic stress-strain relations for the reconstituted silts. It can be seen that the actual and hyperbolic stress-strain relationships are in agreement for both applied confining stresses because the reconstituted silt does not exhibit a strain softening behavior. Table 1 also shows the Mohr-Coulomb and hyperbolic parameters for the partially saturated reconstituted silt. In this case, the modulus exponent is positive, which reflects an increase in stiffness or tangent modulus with increasing applied confining pressure. The hyperbolic stress-strain model provides acceptable agreement with the measured stress difference-axial strain data over a significantly larger strain range for the reconstituted silt compared to the results for the structured/cemented silt.

Fig. 5 quantifies the variation in initial tangent modulus (Fig. 5a) and tangent modulus at an axial strain of 0.5 percent (Fig. 5b) for the four structured/cemented silt specimens trimmed in the vertical orientation and the four reconstituted silt specimens. The initial tangent modulus is significantly higher for the structured/cemented silt at low applied confining pressures. However, as the applied confining pressure increases, the difference in initial tangent modulus decreases. This indicates that at higher confining pressures, the effect of structure/cementation is removed and the soil behavior undergoes a transition towards that of a reconstituted material. Because the preconsolidation pressure of the structured/cemented silt is approximately 958 kPa, the structured/cemented and reconstituted silt are expected to approach a similar value of initial tangent modulus at confining pressures in excess of 958 kPa.

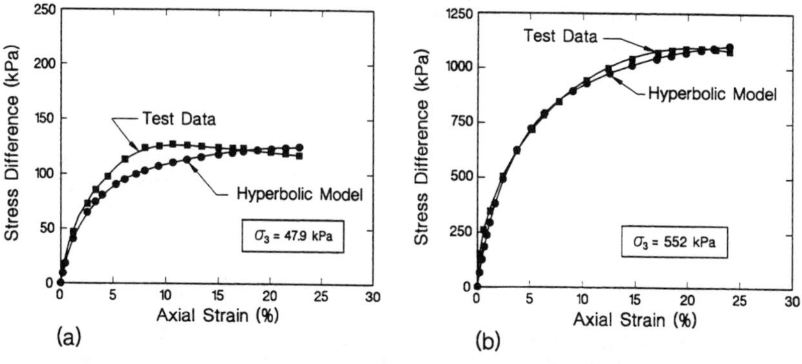

Fig. 4. Comparison of measured and hyperbolic stress-strain relationships of partially saturated reconstituted silt at a confining pressure of a) 47.9 kPa and b) 552 kPa

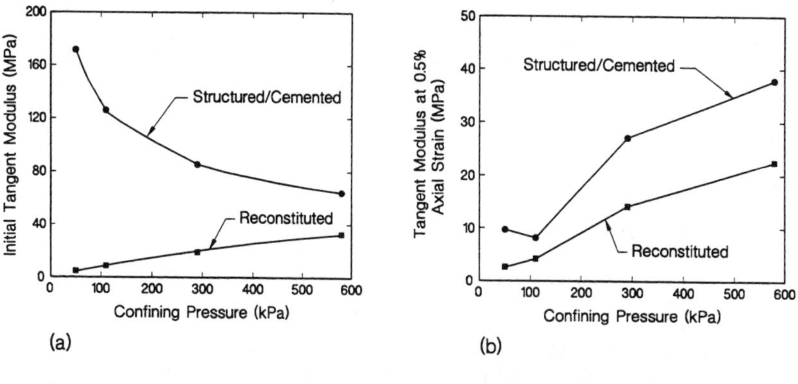

Fig. 5. Variation in tangent moduli for partially saturated structured/cemented and reconstituted silt

The tangent modulus at an axial strain of 0.5 percent increases with increasing applied confining pressure for the structured/cemented and reconstituted silt (Fig. 5 b). The tangent modulus for the structured/cemented silt remains higher than that for the reconstituted silt for confining pressures ranging from 47.9 to 552 kPa.

Anisotropy of Structured/Cemented Silts

Table 1 also presents the Mohr-Coulomb shear strength and hyperbolic stress-strain parameters for partially saturated, structured/cemented silt specimens that were tested in the horizontal orientation. It can be seen that the horizontally oriented silt exhibits similar Mohr-Coulomb shear strength parameters as the vertically oriented silt. However, the horizontal silt exhibits a significantly lower value of modulus number and a greater (less negative) exponent. The modulus number of the vertical silt is approximately four times greater than that of the horizontal silt (Table 1).

The modulus exponent value for the horizontal trimmed silt is -0.10 or nearly zero versus –0.4 for vertical trimmed silt. Therefore, the stiffness or tangent modulus does not decrease as significantly with increasing confining pressure as observed with the vertically oriented specimens. This means a smaller collapse of the structure/cementation occurs in the horizontal orientation. This reinforces the large difference in the modulus numbers, which shows that the structure/cementation is anisotropic. The anisotropy reflects in a stiffer behavior in the vertical orientation.

Stress-Strain Behavior of and Hyperbolic Parameters for Saturated Silt

Silt specimens were saturated to investigate the effect of laboratory saturation on the stress-strain behavior or performance of the structure/cementation in Vicksburg silt. In addition, saturation of the specimens allowed volume change information to be obtained. A series of ICD triaxial compression tests were conducted on structured/cemented silt specimens after the specimens had been saturated in the laboratory. The laboratory saturation technique, referred to as hydrostatic saturation, involved percolating distilled-deaired water through the specimen under a hydraulic head of 0.30 m, or 3 kPa, for a period of 24 hr. A confining pressure of 24 kPa was applied to the specimen prior to the hydrostatic saturation. This saturation process resulted in a B value of 0.95 to 0.99 and therefore it was assumed that little, if any, capillary pressures were present after the saturation process. As a result, the shear strength parameters are referred to as effective stress parameters. After completion of the saturation process, the desired confining pressure was applied. Upon equilibration, the specimen was sheared at an axial displacement rate of 0.02 mm/minute, which corresponds to an axial strain rate of 0.026 percent/minute.

ICD triaxial compression tests were conducted on four saturated/cemented and three reconstituted vertical silt specimens, after hydrostatic saturation. The resulting effective stress cohesion and friction angle were 0 kPa and 33 degrees, respectively, for the structured/cemented specimens, and 0 kPa and 31 degrees, respectively, for the reconstituted specimens (see Table 2). The Mohr-Coulomb shear strength parameters of the saturated structured/cemented silt specimens result in a slightly lower shear strength than that measured for the partially saturated structured/cemented silt ($c = 46$ kPa and $\phi = 28$ degrees). For example, at a normal stress of 192 kPa, a saturated silt specimen exhibited a shear strength that is 15 percent less than that for the partially saturated silt specimen. This finding provides insight to the importance of the various cementation mechanisms in the loess. The two major cementation mechanisms appear to be carbonate and clay/capillary effects (Krinitzsky and Turnbull 1967). Since the partially saturated and saturated triaxial specimens generally yielded similar shear strength and compressibility parameters, it was concluded that the carbonate cementation is resistant to distilled water and is a stronger cementing agent than the clay/capillary effects. The increase in moisture content from approximately 19 to 26 percent during laboratory saturation probably reduced the effect of the clay/capillary bonding. Since the shear strength after laboratory saturation was similar to the shear strength of the loess at the natural water content, the importance of the clay/capillary bonding was assumed to be small. Therefore, the difference between the shear strength of the laboratory saturated structured/cemented and reconstituted specimens is attributed to carbonate cementation.

Table 2 - Mohr-Coulomb Shear Strength and Hyperbolic Stress-Strain Parameters for Saturated Structured/Cemented and Reconstituted Silt

Type of Specimen	Average Initial Total Unit Weight (kN/m^3)	Average Initial/ Final Water Content (%)	Effective Stress Cohesion and Friction Angle (kPa/deg)	Modulus Number and Exponent (K/n)	Failure Ratio (R_f)	Bulk Modulus Number and Exponent (K_b/m)
Structured/ Cemented (Vertical Orientation)	16.7	17.3/ 27.5	0/33	850/ -0.55	0.85	310/ -0.6
Reconstituted	17.2	19.6/ 25.4	0/31	95/ 0.80	0.85	24/ 0.57
Structured/ Cemented (Horizontal Orientation)	17.9	18.2/ 26.0	12/31	305/ -0.10	0.95	50/ -0.4

Table 2 presents the effective stress Mohr-Coulomb and hyperbolic stress-strain parameters for the hydrostatically-saturated, structured/cemented silt (field orientation). The geometric agreement between the data and the hyperbolic stress-strain model was emphasized at axial strains of 5 percent or less because the hyperbolic model does not capture strain softening behavior. Table 2 shows that the modulus exponent is negative, which reflects the degradation of the structure/cementation with increasing confining pressure. The value of modulus number and exponent for the reconstituted structured/cemented silt specimens are 95 and 0.8, respectively (Table 2). These values are within the range of values reported by Stark et al. (1991) for reconstituted clay/silt mixtures and the modulus exponent indicates an increase in stiffness with increasing confining pressure.

Mohr-Coulomb shear strength and hyperbolic stress-strain parameters for saturated, structured/cemented silt, tested in the horizontal orientation, are also presented in Table 2. The horizontally oriented silt exhibits similar effective stress

Mohr-Coulomb shear strength parameters as the vertically oriented silt. However, the horizontal silt exhibits significantly lower values of modulus number and exponent. The modulus number of the vertical silt is approximately three times greater than that of the horizontal silt (Table 2). The modulus exponent value for the nonfield-oriented silt is again -0.10 or near zero. This means there is a smaller collapse of the structure/cementation in the nonfield orientation than in the field orientation even after saturation. Therefore, the anisotropy observed in the partially saturated, structured/cemented silt appears not to be significantly altered by laboratory saturation. This anisotropy should be considered when laterally loading the structure/cemented silt deposit.

Hyperbolic Volume Change Parameters of Saturated Structured/Cemented Silt

Volume change parameters for the laboratory saturated structured/cemented and reconstituted silt specimens, (Table 2) were obtained using the best geometric agreement between measured and hyperbolic stress-strain relationships at low axial strains, i.e., before the specimen exhibited dilation during shearing. Thus, the volume change model is restricted to low strains (e.g., two percent or less). In contrast to the results of the structured/cemented specimens, the volume change model provides a reasonable representation of the volumetric strain relationship for the reconstituted specimens. The reconstituted specimens exhibited no dilation during shearing, which can be attributed to the absence of structure/cementation after the reconstitution process. As expected, the structured/cemented silt specimens exhibited a negative bulk modulus exponent, and the reconstituted specimens exhibited a positive bulk modulus exponent (Table 2). The bulk modulus number of the structured/cemented silt tested in the field/vertical orientation is thirteen times higher than K_b for the reconstituted silt. The bulk modulus number for the horizontal orientation is two times higher than the value of K_b estimated for the reconstituted silt.

Drained Unload/Reload Hyperbolic Stress-Strain Parameters

Four unload/reload isotropically-consolidated triaxial compression tests were conducted to estimate the unload/reload modulus of the structured/cemented silt and the effect of unloading/reloading on the degradation of the structure/cementation. The tests were conducted on partially saturated (i.e., at the natural water content) structured/cemented silt. A partially saturated, structured/cemented specimen was loaded in the vertical orientation to approximately 50 percent of the maximum principal stress difference measured in a previous test (see Figs. 2.a through 2.d) at the same confining pressure. The specimen used in the unload/reload test was sheared to 50 percent of the maximum principal stress difference using an axial displacement rate of 0.2 mm/minute or axial strain rate of 1.7 percent/minute. After reaching 50 percent of the maximum principal stress difference, the specimen was unloaded to a stress difference of zero using an axial displacement rate of 0.2 mm/minute. The specimen was then reloaded to 50 percent of the maximum principal stress difference and unloaded. This was repeated until the specimen had been subjected to four unload/reload cycles. Figs. 6.a and 6.b present the stress difference-axial strain relationships from the ICD unload/reload triaxial compression tests conducted at confining pressures of 47.9 kPa and 552 kPa, respectively. Two other tests (not shown) were conducted in the same fashion at confining pressures equal to 103.4 kPa and 275.8 kPa.

The structured/cemented specimen tested in the ICD unload/reload triaxial test at a confining pressure of 47.9 kPa collapsed unexpectedly after application of a principal stress difference of approximately 36 kPa after the unload/reload cycles (Fig. 6.a). It is important to note that the axial strain scale in Fig. 6.a is exaggerated compared to Fig. 6.b so the specimen collapse could be illustrated. The three remaining structured/cemented specimens were tested at higher confining pressures under unload/reload cycles and were loaded to failure without collapsing (e.g., Fig. 6.b). Therefore, it appears that the structure/cementation is more susceptible to collapse under cyclic loading at low confining pressures.

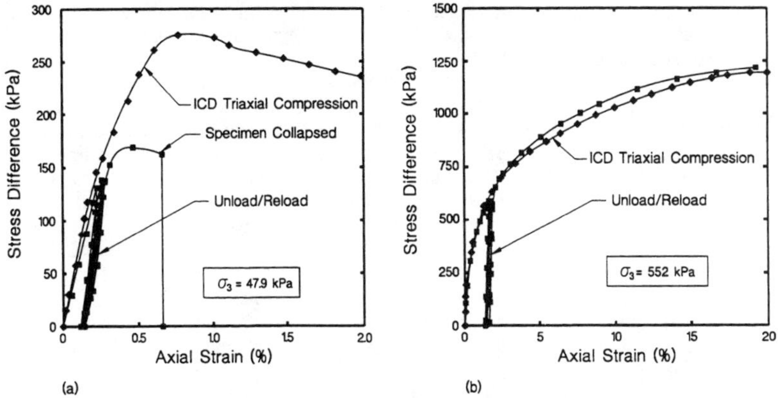

Fig. 6. Comparison of unload/reload and conventional ICD triaxial compression test results on partially saturated structured/cemented silt at a confining pressure of a) 47.9 kPa and b) 552 kPa

The resulting cohesion and friction angle for the cyclically loaded structured/cemented silt are 33.5 kPa and 29 degrees, respectively. The corresponding Mohr-Coulomb shear strength envelope is slightly less than that corresponding to the combined cohesion (46 kPa) and friction angle (28 degrees) measured in conventional ICD triaxial compression tests on partially saturated, structured/cemented silt (Table 1). Therefore, the unloading/reloading of the specimen may have broken or ruptured some of the structure/cementation in the silt. Not all of the structure/cementation was removed during the four unload/reload cycles because the resulting cohesion is greater than the cohesion (c = 0 kPa) measured for the partially saturated reconstituted specimens (Table 1). However, more than four unload/reload cycles or unload/reload cycles with a stress difference greater than 50 percent of the maximum stress difference may result in additional breakage or rupture of the structure/cementation in the silt.

Figs. 6.a and 6.b present a comparison of stress difference-axial strain relationships for unload/reload and conventional ICD triaxial compression tests at confining pressures of 47.9 kPa and 552 kPa, respectively. At a confining stress of 47.9 kPa, the unload/reload cycles resulted in a peak stress difference that is approximately 50 percent lower than that observed in the comparable ICD triaxial compression test (Fig. 6.a). At a confining stress of 552 kPa, the unload/reload cycles did not significantly reduce the peak stress difference (Fig. 6.b). In conclusion, the unload/reload cycles appear to be more detrimental at lower confining pressures. The reduced confinement probably allows some of the structure/cementation to be damaged.

The unload/reload modulus number was estimated using the hysteresis loops obtained from the unload/reload triaxial compression tests and the procedure outlined by Duncan et al. (1980). Values of K_{ur} and n_{ur} equal to 1250 and 0.3, respectively, were estimated from the unload/reload data, some of which is shown in Fig. 6. The value of K_{ur} is approximately equal to the primary loading modulus number (1200) of the partially saturated structured/cemented silt (Table 1). The ratio of K_{ur}/K is approximately unity, which is slightly less than the recommended range of 1.2 to 3 suggested by Duncan et al. (1980).

However, the recommended range of K_{ur}/K is only applicable if n_{ur} equals n (Duncan et al. 1980). Duncan et al. (1980) suggest that, for nonstructured soils, the value of n_{ur} is similar to n, which is the modulus exponent for primary loading. The values of K (1200) and K_{ur} (1250) are similar but the values of modulus exponent differ significantly. Because the difference in exponents is attributed to the structured/cemented nature of the silt, the conclusion that n_{ur} equals n appears unwarranted for structured/cemented soils. As a result, it is recommended that at least one unload/reload triaxial compression test be conducted to estimate the appropriate values of K_{ur} and n_{ur} in structured/cemented materials instead of using K and n to estimate these values. For soil-structure interaction analysis involving the Vicksburg silt/loess, reasonable values of K_{ur} and n_{ur} appear to be 1250 and 0.3, respectively.

Conclusions

Based on the results of this study, the following conclusions appear to be warranted:

(1) The structure/cementation present in Vicksburg, Mississippi loess results in high shear strength and stiffness characteristics. The structure/cementation frequently allows slopes to stand at or near a vertical angle. The two major cementation agents in Mississippi loess appear to be carbonates and clay/capillarity. The carbonate cementation appears to provide the greatest contribution to the overall structure/cementation.

(2) Structure/cementation in the Mississippi loess results in preconsolidation pressures that significantly exceed the effective overburden pressure and preconsolidation pressures measured for reconstituted loess/silt specimens. As a result, this preconsolidation effect is probably not caused by historical prestressing, but by particle cementation.

(3) As the confining pressure in a triaxial compression test approaches the preconsolidation pressure of the Mississippi silt, the effects of the structure/cementation are diminished due to bond breakage and the silt begins to exhibit a stress-strain behavior similar to that of reconstituted silt.

(4) Saturation of the structured/cemented Mississippi loess with distilled water did not significantly alter the compressibility, shear strength, or stress-strain behavior of the material. However, inundation in the virgin compression range in oedometer tests resulted in an increase in axial strain, without a change in applied vertical stress.

(5) The hyperbolic stress-strain parameters for the partially saturated, structured/cemented Mississippi loess differ significantly from the reconstituted values. The modulus number for the structured/cemented silt is three times higher than the value for reconstituted silt. This indicates a higher stiffness, and thus a higher initial tangent modulus. However, the structured/cemented modulus exponent is negative, which indicates that the tangent modulus decreases with increasing effective confining pressure.

(6) Tables 1 and 2 can be used to estimate the Mohr-Coulomb shear strength and hyperbolic stress-strain parameters for partially saturated and saturated (with distilled water) structured/cemented or reconstituted Mississippi loess, respectively.

(7) Horizontally oriented, i.e., 90 degrees from field orientation, Mississippi loess specimens exhibit similar Mohr-Coulomb shear strength parameters as the vertically oriented specimens. However, the horizontally oriented silt exhibits significantly lower values of modulus number and exponent. Therefore, the stiffness of structured/cemented Mississippi loess is anisotropic.

(8) Unload/reload cycles initiated at a principal stress difference corresponding to 50 percent of the maximum principal stress difference in a conventional ICD triaxial compression test resulted in slightly lower values for the Mohr-Coulomb strength parameters compared to the values measured in conventional ICD triaxial compression tests. However, the structured/cemented Mississippi loess appears to be more susceptible to collapse under cyclic loading at low confining pressures. This is reflected in the hyperbolic stress-strain parameters. The average primary and unload/reload modulus numbers of partially saturated, structured/cemented Mississippi loess (vertical orientation) are similar (1200 versus 1250). However, the values of the modulus exponent differ significantly (-0.4 versus 0.3). In summary, unload/reload cycles, especially at low confining pressures, may damage or rupture the structure/cementation.

Acknowledgement

Financial support for this research was provided by the National Science Foundation (Grant No. BCS-93-00043) and the Waterways Experiment Station of the U.S. Army Corps of Engineers. The first author also acknowledges the support provided by the W.J. and E.F. Hall Scholar Award. Permission was granted by the U.S. Army Chief of Engineers to publish this information.

Appendix. References

Duncan, J.M., Byrne, P., Wong, K. S., and Mabry, P. (1978), "Strength, Stress-Strain and Bulk Modulus Parameters for Finite Element Analyses of Stresses and Movements in Soil Masses," Report No. UCB/GT/78-02, University of California, Berkeley, CA.

Duncan, J.M., Byrne, P., Wong, K. S., and Mabry, P. (1980), "Strength, Stress-Strain and Bulk Modulus Parameters for Finite Element Analyses of Stresses and Movements in Soil Masses," Report No. UCB/GT/80-01, University of California, Berkeley, CA.

Duncan, J.M. and Clough, G.W. (1971), "Finite Element Analyses of Port Allen Lock," *Jour. of Soil Mechanics and Foundation Engng. Div.*, ASCE, 97(8): 1053-1068.

Ebeling, R.M. and Mosher, R.L. (1996), "Red River U-Frame Lock No. 1 Backfill-Structure-Foundation Interaction," *Jour. of Geotech. Engng. Div.*, ASCE, 122(3): 216-225.

Krinitzsky, E.L. and Turnbull, W.J. (1967), "Loess Deposits of Mississippi," Special Paper Number 94, Geological Society of America, Boulder, Colorado, 64 p.

Standard Test Method for One-Dimensional Consolidation Properties of Soils; D2435-90. (1994), American Society for Testing and Materials, Philadelphia, Pa.

Stark, T. D., Daly, K. R., and Ebeling, R. M. (1996), "Hyperbolic Stress-Strain Parameters for Structured/Cemented Silts," Tech. Report ITL-96-8, U.S. Army Engineer Waterways Experiment Station, Vicksburg, MS.

Stark, T. D., Ebeling, R. M., and Vettel, J. J. (1994), "Hyperbolic Stress-Strain Parameters for Silts," *Jour. of Geotech. Engng. Div.*, ASCE, 120(2): 420-441.

Stark, T. D., Vettel, J. J., Fitzwilliam, S. M., and Ebeling, R. M. (1991), "Soil-Structure Interaction Parameters for Silts," Tech. Report ITL-91-2, U.S. Army Engineer Waterways Experiment Station,

Nature of Principal Strain Response of Unsaturated Soils Under Multi-Axial Stress States

Laureano R. Hoyos Jr., M. ASCE [1]
Emir J. Macari, M. ASCE [2]

Abstract

A servo-controlled true triaxial (cubical) testing device was modified to test 10-cm cubical specimens of unsaturated soil under varying matric suction conditions. Recompacted silty sand specimens were tested following a multi-stage testing scheme by controlling the major, intermediate, and minor principal stresses. Matric suction states in the specimens were induced and maintained constant during testing by using the axis-translation technique. True triaxial testing data from a series of drained, stress/suction-controlled triaxial compression, triaxial extension, and simple shear tests, were used to evaluate the nature of principal strain response of unsaturated soils under varying matric suction conditions, and for shear stress paths not achievable in a conventional cylindrical apparatus. Normalized test data of octahedral stress versus principal strain, on the basis of matric suction level, indicate consistent stress-strain behavioral patterns. The paper presents these test results and proposes behavioral trends for unsaturated soils as a function of matric suction.

Introduction

Geotechnical laboratory testing of unsaturated soils usually requires special modifications of the conventional oedometer, triaxial, and direct shear apparatuses commonly used to analyze the stress-strain-strength characteristics of saturated soil specimens. These modifications must accommodate the independent measurement or control of the pore-air (u_a) and pore-water (u_w) pressures in the unsaturated soil sample. Most of the volume change constitutive relations, the shear strength

[1] Assistant Professor, University of Texas at Arlington, Arlington, TX 76019.
[2] Chairman and Bingham C. Stewart Distinguished Professor, Department of Civil and Environmental Engineering, Louisiana State University, Baton Rouge, LA 70803.

characteristics, and the elasto-plastic critical state-based constitutive models proposed for unsaturated soils (Alonso et al. 1990; Fredlund and Rahardjo 1993; Wheeler and Sivakumar 1995), involve oedometer, triaxial, and direct shear testing using the axis-translation technique originally proposed by Hilf (1948, 1956). These testing devices allow for the application of load and measurement of stress along limited paths and modes of deformation, that is, under hydrostatic, one-dimensional, or axisymmetric loading conditions. In reality, soils in nature are often subjected to three-dimensional stress states, $(\sigma_{ij} - u_a\delta_{ij})$ and $(u_a - u_w)\delta_{ij}$, as a result of external stress gradients and varying matric suction conditions. It is in this context where a true triaxial (cubical) testing device, capable of controlling the matric suction state imposed to the unsaturated soil specimen, plays a fundamental role in the stress-strain-strength characterization of unsaturated soils under general stress states.

Stress State Variables and Testing Technique for Unsaturated Soils

1. Stress state variables: Fredlund and Rahardjo (1993) presented a theoretical stress analysis of an unsaturated soil on the basis of multiphase continuum mechanics. The analysis concluded that two independent sets of stress state variables, the net normal stress, $(\sigma - u_a)$, and the matric suction, $(u_a - u_w)$, may be used to describe the shear strength and the volume change behavior of an unsaturated soil. For the general stress case, the three principal stresses are known as: net major principal stress, $(\sigma_1 - u_a)$, net intermediate principal stress, $(\sigma_2 - u_a)$, and net minor principal stress, $(\sigma_3 - u_a)$.

2. The axis-translation technique: Difficulties arise in testing unsaturated soils with negative pore-water pressures approaching −1 atm (cavitation). The *axis-translation technique* is commonly used in the laboratory testing of unsaturated soils in order to prevent having to measure pore-water pressures less than zero absolute. The procedure involves a *translation* of the reference or pore-air pressure (u_a) such that the pore-water pressure (u_w) can be referenced to a positive air pressure.

Figure 1 illustrates the application of the axis-translation technique to an unsaturated soil specimen undergoing a matric suction of 101 kPa. The pore-water pressure (u_w) is measured below a saturated high-air-entry porous ceramic disk with an air-entry value of 202 kPa. The water phase of the specimen was originally under 101 kPa tension, i.e., $u_w = -101$ kPa. An air pressure of 202 kPa is then applied directly to the specimen in order to raise the pore-air pressure (u_a) to the same magnitude, i.e., $u_a = 202$ kPa. As a result, the pore-water pressure (u_w) is increased by an equal amount, with a final *positive* value of 101 kPa, i.e., $u_w = 101$ kPa, as shown in Figure 1. Under these conditions, there is no problem associated with the cavitation of the water in the measuring system. Despite the translation of the reference pore-air pressure from 0 to 202 kPa, the matric suction experienced by the soil remains constant at 101 kPa.

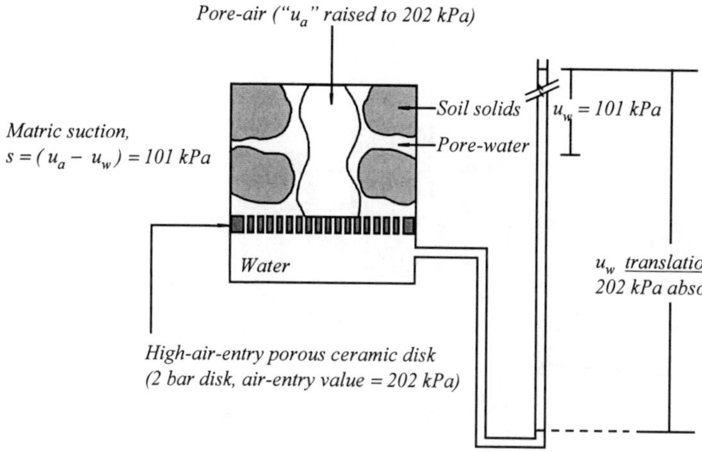

Figure 1. Concept of Axis-translation Technique.

Periodic flushing in the water compartment is required to eliminate the possible presence of entrapped air bubbles originated by diffusion of pore air through the ceramic disk. The axis-translation technique was used in suction-controlled true triaxial tests conducted in the present work.

A Computed-based, Stress/suction-controlled True Triaxial Testing Device

Figure 2 shows a cross sectional view of a computer-based, stress/suction-controlled true triaxial (cubical) testing device implemented at Georgia Institute of Technology to simulate true triaxial stress states in soils under partially saturated conditions. The 10-cm cubical specimens were tested following a multi-stage testing scheme by controlling the major, intermediate, and minor principal stresses. Matric suction states in the specimens were induced and maintained constant during true triaxial testing by using the axis-translation technique.

The device implemented herein is a mixed-boundary type of cell, with the cubical soil specimen seated on top of a water-saturated, 5 bar high-air-entry disk, and between five LATEX/POREX membranes, as shown in Figure 2 (Hoyos Jr. 1998). The device permits the study of stress-strain response of an unsaturated soil under simple-to-complex stress paths and varying matric suction conditions. The complete computer-driven testing setup consists basically of a frame (*1*) featuring six pressure cavities to accommodate five (one top and four lateral) flexible LATEX/POREX membranes (*2*), and a cubical base aluminum piece (*3*) at the bottom to hold the saturated 5 bar ceramic disk (*4*).

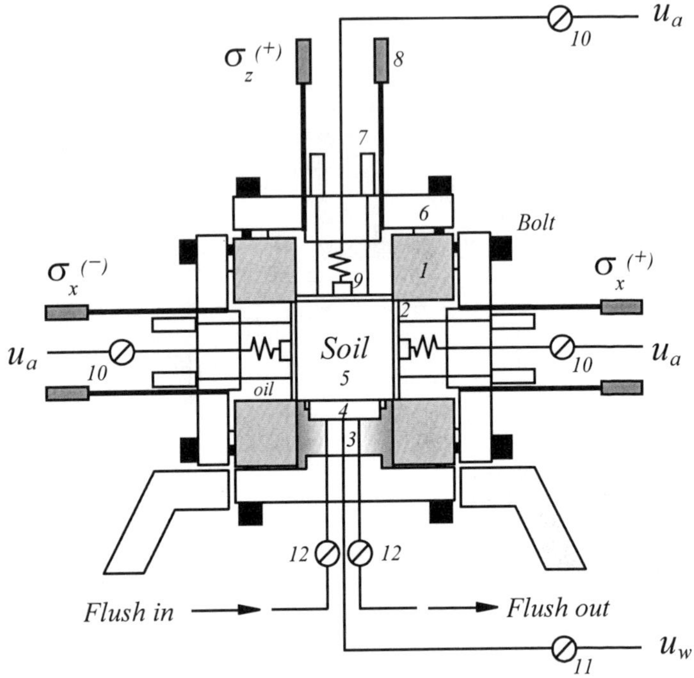

Figure 2. Cross Sectional View of Stress/Suction-controlled Cubical Device.

Once the soil sample (*5*) is mounted, the remaining five walls (*6*) are assembled to the frame. Three LVDT's (*7*) per face are used to monitor soil deformations. Hydraulic fluid is used to pressurize the specimen. The external pressure is transmitted to the fluid through pressure inlet/outlet connections (*8*) on the walls. The pore-air pressure (u_a) is applied to the top and four lateral faces of the specimen via a small block (*9*) attached to the flexible membranes with a threaded stem. The air pressure is applied via flexible nylon tubing (*10*) from the pressure board. The pore-water pressure (u_w) is applied to the bottom of the soil sample through the saturated 5 bar disk (*4*). The water pressure is applied via flexible nylon tubing (*11*) from a water compartment.

A *flushing* mechanism (*12*) has been added to the bottom wall assembly to eliminate any air bubbles that migrate through the 5 bar, high-air-entry ceramic porous disk by diffusion. The tests are entirely computer-driven via a data acquisition and process control system (*DA/PCS*). Figure 3 shows a general view of the complete testing setup and its main external/hardware components.

Test Material and Experimental Program

The material used herein is a remolded, low-plasticity Piedmont residual type of soil, classified as silty sand (SM) according to the USCS. The 10-cm cubical specimens were compacted in 10 layers using the in-place *tamping* compaction method, at a moisture content of 4 % less than the optimum from the Standard Proctor. The in-place *tamping* compaction corresponds to a compactive effort considerably less than the Standard Proctor dynamic compaction test, and resulted in an average dry density, $\gamma_d = 1.08$ Mg/m^3, and an average degree of saturation, $S_r = 48$ %.

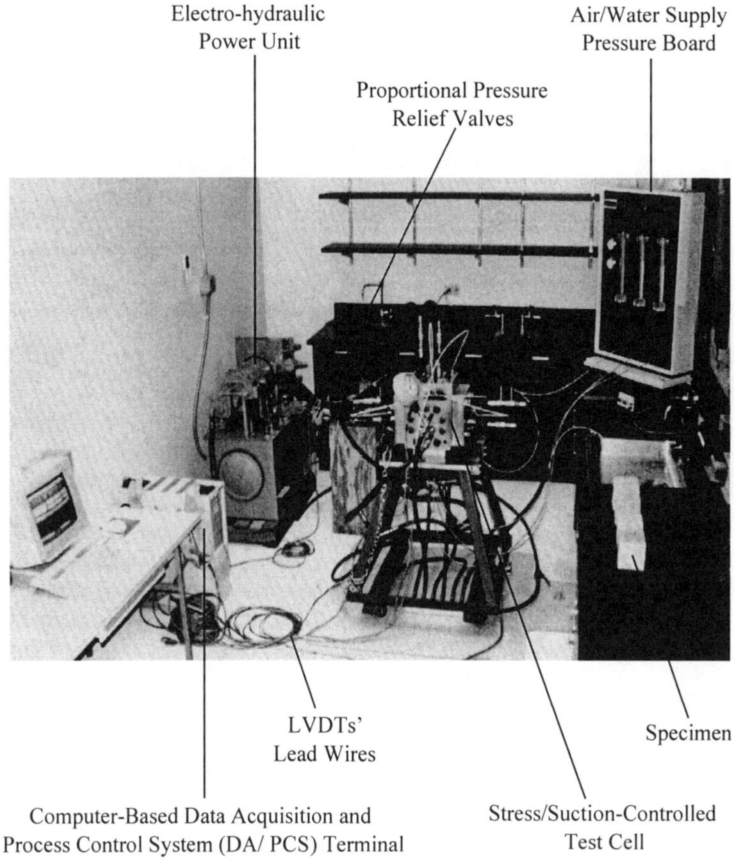

Figure 3. Photograph of Complete Laboratory Testing Setup.

The series of drained (constant suction), stress/suction-controlled tests conducted in this study included: **(1)** Hydrostatic compression (HC) tests; **(2)** Triaxial compression (TC) tests *(b = 0)*; **(3)** Simple shear (SS) tests *(b = 0.5)*; and, **(4)** Triaxial extension (TE) tests *(b = 1)*. All tests were conducted at a loading rate of 10 kPa/hr, to allow for proper excess pore-air and pore-water pressure dissipation, and at three different levels of net octahedral normal stress, σ_{oct} = 50 kPa, 100 kPa, and 200 kPa, for three different values of matric suction, s = 50 kPa, 100 kPa, and 200 kPa. Each specimen followed a multistage testing procedure where the pre-established matric suction, $s = (u_a - u_w)$, was always kept constant, as depicted in Figure 4, where the stress ratio, b, is defined as,

$$b = \left(\frac{\sigma_2 - \sigma_3}{\sigma_1 - \sigma_3} \right)$$

During constant-suction (drained) true triaxial testing, any excess pore-air and pore-water pressures, caused by the applied loads, are dissipated by allowing the pore fluids (air and water) to flow in or out of the soil specimen. In most cases, the matric suction induced in the pore-water of the specimen was achieved by applying a constant air pressure via the top and lateral pore-air-pressurizing units (block, stem, and POREX membranes), and a variable water pressure in the grooved water compartment beneath the 5 bar, high-air-entry ceramic disk, as depicted in Figure 2.

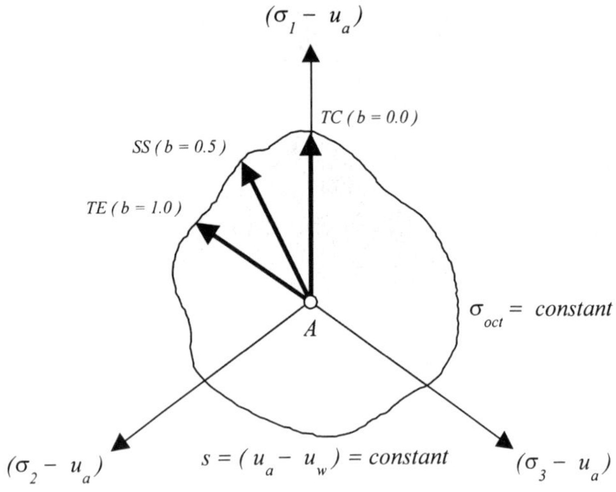

Figure 4. Experimental Suction-controlled Stress Paths in the Octahedral Plane.

Nature of Principal Strain Response

Figures 5, 6, and 7, show the principal strain response of recompacted silty sand from the series of TC, SS, and TE tests conducted herein. Figure 5 presents test results for $\sigma_{oct} = 50$ kPa. Figure 6 presents test results for $\sigma_{oct} = 100$ kPa. Finally, Figure 7 presents test results for $\sigma_{oct} = 200$ kPa. All test data are presented on the basis of matric suction level ($s = 50$, 100, and 200 kPa), in terms of the normalized octahedral shear stress, $\tau_{oct} / (\tau_{oct})_{max}$, versus the major, intermediate, and minor principal strains (ε_1 , ε_2 , ε_3). The maximum octahedral shear stress, $(\tau_{oct})_{max}$, corresponds to a total shear strain (ε_q) of approximately 12 % in all cases. The net octahedral normal stress (σ_{oct}) and the octahedral shear stress (τ_{oct}) were calculated as follows,

$$\sigma_{oct} = \frac{\sigma_1 + \sigma_2 + \sigma_3}{3} - u_a$$

$$\tau_{oct} = \frac{1}{3}\sqrt{(\sigma_1 - \sigma_2)^2 + (\sigma_2 - \sigma_3)^2 + (\sigma_1 - \sigma_3)^2}$$

Figures 5–7 show a paramount influence of matric suction, $s = (u_a - u_w)$, on the stress-strain-strength response of recompacted silty sand specimens under general stress states. For high values of matric suction (i.e., $s = 200$ kPa), the maximum octahedral shear stress, $(\tau_{oct})_{max}$, is reached at lower deformation levels. This is a clear indication of the direct effect of matric suction on the stiffness of the unsaturated soil material. In general, all normalized curves show similar trends, regardless of the suction level induced in the sample. Nevertheless, the level of confinement minimize the influence of matric suction, as the normalized curves tend to be closer to each other for increasing levels of net octahedral normal stress (σ_{oct}).

During constant suction TC tests, as the intermediate and the minor principal stresses were equally decreased, the corresponding strains were found to be expansive (–). Instead, the major principal stress was increased during testing, and consequently, its corresponding strain was compressive (+). **During constant suction SS tests**, the major principal stress was increased, and, at the same time, the minor principal stress was decreased; while the intermediate principal stress was maintained constant. Consequently, the major principal strain was compressive (+), and the minor principal strain was expansive (–). As the intermediate principal strain (ε_2) was very small, the SS type of test condition closely corresponds to a constant matric suction, plain strain condition. **During constant suction TE tests**, the intermediate and major principal stresses were equal and equally increased, while the minor principal stress was decreased. Consequently, the intermediate and major principal strains were found to be compressive (+), and the minor principal strain was expansive (–). Table 1 summarizes all experimental values of the maximum octahedral shear stress, $(\tau_{oct})_{max}$.

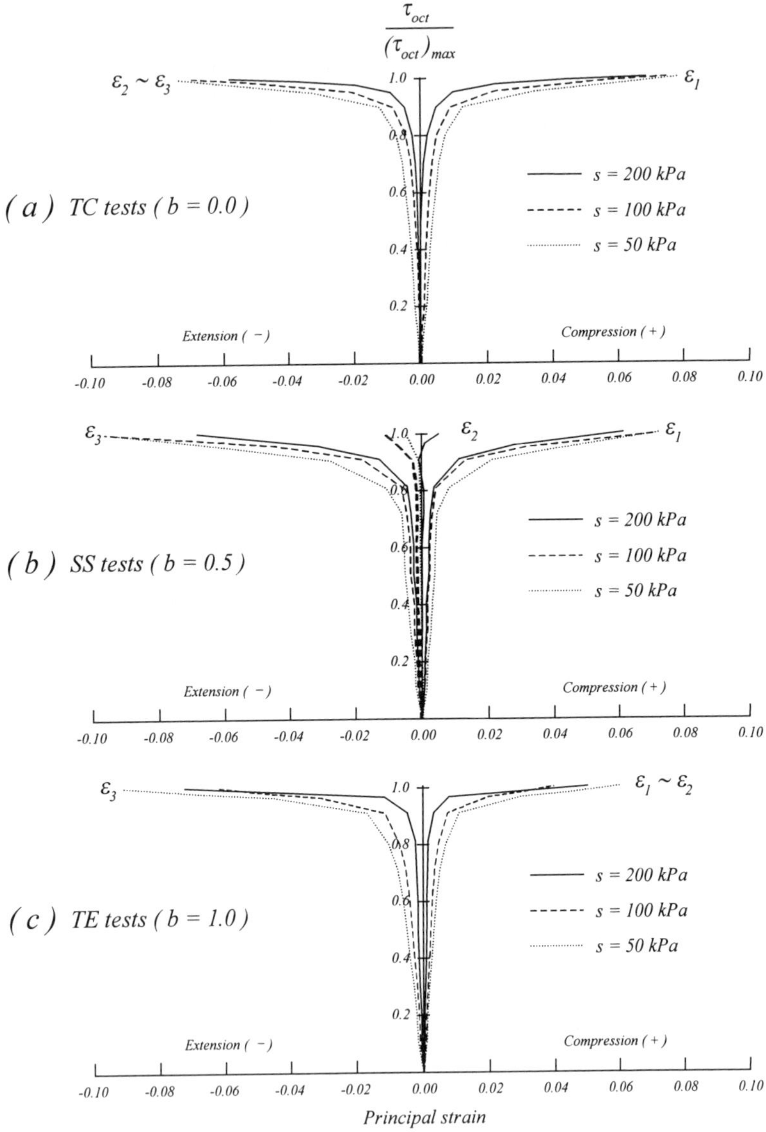

Figure 5. Principal Strain Response for, $\sigma_{oct} = 50$ kPa.

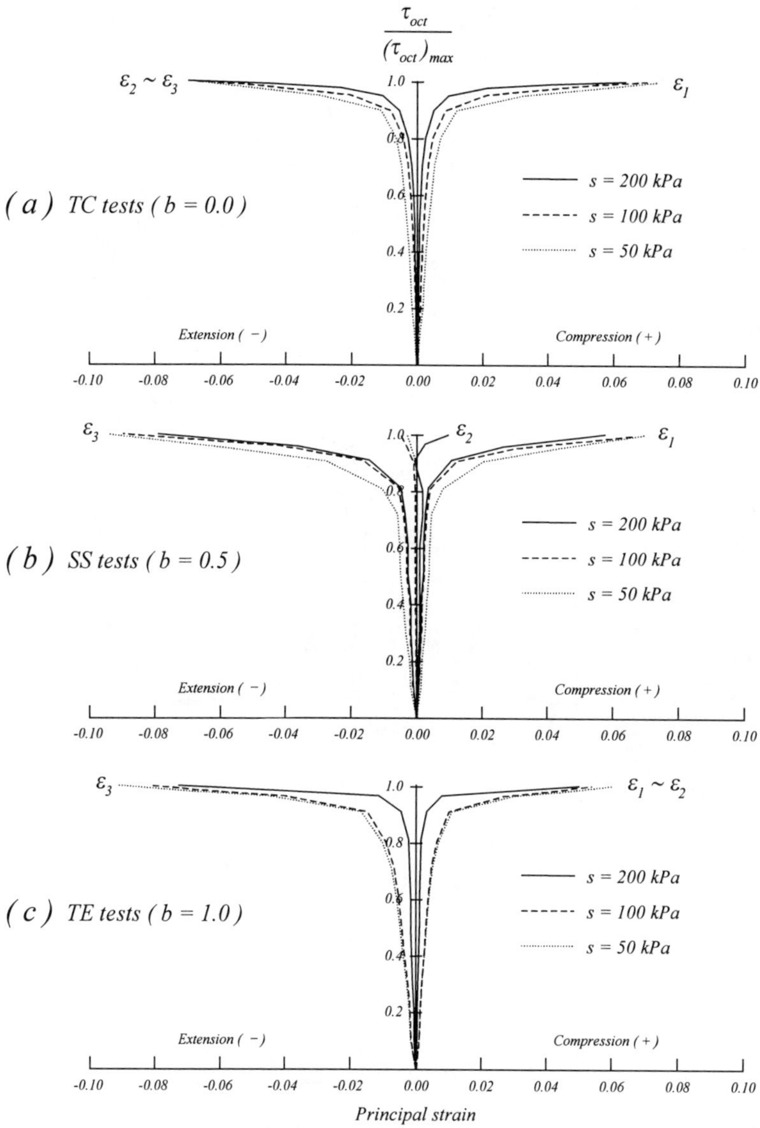

Figure 6. Principal Strain Response for, $\sigma_{oct} = 100$ kPa.

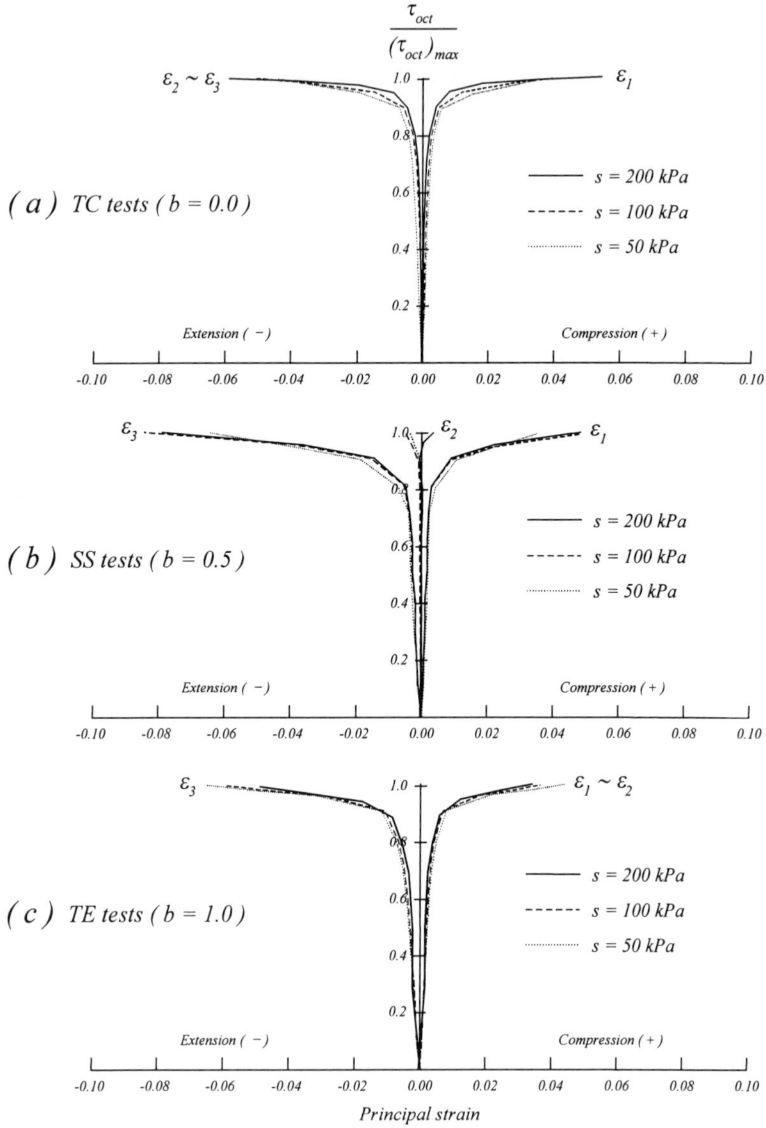

Figure 7. Principal Strain Response for, $\sigma_{oct} = 200$ kPa.

Table 1. Experimental Values of Maximum Octahedral Shear Stress, $(\tau_{oct})_{max}$: kPa.

Test Type	Matric suction, $(u_a - u_w)$: kPa	Net octahedral normal stress, σ_{oct} : kPa		
		50	100	200
TC	50	56	80	128
	100	88	118	168
	200	154	180	222
SS	50	40	70	96
	100	80	105	134
	200	120	148	182
TE	50	47	64	97
	100	80	100	133
	200	120	148	180

References

Alonso, E. E.; Gens, A.; and Josa, A. (1990). "A constitutive model for partially saturated soils." Géotechnique, **40**(3), 405-430.

Fredlund, D. G.; and Rahardjo, H. (1993). "*Soil Mechanics for Unsaturated Soils.*" John Wiley and Sons, Inc., New York, NY, 57 p.

Hilf, J. W. (1948). "Estimating Construction Pore Pressures In Rolled Earth Dams." Proc., 2nd International Conference on Soil Mechanics and Foundation Engineering, Rotterdam, Vol. III, 234-240.

Hilf, J. W. (1956). "An Investigation of Pore-Water Pressure in Compacted Cohesive Soils." Technical Memorandum No. 654, U.S. Department of the Interior, Bureau of Reclamation, Design and Construction Division, Denver, CO.

Hoyos Jr., L. R. (1998). "Experimental and computational modeling of unsaturated soil behavior under true triaxial stress states." Ph.D. dissertation, Georgia Institute of Technology, 352 p.

Wheeler, S. J.; and Sivakumar, V. (1995). "An elasto-plastic critical state framework for unsaturated soils." Géotechnique, **45**(1), 35-53.

SWELLING BEHAVIOR UNDER OEDOMETRIC AND TRIAXIAL LOADING CONDITIONS

Mosleh A. Al-Shamrani[1] and Abdullah I. Al-Mhaidib[2]

ABSTRACT

Preconstruction treatment or the selection and design of a foundation system both rely on accurate estimates of the potential heave of the supporting expansive soil. Swell percentage of expansive soils is commonly obtained from oedometer swell tests under fully lateral restraint conditions. In situ heave, however, is not a one-dimensional volume change but more likely a three-dimensional process. Consequently, surface heave predicted from results of oedometer tests is usually conservative and largely differs from heave actually observed in the field. To account for the discrepancy between oedometer and in situ boundary conditions, some researchers have suggested a lateral restraint factor to be applied to heave predictions evaluated based on parameters obtained from oedometer swell tests.

This paper presents a laboratory testing procedure by which the vertical and the volumetric swells of compacted expansive soil samples are concurrently measured using a hydraulic triaxial stress path apparatus. A series of triaxial swell tests is conducted and the results are compared with the volume change measured in specimens tested under one-dimensional loading conditions in the oedometer. The percentages of vertical swell obtained from the triaxial swell tests are found to be considerably lower than the corresponding values measured in the oedometer tests. For the triaxial swell tests, the ratio of swell in the vertical direction to the volumetric swell, SR, is evaluated. The test results indicate that SR increases with increasing applied confining pressure, and for a particular swell test, the value of SR is not constant but rather changing with elapsed swelling time.

INTRODUCTION

Expansive soils increase in volume when subjected to variance in moisture content, and hence causing serious engineering problems on structures built on these soils. A substantial literature have unequivocally demonstrated and spelled out the severity and extent of damages inflicted by soil deposits of swelling nature in many urban areas throughout the world. Damages related to swelling soils range from minor cracking of

[1] Assoc. Professor, Department of Civil Engineering, College of Engineering, King Saud University, P.O. Box 800, Riyadh 11421, Saudi Arabia.

[2] Assis. Professor, Department of Civil Engineering, College of Engineering, King Saud University, P.O. Box 800, Riyadh 11421, Saudi Arabia.

pavement or interior finishes in buildings, which is very common, to irreparable displacement of footings and superstructure elements. Vertical ground movements associated with the expansive soil deposits accounts for more expense annually than the combined effects of flooding, earthquakes, hurricanes and tornadoes (Jones and Holtz, 1973).

The high cost of repairing structural damages associated with swelling soils has drawn attention to the need for more dependable evaluations of various engineering characteristics of such problematic soils. After identifying a soil as being expansive, the next most important step is to accurately quantify the amount of anticipated heave the soil will undergo when subjected to variance in moisture content. Reliable estimate of surface heave is required for the selection of treatment alternatives to minimize volume change or in the preparation of a foundation design to accommodate the anticipated volume change.

The oedometer swell tests are the methods most widely used to identify soils that might swell and to evaluate the amount of swell that may occur. This is mainly because of simplicity of its operation and the availability of conventional consolidometer in most soil mechanics laboratories (Kassiff and Ben-Shalom, 1971). However, a large discrepancy is usually found between heave predicted using parameters obtained from oedometer swelling tests and that actually realized in the field. Comparison of field and laboratory data obtained from oedometer tests revealed that the laboratory results from oedometer tests overestimate the in situ heave by a factor of about 3 (Gizienski and Lee, 1965; Richards, 1967; Johnson and Snethen, 1979; Dhowian et al., 1990; Erol, 1992). This suggests that about one-third of the volume change is reflected as a surface heave, the remainder will be laterally. Accordingly, to convert the potential volume change to the anticipated vertical heave, some researchers have suggested a lateral restraint or a correction factor in the order of one-third to be applied to swell percentages obtained from the oedometer laboratory measurements (Johnson and Snethen, 1979; Dhowian et al., 1990). However, as indicated by Crilly et al. (1992) the ratio of vertical to volumetric strain is not constant throughout the swelling layer, and the use of a single factor would tend to overestimate movement near ground surface and underestimate it at depth.

Evidently, the vertical movement is more relevant to the design of foundation systems on expansive soils than the total volume change since it is the main cause of deterioration in light buildings or pavements. Oedometer testing methods do provide a measure of the anticipated vertical heave but they significantly overestimate it because of the lack of similarity between the states of stress conditions in the oedometer testing as compared to the actual states in the field. The conventional triaxial swell tests, on the other hand, furnish more realistic loading conditions and offer a more satisfactory way of measuring swelling of expansive soils. However, in these studies in which expansive soil samples were tested in the triaxial apparatus under multi-axial state of stress, it has been possible to measure only the total volume change of the tested samples (e.g., Blight, 1967; Dakshanamurthy, 1979; Abduljauwad and Al-Sulaimani, 1993).

This paper presents a testing procedure through which the vertical swell and the total volume change were simultaneously measured from swell tests carried out on statically compacted samples subjected to triaxial loading conditions. The swell tests were conducted in a hydraulic stress path cell complemented by a manometer for measurement of sample volume change. The presented testing technique was specifically designed to first compare the vertical and volumetric swells measured under triaxial loading conditions with the volume change (which is also the vertical swell) of samples prepared under the same initial conditions of dry density and moisture content and tested in the oedometer. Second, to examine the ratio of vertical swell to volumetric swell, hereafter called the swell ratio or for short SR, of expansive soil samples subjected to triaxial state of stress, and how this ratio is related to the magnitude of the applied confining pressure.

TESTED MATERIAL AND TESTING PROCEDURE

Material

The material used in this study was brought from the town of Al-Ghatt, located 270 km Northwest of Riyadh, the capital of Saudi Arabia. The soil formation in the region represents typical expansive shale 8 to 10 meters thick and with relatively high swelling parameters (i.e., volume change and swell pressure). Soil samples for laboratory testing were obtained from a test pit at a depth of about three meters. The pit was excavated in an area where many residential buildings, sidewalks, and pavements have been severely damaged due to differential heave of the supporting soft rock shales.

Laboratory testing on extracted samples involved the performance of routine classification and mechanical property tests. The physical properties of the soil recovered from the test pit are summarized in Table 1. Atterberg limits, specific gravity, and particle size were determined in accordance with the ASTM test standards. The soil is classified as a CH soil as per the Unified Soil Classification System (USCS).

Table 1. Properties of Al-Ghatt Shales.

Parameters	Values
Sand content (grain size >75μ)%	0.0
Silt content (grain size 2 to 75μ), %	28.0
Clay content (grain size <2μ), %	72.0
Liquid limit, LL, %	60.0
Plastic limit, PL, %	30.0
Shrinkage limit, SL, %	16.0
Natural water content, %	22.0
Dry unit weight, kN/m^3	18.0
Specific gravity of solids	2.8
Activity	0.4
pH	8.0

In terms of clay activity, the shale has low activity. The X-ray diffraction analyses showed that the clay fraction of the shale consists primarily of illite and kaolinite, with no trace of smectite (Dhowian et al., 1990). Both minerals are known for being low to moderately expansive materials. However, the properties shown in Table 1 put the soil into the "high' to "very high" categories of four of the most commonly methods of determining a soil's intrinsic expansiveness (Holtz and Gibbs, 1956; Seed et al., 1962; Van Der Merwe, 1964; Dakshanamurthy and Raman, 1973). Furthermore, high magnitudes of swell parameters obtained in oedometer tests clearly reflect high expansive nature of Al-Ghatt shale formation (Erol et al., 1987). The relatively high expansion in these deposits was primarily attributed to the destruction of capillary stresses resulting from extreme desiccation, coupled with the destruction in the stratified soil structure through water infiltration (Dhowian et al., 1990). In addition, the natural moisture contents were well below the plastic limit of the soil, resulting in a considerable water deficiency and high water intake potential.

Sample Preparation

In order to minimize variations in test data, remolded samples were used in the testing program. To ensure uniformity of testing, the block samples were dried, pulverized into a powder in batches and screened through sieve No. 40. The batches were thoroughly mixed to obtain a uniform bulk sample. Individual samples were oven-dried for about 24 hours and then thoroughly mixed to the chosen molding moisture content. They were then stored in air-tight plastic bags and sealed in plastic wraps to avoid loss in moisture content, and were allowed to cure at room temperature for 24 hours to allow uniform distribution of moisture content.

The 70 mm-diameter, 19 mm-thick oedometer specimens were statically compacted in two layers in the consolidation ring, which was enclosed in a special compaction mold. The 38 mm-diameter, 76 mm-long specimens for the triaxial swell tests were statically compacted to the desired density in a vertically-split compaction mold. Static compaction method was chosen as it provides better control of the specimen condition. In addition, to maximize overall uniformity, the soil was compacted in three layers each with about 25-mm thickness. The required amount of soil for the first layer was poured into the compaction mold. The soil was then compacted by inserting into the mold a 35 mm-diameter aluminum rod whose length is 25 mm shorter than the height of the mold. Using a jack, the rod was pushed against the soil sample until a flange at its top touched the rim of the compaction mold. The same procedure was repeated for the other two layers, except that the rods used for compacting the second and the third layers were, respectively, 25 mm and 50 mm shorter than the first rod.

Testing Procedure

Oedometer Swell Tests

The amount of swell measured in oedometer tests depends on test procedure with regard to stress path and soaking conditions. Different oedometer methods and testing procedure have been used for the measurement and prediction of heave of expansive soil, namely the "free swell", "constant volume swell", and "swell overburden". The swell overburden method has the important advantage of being more representative of actual field loading and wetting conditions. It was, therefore, used and the conducted tests were performed according to ASTM D4546 standard.

After preparation, the soil specimen was then transferred to a standard rear loading oedometer having a diameter of 70 mm. The specimen was seated against a saturated porous stone and a second similar porous stone was placed on top of the specimen. The expansion of the specimen was measured until the change was negligible, at which time the test was terminated and the final water content was measured.

Triaxial Swell Tests

Equipment

The triaxial swell tests were carried out in a hydraulic triaxial stress path cell of the type reported by Bishop and Wesley (1975). A schematic diagram of the layout of the testing system is shown in Fig. 1. The cell is self-contained, portable, and requires no loading frame. The axial load is applied to the sample by pressurizing the lower chamber at the bottom of the cell. The piston pushes up a loading ram, at the top end of which is the pedestal on which the soil sample is mounted. Thus, in the conventional triaxial apparatus the loading ram is pushed downward against the top of the sample. In the stress path cell, however, it is the sample that is pushed upward against a stationary submersible load cell. This is a salient feature of the stress path cell that makes it possible to measure not only the vertical swell of the tested sample but also its total volume change. In the conventional triaxial apparatus it has been possible to monitor only the total volume change of the tested soil samples.

Test Procedure

Immediately after compacting, the specimen was removed from the mold and with the aid of a stretcher it was enclosed into a single watertight rubber membrane and was mounted on the base of the triaxial cell. The base platen was lightly coated with a film of thin grease prior to attaching the membrane. The specimen was seated against a saturated porous stone and a second similar porous stone was placed on the top of the sample. The membrane was then sealed to the top loading cap and the bottom platen with O-ring seals. Subsequently, valves 2 and 7 were opened, and the perspex cell was filled with de-aired tap water flowing through valve 2. The air release valve 4 on the triaxial cell was opened to release trapped air. The rate of filling was reduced when the level was almost full and valve 4 was closed when water began to flow. Then valve 1 was opened and the chamber pressure was incrementally increased until the loading ram, along with the sample, began to move upward. The chamber pressure required to move the ram upward is easily determined by knowing the combined weight of the loading ram and the specimen.

The upward movement of the loading ram slowly continued until the top loading cap touched the stationary load cell, which recorded the load, and then valve 2 was closed. Utmost care was taken with the alignment of the top loading cap with the load cell. It was important that the loading cap gently touches the load cell so that the sample will not be subjected to axial compression. In the meantime, it was essential to ensure a full contact between the top loading cap and the load cell so that the entire vertical swell will be detected and measured by the dial gauges shown in Fig. 1. If a space is left between the top loading cap and the load cell, the specimen will expand upward, and the dial gauges will not detect this expansion. Therefore, to maintain contact between the top loading cap and the load cell a nominal axial pressure of about 2 kPa was applied to the sample.

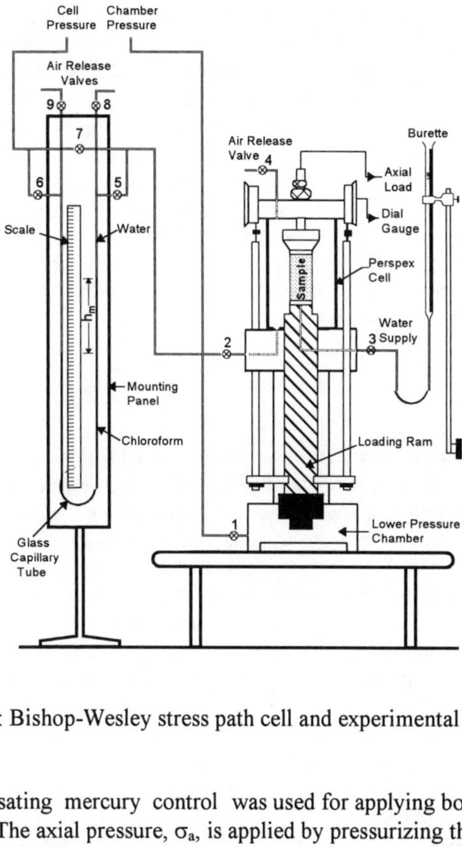

Figure 1: Bishop-Wesley stress path cell and experimental set-up.

A self-compensating mercury control was used for applying both the lateral and chamber pressures. The axial pressure, σ_a, is applied by pressurizing the lower pressure chamber. However, in the stress path apparatus the axial stress, σ_a, and cell pressure, σ_c, are interrelated. The value of σ_a depends on both σ_c and the pressure in the lower pressure chamber, *p,* and the key to the operation of the cell lies in the relationship between σ_a, σ_c, and *p*. Therefore, after deciding the desired cell pressure for the given test, the chamber pressure was determined accordingly. For any particular test, a nominal cell pressure is initially applied and the value of *p* is gently tuned to balance the cell pressure, σ_c, plus the weight of loading ram and soil sample. Considering the equilibrium of the loading ram shown in Fig. 1, the specific value of *p* necessary to produce the required σ_a, for a given value of σ_c, can easily be determined (Bishop and Wesley, 1975).

With valves 1, 2 and 4 closed, the chamber pressure, p, and cell pressure, σ_r, were increased to the specified values. Their values were monitored by pressure transducers connected to electronic readout devices. After reaching the specified values for the chamber and cell pressures valves 1 and 2 were simultaneously opened (valve 7 was also opened). In case the initial setting of the dial gauges changes, fine touching was obtained by a screw arrangement at the top of the cell so that the top loading cap remained in full contact with the load cell. Then valve 7 was closed and valves 5 and 6 were opened. Accumulated air through the manometer is released through the opening of valves 8 and 9 to bring the null indicator back to its original position.

The specimen was allowed to consolidate under the applied isotropic confining pressure for a period of about 24 hours before it was given free access to water. Initial manometer and dial gauge readings were taken, and water was introduced to the bottom of the sample by opening valve 3 which is connected to a 50 ml single-tube drainage burette as shown in Fig. 1. Readings were taken at convenient time intervals until the change in vertical and volumetric swells under the applied confinement was negligible, at which swelling was assumed to be completed and the test was terminated. The sample was then removed from the cell, weighed, and its length and average diameter were measured. Three slices were taken from the bottom, middle, and top of the specimen and used to find the final water contents.

Swell Measurements

The vertical swell of the tested samples was measured with two micrometer dial gauges, mounted on the top of the cell and deflected by two vertical extension rods that pass through clearance holes in the cell base and connected to the loading ram. The volume change was measured with a manometer, which consists of a transparent U-tube of 6 mm-bore glass tubing. The manometer is mounted on a supporting frame and is fitted with a scale marked in millimeters. A water-filled rigid nylon tubing of small bore connects one limb of the manometer to the base of the triaxial cell through valve 2 and the other limb is connected to a pressure gauge. The basic layout and the principal features of the volume measurement system are shown diagrammatically in Fig. 1.

The volume change of the sample was measured by observing the quantity of water entering or exiting the cell to compensate for the change in volume of sample. To make this observation, it was necessary to measure the displacement of a free surface between the water supply to the cell and some other fluid having different density. Such surface was first provided by the boundary between the water and mercury. However, because of the high specific gravity of mercury, a significant head difference arises as the volume of the sample changes and the mercury is displaced from one limb of the manometer to the other.

Therefore, the lower part of the U-tube of the manometer was filled with Chloroform (CHCL3). The difference between the specific gravity of water (1.0) and of Chloroform (1.5) is sufficiently small for volume change to result in less than 5.0 kPa pressure per meter difference in the level of the surfaces. Potassium lodide was added to the Chloroform to define clearly the meniscus between it and the water. The position of the Chloroform in the limbs of manometer can be adjusted by opening valves 8 and 9, which remain closed during the measurement of volume. It is essential that the measuring system is thoroughly de-aired and is watertight without the slightest leak.

Observation of the sample total volume change is dependent on whether the sample is allowed to swell vertically or not. If in a certain test the vertical swell was not permitted then when a volume increase occurs in the sample due to swelling, and if the triaxial cell containing the soil is completely filled with de-aired water before the swelling test is started, water will flow from the cell to the manometer causing the Chloroform to move from one limb to another. The sample volumetric swell, S_v, is then given by:

$$S_v = A_m h_m \tag{1}$$

Where A_m is the manometer cross-sectional area and h_m represents difference in the levels between the Chloroform in one limb measured initially and again after expansion has taken place after known time t. Readings are taken at the interface meniscus inside the indicator.

If the sample is allowed to swell vertically, the load piston is pushed downward out of the cell and the movement of water from manometer to the cell or vice versa will depend on the relative magnitudes of vertical and total volumetric swell of the tested sample. Recorded data should indicate whether the left-hand or right-hand limb was observed (i.e. if inflow or outflow). In all tests conducted water moved into the cell, and the sample volumetric swell is obtained from:

$$S_v = A_p \Delta h - A_m h_m \tag{2}$$

Where A_p is the piston cross-sectional area and Δh is the change in sample height at time t.

The accurate measurement of volume change is particularly important when the cell pressure is large and the swelling is small. A small 6-mm diameter tube was used to give greater accuracy in the volume readings. The size of the tube has been found convenient for most tests on the 76-mm high by 38-mm diameter samples. If the cumulative flow exceeds the null indicator capacity, valve 2 is then closed and the Chloroform column is brought back to its original location, or to another location which becomes a new datum for subsequent readings, by opening valve 8 and valve 9. The clearance is 70 mm (i.e., maximum h_m) and that provides capacity for the anticipated volume change of the tested samples often without needing reversal of flow.

SWELL TESTS

The testing program was specifically designed to: (1) compare the vertical and volumetric swells measured under triaxial loading conditions with the volume change of samples prepared at the same initial conditions and tested under one-dimensional loading conditions in the oedometer, (2) examine the ratio of vertical swell to volumetric swell, SR, of expansive soil samples subjected to triaxial state of stress, and how this ratio is related to the magnitude of the applied confining pressure.

In order to examine the reliability of the triaxial testing equipment and procedure, two replicate tests designated as TST-a and TST-b were first conducted under a confining pressure of 25 kPa on samples compacted at an initial water content of 18% and a dry density of 16 kN/m^3.

The second group of swell tests was aimed at studying the vertical swell under one-dimensional and three-dimensional loading conditions. A series of four oedometer swell tests designated as OST-1, OST-2, OST-3, and OST-4 were conducted, under a range of confining pressures (25, 50, 100, and 200 kPa). Another set of triaxial swell tests identified as TST-1, TST-2, TST-3, and TST-4, was conducted in the stress path triaxial apparatus and subjected to the conventional isotropic triaxial state of stress under the same range of confining pressures used for the oedometer swell tests (i.e., 25, 50, 100, and 200 kPa). The samples for all of the eight tests were prepared at 14% initial water content and a dry density of 18 kN/m^3.

The third group of tests comprises four triaxial swell tests named TST-5, TST-6, TST-7, and TST-8 conducted to examine the effect, if any, of the sample initial water content on SR. The tests were performed on samples statically compacted at a dry density of 18 kN/m^3. For tests TST-5 and TST-6 the samples were prepared at initial water content of 18%. The former was tested under a confining pressure of 25 kPa while the applied confining pressure for the latter was 50 kPa. Tests TST-7 and TST-8 were, respectively, similar to TST-5 and TST-6, except that the initial water content for the samples was 22%.

The sample initial and loading conditions for all tests performed are summarized in Table 2.

Table 2: Test conditions and specimen characteristics of swell tests

Test No.	Initial Water Content (%)	Initial Dry Density (kN/m^3)	Confining Pressure (kN/m^2)
Triaxial			
TST-a	18	16	25
TST-b	18	16	50
TST-1	14	18	25
TST-2	14	18	50
TST-3	14	18	100
TST-4	14	18	200
TST-5	18	18	25
TST-6	18	18	50
TST-7	22	18	25
TST-8	22	18	50
Oedometer			
OST-1	14	18	25
OST-2	14	18	50
OST-3	14	18	100
OST-4	14	18	100

RESULTS AND DISCUSSIONS

The measured percentages of vertical and volumetric swells for the replicate tests, TST-a and TST-b, are plotted against time in Fig. 2. It can be seen that the results of the two tests are comparable with only minor discrepancy in the values of the volumetric swell towards the end of the tests. The percentage of vertical swell is defined as the ratio of the increase in height of the sample to its initial height before it was given free access to water. The increase in volume of the sample after wetting expressed as a percentage of initial sample volume before wetting is the sample volumetric swell or total volume change.

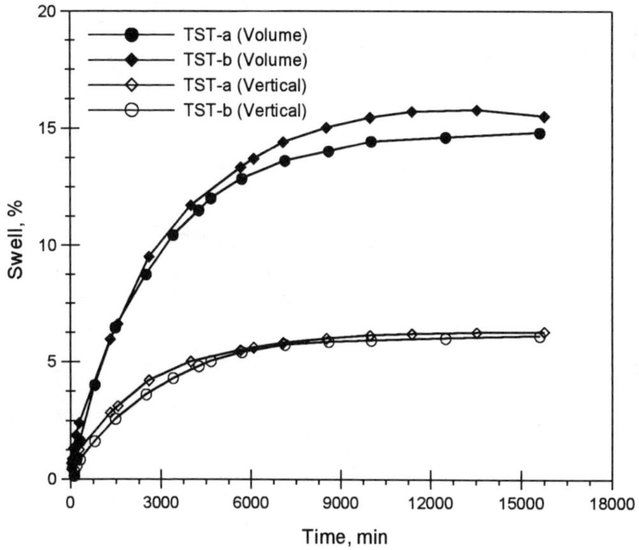

Figure 2: Swell-time relationships for two replicate triaxial swell tests.

Uniaxial and Triaxial Swellings

Typical variation of the percentage of vertical swell with elapsed time is shown in Fig. 3 for tests OST-1 and TST-1, and in Fig. 4 for tests OST-4 and TST-4. As expected, the vertical swells obtained from oedometer tests are considerably larger than the triaxial test measurements. It is also noted that the swelling takes place at a faster rate in the oedometer as compared to the triaxial swell tests. Similar patterns have also been observed for the triaxial and oedometer tests tested under the other two values of applied pressure (i.e., 50 and 100 kPa).

The time required for complete swelling of a specimen after free water is provided on the surface, depends upon the permeability of the compacted samples and the distance through which the water must be drawn through the sample. The eight samples for the oedometer and the triaxial tests are assumed to have approximately the same initial structure, and as such equal permeability. Therefore, the difference in the rate of swelling between the two types of tests is attributed to the difference in sample thickness. For the oedometer tests, samples are thin and saturation for the entire specimen takes place in a quite short period of time leading to a fast rate of swelling. In the case of triaxial tests, the specimen is relatively thick, in comparison to the oedometer specimen, and some time has to pass before the soil at some distance from the bottom of the sample can begin to swell after it becomes saturated. Not only that, but it has also be noted that, regardless of the value of the applied pressure, the final water contents for samples tested in the oedometer are higher than the corresponding values for samples tested in the stress path cell (Al-Shamrani and Al-Mhaidib, 1999). In other word, the moisture deficiency or moisture intake potential, which is the difference between final and initial moisture contents, is higher for oedometer tests. This supports the suggestion made by Erol et al. (1987) that not only a lateral restraint factor but also a moisture factor should be considered in the analysis of predicted heave when using the results of oedometer tests.

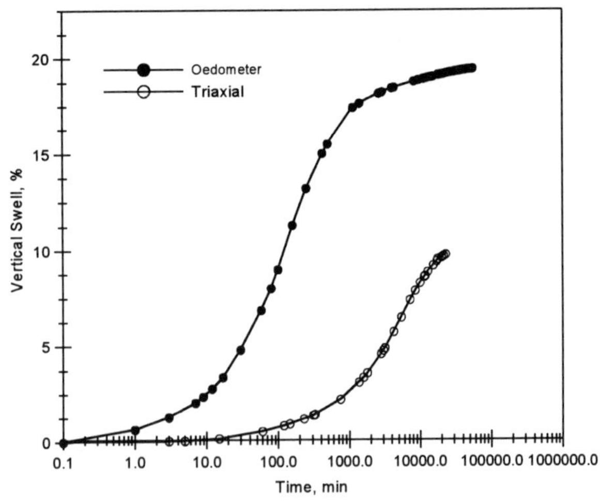

Figure 3: Percentage of vertical swells obtained from oedometer and triaxial swell tests (OST-1 and TST-1).

The percentage of ultimate vertical swells measured in the oedometer and the stress path equipments are plotted against the logarithm of applied pressures in Fig. 5. The data from both the triaxial and oedometer tests are approximately fitted to a straight line. As the confining pressure increases, the ultimate volumetric swells obtained from the triaxial tests decrease at almost the same rate as the ultimate vertical swells measured in the oedometer tests. It is noticed, however, that the change in the triaxial ultimate vertical swell with increasing confinement is smaller as compared to the change in the ultimate vertical swell measured in the oedometer.

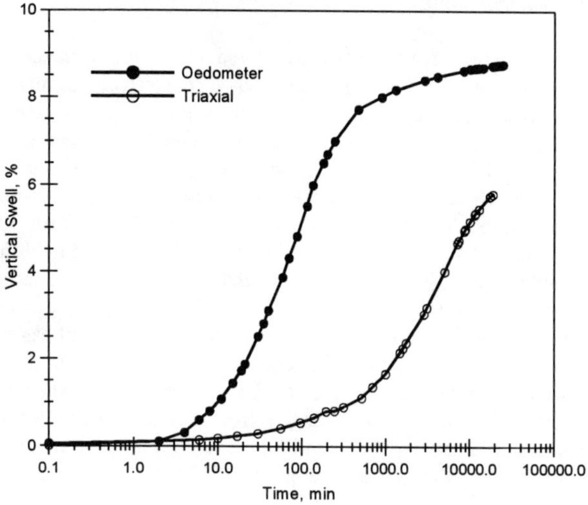

Figure 4: Percentage of vertical swells obtained from oedometer and triaxial swell tests (OST-4 and TST-4).

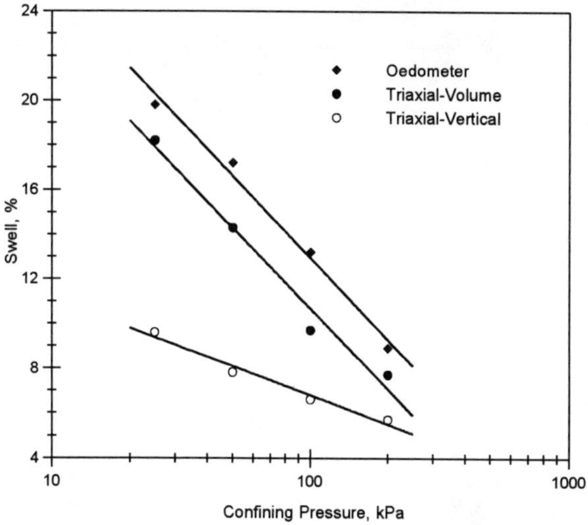

Figure 5: Comparison of ultimate swell from oedometer and triaxial swell tests for the samples having initial water content = 14%.

Vertical and Volumetric Swells

The variation of the percentage of vertical swell and total volume change against time is shown in Fig. 6 and Fig. 7 for TST-1 and TST-4, respectively. Similar patterns have also been observed for TST-2 and TST-3 swell tests. It is noted from the results of Fig. 6 and Fig. 7 that the swell ratio, SR, is not constant throughout the test but rather changing with elapsed swelling time. It can be seen, for instance, from Fig. 6 that the ratio increased from about 0.45 at the start of the test to around 0.54 at the time when the test was terminated. The average value for SR for this test is about 0.50. However, for TST-4 where the confining pressure was 200 kPa, it seen from Fig. 7 that SR ranged between 0.49 and 0.78 at the commencement and termination of the test, respectively, and the average ratio is about 0.64.

It was found that the value of the sample initial water content has almost little or no effect on the vertical swell to volumetric swell ratio. However, depending on the magnitude of confining pressure the value of SR at the beginning of the performed swell tests lie between 0.40 and 0.49 and ranged from 0.54 to about 0.78 at the end of the tests. The relationship between the logarithm of confining pressure and SR values at ultimate swell for TST-1 through TST-8 is shown in Fig. 8. The average value of SR for each test was also evaluated and drawn in Fig. 8 against the logarithm of confining pressure. It is noted that the average swell ratio as well as the swell ratio at ultimate swell approximately assumes a linear relationship with the logarithm of confining pressure. If only the ultimate state is considered, the average value of SR is around 0.64. However, if the swell ratios for all tests at all time intervals are considered, the average value of SR is about 0.55.

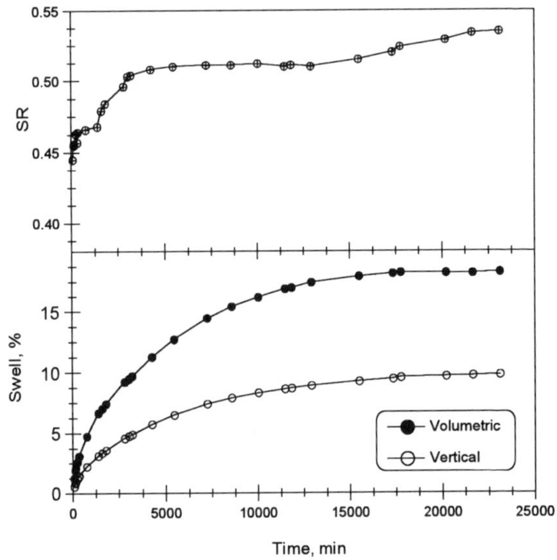

Figure 6: Variation with time of vertical and volumetric swell for test TST-1.

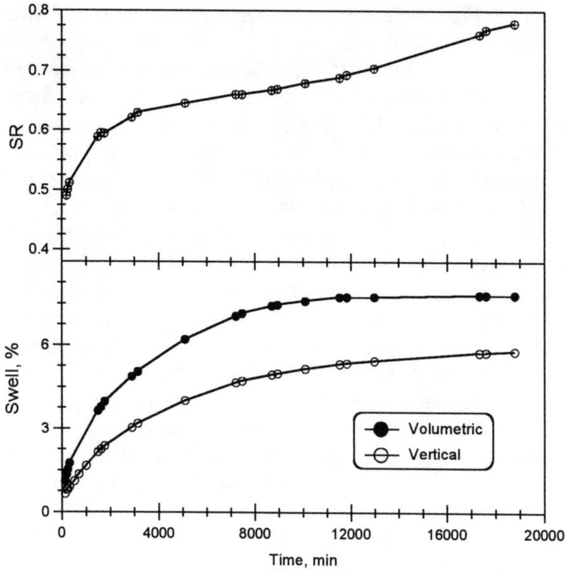

Figure 7: Variation with time of vertical and volumetric swell for test TST-4.

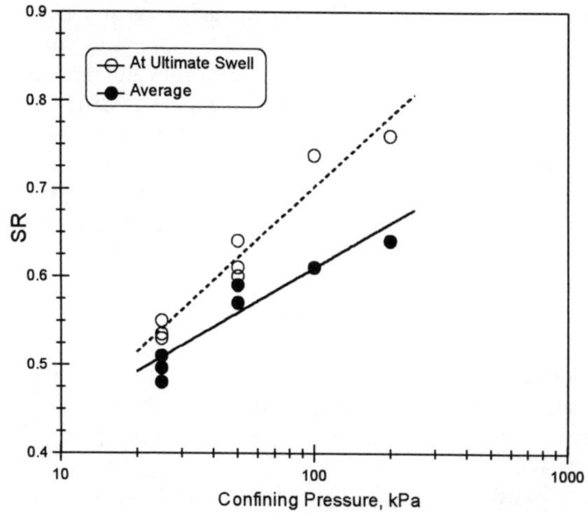

Figure 8: Relationship between swell ratio and applied confining pressure.

The results of Fig. 8 also indicate that the ratio of vertical swell to volumetric swell increase as the confining pressure increases. This implies that the level of confinement has relatively more effect on the total volume change than on the vertical swell of expansive soil samples. The increase, for instance, in confining pressure from 25 kPa in test TST-1 to 200 kPa in test TST-4 reduced the percentage of ultimate vertical swell and volumetric swell by about 41% and 57%, respectively. This reality is more obvious when the percentages of ultimate vertical and volumetric swells for TST-1 through TST-4 are plotted against the logarithm of applied pressure as shown in Fig. 5. It is noted that the ultimate volumetric swell decreases at a higher rate in comparison to the swell in the vertical direction.

Therefore, upon saturation, expansive soils expand in all directions. However, as one would intuitively expect, the higher is the applied confining pressure the more volume change would be reflected as a surface heave. The maximum percentage of vertical swell takes place under the fully lateral restraint condition maintained in the different forms of oedometer swell tests. In these tests, the soil is prevented from swelling in the horizontal direction and volume change occurs in the vertically upward direction. This results in the percentages of vertical swell and volume change measured in the oedometer tests being identical; and hence SR is equal to unity. Therefore, the ratio of vertical swell to volumetric swell could be as low as about 0.40 for samples subjected to low confinement under multi-axial state of stress to exactly one for samples tested under one-dimensional loading conditions in the oedometer.

SUMMARY AND CONCLUSIONS

Selection of a treatment technique to minimize the volume change and the design of a foundation system to accommodate the anticipated vertical movement, both rely on accurate estimates of the potential heaves of the supporting soil. The various one-dimensional swell tests are usually found to overestimate the actual field heaves of expansive soils. This may lead to a conservative and unnecessarily costly design.

In this paper, it was demonstrated that the hydraulic triaxial stress path cell is versatile equipment that can be used effectively for measuring swell parameters of expansive soils. Complementing the cell with a manometer for the measurement of volume change, it has been possible, in this study, to measure both the vertical swell and the total volume change of the tested samples. This is to the writers' knowledge the first attempt in which both vertical and volumetric swells of expansive soil specimens were directly and concurrently measured in the same triaxial swell test.

Based on the experimental results compiled in this study, it was found that the vertical swells obtained from oedometer swell tests are considerably higher than the triaxial test measurements. This is attributed to the differences in loading and wetting conditions between the two equipments. However, irrespective of the test loading condition, the relationship between the percentage of ultimate vertical swell and the logarithm of confining pressure is approximately linear, yet the rate of decrease in the ultimate vertical swell with increasing applied pressure was higher for the oedometer tests.

Under triaxial state of stress, the value of the ratio of vertical swell to volumetric swell, SR, was not constant but changed with elapsed swelling time. Furthermore, SR linearly increased with increasing applied confining pressure. The minimum and maximum values for SR were about 0.40 and 0.78, respectively. The overall average swell ratio for all tests at all time intervals was around 0.55. The ratio of vertical swell to volumetric swell could, therefore, be as low as about 0.40 for samples subjected to relatively low triaxial confining pressures to exactly one for samples tested under one-dimensional loading conditions.

The presented results are limited to a relatively small number of tests on a single soil type. Further work is still required to determine whether these findings are more generally applicable to other reconstituted and natural soils, having different initial characteristics and tested under different loading conditions.

REFERENCES

Abduljauwad, S. N. and Al-Sulaimani, G. J., (1993), "Determination of swell potential of Al-Qatif Clay," *Geotechnical Testing Journal*, ASTM, Vol. 16, No. 4, pp. 469-484.

Al-Shamrani, M. A. and Al-Mhaidib, A. I. (1999), "Prediction of potential vertical swell of expansive soils using a triaxial stress path cell," *Quarterly Journal of Engineering Geology*, Vol.32, pp. 45-54.

Bishop, A. W. and Wesley, L. D., (1975), "A hydraulic triaxial apparatus for controlled stress path testing," *Geotechnique*, Vol. 25, pp. 657-670.

Blight, G. E., (1967), "Effective stress evaluation for unsaturated soils," *Journal of the Soil Mechanics and Foundation Division*, ASCE, No. SM2, 125-148.

Crilly, M. S., Driscoll, R. M. and Chandler, R. J., (1992), "Seasonal ground and water movement observations from an expansive clay site in the UK," *Proceedings of the 7th International Conference on Expansive Soils*, Dallas, Texas, 313-318.

Dakshanamurthy, V. and Raman, V. (1973), "A simple method of identifying expansive soil," *Soils and Foundations*, 13, 1, pp. 97-104.

Dakshanamurthy, V., (1979), "A stress-controlled study of swelling characteristics of compacted expansive clays," *Geotechnical Testing Journal*, Vol. 2, No. 1, pp. 57-60.

Dhowian, A. W., Erol, A. O. and Youssef, A., (1990), "Evaluation of expansive soils and foundation methodology in the kingdom of Saudi Arabia," Final Report, King Abdulaziz City for Science and Technology, KACST, AT-5-88.

Erol, A. O., Dhowian, A., and Youssef, A. (1987), "assessment of oedometer methods for heave prediction," *Proceedings 6th International Conference on Expansive Soils*, New Delhi, India, pp. 99-103.

Erol, A. O., (1992), "In situ and laboratory measured suction parameters for prediction of swell," *Proceedings of the 7th International Conference on Expansive Soils*, Dallas, Texas, pp. 30-32.

Erol, A. O., Dhowian, A., and Youssef, A. (1987), "Assessment of oedometer methods for heave prediction," *Proceedings 6th International Conference on Expansive Soils*, New Delhi, India, pp. 99-103.

Gizienski, S. F. and Lee, L. J., (1965), "Comparison of laboratory swell tests to small scale field tests, engineering effects of moisture changes in soils," *Concluding Proceedings, International Res. and Engrg. International on Expansive Clay Soils*, Texas A&M Press, Texas, pp. 108-119.

Holtz, W. G. and Gibbs, H. J. (1956), "Engineering properties of expansive clays," *Proceedings*, ASCE, 80, Separate No. 516, pp. 1-28.

Jones, D. E. and Holtz, W. G. (1973),"Expansive soils-the hidden disaster," *Civil Engineering*, ASCE, Vol. 43, No. CE8, p. 49.

Johnson, L. D. and Snethen, D. R., (1979), "Prediction of potential heave of swelling soils," *Geotechnical Testing Journal*, ASTM, Vol. 1, No. 3, pp. 117-124.

Kassiff, G. and Ben-Shalom, A. (1971), "apparatus for measuring swell potential under controlled moisture intake," *Journal of Materials*, JMLSA, 6, 1, pp. 3-15.

Richards, B. G., (1967), "Moisture flow and equilibria in unsaturated soils for shallow foundations," *Permeability and Capillarity of Soils*, STP 417, ASTM, Philadelphia, PA, pp. 4-34.

Seed, H. B., Woodward, R. J., Jr., and Lundgren, R. (1962), "Prediction of swelling potential for compacted clays," *Journal of the Soil Mechanics and Foundation Division*, ASCE, 88, 3, pp. 53-87.

Van Der Merwe, D. H. (1964), "The prediction of heave from the plasticity index and percentage clay fraction of soils," *Civil Engineers in South Africa*, 6, 6, pp. 103-107.

Mineralogy-Swelling Potential Relationships For Expansive Shales

Harold W. Olsen[1], Lisa Krosley[1], Karl Nelson[1],
Sabine Chabrillat[2], Alexander F. H. Goetz[2, 3], and David C. Noe[4]

Abstract

This study examines the extent to which mineralogy and swelling potential can be correlated in the expansive clays and shales along the Colorado Front Range Urban Corridor. 182 undisturbed samples were collected from 20 sites from Boulder to Pueblo. Sites were selected in Cretaceous shales, including the Pierre Shale, that have been uplifted into steeply dipping strata near the foothills of the Rocky Mountains, and that are well known to be hazardous to residential and light commercial developments in this region. For each sample, mineralogy was determined by x-ray diffraction and swelling potential was obtained from moisture content (w), suction (h), and clod volume (V) measurements in terms of the suction potential (dh/dw) and suction-compression index (dV/dh) parameters used in the classification scheme McKeen proposed in 1992. Swelling potentials were also obtained on more limited suites of samples with conventional and labor-intensive schemes including Seed and Chen's schemes based on tests for grain-size distribution and Atterberg limits, and with swell-consolidation measurements in response to saturation, consolidation, and rebound in an oedometer.

The results show the percent total smectite provides a useful index of the swelling potential concept defined by Seed and correlates reasonably well with the swelling potential indices developed by Seed, Chen, and McKeen. However, the percent total smectite does not correlate well with conventional swell-consolidation test indices. The causes for this lack of correlation appear to be placement and environmental factors such as the initial moisture content, stress history, and surcharge loading that are beyond the scope of Seed's swell potential concept and Seed, Chen, and McKeen's swelling potential indices.

Introduction

It has long been recognized that expansive clays and clay-shales cause billions of dollars of damage every year, more than all other natural hazards

[1] Colorado School of Mines, Golden, CO 80401
[2] Center for the Study of Earth from Space/CIRES, University of Colorado, Boulder, CO 80309-0449
[3] Department of Geological Sciences, University of Colorado, Boulder, CO 80309-0449
[4] Colorado Geological Survey, Denver, CO, 80203

combined (Jones and Holtz, 1973; Chen, 1988). One example of this problem is the current situation in the Denver Metropolitan Area. This area is underlain by Cretaceous clay-shales, including the Pierre Shale, that also underlie vast areas of the United States and Canada, including the following Rocky Mountain States: Montana, North Dakota, South Dakota, Wyoming, Nebraska, Colorado, Kansas, Missouri, and Arizona. The sedimentary strata are generally flat lying, except near the foothills of the Rocky Mountains, where they have been uplifted into steeply dipping strata. These conditions underlie several cities along the Front Range Urban Corridor in Colorado that extends from Pueblo to the Wyoming border. This corridor is expected to provide land for much of the housing to be built for a substantial increase in Colorado's population in the next 25 years (Noe and Dodson, 1995, 1999; Gill et al, 1996).

For more than four decades, geotechnical engineering practice for mitigating damage from expansive soils in the Denver metropolitan area has involved (1) locating potentially expansive soils in the subsurface with vertical drill holes sampled at 5 foot intervals, (2) characterizing the expansive potential of relatively undisturbed samples in terms of laboratory geotechnical index tests, and (3) supporting structures on drilled piers anchored in bedrock. The geotechnical index tests include the Atterberg limits and one-dimensional swell-consolidation tests. The latter commonly involves loading a relatively undisturbed sample in an oedometer to a pressure of 24 or 48 kPa (500 or 1000 pounds per square foot), flooding it with water and monitoring the swell induced thereby, and subsequently consolidating the sample to determine the load needed to bring the sample back to its initial void ratio. Generally, the "percent swell" and/or the “swell pressure” determined from this swell-consolidation test is used as an empirical index for classifying the severity of potential differential vertical deformations at the site, and as a basis for predicting when drilled piers, structural floors, and other stabilization measures are needed.

The study described in this paper is part of a larger project aimed at determining the feasibility of using reflectance spectroscopy to identify and characterize the expansive clays and clay shales along the Colorado Front Range Urban Corridor. This remote sensing technology is of interest because it permits rapid identification of constituent minerals on in situ earth materials with a hand-held spectrometer at specific sites such as outcrops or trench walls. This technology also allows reconnaissance studies of mineralogy with spectrometers mounted in aircraft or satellites. The question arises whether this technology could become a useful tool for identifying and evaluating expansive soils at either specific sites or on a regional basis.

The development of reflectance spectroscopy for identifying and characterizing expansive soils requires the correlation of clay mineralogy with swelling potential indices determined from engineering tests, and the expansion of existing correlations of reflectance spectra with clay mineralogy determined by x-ray diffraction. This paper examines the extent to which clay mineralogy can be correlated with swelling potential for the clays and shales along the Denver Front Range Corridor. Our progress in correlating clay mineralogy with reflectance spectra is being reported elsewhere (Chabrillat et al, 1997, 1999, Chabrillat and Goetz, 1999).

Methods

Sampling in the Colorado Front Range Corridor. 182 undisturbed samples were collected from 20 different locations along the Colorado Front Range Urban Corridor from Boulder To Pueblo. 145 samples were taken at 30-50 cm depth beneath the ground surface. 37 samples were taken at 30-50 cm depth beneath the floor of an excavation for a large building, approximately 8 feet beneath the ground surface. The latter samples were taken along a 30-meter profile perpendicular to the strike of the underlying steeply dipping strata. Sites were selected in Cretaceous shales most of which are uplifted into steeply dipping strata near the foothills of the Rocky Mountains, and that are well known to be hazardous to residential and light commercial developments in this region (Figure 1). The sites were all located on Upper Cretaceous sediments including the Benton, Smoky Hill, and Pierre Shales. The great majority of samples obtained were clay-shales, consisting predominantly of mixed-layer illite and smectite with frequent thin bentonite beds, 10-30 cm thick, that consisted of almost pure smectite.

Mineralogical Composition. Quantitative mineralogy was derived from XRD patterns of oriented preparations using a Millipore filter peel transfer onto glass slides (Drever, 1973). The calculations were carried out following the equations of Schultz (1964). When possible, mixed-layer illite/smectite (I/S) expendabilities were determined by the °2θ method of Moore and Reynolds (1989). When peak interference from other phases made this determination difficult, ordering of the sample allowed an educated guess of the I/S expandability. In this case we estimated that the incertitude on the expandability value led to an error in the percent of total smectite (smectite phase + smectite in the I/S layers) and total illite (illite phase + illite in I/S) within ± 10 %. The mineralogical composition determined with XRD is is expressed in clay fraction, i.e. reported to 100% clays. We calculate the mineralogical composition related to the whole sample by correcting it with the percent clay of the sample determined by particle size analysis.

Swell Potential. The principal correlation between swelling potential and clay mineralogy in this study is based on the scheme introduced by McKeen (1992). This scheme was derived from the fundamental suction, water content, and volume change relationships that were developed to provide a quantitative basis for airport pavement design on expansive soils (McKeen 1978, 1981, 1985, 1990). For this study, the practical advantage of McKeen's scheme is that it provides a rapid and inexpensive quantitative measure of swell potential. This advantage was of fundamental importance for obtaining data on a sufficient number of samples to develop a significant correlation with clay mineralogy.

This study also includes measurements of the classic geotechnical indicators of swell potential that are the basis for geotechnical engineering practice in the Denver Metropolitan Area. These indicators include the classification schemes introduced by Seed (1962) and Chen (1988) that are based on measurements of the Atterberg Limits (liquid limit, plastic limit, plasticity index) (ASTM D4318-93), and particle size distribution (ASTM D 422-63). The classic indicators also include the percent swell, swell pressure, and coefficient of rebound that can be obtained in the swell-consolidation test (ASTM D4346-90). The purpose of these tests was to evaluate the consistency of McKeen's classification scheme with the classic schemes proposed by Seed and Chen, and also with the swell-consolidation test. Because the classic indicator tests are far more time consuming and labor intensive than the moisture content and suction tests involved in McKeen's scheme, they were run on limited suites of samples exhibiting the full range of swelling potentials according to McKeen's classification scheme.

McKeen's Swell Potential Classification Scheme. Figures 2-4 illustrate the suite of suction-related data obtained in this study on the undisturbed sample MR97001, the assumptions underlying McKeen's classification scheme, and how the data are interpreted with McKeen's scheme.

In Figure 2, four data points denote the suction of sample MR97001 at varying degrees of wetness. These data were obtained on one split of the sample with soil suction measurements using the filter paper method (ASTM D5298-94) in conjunction with water content measurements (ASTM D 2216-92). The solid line represents the suction vs. water content relationship and the slope of this line, dh/dw, is the suction potential of the sample. This line is based on two assumptions introduced by McKeen. First, the suction vs. water content relationship is linear between suction values of 5.0 log kPa and 2.0 log kPa. Second, an extension of this line passes through a benchmark intercept of 5.25 log kPa on the suction axis where the water content of the soil equals zero. The slope of the line shown is the average of the slopes of individual lines connecting the individual data points with the benchmark intercept.

In Figure 3, the seven data points show the variation of volumetric strain with changes in moisture content on another split of sample MR97001. These data were obtained using the clod test developed by McKeen (1981, 1985, and 1990). In this test, an undisturbed clod is first moistened and then encapsulated with a saran resin that allows the pore water to evaporate and the clod to shrink slowly over a period of several days. Periodically, the weight of the clod is measured and its volume obtained by water displacement. These data were used to calculate the volumetric strain and water content at each step in the clod test.

Figure 4 shows the variation of volumetric strain with suction for Sample MR97001. The points were derived from the data points in Figure 3 together with suction values estimated in Figure 2 at water contents corresponding to those for the data in Figure 3. The solid line represents the volumetric strain vs. suction relationship for the sample, and the slope of this line is the suction compression index, $dV/V_odh = C_h$. McKeen assumes this relationship is linear between upper and lower suction levels at which volume change ceases. He notes the upper suction level of 4.5 log kPa is that associated with the shrinkage limit, and the lower suction level of about 1.5 log kPa is that associated with field capacity when water tends to run off rather than infiltrate into soil.

McKeen developed his classification scheme by correlating the suction potential index and the suction compression index using his database on Texas soils.

In this correlation, he defines a linear relationship located such that 85% of the C_h values are larger than would be predicted by the equation of the relationship.

$$C_h = (-0.02673)(dh/dw) - 0.38704. \quad (1)$$

With this correlation, McKeen shows that the suction potential, which requires relatively simple suction and water content measurements, provides a useful index for estimating the more fundamental suction compression index, which also requires the clod test. McKeen further develops quantitative criteria for using both suction potential and suction compression index values to differentiate five categories of swell potential ranging from very high to non-expansive, as shown in Table 1.

Table 1 – McKeen's Swell Potential Categories

Category	Swell Potential	Suction Potential dh/dw	Suction Compression Index, dV/V_0dh
I	Very High [McKeen calls this category "Special Case"]	>-6	<-0.227
II	High	-6 to –10	-0.227 to -0.120
III	Moderate	-10 to –13	-0.120 to -0.040
IV	Low	-13 to –20	-0.030 to non-expansive
V	Non-expansive	<-20	Non-expansive

For example, Figures 2 and 4 show suction potential and suction compression index values of -4.41 and –0.32, respectively, for Sample MR97001. In this case, the value for the suction compression index was obtained directly from a clod test rather from equation 1. According to Table 1, both indices show the sample falls in category I, and hence has a very high swell potential.

Seed's Swell Potential Classification Scheme. Seed's swell potential classification scheme is based on well-defined experimental relationships that were established using artificially prepared soils (Seed et al, 1962). Twenty-three experimental soils were prepared by mixing commercially available kaolinite, illite, and Wyoming bentonite with sand in different proportions to obtain specimens having a wide range of soil characteristics. Each of the experimental soils was tested in the laboratory to determine its liquid limit, plastic limit, shrinkage limit, grain size distribution, and its expansion or percentage swell under the following standard conditions. The percentage swell was measured on specimens under a 1-psi surcharge that were compacted at optimum water content to maximum density in the Standard AASHO Compaction Test. The experimental relationships developed from these data were used to establish equation 2, which expresses the percentage of swell (S) in terms of the percentage of clay sizes (C) and the activity of the clay (A).

$$S = (3.6x10^{-5})(A^{2.44})(C^{3.44}) \quad (2)$$

Seed correlated his results to the extent possible with the U.S. Bureau of Reclamation (USBR) standards for expansion characteristics, and their criteria for differentiating low, moderate, high, and very high categories of swell potential. The boundaries between these categories, calculated from equation 2, are shown in Table 2.

Table 2 – Seed's Swell Potential Categories

Category	Very high	High	Moderate	Low
Seed's percentage of swell (Calculated from equation 2)	>25%	5-25%	1.5-5%	0-1.5%

For example, consider sample MR 97001 which fell in McKeen's very high category for swelling potential. For this sample the activity, A, is 1.1 and the percent of clay sizes, C, is 32.3%. The percentage of swell calculated from equation 2 is 7.1%, which falls in Seed's high category for swell potential. Note that this result differs somewhat from that obtained above for the same sample using McKeen's classification system.

Chen's Swell Potential Classification Scheme. Chen's classification system is based solely on the plasticity index, as shown in Table 3. This system is consistent with the conclusion reached by Seed et al (1962) that the plasticity index alone can provide an assessment of swelling potential that is probably correct to within ± 35%, and that "this method of prediction might be best suited for practical purposes."

Table 3 – Chen's Swell Potential Categories

Category	Very high	High	Moderate	Low
Plasticity Index	>35	20-55	10-35	0-15

For example, consider sample MR 97001 which fell in McKeen's very high category and in Seed's high category for swelling potential. The plasticity index for this sample was 36%. According to Chen's criteria, this sample falls either in the middle of the high, or the bottom of the very high swell potential categories.

The Swell-Consolidation Test. Figure 5 illustrates the data obtained from the swell-consolidation test on each sample in this study. The data shown are for Sample MR97001. The test procedure is an expanded version of Method B in the Standard ASTM Test Method for One-Dimensional Swell or Settlement Potential of Cohesive Soils (ASTM D 4546-90). Method B includes the initial application of a 500-psf surcharge, the swelling of the sample in response to inundation, and the subsequent reconsolidation of the sample to its void ratio prior to inundation. The expanded aspects include an extended consolidation segment and a subsequent rebound segment similar to those for Method C. Four indices of swell potential are shown in Figure 5: (1) the percent swell of the sample in response to inundation, (2) the swell pressure which is the pressure required to reconsolidate the sample to it void ratio prior to inundation, (3) the compression index, and (4) the rebound index. The percent swell is the predominant index currently being used in practice in the Denver Metropolitan Area. Its relationship to slab performance risk categories are shown in Table 4 (HBA Soils Task Force, 1996). The same criteria for the swell potential classification of samples tested under 48 kPa (1000psf) surcharge are documented by Thompson (1997).

Table 4 – Recommended Representative Swell Potential Descriptions and Corresponding Slab Performance Risk Categories

Slab Performance Risk Category	Representative Percent Swell under 24 kPa (500 psf) Surcharge	Representative Percent Swell under 48 kPa (1000 psf) Surcharge
Low	0 to <3	0 to <2
Moderate	3 to <5	2 to <4
High	5 to <8	4 to <6
Very High	8 and >8	6 and >6

For example, Figure 5 shows the percent swell for sample MR97001 equals 2.2% which falls in the low category in Table 4. Note that this result differs substantially from those obtained with McKeen's, Seed's, and Chen's classification schemes.

Results

The mineralogy of the clay fraction of all the samples ranges from pure smectite to varying degrees of mixed-layer smectite/illite and smectite to very low smectite. Illite content can be up to about 60%. Kaolinite content is usually less than 10% in these shale formations but can be up to about 50%. In general, the Perry Park samples are very rich in smectite. The Canon City samples are rather evenly distributed with respect to smectite content. In the following, the term "total smectite" includes both pure smectite and smectite in the mixed-layer smectite/illite.

The clay mineral of particular interest is smectite, because of its well-known swelling behavior and because the other clay minerals present are non-swelling. Figure 6 presents, for 178 samples, the percent total smectite in each whole sample plotted versus the percent clay in each sample. This figure shows that smectite is generally the most abundant clay mineral in the clay fraction of the samples, and that the percent total smectite and the percent clay both vary from nearly zero to 100 percent.

The suction vs. water content data obtained on 182 samples are presented in Figure 7 together with McKeen's swell potential categories. The latter are differentiated with linear relationships that begin at the benchmark intercept illustrated in Figure 2, and that have the slopes or suction potentials, dh/dw, defined in Table 1. In Figures 8, the suction potential (dh/dw) data are plotted versus the suction compression index (C_h) data for all of the samples. Also shown are the suction potential values that differentiate McKeen's swell potential categories, according to Table 1. These figures show the samples are widely distributed throughout all McKeen's categories. Figure 8 also shows the suction potential and the suction compression index values from 93 samples are related with a correlation coefficient, r, = -0.71.

Activity versus clay fraction data on 64 samples are presented in Figure 9 together with Seed's swell potential categories. For the same 64 samples, the plasticity index data are presented in Figure 10, together with Chen's swell potential categories. The data are distributed fairly uniformly throughout the categories, consistent with the distribution of the data throughout McKeen's categories in Figure 7.

The properties obtained from swell-consolidation tests on 44 samples are plotted versus the percent total smectite in Figure 11. These properties, defined in Figure 5, include the swell pressure, compression index, rebound index, and percent swell. These plots show extensive data scatter with very small correlation coefficients for swell pressure, compression index, and percent swell versus percent total smectite. Although the correlation coefficient for the rebound index and smectite data is appreciably larger, r = -0.36, it does not indicate a strong correlation. Finally, note that the percent swell values are in the low category for slab performance risk, according to Table 4.

Figure 12 shows the extent to which the percent total smectite in each sample correlates with the swelling potential indices introduced by McKeen, Seed, and Chen. For 177 samples, the correlation coefficients for McKeen's suction compression index and suction potential vs. percent total smectite are 0.61 and 0.58, respectively. For 59 samples the correlation coefficient for percent total smectite with Seed's swell potential index and Chen's plasticity index were 0.76 and 0.74 respectively.

Discussion

When comparing different indices of swelling potential, it is important to keep in mind the difference between "swelling potential" and amount of "potential swell." Seed (1962) defined and differentiated these concepts by dividing the factors that influence swell into two groups. With the first group he defined "swelling potential" in terms of those factors that depend on the nature of the soil particles, and that govern the capacity of a soil to swell under any conditions. These factors are the type of clay present in the soil and the amount of clay. With the second group, Seed defined "potential swell" in terms of the placement and environmental factors that govern the amount a soil will swell, but are unrelated to the nature of the soil particles. These factors include the soil structure, the electrolyte concentration in the water, the dry density to which a soil is compacted, the water content at compaction, the method of compaction, the availability of water, and the surcharge pressure.

Although the scope of Seed's classic study was limited to compacted soils, it is proposed that his concepts are also applicable to natural and undisturbed soils. In Seed's first group, the factors governing swelling potential that depend on the nature of the soil particles are equally applicable to either compacted or natural soils. In Seed's second group, the factors governing the magnitude of potential swell can readily be expressed in terms of the placement and environmental factors for natural soils. These factors include the density and state of stress for a soil in-situ, its geologic origin and history, and its structure and pore-fluid chemistry.

The swelling potential indices in this study differ in the degree to which they conform to Seed's swelling potential and potential swell concepts. These differences are explained in the following.

Chen's scheme involves the same distinction between swelling potential and potential swell concepts, but uses a single property, plasticity index, to characterize swelling potential. This provides an approximation to Seed's percentage swell index because both the type and percentage of clay in the soil govern its plasticity index.

Regarding the total percent smectite, this index differs from Seed's percentage swell index only in how they characterize the clay type. The smectite index is based on the percentage of clay in the whole sample and the amount of smectite in the clay fraction. Seed's percentage swell index also depends on the percentage of clay in the whole sample but uses an indirect measure of the mineralogy of the clay fraction, namely, the activity which is the plasticity of the clay fraction. The advantage of Seed's index is that the activity reflects the influence of all the constituents of the clay. The advantage of the smectite index is that it is based on the dominant source of expansive soil behavior, namely smectite. A possible limitation of this index, as defined in this study, is that it does not take into account the composition of exchangeable cations in the smectite, particularly calcium (Ca^{2+}) and sodium (Na^{+}).

McKeen's suction potential and suction compression indices differ from Seed's concepts in that they are obtained on undisturbed clods of soil. Therefore, McKeen's swelling potential indices are governed not only by the type and amount of clay in a soil, but also by the structure and pore-fluid chemistry of the soil, that result from its geologic origin and history. In the Colorado Front Range Corridor, the origin and geologic history of the clays and clay shales are generally similar in that they are primarily Cretaceous sediments. Nevertheless, it appears reasonable to expect some variations in the structure and pore-fluid chemistry of the samples collected in this study. Such variations could be contributing to the scatter in the correlations of McKeen's suction parameters with percent total smectite, and hence account for their somewhat lower correlation coefficients (0.6) compared with those (0.75) involving Seed and Chen's swelling potential indices.

In the swell-consolidation test, the most commonly used index in the Denver Metropolitan Area is the initial percent swell. The significance of this index differs from those discussed above in that it is governed not only by the nature of the material, but also by environmental and placement factors, such as the initial water content, stress history, and surcharge loading. That these factors may be controlling the behavior of the samples in the swell-consolidation tests is suggested by the low degrees of correlation shown in Figure 12 between the swell-consolidation test parameter data and the percent total smectite data. The importance of these factors is also suggested by the relatively large correlation coefficient for the rebound index and percent total smectite data compared with the very small correlation coefficients for the initial swell, swell pressure, and compression indices. The steps preceding the rebound phase of a swell-consolidation test minimize the influence of the initial water content and stress history by saturating and compressing the samples.

Conclusions

In this study the clay mineral of particular interest is smectite because of its well-known swelling behavior and because the other clay minerals present were non-sswelling. The percent total smectite provides a useful index of the swelling potential concept defined by Seed and correlates reasonably well with the swelling potential indices developed by Seed, Chen, and McKeen. Correlation coefficients are in the range of 0.6 to 0.75. However, the percent total smectite does not correlate well with conventional swell-consolidation test indices. The causes for this lack of correlation appear to be placement and environmental factors such as the initial moisture content, stress history, and surcharge loading that are beyond the scope of Seed's swell potential concept and Seed, Chen, and McKeen's swelling potential indices. The latter are only governed by factors that characterize the composition of a soil, and that govern its capacity to swell under any conditions.

References

ASTM. (1993). *Annual Book of ASTM Standards*, 04.08, American Society. of Testing and Materials, Philadelphia, Pennsylvania.

Chabrillat, S., Goetz, A. F. H., Olsen, H. W., and Noe, D. C. 1997, "Field spectrometry techniques for identification of expansive soils," *Proceeding of the 12th Conference on Applied Geologic Remote Sensing*, 17-19 Nov. 1997, I:141-148.

Chabrillat, S., Goetz, A. F. H., Olsen, H. W., Krosley, L., and Noe, D. C. 1999, "Use of aviris hyperspectral data to identify and map expansive clay soils in the front range urban corridor in colorado," *Proceedings of the Thirteenth International Conference on Applied Geologic Remote Sensing,* Vancouver, British Columbia, Canada, 1-2 March, 1999.

Chabrillat, S. and Goetz, A. F. H., 1999, "The search for swelling clays along the colorado front range: the role of aviris resolution in detection," *Proceedings AVIRIS Workshop*, February 6-8, 1999.

Chen, F. H., 1988, *Foundations on Expansive Soils*, 2nd Edition, Elsevier, New York.

Drever, J. I., 1973, "The preparation of oriented clay mineral specimens for x-ray diffraction analysis by a filter-membrane peel technique," *American Mineralogist*, Vol. 58, pp. 553-554.

Gill, J.D., West, M.W., Noe, D.C., and Olsen, H.W., 1996, "Geologic control of severe expansive clay damage to a sub-division in the Pierre Shale, Southwest Denver Metropolitan Area," *Clays and Clay Minerals,* v. 44, no. 4, p. 530-539.

HBA Soils Task Force, (1996), "Soils task force report, draft for review," *Home Builders Association of Metropolitan Denver,* November 1, 1996.

Jones, D. E. and Holtz, W. G., 1973, "Expansive soils," - The Hidden Disaster, *Civil Engineering,* ASCE: v. 43, no. 8, August p. 49-51.

McKeen, R. G., 1992, "A model for predicting expansive soil behavior," *Proceedings of the International Conference on Expansive Soils,* Vol. 1, p. 1-6, 1992.

McKeen, R. G. and Nielsen, J. P., 1978, *Characterization Of Expansive Soils For Airport Pavement Design,* U. S. Department of Transportation, Federal Aviation Administration, FAA Report No. FAA-RD-78-59.

McKeen, R. G., 1981, *Design of Airport Pavements on Expansive Soils,* DOT/FAA/RD-81/25, Systems Research and Development service, Federal Aviation Administration, Washington, D.C.

McKeen, R. G., 1985, *Validation of Procedures for Pavement Design on Expansive Soils,* DOT/FAA/PM-85/15, Program Engineering and Maintenance Service, Federal Aviation Administration, Washington, D.C., July.

McKeen, R. G. and Johnson, L D., 1990, "Climate-controlled soils design parameters for mat foundations," *Journal of Geotechnical Engineering,* ASCE, Vol. 116, No 7, July, pp. 1073-1094. Moore, D. M., and Reynolds, R. C., Jr. (1989). *X-ray Diffraction and the Identification and Analysis of Clay Minerals*, Oxford University Press, Oxford, New York.

Noe, D.C. and Dodson, M.D., 1995, "The Dipping Bedrock Overlay District (DBOD): an area of potential heaving bedrock hazards associated with expansive, steeply dipping bedrock in Douglas County, Colorado," *Colorado Geological Survey Open-File Report 95-5*, Colorado Geological Survey, Denver, CO, 32p., 1 plate, scale 1:500.000.

Noe, D.C. and Dodson, M.D., 199, "Heaving-bedrock hazards associated with expansive, steeply dipping bedrock in Douglas County, Colorado," *Colorado Geological Survey Special Publication 42*, 80p. 2 plates, scale 1:24000.

Schultz, L. G., (1964). "Quantitative Interpretation of Mineralogical Composition from X-ray and Chemical Data for the Pierre Shale". *United States Geological Survey, Professional Paper 391-C.*

Seed, H. B., Woodward, R.J., Jr., and Lundgren, R., 1962, "Prediction of swelling potential for compacted clays," *Journal of The American Society of Civil Engineers, Soil Mechanics and Foundations Division,* Vol. 88, No. SM3, pp. 53-87.

Thompson, R. W., 1997, "Evaluation Protocol for Repair of Residences Damaged by Expansive Soils," Special Publication No. 68, *Proceedings on Unsaturated Soils at Geo-Logan '97,* sponsored by the Geo-Institute of ASCE, Logan Utah, July 1997, pp. 255-276.

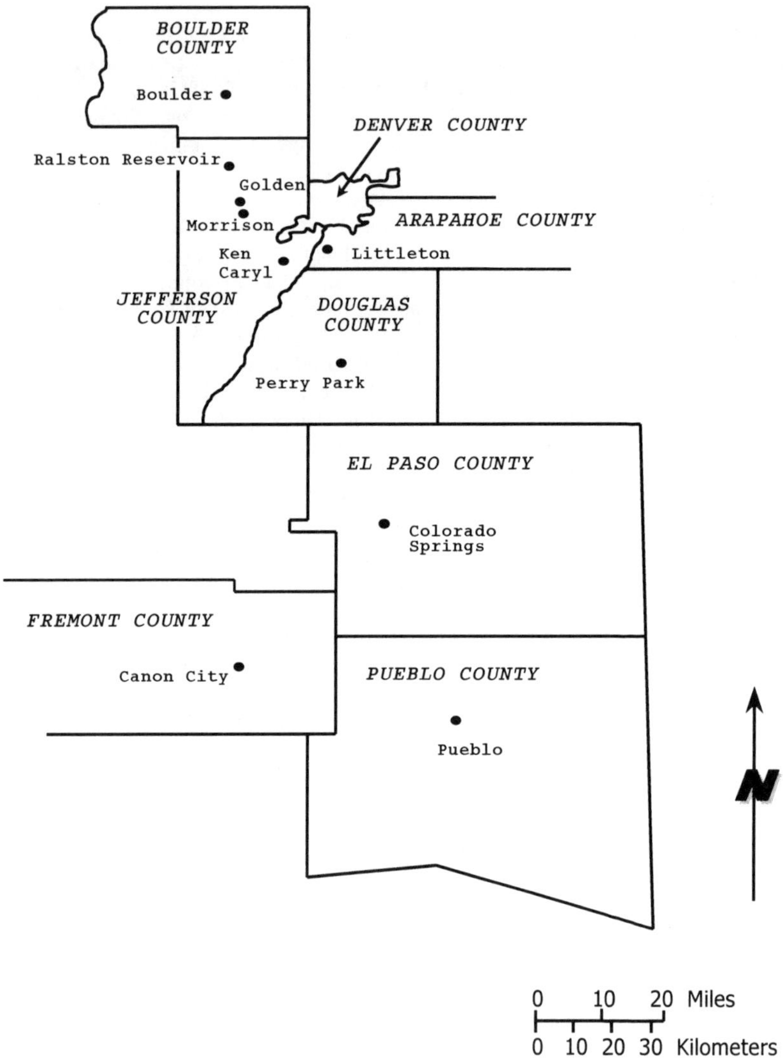

Figure 1 Counties (upper case) and cities along the Colorado Front Range Urban Corridor where samples were obtained.

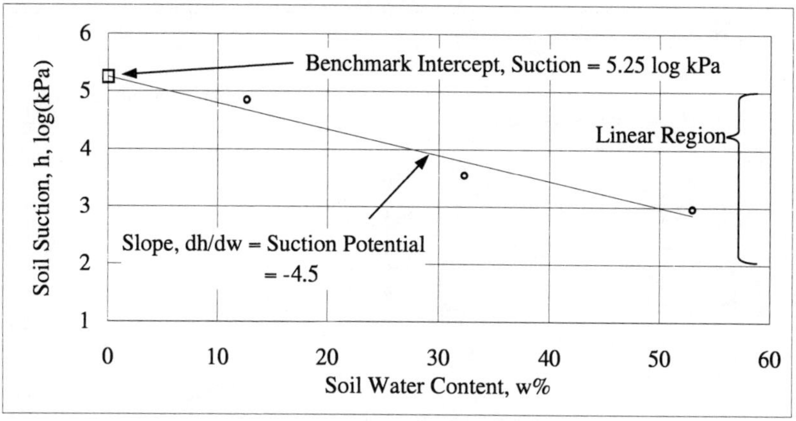

Figure 2 Suction vs. water content data for Sample MR97001.

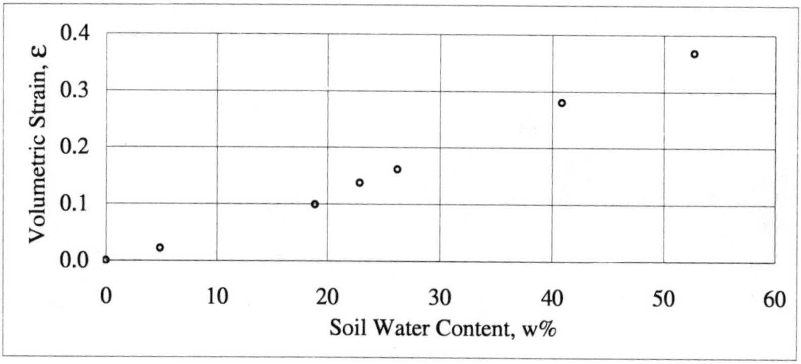

Figure 3 Volumetric strain vs. water content data for Sample MR97001.

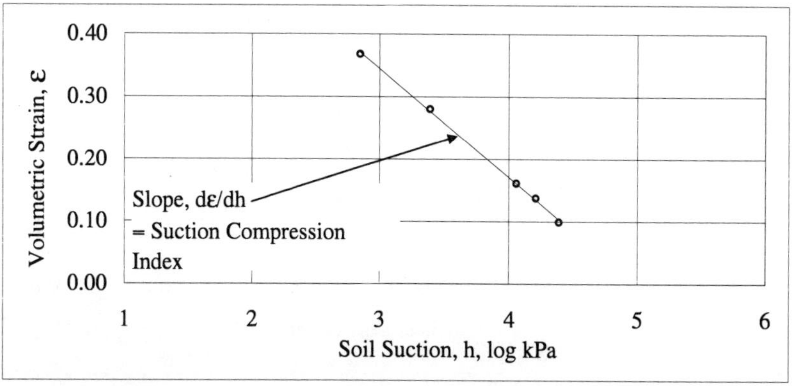

Figure 4 Volumetric strain vs. suction data for Sample MR97001.

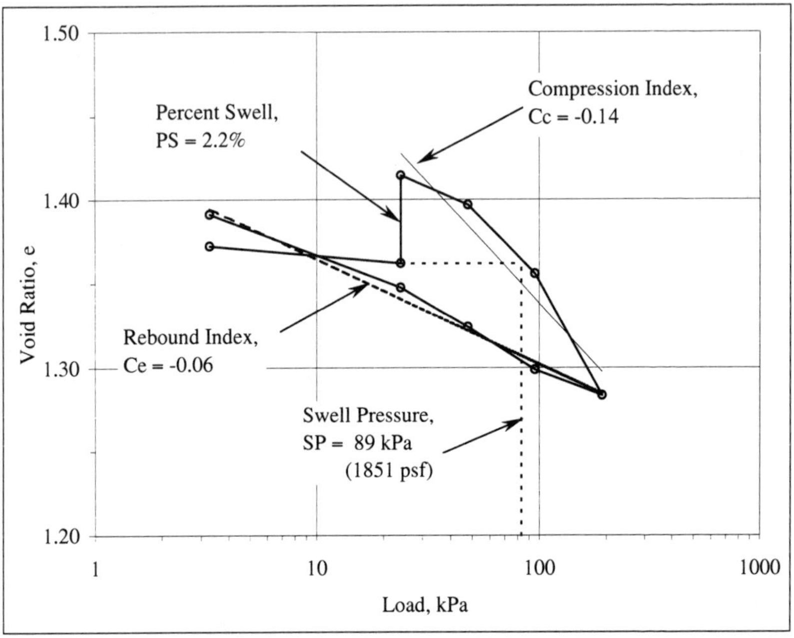

Figure 5 Swell consolidation test data for Sample MR97001.

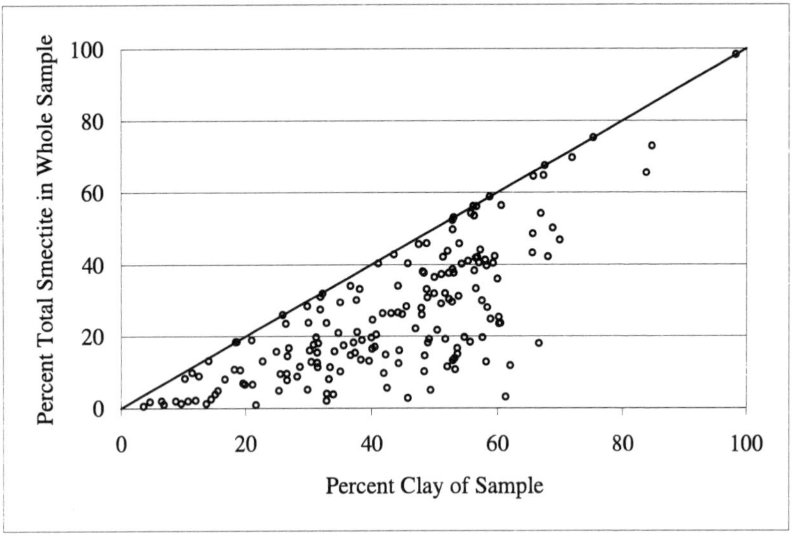

Figure 6 Percent total smectite vs. percent clay in 178 samples.

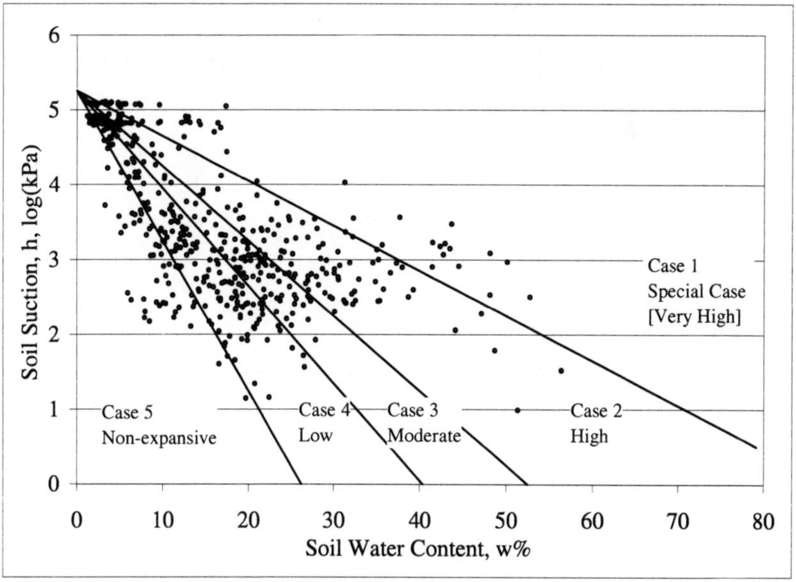

Figure 7 Suction vs. water content data for 182 samples together with McKeen's classificaiton categories in Table 1.

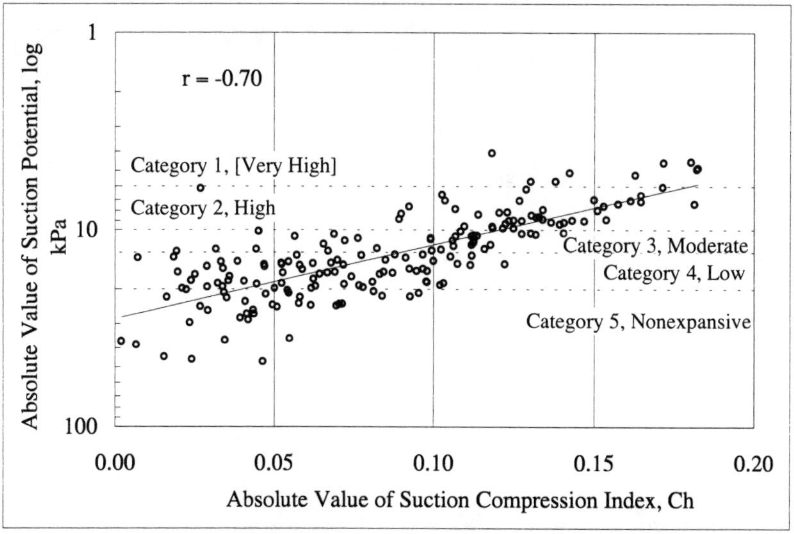

Figure 8 Suction potentials vs. suction compression index for 182 samples together with McKeen's classificaiton categories in Table 1.

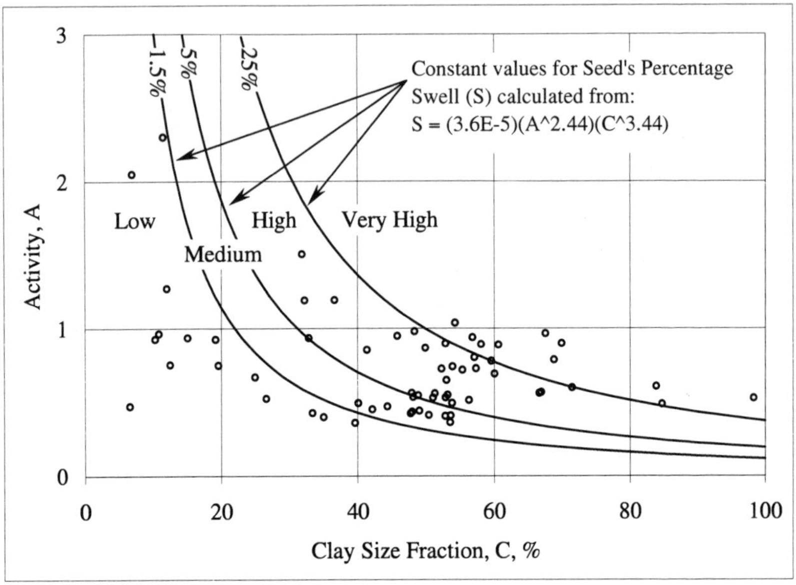

Figure 9 Activity vs. clay fraction data for 64 samples together with Seed's classification categories in Table 2 and Figure 4.

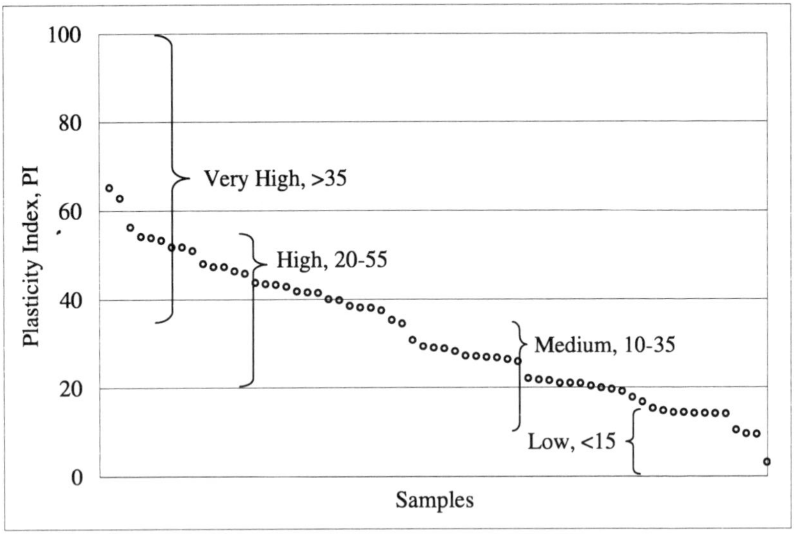

Figure 10 Plasticity index data for 64 samples together with Chen's classification categories in Table 3.

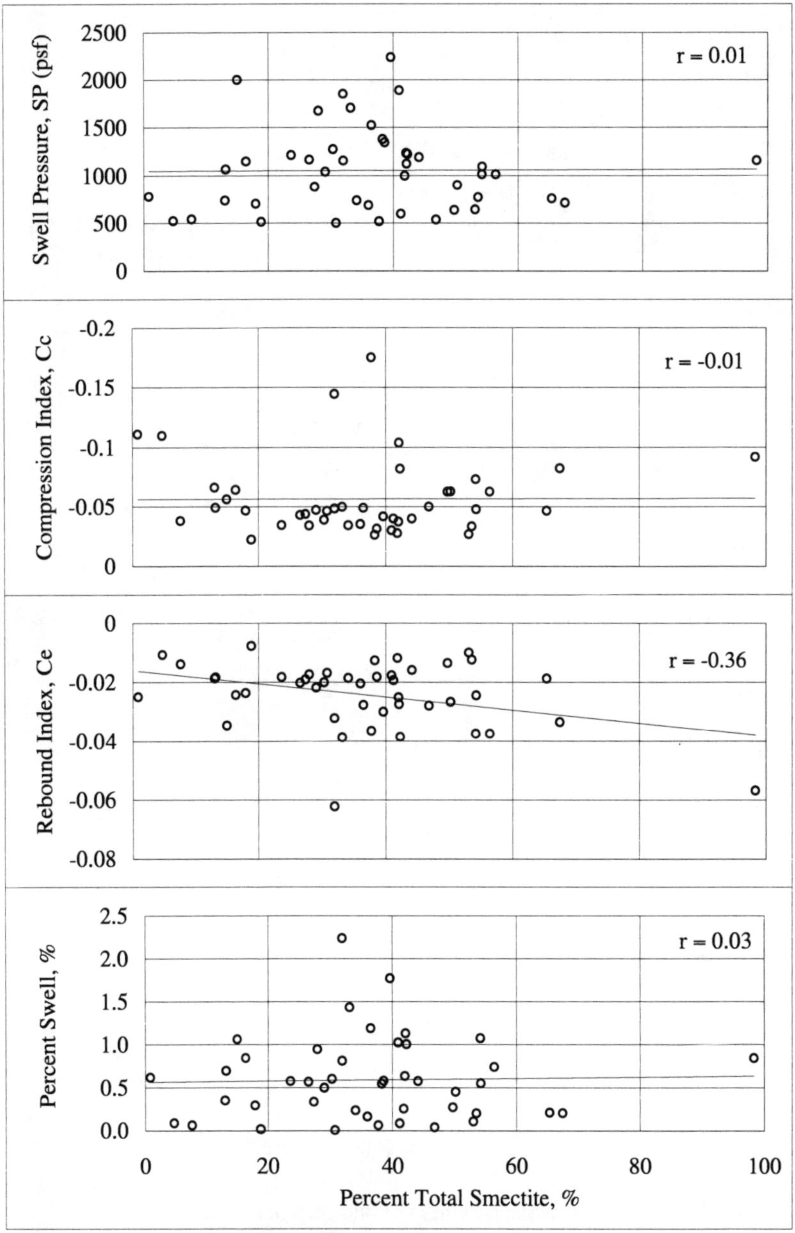

Figure 11 Swell-consolidation test parameters vs. total smectite for 44 samples.

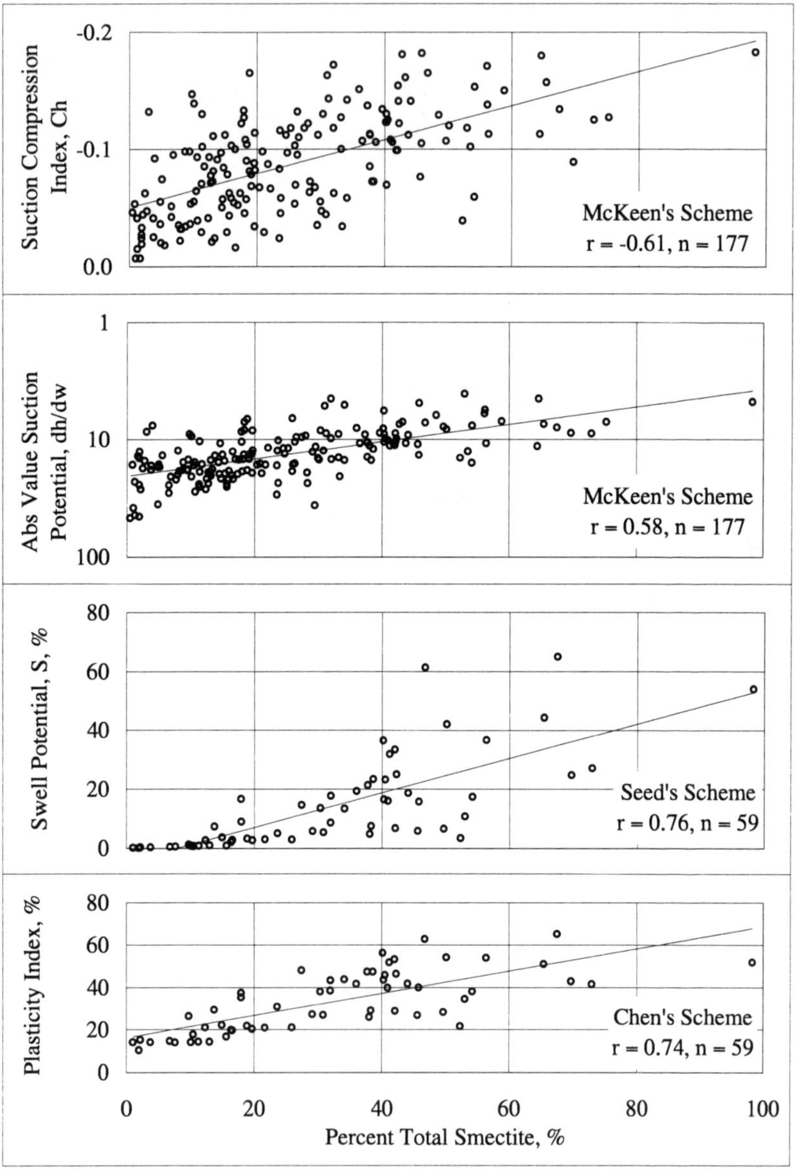

Figure 12 Swelling potential schemes vs. total percent smectite.

Suction-Swelling Relations

Erdal Cokca[1] and Altay Birand[2]

Abstract

In this paper, thermocouple psychrometric technique is used to measure soil suction and reliability of swell percentage prediction method based on soil suction data is evaluated. Kaolinite-bentonite clay mixtures were prepared to obtain expansive soils in a wide range of plasticity indices. Suction measurements using thermocouple psychrometer technique and free swell tests in oedometers were made on statically compacted specimens. Soil suction-water content relationships were evaluated. The results indicated a linear relationship between log suction and swell percentage measured in oedometer. It has been shown that the predicted swell percentages by using suction test results underestimate the ultimate swell percentages in oedometer free swell tests.

Keywords: clay, expansive soil, soil suction, swell percentage, psychrometer

Introduction

Expansive soil is a term applied to any soil that has a potential for swelling under changing moisture conditions. The unpredictable swell of unsaturated expansive soils resulting from wetting may cause undesirable cracks and damages to many structures built in arid or semi-arid areas. Since engineers have been confronted with the problems of expansive soils, many techniques have been developed to enable a reliable prediction of swell percentage (e.g. oedometer tests).

Recently psychrometers, which measure total suction, have received increased usage in measuring the suction of a soil in the laboratory. Johnson and Snethen (1978) states that, the soil suction concept and associated testing and prediction procedures provide a reliable estimate of swell percentage. Suction

[1] Assoc.Prof..Dr. and [2]Prof.Dr. of Civil Engineering, Middle East Technical University, Department of Civil Engineering, 06531 Ankara Turkey, email: ecokca@metu.edu.tr , abirand@metu.edu.tr, Fax: 90-312-2101262

methods for prediction of swell percentage of expansive soils have advantages over conventional oedometer tests. In suction tests there is no restriction for sample geometry, therefore difficulties met in sampling and trimming of oedometer test specimens are eliminated.

Definition of Soil Suction

Although not technically correct, soil suction can be described in layman's terms as a measure of a soil's affinity for water. In general, the drier the soil, the greater is the soil suction (Wray 1984). Snethen (1980) defines suction as a pressure term that is a measure of the pulling force (tension) exerted on the water. The total soil suction, τ may be alternatively defined as the free energy present in soil water with respect to a pool of pure water located outside of the soil at the same elevation. This energy is the source of work that is done when the pool of pure water comes in contact with the soil. At equilibrium following contact with the pure water, the free energy is dissipated in the form of work that is done to pull the pure water into soil, counter the friction, and expand the soil lattice.

Total suction τ is comprised of two component suctions; matrix suction, τ_m, which is negative pressure which will hold pore water in equilibrium through a porous membrane with the same soil water within a sample of soil; and osmotic (or solute suction), τ_s, which is the negative gauge pressure which will hold pore water in equilibrium, arises from the presence of soluble salts in the soil water provided the type and concentration of the salts in the soil differ from that of free water (Snethen 1980).

According to Nelson & Miller (1992) within the range of water contents encountered in most practical problems in geotechnical engineering, significant changes in osmotic suction do not occur, changes in total suction that will occur are those that are due only to changes in matrix suction.

Evaluation of Thermocouple Psyhrometers

Here soil suction is evaluated from measurements of relative humidity in soils determined with thermocouple psychrometers (according to Edil et al. (1981) their use is restricted to soil suctions greater than 100 kPa). Details of the procedure for measuring soil suction using psychrometers is presented in the literature (Snethen 1980) but for the sake of completeness a summary description is also given in this paper.

The thermocouple used in geotechnical engineering is a miniature thermocouple enclosed in protective casing made of ceramic or steel mesh (Fig.1), it measures the relative humidity in the soil by a technique called Peltier cooling. By causing a small direct current of about 4 to 8 milliampers to flow through the thermocouple junction for about 15 seconds in the correct direction, this junction will cool and water will condense on it when the dew point temperature is reached (Snethen 1980). Condensation of this water inhibits further cooling of the junction, and the voltage developed between the thermocouple and reference junction is measured by microvoltmeter.

Correction for the readings taken at temperatures other than 25°C was made by using the following formula:

$E_{25}=E_T/(0.325+0.027*T)$ (1)

where

E_{25} =the equivalent output at the calibration temperature of 25°C,

E_T = the psychrometer voltage output, at T °C (in microvolts)

T = temperature in °C.

Before using the psychrometers in soils, the voltage outputs of the psychrometers are calibrated by tests with salt solutions.

Experimental Works

The tests presented in this paper were performed to compare the swell percentage measurement by soil suction test and by oedometer test. Soil suction tests using thermocouple psychrometers were performed to obtain the soil suction versus water content relationship, also free swell oedometer tests were performed to measure swell percentages.

Material

Commercially processed kaolinite and bentonite mineral clays were mixed in preselected proportions to obtain clays possessing a wide range of plasticity index. Kaolinite and bentonite mixture ranges are shown in Table 1. The index properties of kaolinite and bentonite mixtures were found and hydrometer tests were performed to find the clay percent of soil mixtures (Table 1).

Suction Tests Using Thermocouple Psychrometer Method

Equipment

The psychrometric technique has been used successfully for laboratory measurements. Thermocouple probes and readout apparatus are commertially available. In this study, the apparatus manufactured by Wescor Inc. (Wescor type HR-33T Dew point Microvoltmeter) of the United States (it is accurate in the range of pF3.2-4.8) was used. A glass jar as a sample container, rubber stoppers, plastic mud, a thick cardboard box filled with plastic foam, a medicine dropper and NaCl solutions with different molalities of 0.3, 0.55 and 0.7 were used.

Calibration

NaCl solutions with molalities of 0.3, 0.55 and 0.7 and having water potentials of -13.68, -25.1 and -32.10 bars respectively were used to calibrate the psychrometers. The psychrometer measurements on sodium chloride solutions reflect the osmotic (and the total) suction equivalent of the solution. Four jars were filled with NaCl solutions of the same molality and then the jars were closed with the rubber stoppers with psychrometer wires passing through. Finally, plastic mud was used to seal the openings between the psychrometer wire and rubber stopper in order to prevent air entrance.These jars were placed in the thick cardboard box filled with plastic foam in order to minimize the temperature variation of the sample container (Fig. 2). After two or three hours, the psychrometer voltage output readings stabilized. Thus the vapor equilibrium was reached. After reaching the equilibrium, the psychrometer voltage output and temperature readings were taken. The psychrometer voltage outputs E_T (in microvolts) were converted to the equivalent output at the calibration temperature of 25°C, E_{25}, by using Equation (1).

The psychrometer output E_{25} is then plotted against water potentials (i.e.suction), and this procedure is repeated for the remaining two molalities to obtain the calibration line for the specific psychrometer (Fig. 3). A typical example of a calibration line for the thermocouple psycrometers used to collect the soil suction data in this study is:

$$\tau=2.3636*E_{25}-1.7188 \qquad (2)$$

Test Procedure

From each clay mixture four disc shaped specimens having diameter of 36 mm and height of 15 mm were prepared at a dry density of 1.64Mg/m^3 and a water content of 10%, using static compaction. After preparing four compacted samples, one of them was left to stay at the initial water content level. The remaining three samples were wetted with varying amounts of water for varying lengths of time to establish a range of water content conditions. After this gradual wetting process, all four specimens were left to swell in the desiccator for 2 to 3 days to attain equilibrium water contents.

At the end of this period water content and specific volume of the specimen sets were calculated to find the volumetric compressibility factor for the specimen sets. The volumetric compressibility factor is the slope of the relationship between the specific volume and water content of the specimens. The specific volume was calculated from the changes in vertical and radial dimensions of the specimens. Both the radial and vertical dimension readings were taken from different points of the specimen and in calculations mean values of the readings were utilised.

Then all four specimen were waxed. A hole was drilled on the waxed covered sample so that the ceramic shield of psychrometer could be placed in the hole to measure the final equilibrium soil suction values. Sides of the psychrometer were isolated by plastic modelling clay to prevent any water content change. These four samples were placed in jars and the jars were closed with the rubber stoppers which had been affixed to the psychrometer wire (Fig.2). The jars were placed in a cardboard box filled with plastic foam packing material for better heat insulation and left to reach equilibrium (Fig.2). Usually it took 24 hour to reach equilibrium. Psychrometer voltage outputs and temperature readings are taken.

This procedure is repeated at least three times in a day until the successive readings stabilize, meaning that vapor pressure equilibrium is indicated.

Data Reduction and Interpretation

The results of the volumetric swell measurements showed that volumetric compressibility factor α, which were determined by calculating the slope of the specific volume versus water content relationship (Fig.4) is a function of plasticity index. The data trends indicated that as the plasticity index increases α also increases.

The psychrometer output (microvolt) at the measured temperature is converted to an equivalent output using Equation (1). The psychrometer output is then converted to soil suction using the calibration line for the specific psychrometer. The data is then plotted versus water content on a semi log plot to establish the log soil suction versus water content relationship (Fig.5), which is linear and has the form

$$\log\tau_m = A - B*w \quad (3)$$

where τ_m= matrix soil suction without surcharge pressure

A = y-intercept of soil suction versus water content curve, i.e. the soil suction value (logarithmic scale) at zero moisture content.

B= slope of soil suction versus water content curve

w= water content

Calculation of Swell Percentage

Johnson and Snethen (1978) gives a swell prediction method based on soil suction, in this method, the swell is given by the following equation:

$$\Delta H/H = (C_\psi/(1+e_0))*\log(\tau_i/(\tau_f+\alpha*\sigma_v)) \quad (4)$$

where

τ_i= initial soil suction,

τ_f= final soil suction,

σ_v= vertical stress,

e_0= initial void ratio

C_ψ= suction index,
$C_\psi = \alpha * G_s/(100*B)$ (5)
where α= Volume compressibility factor, B= Slope of log soil suction versus water content curves, G_s= Specific gravity of the soil particles

Oedometer Method

Standard free swell tests (ASTM D4546-90) were conducted on statically compacted samples of the kaolinite-bentonite mixtures (initial water content of 10% and having dry density values of 1.64Mg/m^3) in conventional oedometer cells; and the swell percentages of each specimen, possessing plasticity indices comparable to the ones obtained in the suction measurement tests, were directly measured. The sample is confined in the consolidation ring, which is placed in the oedometer under a small surcharge of about 1.94 kPa, and water is than added to sample and allowed to swell freely. Because of the high swell potential of bentonite, new consolidation rings and free swell test apparatus were made for this test method. The height of the rings were 6.0 cm. All the laboratory free swell tests were laterally confined. As the sample swells the deflections of the dial gauge is recorded. At some point the sample has no further tendency to swell and the maximum deflection recorded is used for the calculation of the free swell. The percentage of free swell can be expressed as;
Free swell= $\Delta H/H*100$ (6)
where ΔH= Change in initial height (H) of the specimen, H= Initial height of the specimen
The results of swell percentage tests performed under various bentonite contents are shown in Table 1.

Test Results

A linear relationship is assumed between the log soil suction, and water content (see Fig. 5), which has the form: $\log \tau = A\text{-}B.w$

Table 1 summarises the intercept A, which is the soil suction value (logarithmic scale) at zero moisture content, slope B of log soil suction versus water content behaviour, volumetric compressibility factor α, which relates the change in specific volume to the corresponding change in water content, and suction index C_ψ values for different plasticity indices.

The intercept A and slope B of log soil suction versus water content behaviour decrease with increasing PI. The steep slope (large B) for a soil sample is indicative of a relatively low swelling capability (e.g. sample with PI=22%), while the flat slope (small B) for a soil sample is indicative of a relatively large swelling capability (e.g. sample with PI=225%).

Suction index, C_ψ which reflects the rate of change of void ratio with respect to soil suction, increases with increasing plasticity index.

The swell percentage predicted by Equation (4) are compared with the direct measurements from the oedometer free swell test, as shown in Table 1. Swell percentage prediction by suction method greatly underestimated the swell percentage.

The regression analysis made on the swell percentage (oedometer test)-initial soil suction data has revealed the following correlation (Fig.6):

$$\text{swell } \% = -1026.4 + 661 \log(\text{suction}) \qquad (7)$$

As it can be seen from Figure 6, that there is a strong relationship between the swell percent and initial soil suction. As the swell percent increases, initial soil suction also increases.

Conclusions

A simple method for prediction of swell percentage by suction measurements using thermocouple psychrometers has been described. Swell oedometer tests and suction measurements have been performed on compacted expansive soil samples. A linear relationship between the log soil suction, τ and water content w has been observed:

$$\log \tau = A - B.w$$

and the intercept A and slope B of log τ versus water content behaviour decrease with increasing plasticity index.

The volumetric compressibility factor α which is strongly dependent on plasticity index, increases as the plasticity index increases.

Suction index , C_ψ, increases with increasing plasticity index.

The experimental relationship, which directly relate initial soil suction to the swell percent is established, the coefficient of correlation of this relationship is 0.97, but this relationship is valid for the given initial dry density and water content.

The swell percentages measured in oedometers were compared with the swell percentages estimated by the suction method. The comparison of measured swell percentage with the ones predicted by Equation (4) reveal that the method proposed by Johnson & Snethen (1978) underestimates the swell percentages of compacted clays. The discrepancy between measured and predicted swell percentages can be attributed to the methods used to characterize the relationship between soil suction and volume change. Both compressibility factor α and the parameter B that reflects the suction-water content relationship contribute to the underestimation, these parameters were determined from volumetric swell measurements, whereas oedometer Free Swell percentage values are obtained from one-dimensional swell.

References

ASTM (1993). Standard Test Method for One-Dimensional Swell or Settlement Potential of Cohesive Soils, Annual Book of ASTM Standards, D4546-90, **04.08**, 853-859.

Edil, T.B., Motan, S.E. & Toha, F.X. (1981). Mechanical behavior and testing methods of unsaturated soils, *American Society of Testing and Materials*, Special Technical Publication, 114-129.

Johnson, L.D. & Snethen, D.R. (1978). Prediction of potential heave of swelling soils, *Geotechnical Testing Journal*, GTJODJ, **1**, No.3, sept.,117-124.

Nelson, J.D. & Miller, D.J. (1992). Expansive Soils, Problems And Practice in Foundation and Pavement Engineering, John Wiley and Sons Inc., NewYork.

Snethen, D.R. (1980). Characterization of expansive soils using soil suction data, *4th International Conference on Expansive Soils*, pp.54-75.

Wray, W.K. (1984). The principle of soil suction and its geotechnical engineering applications, *Fifth International Conference on Expansive Soils*, Adelaide, South Australia, pp.114-118.

TABLE 1. Soil Properties, A, B, α and C_ψ Values, Measured and Predicted Swell Percentages

Sample* (%)	G_s	Clay Content (%)	LL %	PL %	PI %	Soil Class**	A	B	α	C_ψ	Measured Swell %	Predicted Swell %
100%K	2.62	47.5	42	20	22	CL	2.909	0.1214	0.16	0.0345	23.26	6.70
95%K 5%B	2.61	48.0	48	20	28	CL	2.640	0.0933	0.16	0.0448	26.15	7.80
90%K 10%B	2.59	48.8	71	20	51	CH	2.473	0.0781	0.17	0.0564	40.41	10.30
85%K 15%B	2.58	49.9	93	21	72	CH	2.479	0.0756	0.19	0.0648	59.00	12.84
80%K 20%B	2.56	51.0	112	23	89	CH	2.332	0.0605	0.22	0.0931	71.09	17.89
75%K 25%B	2.55	52.0	135	24	111	CH	2.265	0.0519	0.27	0.1330	85.04	25.12
70%K 30%B	2.54	53.2	155	26	129	CH	2.159	0.0427	0.28	0.1666	103.63	31.04
65%K 35%B	2.53	54.0	180	26	154	CH	2.132	0.0371	0.28	0.1909	118.24	35.62
60%K 40%B	2.52	54.8	205	28	177	CH	2.101	0.0317	0.30	0.2385	132.74	43.30
55%K 45%B	2.51	56.0	230	28	202	CH	2.112	0.0299	0.32	0.2686	143.18	48.95
50%K 50%B	2.50	57.5	255	30	225	CH	2.099	0.0270	0.33	0.3056	168.57	56.10

*K: Kaolinite, B: Bentonite

**Unified Soil Classification System

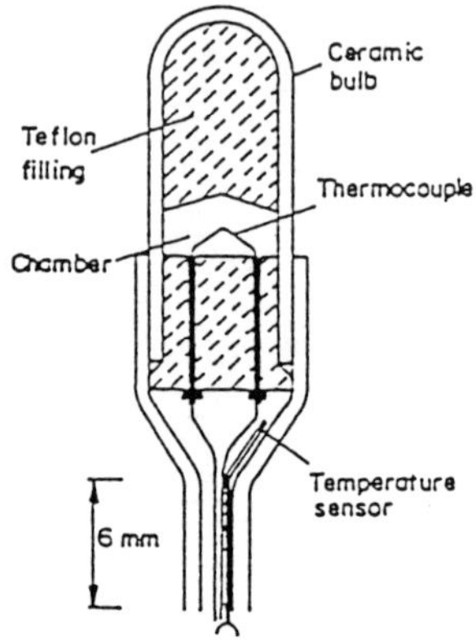

Fig. 1. Ceramic shielded thermocouple psychrometer

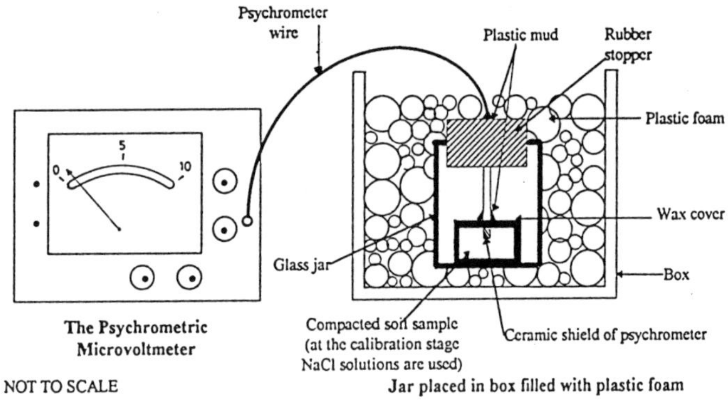

Fig. 2. The equipment used to evaluate the soil suction using thermocouple psychrometer

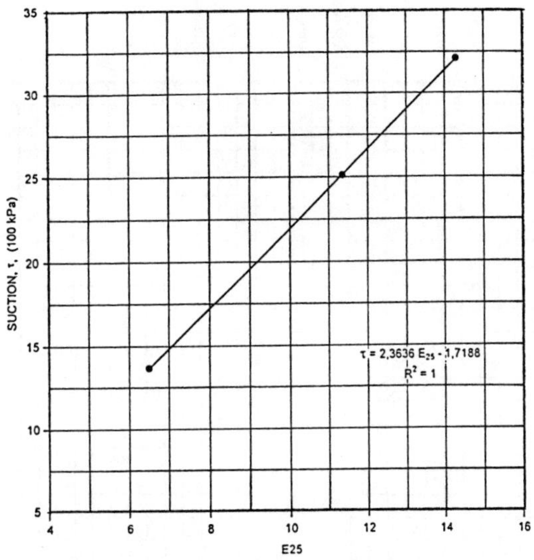

Figure 3. Calibration line for No. 36670 psychrometer

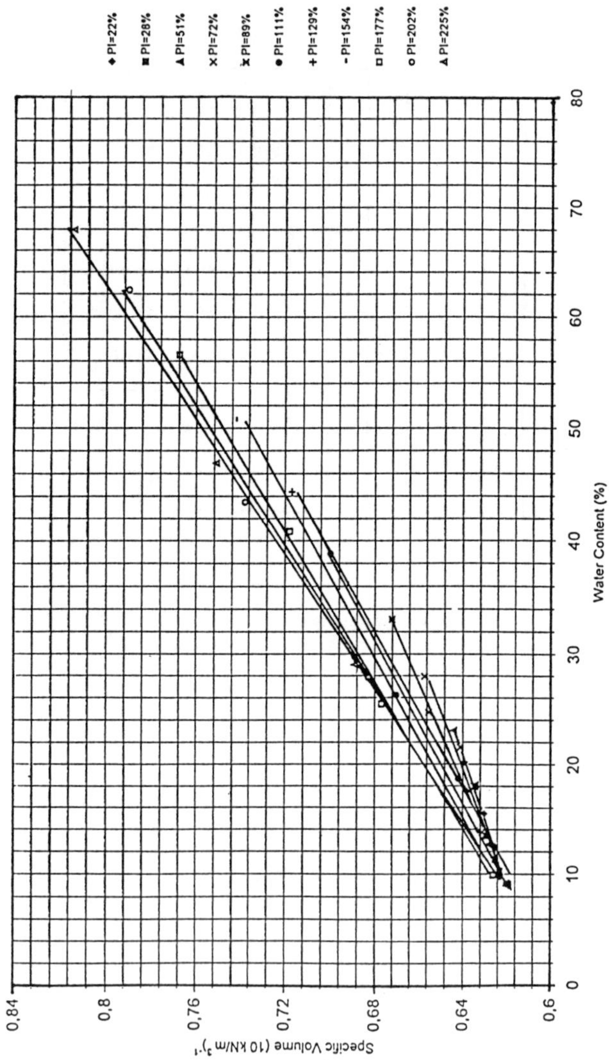

Figure 4 The relationship between the specific volume and water content

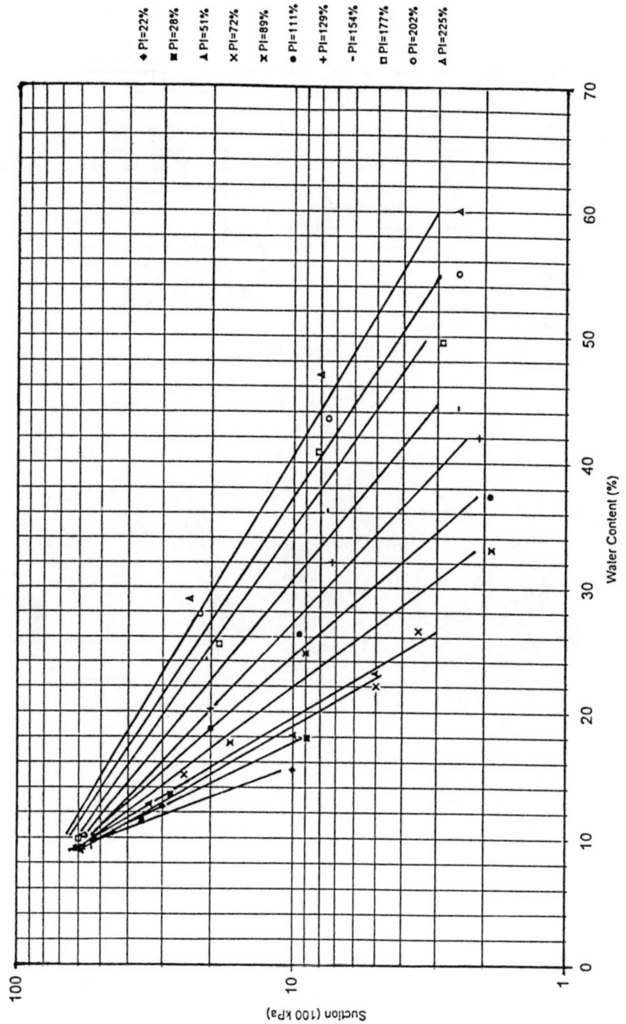

Figure 5 The relationship between the suction and water content

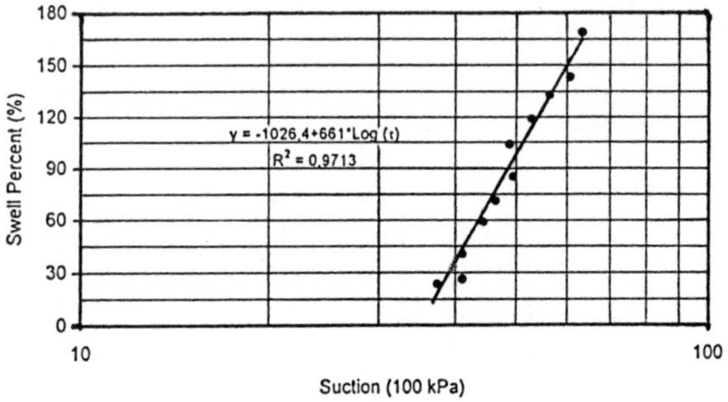

Figure 6 Swell percent versus suction relationship

Suction Compression Index based on CLOD Test Results

Howard A. Perko, PE, Associate Member[1], Robert W. Thompson, PE, Fellow[2], and John D. Nelson, Ph.D., PE, Fellow[3]

Abstract

Determination of the suction compression index, C_h, is necessary for predicting expansive soil heave based on soil suction. The CLOD Test is a convenient laboratory procedure for determining C_h, however it is not widely used. Rather, engineers often rely upon empirical correlations to find C_h. A laboratory investigation was conducted to verify previously established empirical relationships. The investigation involved CLOD tests on relatively undisturbed drive samples of clay and claystone bedrock from the Denver, Colorado area. Test results generally matched previous studies. Samples utilized in this investigation spanned a large range of suction and moisture characteristics. A new empirical relationship for determining C_h over a broader range of materials is presented.

In an effort to expedite the CLOD test, studies were performed to ascertain the effect of various procedural simplifications on the test results. Information obtained by these studies is used to suggest a modified procedure for professional practitioners. Filter paper suction, one-dimensional swell, Atterberg Limits, and gradation tests were also performed on numerous select samples. Swell prediction based on soil suction and the CLOD test results compared well with direct laboratory swell measurements. The average C_h measured in the Denver area is shown to be a function of bedrock geology. A semi-empirical relationship is also presented between Atterberg limits, percent passing the No. 200 sieve and C_h.

---

[1]Research Fellow, Civil Eng. Dept., CSU and Project Eng., CTL/Thompson, Inc.
727 Laporte Ave., Ft. Collins, CO 80521
howie@engr.colostate.edu, phone: 970-472-5068, fax: 970-472-0264

[2]CEO, CTL/Thompson, Inc.
1971 W. 12th Avenue, Denver, CO 80204
Expdirt@aol.com, phone: 303-825-0777, fax: 303-825-1471.

[3]Professor, Civil Eng. Dept., Colorado State University
CSU Campus, Civil Engineering Department, Fort Collins, CO 80523
john@engr.colostate.edu, phone: 970-491-6081, fax: 970-491-3584.

Introduction

The more traditional method of determining expansive soil heave in practice today in the Denver, Colorado area utilizes conventional, one-dimensional, swell-consolidation measurements. This method has an established heritage and is based on representing soil suction changes through "equivalent" changes in effective stress. Another method for estimating heave is based directly on soil suction. This method considers soil suction as the primary stress state variable and relates heave directly to suction changes. There are many methods of representing soil heave in terms of soil suction, some of which are described in Matyas and Radhakrishna (1968), McKeen and Nielsen (1978), Fredlund and Rahardjo (1993), and Hamberg and Nelson (1984). The subject of this particular study is a formalism presented by McKeen (1992) who incorporated a lateral restraint factor and combined the effects of changes in total suction and effective confining stress into one term and presented the equation

$$\Delta H = C_h \, \Delta h \, \Delta t \, f \, s \tag{1}$$

where ΔH is the vertical movement of the layer being considered, Δh is total suction change, Δt is layer thickness, C_h is the suction compression index, f is the lateral restraint factor (between 0.5 and 0.8 for clays), and s is the load effect coefficient (typically 0.9). The latter equation has been used by engineers in the Denver, Colorado area (Thompson 1997, and Thompson and McKeen, 1995, and McOmber and Thompson, 2000).

Application of the McKeen method requires direct measurement of soil suction. This is typically accomplished by filter paper tests (McQueen and Miller, 1968 and ASTM D5298). The constitutive parameter used to relate total suction to volume change, called the suction compression index, C_h, can be measured using the CLOD test (McKeen, 1985 and Hamberg, 1985) but is often found by an empirical relationship (McKeen, 1992)

$$C_h = -0.02673\left(\frac{\Delta h}{\Delta w}\right) - 0.38704 \tag{2}$$

where Δw is the change in soil water content. This relationship is based on a line constructed at the 85th percentile for laboratory tests on Shelby tube and California drive samples.

The objectives of the present investigation are to augment the validity of the empirical suction compression index relationship, and expand its applicability to relatively undisturbed samples from the Denver, Colorado area. Laboratory procedures for measuring C_h are reviewed and a few methods for expediting the test are presented. Effects of plastic limit, gradation, and geology on C_h are discussed.

Investigation

A total of 89 relatively undisturbed soil and bedrock samples from 22 sites around the Denver, Colorado area were tested. Samples were obtained using a California type (6.4 cm or 2.5 in outside diameter) split barrel sampler. Sample depth generally ranged from 1.5 to 7.5 m (5 to 25 ft). Most samples were separated into two pieces, and filter paper suction and CLOD tests were conducted. Some samples were separated into three pieces, and conventional swell tests were performed in addition to suction and CLOD tests. Atterberg limits and gradation information were obtained from nearby samples of representative soils and bedrock. Silt and clay content (percent passing the No. 200 sieve) ranged from 50 to 100 percent. Liquid limit and plasticity index ranged from 35 to 165 percent and 21 to 124 percent, respectively.

Procedures

All tests were performed by personnel employed by CTL/Thompson, Inc. Filter paper suction, one-dimensional swell, Atterberg limits and gradation tests were conducted according to standard procedures. Suction measurements were made by suspending the filter papers over the samples so as to measure total suction. Osmotic suction has been shown conclusively to be a valid stress state variable but its impact on soil behavior is subtle in comparison with soil response to matric suction (Miller and Nelson, 1993). It was assumed that osmotic suction changes were negligible during these tests and that total suction changes are equal to matric suction changes.

Standard test methods have not been developed for the CLOD test. McKeen (1985) originally formulated the CLOD test method based on the COLE (Coefficient of Linear Extensibility) procedure presented by the U.S. Department of Agriculture, Soil Conservation Service (1972). The method is also described in Nelson and Miller (1992) and in Hamberg (1985). Procedures for performing the CLOD test involve coating irregularly shaped soil specimens with liquid resin and monitoring weight and volume change over time while either air drying or wetting. Volume measurements are obtained by Archimedes principle which involves weighing the samples underwater, as shown in Fig. 1. The resin coating is essentially watertight, but is permeable to water vapor. Through these procedures a relationship between sample volume and water content is obtained. After the drying or wetting process, the specimens are oven dried and the suction compression index, C_h, is determined.

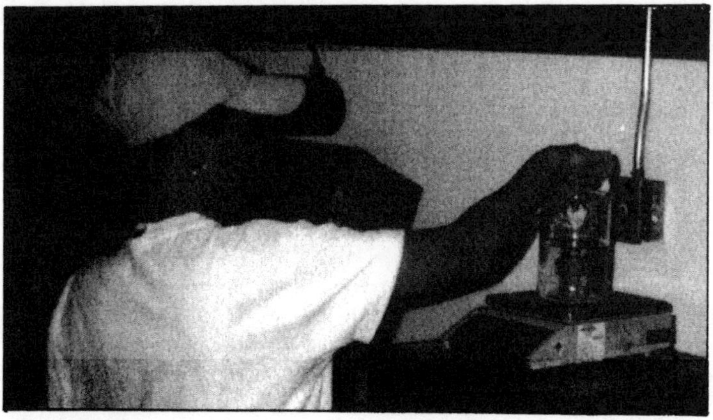

Fig. 1 Specimen Volume Change Measurement by Submersion

The precision of the CLOD test method was evaluated by testing 12 nearly identical remolded specimens from a homogeneous clay material. Deviations in CLOD test results for these samples are shown on the left side of Fig. 2. From these results, the limit of precision on the suction compression index of the CLOD test appears to be ±0.02. In heave calculations, this corresponds to about ±1.3% heave per 1 pF change in suction.

For this procedure to be practical, it is desirable to perform the tests as expeditiously as possible without sacrificing accuracy. Techniques of simplifying the CLOD test procedure were tried. One such technique was the elimination of the intermediate wetting/drying process, since the primary objective was only to determine the final value of C_h. To evaluate the effect of this procedural change, four soil and bedrock samples were tested by exposing half of each sample to a 10 day air drying process followed by 48 hr low temperature (75 deg C, 170 deg F) oven drying, while the other half of each sample was placed immediately in a high temperature oven for 48 hrs (105 deg C, 220 deg F). The results of these tests are shown on the right side of the bar graph in Fig. 2. In general, deviations in C_h between the pairs of samples were small and within the anticipated test accuracy. All four specimens exposed to the higher temperature oven exhibited slightly more negative results, indicating larger overall volume change despite the much shorter drying period.

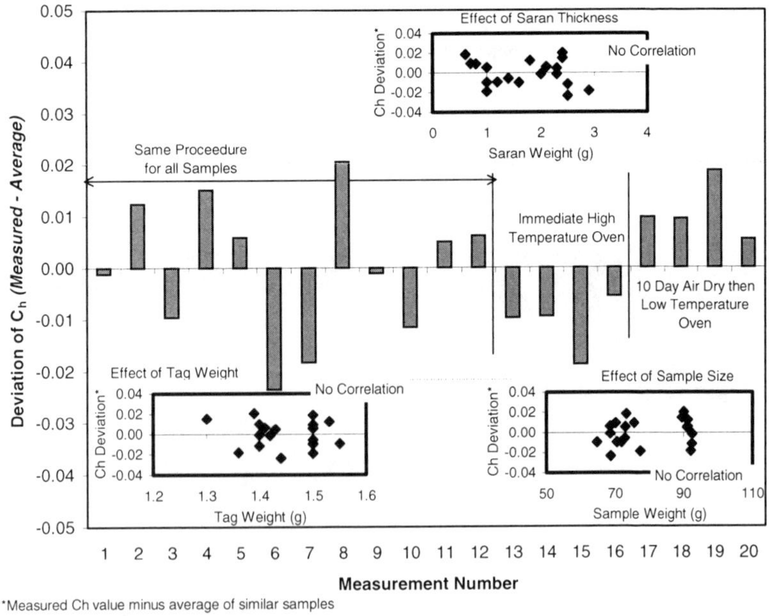

Fig. 2 Deviation of CLOD Test Measurements

The McKeen (1985) CLOD test procedure involves coating samples three times separated by resin drying times of various durations. An attempt was made to evaluate the effect of Saran/resin thickness on measurements of C_h. Results are shown in the smaller graph at the top of Fig. 2. Samples were coated with variable numbers of resin layers indicate that there is no significant correlation between resin thickness and C_h. However, it was observed that cracking of the samples during drying occurred more abundantly for those specimens with thinner coatings of resin. This cracking has an apparently insignificant effect of the final C_h value when only one submerged measurement after drying is required. A reduction in number of coatings appears to be permissible. It is recommended that McKeen's method be followed when intermediate measurements are to be taken in order to establish the full drying/wetting function. CLOD tests were also attempted on two samples without resin coating. It was postulated that if the volume measurement was conducted quickly there would be insufficient time for the clays to absorb additional water and the procedure could be further simplified. Unfortunately, the oven dried clays were so reactive to the water that their exteriors crumbled almost immediately upon immersion, thereby rendering the results useless.

Since suction and CLOD tests were performed on separate halves of the same California drive samples, it was found to be convenient to utilize specimens weighing between 70 and 90 g rather than the 120±20 g size recommended by McKeen (1985). As shown in the smaller graph at the lower right hand corner of Fig. 2, sample size had no discernable effect on the outcome of the tests. However, 140 g size samples were not tested. Common sense dictates that precision should improve with larger samples. As an aside, it was also shown that the weight of the identification tag had a negligible effect on test results, as shown in the smaller graph at the lower left corner of Fig. 2.

The time required for oven drying was also examined to determine if any reduction in drying time was permissible. Results of weight measurements in-air are plotted with respect to oven drying time for 16 samples of clay and claystone bedrock in Fig. 3. The weight measurements were normalized with respect to final oven dried values so that the specimens could be compared at the same scale. It can be seen from the figure that equilibrium oven-dried conditions appear to occur after 48 hours. This time duration matches the recommendation by McKeen (1985).

Considering time efficiency and the results of the test bias experiments, a slightly modified method of performing the CLOD test to determine C_h is shown in Table 1. These procedures were followed for the majority of samples tested in this study.

Calculations

Initial water content is computed from CLOD test measurements as follows (Hamberg, 1985)

$$w_i = \frac{W_1 - W_s}{W_s} \times 100\% \tag{3}$$

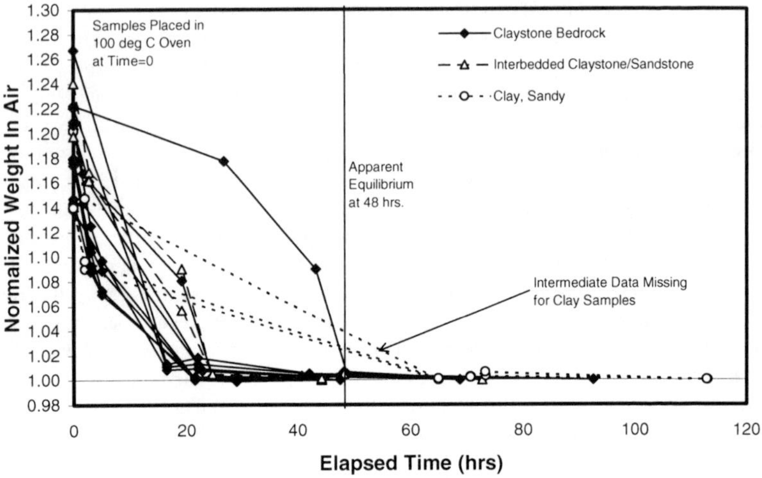

Fig. 3 Time Duration for Sample Oven-Drying

Table 1. Expedited CLOD Test Method
(modified from Hamberg, 1985)

1	Obtain specimens (clods) of undisturbed representative soil samples. The clods may be irregularly shaped. Take necessary precautions to maintain field moisture content of the specimens.
2	Separate the specimens into two halves. Perform a filter paper suction test on one half and reserve the other for immediate CLOD testing.
3	Measure the weight of each specimen (W1). Sample weight should be between approximately 70 and 140 g.
4	Tie a fine wire around each sample and attach a tag for handling and identification. Measure the weight of each tagged specimen (W2).
5	Coat the specimens with liquid resin similar to 1 part DOW Saran© F310 dissolved in 7 parts methyl ethyl keytone or acetone. The coating procedure should be done under a fume hood as follows: dip in resin, dry five minutes in air, then dip again and dry an additional fifty-five minutes in air. Weigh each coated specimen immediately after air drying (W3).
6	Set a container of water on an electronic balance and "tare out" its weight. The container should be large enough to facilitate specimen submersion without touching the sides. Momentarily submerge each specimen under water and record its buoyant force (W5).
7	Oven dry the specimens for 48 hrs at 105 deg C. Weigh the oven-dried specimens in air and submerged under water to determine W7 and W8, respectively.

where W_s is the weight of solids given by

$$W_s = W_7 - 0.85(W_3 - W_2) - (W_2 - W_1) \tag{4}$$

The weight measurements, W_1 through W_8, are defined in Table 1. The factor of 0.85 is applied to account for losses during oven drying due to volatility of the resin (McKeen, 1981 and Hamberg, 1985).

The suction compression index is obtained from (McKeen, 1992)

$$C_h = \frac{V_i - V_f}{\Delta h(V_i)} \tag{5}$$

where Δh is the change in soil suction (in pF units), V_i is initial specimen volume, and V_f is oven dried specimen volume. It is generally accepted that a suction value of about 5.5 pF corresponds to the shrinkage limit (McKeen and Nielsen, 1978). Thus, the change in soil suction, Δh, during the CLOD test is the filter paper suction measurement from the other half of the specimen minus 5.5 pF. The initial and final specimen volumes are calculated from CLOD measurements as follows (Hamberg, 1985).

$$V_i = \frac{W_3}{\gamma_w} - \frac{(W_3 - W_2)}{\gamma_r} \tag{6}$$

$$V_f = \frac{W_8}{\gamma_w} - \frac{0.85(W_3 - W_2)}{\gamma_r} \tag{7}$$

The parameter γ_w is the unit weight of water (1 g/cm^3) and γ_r is the density of resin (1.3 g/cm^3 for DOW Saran F310 at 1:7 solute:solvent).

In order to correlate with previous studies, computation of the ratio of suction change to water content change, Δh/Δw, was necessary. Here, suction change is taken as the difference between the field suction value and the commonly accepted oven-dried intercept suction value of 6.25 pF. Water content change is simply the initial water content, w_i, from Eq. (3), since the water content of oven dried soil is presumed to equal zero.

Results

Suction compression index results from CLOD tests are plotted against the ratio Δh/Δw in Fig. 4. The diagonal dashed line shown in the figure represents the empirical relationship previously derived by McKeen (1992), Eq. (2). It can be seen that data from undisturbed drive samples tested in this investigation generally match McKeen's relationship, thereby supplementing its validity and versatility. Soil and bedrock specimens tested in this study had a greater range in plasticity compared to those tested by McKeen. The outlying points at more negative Δh/Δw values consisted of low-plasticity, sandy clay samples, whereas the outlying point at the least negative value of Δh/Δw consisted of very high plasticity claystone bedrock (Plastic Index of 124).

Although McKeen's empirical relationship fits the data for the samples in mid-range, it is less dependable at the extremes. A qualitative assessment of the practical limits of C_h can be made by hypothetically assuming that soil suction changes by approximately -2.15 pF between field moisture and oven-dried conditions (i.e. 4.1 pF - 6.25 pF) for all types of soils. For example, the field moisture content of silty sands might be 10 percent, so $\Delta h/\Delta w$ would be on the order of -21.5. Cleaner sands should have even lower moisture contents, and consequently, lower (more negative or absolute) values of $\Delta h/\Delta w$. One would expect materials in this extreme to exhibit very small volume change due to suction, therefore C_h should approach zero for very negative values of $\Delta h/\Delta w$. On the other hand, the field moisture content of highly plastic clays might be 30 percent, so $\Delta h/\Delta w$ would be on the order of -7.2. These materials are expected to exhibit large changes in volume due to changes in soil suction, thus C_h should continue to become lower (more negative) as $\Delta h/\Delta w$ approaches zero. Given the new data and the foregoing discussion on limits, a relationship shown by the solid curve in Fig. 4 is suggested. Derivation of this logarithmic statistical fit is described in the next section.

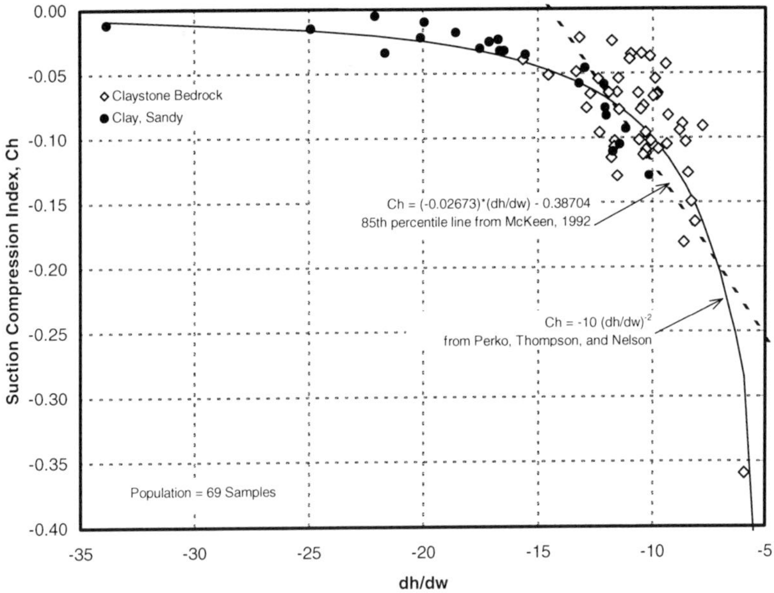

Fig. 4 Suction to Water Content Ratio and Suction Compression Index Data

Heave values predicted using McKeen's method, Eq. (1), and measured suction compression index data are compared with measured one-dimensional swell in Fig. 5. Some uncertainty exists as to the value of soil suction to be used in Eq. (1) that appropriately describes the final conditions in a conventional swell test. Since the sample is completely submerged during this test, there has been some speculation that the specimens approach their field capacity at a suction value of 2.5 pF. At this value, model predictions severely overestimate actual sample swell. By trial and error, the best possible statistical fit between swell based on suction and swell measured conventionally was obtained by setting the final suction value equal to 3.5 pF. It is believed that the difference between field capacity and the final suction value suggested by the best fit is the result of a high osmotic suction component for soils in Denver, Colorado. CTL/Thompson, Inc. is currently conducting a subsequent investigation into the osmotic component of soil and bedrock. Initial evidence indicates that the difference between total and matric suction is almost always on the order of 1.0 pF in the Denver area.

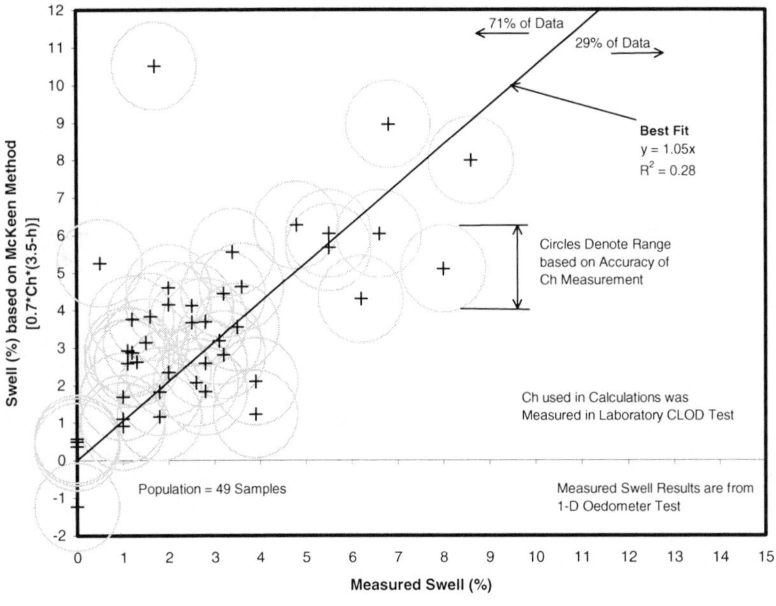

Fig. 5. Comparison Between McKeen Model and Measured Swell

Due to geologic uplifting and subsequent weathering of bedrock along the toe of the Rocky Mountains, a number of different geologic units appear from east to west across the Colorado Front Range. Geologic information was available for a number of the sites where samples were obtained. Suction compression indices were averaged for each major geologic unit. Results are shown in Fig. 6 along with a local bedrock stratigraphic column. It is observed that the suction compression index is strongly related to geology. In fact, the volume change responses of older, Cretaceous system formations are significantly greater than that of the younger, Tertiary system formations. This conclusion is consistent with local philosophy.

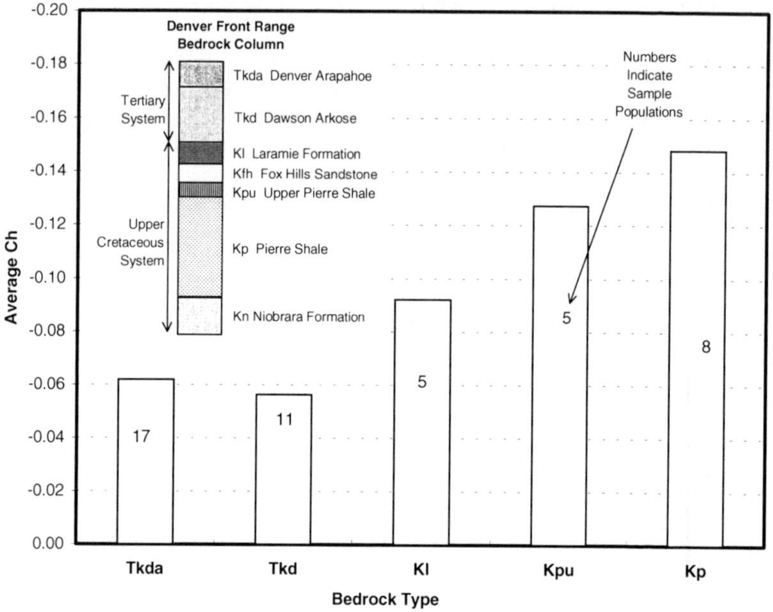

Fig. 6 Suction Compression Indices for Various Colorado Bedrock Types

Analysis

Through iterations of empirical data analysis, the relationship shown in Fig. 7 was eventually found between C_h and the $\Delta h/\Delta w$ ratio. In this figure, the logarithm of the absolute values of $1/C_h$ and $\Delta h/\Delta w$ are plotted with respect to one another. A statistical linear regression was performed on the data and the resulting slope and intercept were astonishingly close to integers.

The equation for the straight line in Fig. 7 is

$$Log(\frac{-1}{C_h}) = 2.0 \; Log(\frac{-\Delta h}{\Delta w}) - 1.0 \tag{8}$$

which, by algebraic manipulation, simplifies to

$$C_h = -10 \; (\frac{\Delta h}{\Delta w})^{-2} \tag{9}$$

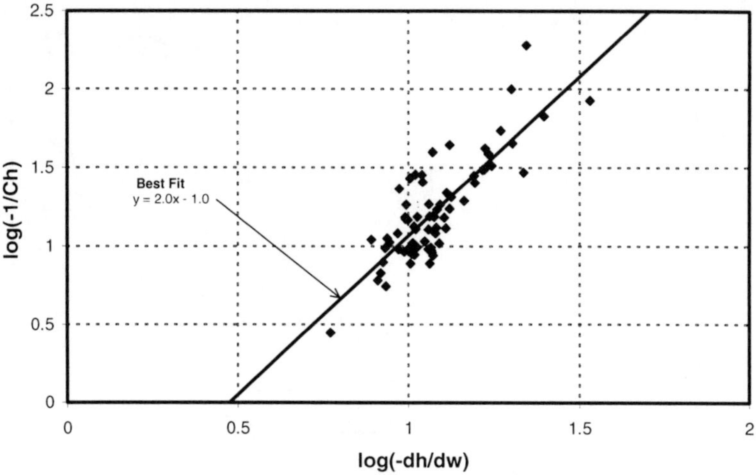

Fig. 7 Logarithmic Fit Analysis

This is the equation for the solid curve plotted in Fig. 4, as described previously. For purposes of discussion, Eq. (9) will be referred to as the PTN relationship.

The PTN relationship (Eq. (9)) is compared with the presently used McKeen linear empirical relatonship (Eq. (2)), in Fig. 8. Expansive soil classifications of low, moderate and high heave potential based on McKeen (1992) are shown at the top of the graph. The differences between the suction compression indices determined by the two methods are plotted along the y-axis. Above the zero baseline (i.e. the line depicting where the two methods are equal), the shaded area indicates that the PTN relationship yields more negative values than does the McKeen relationship, indicating that it is more conservative. Conversely, the shaded area below the baseline indicates that the McKeen relationship is more conservative than the PTN relationship.

A significant observation is that, for moderate expansive soils, both methods are essentially the same, within the expected accuracy of the CLOD test. For low expansive soils, the McKeen relationship predicts positive values of C_h indicating compression, whereas the PTN relationship tends toward zero. For these low expansive soils, the differences between McKeen and PTN are not particularly relevant except that the PTN relationship prevents misleading predictions of soil compression. At the other extreme, the PTN relationship again tends toward more negative values of C_h than those of the McKeen relationship. Consequently, the PTN relationship tends toward higher estimates of soil heave.

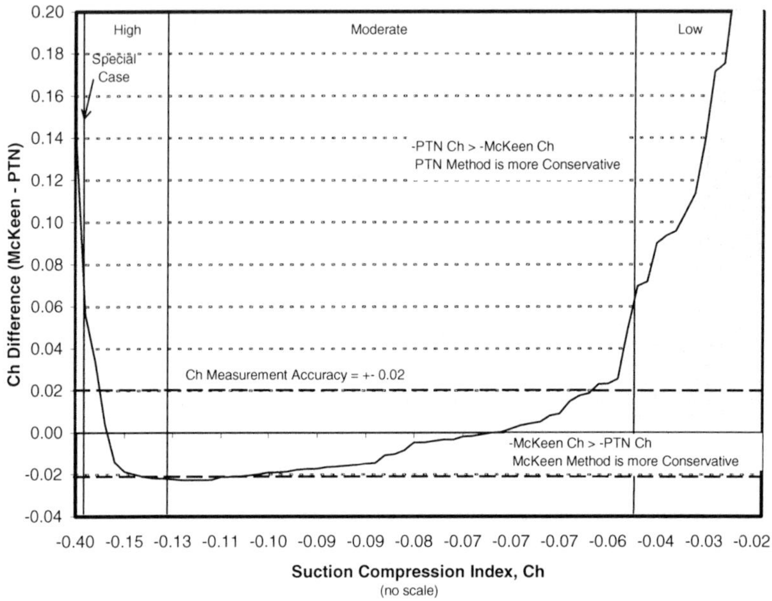

Fig. 8 Comparison between McKeen and PTN Empirical Relationships

Further analysis of laboratory CLOD test data qualitatively indicated that C_h values are exponentially related to soil plasticity. There also appears to be some dependence on silt and clay content, as should be expected. It is desirable to better relate suction compression index to the Atterberg limits and gradation of a soil specimen. Statistical regressions were performed on the data, but a convenient representation was not found. Instead a semi-empirical approach was taken. The method is called semi-empirical because it is theoretically derived based on the empirical PTN relationship.

It is believed that the plastic limit of a soil generally corresponds to a soil suction value of between 3.0 and 3.5 pF (McKeen, 1992). As mentioned previously, the suction of oven-dried soil has been found experimentally to be 6.25 pF. The change in water content between the plastic limit and oven-dried conditions is simply the plastic limit. Therefore, the suction compression index according to the PTN relationship should be approximately equal to

$$C_h \approx -10\left(\frac{3.25-6.25}{PL}\right)^{-2} \approx -\frac{10}{3}PL^2 \tag{10}$$

where PL is the plastic limit (%). However, this equation does not account for the sand fraction of the soil, which has a C_h of nearly zero. The expression in Eq. (10) can be adjusted to account for the coarse fraction by performing a simple correction similar to a "rock correction" in soil density testing. Subtracting the coarse volume, C, from the total volume in the definition of C_h given by Eq. (5) yields

$$-\frac{10}{3}PL^2 = \frac{\Delta V}{(V_i - C)\Delta h} = \frac{\Delta V}{V_i\Delta h - C\Delta h} \quad \textbf{(11)}$$

Taking the inverse of Eq. (11) results in

$$-\frac{3}{10PL^2} = \frac{V_i\Delta h}{\Delta V} - \frac{C\Delta h}{\Delta V} \quad \textbf{(12)}$$

The volume of coarse material in a soil is

$$C = \frac{W_c}{\gamma_w G_s} \quad \textbf{(13)}$$

where G_s is soil mineral specific gravity, and W_c, the weight of coarse grained solids, is simply

$$W_c = V_i\gamma_d(1-F) \quad \textbf{(14)}$$

where γ_d is the dry unit weight of soil and F is the weight percent passing the No. 200 sieve. Substituting Eq. (14) into Eq. (13) and then the result into Eq. (12) produces

$$-\frac{3}{10PL^2} = \frac{V_i\Delta h}{\Delta V} - \frac{\gamma_d(1-F)}{\gamma_w G_s}\frac{V_i\Delta h}{\Delta V} \quad \textbf{(15)}$$

Recognizing that $\gamma_d/(\gamma_w G_s) = 1/(e+1)$ where e is void ratio, and that $V_i\Delta h/\Delta V$ is the inverse of C_h, the semi-empirical relationship between suction compression index, plastic limit and percent passing the No. 200 sieve is found by solving Eq. (15) for C_h

$$C_h \approx \frac{-10}{3}PL^2\left(\frac{e+F}{e+1}\right) \quad \textbf{(16)}$$

The validity of Eq. (16), hereafter termed the PL-200 relationship, was evaluated by comparing C_h values with those determined using the McKeen and PTN relationships, as shown in Fig. 9. Ratios of predicted to measured C_h are shown in the larger graph. In the smaller graph, C_h values were used to categorize samples according to the expansive soil classification provided by McKeen (1992). Soil sample classifications based on the measured swell from one-dimensional, oedometer type tests are also shown.

Inspection of Fig. 9 reveals that the McKeen relationship poorly matches measured C_h for low swelling soils at the larger (least negative) values of C_h, but otherwise exhibits good correlation. The PL-200 relationship tends to underestimate C_h in the midrange, but appears to be a worthwhile method especially considering that the plastic limit and percent passing the No. 200 sieve information was obtained from representative samples and not from the actual samples tested. Predictions based on the PTN relationship most closely fall within the boundaries of accuracy anticipated for the laboratory CLOD test method. All three C_h relationships are valid for classification of samples according to low, moderate or high swell potential.

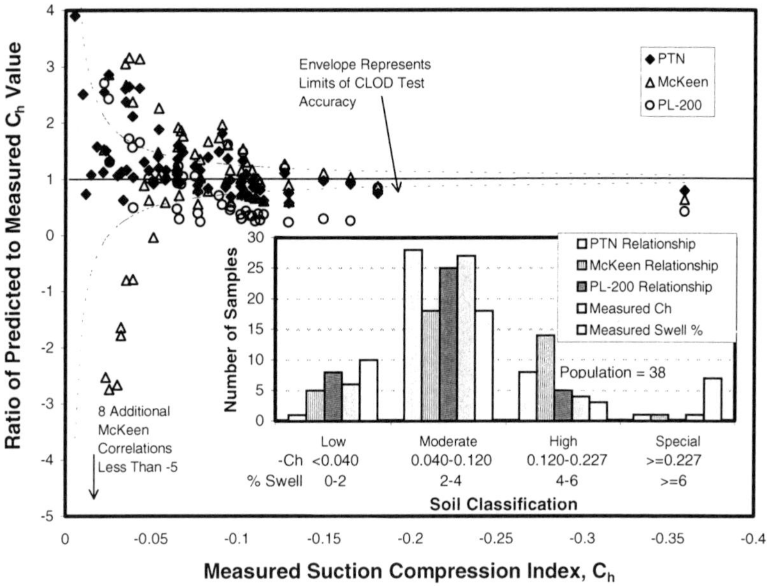

Fig. 9 Summary Comparison Between Various C_h Prediction Models

Conclusions

A laboratory investigation was conducted into the determination of the suction compression index, C_h (McKeen, 1992), using the CLOD test method (McKeen, 1985, Nelson and Miller, 1992, Hamberg, 1985). A new empirical correlation (termed the PTN relationship) was developed to relate C_h to the ratio of change in suction to change in water content. The PTN relationship agrees closely with an empirical correlation previously presented by McKeen, but is more realistic and accurate at very low or very high ratios of suction to water content change. A semi-empirical relationship was also developed that relates C_h to the Plastic Limit. This method appears to correlate fairly well with the other methods for the soils tested. Heave prediction based on suction also correlated well with conventional one-dimensional swell measurements. A procedure for expediting the CLOD test is presented.

Acknowledgments

The funding for this research program was provided by CTL/Thompson, Inc. Editing and advice provided by Dr. Debra J. Hamberg Miller and Dr. R. Gordon McKeen are much appreciated. Paul Hunsader and Keeley Phelps of CTL/Thompson, Inc. are recognized for their participation in the actual laboratory testing.

Nomenclature

γ_w	unit weight of water (g/cm^3)
γ_r	unit weight of resin (g/cm^3)
γ_h	compressibility coefficient (Pa^{-1})
γ_d	unit weight of soil (g/cm^3)
ΔH	soil heave (m)
Δh	change in soil suction (pF)
Δt	soil layer thickness (m)
Δw	change in specimen water content (%)
σ	total stress (Pa)
D_s	suction-volume change constitutive parameter
D_t	effective stress-volume change constitutive parameter
f	lateral restraint factor
F	weight fraction passing the No. 200 sieve (%)
PL	plastic limit (%)
s	load effect coefficient
u_a	pore air pressure (Pa)
V_i	initial volume of CLOD specimen (cm^3)
V_f	oven dried volume of specimen (cm^3)
w_i	initial specimen water content (%)
W_s	weight of solids (g)
W1	initial weight of CLOD specimen (g)
W2	weight of tagged specimen (g)
W3	weight of resin coated, tagged specimen (g)
W5	initial specimen buoyant force (g)
W7	weight of oven dried specimen (g)
W8	oven dried specimen buoyant force (g)

References

Fredlund, D.G. and Rahardjo, H. (1993). Soil Mechanics for Unsaturated Soils, John Wiley & Sons, Inc., New York.

Hamberg, D.J. and Nelson, J.D. (1984). "Prediction of Floor Slab Heave", Proceedings of the 5th International Conference on Expansive Soils, Adelaide, South Australia, pp. 137-140.

Hamberg, D.J. (1985). "A Simplified Method for Predicting Heave in Expansive Soils", Masters Thesis, Colorado State University, Fort Collins, Colorado

Matyas, E.L. and Radhakrishna, H.S. (1968). "Volume Change Characteristics of Partially Saturated Soil", *Geotechnique*, Vol. 18, No. 4, pp. 432-448.

McOmber, R.M., and Thompson, R.W. (2000). "Verification of Depth of Wetting for Potential Heave Calculations", Proceedings of Sessions on Unsaturated Soils at Geo-Denver 2000, The Geo-Institute, ASCE, Denver, Colorado (this conference).

McKeen, R.G., and Nielsen, J.P. (1978). "Characterization of Expansive Soils for Airport Pavement Design", U.S. Department of Transportation, Federal Aviation Administration, Interim Report No. FAA-RD-78-59, Systems Research & Development Service, Washington, D.C. 20590.

McKeen, R.G. (1981). "Design of Airport Pavements for Expansive Soils", U.S. Department of Transportation, Final Report No. DOT/FAA/RD-81/25, Systems Research & Development Service, Washington, D.C. 20590.

McKeen, R.G. (1985). "Validation of Procedures for Pavement Design on Expansive Soils", U.S. Department of Transportation, Federal Aviation Administration, Report No. DOT/FAA/PM-85/15, Program Engineering and Maintenance Service, Washington, D.C. 20591

McKeen, R.G. (1992). "A Model for Predicting Expansive Soil Behavior", Proceedings of the 7th International Conference on Expansive Soils, pp. 1-6

McQueen, I.S., and Miller, R.F. (1968). "Calibration and Evaluation of a Wide-Range Gravimetric method for Measuring Moisture Stress", Soil Science, Vol. 106, No. 3, pp. 225-231.

Miller, D.J. and Nelson, J.D. (1993) "The Role of Osmotic Suction in Stress State", ASCE Annual Meeting Session on Unsaturated Soil, Dallas, TX, ASCE Geotechnical Special Publication No. 39

Nelson, J.D. and Miller, D.J. (1992) Expansive Soils: Problems and Practice in Foundation and Pavement Design, John Wiley and Sons, New York, NY

Thompson, R.W. (1997) "Evaluation Protocol for Repair of Residences Damaged by Expansive Soils", Special Publication No. 68, Proceedings of Sessions on Unsaturated Soils at Geo-Logan '97, The Geo-Institute, ASCE, Logan, Utah.

Thompson, R.W. and McKeen, R.G. (1995) "Heave Prediction using Soil Suction: A Case History", ASCE Soil Suction Applications in Geotechnical Engineering Practice, Geotechnical Special Publication, No. 48, pp. 1-12.

U.S. Department of Agriculture (1972) "SCS, Soil Survey Laboratory Methods and Proceedures for Collecting Soil Samples", Soil Survey Investigations Report No. 1, Soil Conservation Service

Verification of Depth of Wetting for Potential Heave Calculations

By Ronald M. McOmber, P.E.[1]
Robert W. Thompson, P.E.[2]

Abstract

A protocol has been developed for calculation of potential future heave of basement floors in residential structures in the Colorado Front Range area. Data from swell/consolidation tests commonly used in the area and soil suction tests are used to predict potential heave. The calculations are very dependent upon the anticipated depth of wetting. Pre-construction soil suction data were compared to post-construction suction data. The results indicate the assumptions of depth of wetting developed for the protocol are appropriate in most instances. Unusually deep wetting may occur in areas where geologic conditions allow water to penetrate more easily into soils and sedimentary bedrock.

Introduction

The widespread presence of expansive soils and bedrock is well known in the Front Range area of Colorado. The existence of the swelling materials creates risk for man-made improvements constructed on these soils and bedrock. The risks can be mitigated through proper engineering design, but not eliminated. The historical geotechnical practice in the Denver area has been to judge potential heave based upon one-dimensional swell/consolidation tests, other laboratory tests, soil profile, ground water conditions and other factors, and to qualitatively assess the likely behavior in terms of low to high swell potential. Swell testing has emerged as the primary tool used in the assessment of likely structural performance (Thompson, 1992). The Colorado

---

[1] Executive Vice President, CTL/Thompson, Inc., 1971 W. 12th Ave., Denver, CO., 80204; e-mail rmcomb@aol.com

[2] Chairman, CEO, CTL/Thompson, Inc.

Association of Geotechnical Engineers has described this assessment (1996). Geotechnical engineers have been reluctant to predict potential heave because it is known that actual movements are influenced by many factors that may be controlled by the environment and activities of an owner.

From 1993 through 1997, a series of lawsuits occurred in the Denver area. These cases were filed on behalf of home owners against developer/builders, and were largely based upon performance of slab-on-grade basement floors. A method, termed protocol in this paper, was developed to determine the eligibility of individual homes for repair, and to establish whether slab-on-grade floors would be repaired or replaced by new slabs or by structurally supported floors. The development of the protocol was described by Thompson (1997) and can involve the use of either swell/consolidation testing, or soil suction testing to calculate potential future heave.

The authors' firm has performed evaluations of potential heave of basement floors using the protocol for several years for many residences. Starting in 1996, samples obtained during design-level soils investigations have been tested for both soil suction and swelling characteristics. This has provided better understanding of the variability of soil suction and the probable depth of wetting, and a broader basis for comparison of probable heave predictions using swell and suction test results. In the 1997 paper, Thompson proposed the sharing of data related to wetting depth and the performance of slabs and foundations. The goal was to improve future performance of buildings and the local geotechnical practice. The geotechnical engineering design practice in Denver, Colorado at this time does not routinely include heave prediction. Hopefully, sharing of the data will assist other engineers to understand the basis for the calculations, and to accept quantitative assessment of potential heave as an important supplement to engineering judgement necessary to provide design recommendations.

Geologic conditions affect the behavior of expansive soils and sedimentary bedrock in response to wetting. Along the foothills of the Front Range of Colorado, sedimentary rocks have been turned upwards by regional geologic activity, resulting in steeply dipping beds of expansive clay shales. Comparison of suction data obtained before and after construction in the area underlain by steeply dipping bedrock indicates depth of wetting in this area is higher than for areas underlain by more flat-lying sedimentary bedrock.

Potential Heave Prediction

Calculations necessary for the prediction of potential heave have been described by Thompson (1997) and will not be repeated here in detail. In general, heave predictions can be done using data obtained from swell/consolidation testing or soil suction testing. The methodology involves dividing the soil profile into layers. Laboratory tests are performed to characterize the stresses and soil properties of each soil layer. In the protocol, a depth of wetting of about 20 feet, or 14 feet below residential basement floors, was used. This depth was established after consultation with engineers representing the home owners and engineers representing the developer/builders, and was believed to represent likely depth of wetting in areas underlain by comparatively flat-lying bedrock.

Swell/consolidation tests are performed at pressures that approximate the overburden weight of soils/bedrock at the sample depth. The measured swell of the specimens is used in the calculation of heave for each soil layer. In the protocol, the computed swell is reduced by an "adjustment factor" of 70 percent to predict probable layer heave.

Suction tests can also be used as a basis of computing potential heave. Several methods can be used to perform this calculation. The procedure described by McKeen (1992) was incorporated into the protocol. Soil suction can be expressed in various units of measurement, including a "pF". The pF unit corresponds to the logarithm of the hydraulic pressure at the bottom of a column of water measured in centimeters. All suction valves in this paper are presented in pF units, along with corresponding pressures in kPa units. In general, the drier the expansive soil, the higher the suction value.

Local Soil Suction Data

The authors' firm has been collecting soil suction data on numerous projects since about 1982 using laboratory techniques involving both psychrometer and filter paper measurements. Most recently, filter paper measurements have been made in accordance with ASTM 5298. Some of the suction data for flat-lying bedrock areas compiled since 1996 are shown on Figures 1 and 2. The data include measurements on samples from 51 residence sites prior to construction (Figure 1), and 41 sites after construction (Figure 2). All of the sites tested after construction had some form of reported distress in either foundations or basement slab-on-grade concrete floors. Therefore, the data is considered biased and may not represent typical conditions on sites where distress has not been reported. In general, the data show that soil suction values typically fall between 3.5 to 4.5 pF (300 to 3000 kPa), with a few data points outside these limits. Soil suction values less than about 3.0 pF (100 kPa) have not been measured on potentially expansive soils, even under extreme wetting conditions.

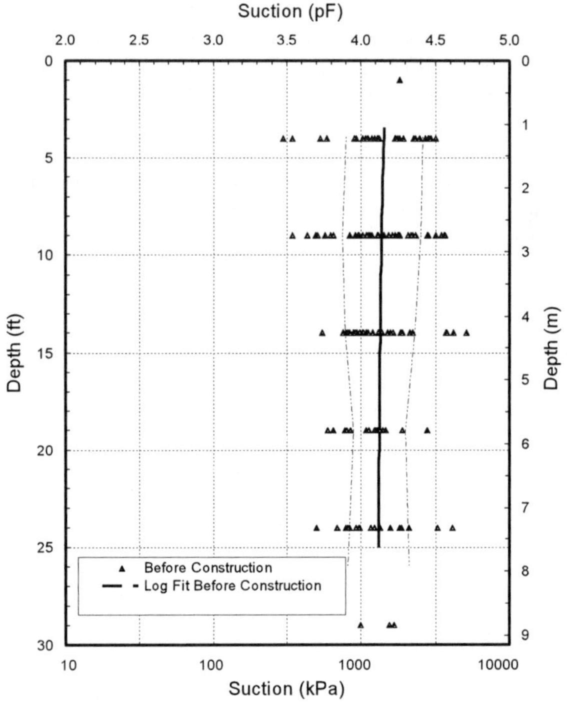

Figure 1. Measured Total Suction (Before Construction)

Figure 1 shows the suction measurements from sites sampled before construction. The bold line represents a logarithmic fit of the data. The line equation was determined by regression analysis of suction as a function of depth with a natural logarithm relationship using Microsoft Excel (copyright). This analysis indicates typical soil suction is relatively constant with depth having a pF value of about 4.1 to 4.2 (about 1200 to 1500 kPa). This is consistent with other measurements of soil suction made by the authors' firm over the past ten to fifteen years. Specimen depths were influenced by typical sampling depths used in design-level investigations. The dashed lines represent a variation of one standard deviation off the logarithmic fit at the depths sampled.

Figure 2 shows data from borings drilled 1 to about 12 years after construction. The variations were determined by calculating the standard deviation off the logarithmic fit for various depth intervals. The logarithmic fit results in a pF value of about 3.2 (150 kPa) at the ground surface provided the line is extended beyond the limits of the data. The value of 3.2 appears reasonable considering the impact of home owner applied irrigation.

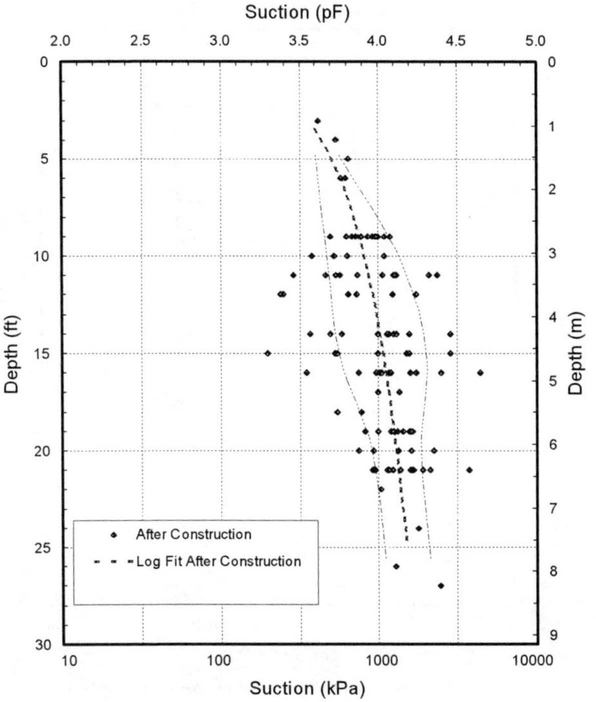

Figure 2. Measured Total Suction (After Construction)

Figure 3 shows the suction data obtained before and after construction, along with a line representing the "final" soil suction profile used in the protocol. The protocol was developed based upon a depth of wetting of 20 feet (6.1 m). The soil suction data agree well with this depth of wetting. The "before construction" and "after construction" regression fits intersect at a depth of about 21 feet (6.4 m). In addition, the final suction profile assumed in the protocol falls below, i.e., to the left of about 80 percent of the "after construction" soil suction data within the range of 6 to 20 feet (1.8 to 6.1 m). The data from distressed residences imply that the final suction profile used in the protocol provides a reasonably good basis for conservative heave estimates.

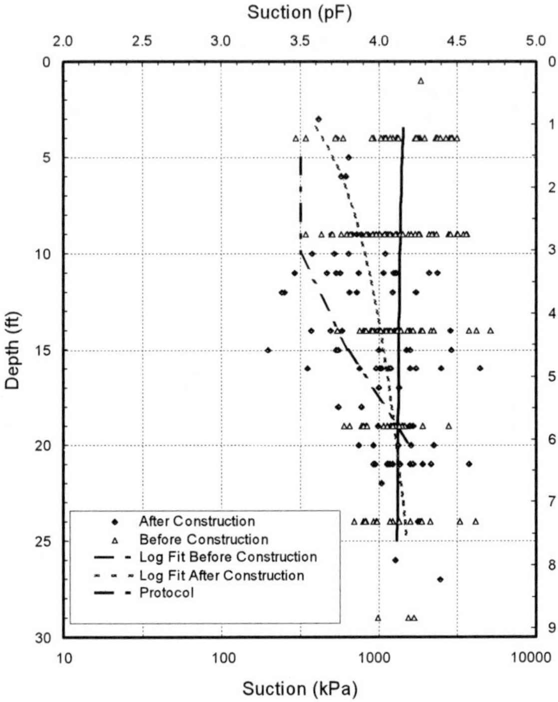

Figure 3. Comparison of the soil suction measurements to the protocol "final" soil suction profile (Thompson, 1997)

The micro-climate at a particular site produces a flux boundary condition that influences soil suction of the site. Several measurements or estimates are reported of the influence of seasonal moisture variation on soil suction, and on heave resulting from construction. The depth of seasonal moisture variation is also called the "active zone" (O'Neill,1980). Figure 4 shows a plot of monthly precipitation over three years at a weather station in the Denver area, along with a monthly average over the same time period. The plot indicates that more precipitation typically falls within the six-month period of March through August than during September through February. In addition, home owners generally irrigate their lawns during the period of April through September or early October.

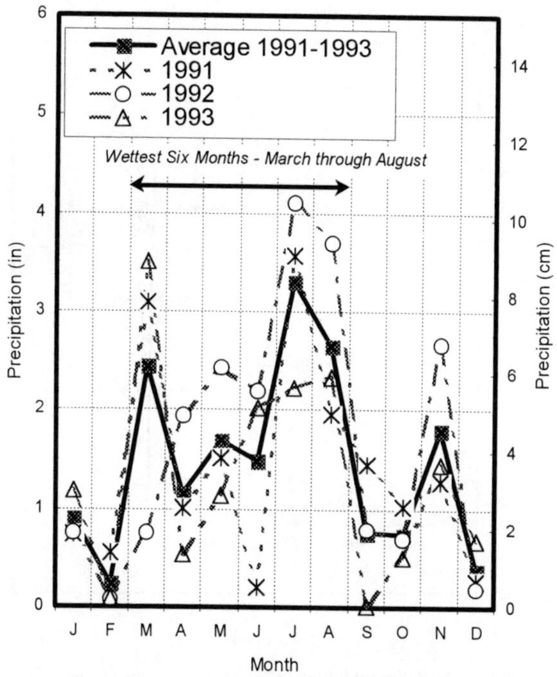

Figure 4. Monthly precipitation for the period 1991 through 1993

The soil suction data from "before construction" were separated into two, six-month time periods based on date of sampling to demonstrate potential seasonal wetting changes at depth, and the results are shown in Figure 5. There is considerable scatter, but the data imply an "active zone" of about 15 to 20 feet (4.6 to 6.1 m). O'Neill (1980) reported an active zone of 10 to 15 feet (3 to 4.6 m). Figure 6 shows a similar comparison for the "after construction" data. The plot of average soil suction values indicates less, and almost negligible, seasonal influence "after construction". This indicates that once the soils and bedrock in the Denver area become wetted, they do not dry significantly. This is consistent with local experience, particularly in areas where most of the ground is irrigated and remaining areas are covered by structures or pavements.

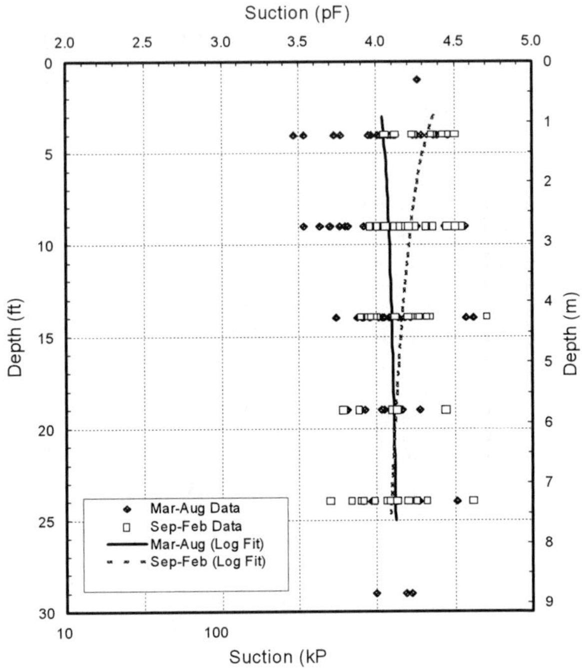

Figure 5. Seasonal Inlfuence on Suction (Before Construction)

Comparison of Heave Calculations using Swell and Suction Data

The protocol (Thompson, 1997) established two alternative methods for calculating potential future heave by using soil suction data and swell/consolidation tests performed after wetting specimens under approximate overburden pressures. Measured laboratory swell is reduced by a factor of 0.7 in the calculation. The data were analyzed to judge if calculated heave using suction data was about 0.7 times heave calculated from swell test results. This comparison was not intended to establish either swell or suction tests as a reference standard; rather, just to evaluate consistency between the two approaches. Side-by-side calculations of basement floor slab heave were made using a wetting depth of 14 feet below the basement floor (i.e. 20 feet or 6.1 m below ground surface). The swell test data were not adjusted, as described previously, and the results are plotted on Figure 7. A linear regression fit was performed on the calculated values of potential heave, with an intercept forced to go through zero. Heave calculated using suction tests was 0.65 times the heave calculated using swell tests. This agrees reasonably well with the factor (0.7) used in the protocol. The data shows that there can be some variation.

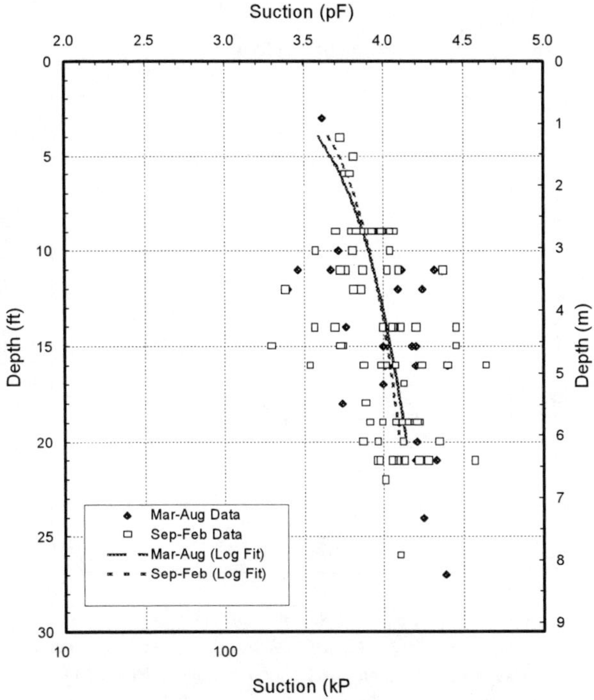

Figure 6. Seasonal Influence on Suction (After Construction)

Application of Heave Prediction to Foundation Design and Performance

The authors' firm has also used heave calculations to evaluate potential future movements in foundations of distressed structures. For footing-type foundations, the calculations have typically been done using procedures similar to those described in the protocol. For footings at shallow depth (i.e. above basement level), potential heave calculations are modified to include the soil above the basement level. It is possible to include the influence of foundation loads in selecting the confining pressures used during swell tests to account for increased confinement. Calculations using soil suction data can also include the effect of applied foundation loading in an empirical manner. Heave calculations can also be performed as part of design for new structures on footing-type foundations.

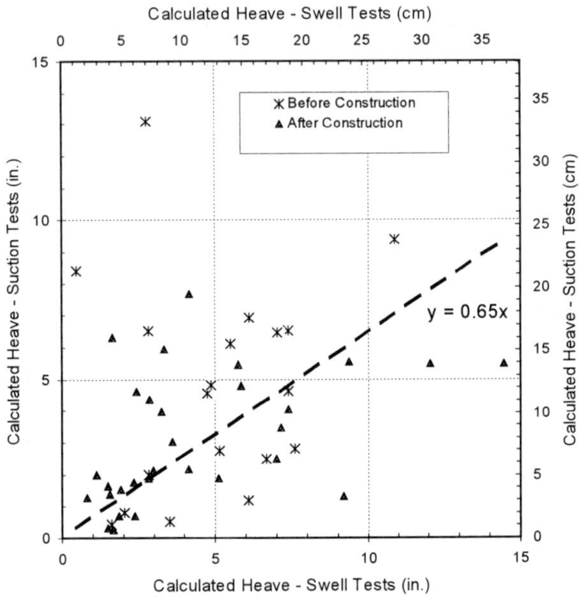

Figure 7. Comparison of calculated heave of basement floo slab using suction and swell tests assuming 14 feet (4.3 m) of wetting below the basement floor

The complex interaction between drilled pier foundations and expansive soils and bedrock is much more difficult to model for heave predictions. At this time, evaluations of potential pier movements requires considerable experience, judgment and intuition. The data can be used to illustrate the complexity of soil/pier interaction.

The heave estimates using changes in soil suction from the "before construction" state and the "after construction" state, and from "before construction" to the protocol "final suction" profiles were used to calculate the variation of heave with depth. The purpose of this analysis was to illustrate potential soil movement at various depths around a drilled pier. The results are shown in Figure 8.

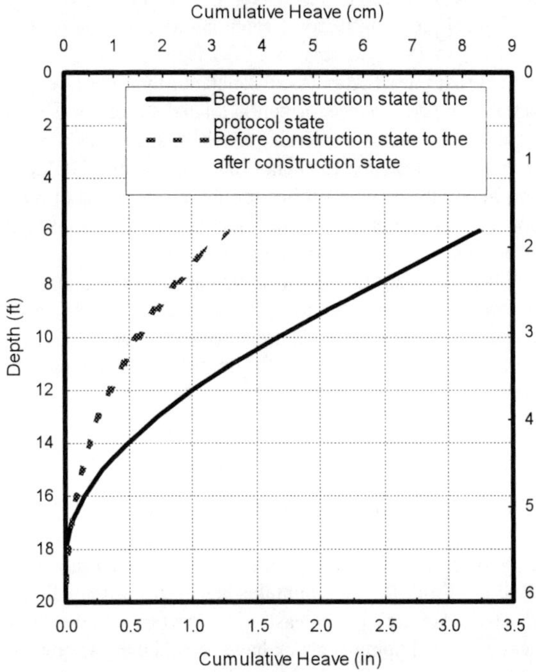

Figure 8. Computed cumulative heave below basement level based on soil of wetting

Footing foundations are commonly designed for total settlements on the order of about 1 inch (2.5 cm), and differential settlements on the order of 0.5 to 0.75 inch (1.3 to 1.9 cm). It seems reasonable to evaluate potential drilled pier heave using similar criteria. Experience indicates that differential heave between piers can approach total heave, because one pier may heave while an adjacent pier does not heave. Assuming 20 feet (6.1 m) of wetting, the calculations imply that comparison of typical changes in soil suction from "before construction" to "after construction" indicate 0.75 inches (1.9 cm) of cumulative heave of soil and bedrock around a pier would occur about 8 feet (2.4 m) below the ground surface. One-half inch (1.3 cm) of heave would occur at a depth of about 11 feet (3.4 m). Using the higher calculated heave resulting from a change in state from the "before construction" data to the protocol "final" suction, heave of 0.75 and 0.5 inches (1.3 and 1.9 cm) occurs at about 12 and 14 feet (3.7 and 4.3 m) below the ground surface; respectively.

Drilled piers have been used in the Denver area with comparatively good success for many years. Typical pier lengths for residential construction have increased from about 12 feet (3.7 m) in 1970, to lengths of 25 feet (7.6 m) or more now. It has historically been held that piers that perform satisfactorily do not heave, but it is more likely that satisfactory pier performance for lightly loaded structures is the result of small total and differential heave which are tolerable for these structures. In the general case, the data suggest that piers designed to anchor below a "depth of influence" of 8 to 11 feet (2.4 to 3.4 m) below the ground surface, or about 6 to 9 feet (1.8 to 2.7 m) below shallow foundation excavation depths or 2 to 5 feet (0.6 to 1.5) below typical basement depth, should perform satisfactorily. In the more conservative case using "final" suction from the protocol, the calculations support a typical "depth of influence" of 12 to 14 feet (3.7 to 4.3 m) below the ground surface. Considering the variability of the data used to develop these values, the designer should use judgement to assess the influence of site-specific soil and bedrock conditions in design for uplift due to swelling soils and bedrock around drilled piers.

Influence of Geologic Conditions on Depth of Wetting

The performance of residence built over areas of steeply dipping bedrock along the eastern flank of the Front Range mountains has been substantially different than in areas of more flat lying bedrock to the east. The frequency and severity of damages resulting from heave of the dipping bedrock have been higher (Thompson, 1992). It is believed that fractured zones developed within the bedrock during uplift and unloading. These fractures can allow water to penetrate more easily into the formation. In many instances, differential heave of 6 to 12 inches (15 to 30 cm) has occurred over horizontal distances of about 30 to 40 feet (9.1 to 12.2 m). In extreme cases, differential heave of 24 to 27 inches (61 to 69 cm) has occurred. This movement has damaged foundations and concrete flatwork, streets, and buried utilities. Damage was severe enough that Jefferson County (located in the west Denver metro area) identified the area underlain by steeply dipping bedrock as a geologic hazard in April 1995, and developed mitigation requirements in County Land Development Regulations and building code.

Soil suction data from 32 exploratory borings drilled in areas underlain by steeply dipping expansive bedrock before residential construction and 21 borings drilled at residence sites where distress has occurred are shown on Figure 9. Logarithmic fits of the before construction data and after construction data were prepared using methods described previously. Comparison of the data indicates that a depth of wetting on the order of 28 to 29 feet (8.5 to 8.8 m) is typical. Based upon experience in the area and

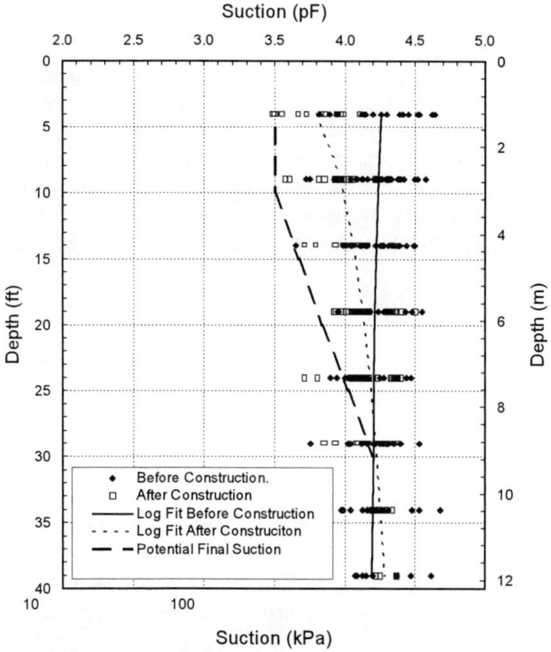

Figure 9. Comparison of soil suction measurements for sites underlain by steeply dipping expansive bedrock

preliminary review of the data, the authors' firm has used a potential final suction profile similar to the profile shown on Figure 9 for calculations of potential heave in areas underlain by steeply dipping expansive bedrock.

Conclusions and Recommendations

Heave prediction can be performed using data from swell/consolidation tests and soil suction tests. A protocol for performing the calculations has been developed (Thompson, 1997) for the Front Range area of Colorado. Data compiled since 1996 suggest the depth of wetting and final suction profile assumptions in the protocol are reasonable. The heave calculations can also be used to study the likely performance of foundations. The potential influence of geologic conditions should be considered during use of the calculation methods.

The soil suction versus depth measurements from "before" and "after" construction states provide information regarding the depth of wetting for calculating heave. In order to advance the understanding on the likely depth of wetting in the Denver area, more information regarding site-specific changes over time would be useful. Considerably more performance measurements need to be made and analyzed in order to increase knowledge on the use and reliability of heave predictions as part of design. The increase in knowledge should provide the confidence necessary for routine design and maintenance procedures.

Appendix I. References

Colorado Association of Geotechnical Engineers (1996), "*Guideline for risk assessment for basement slab-on-grade floors in residential construction*," Denver, CO.

McKeen, R.G. (1992), "A model for predicting expansive soil behavior," *Proceedings of the 7th International Conference on Expansive Soils, Dallas, Texas (USA)*, 1-6.

O'Neill, M.W., and Poormoayed, A.M. (1980), "Methodology for foundations on expansive clays," *Journal of the Geotechnical Engineering Division*, ASCE, GT12, 1345-1367.

Thompson, R.W. (1992), "Swell testing as a predictor of structural performance," *Proceedings of the 7th International Conference on Expansive Soils, Dallas, Texas (USA)*, 84-88.

Thompson, R.W. (1992), "Performance of Foundations on Steeply Dipping Claystone", *Proceedings of the 7th International Conference on Expansive Soils, Dallas, Texas (USA)*, 438-442.

Thompson, R.W. ((1997), " Evaluation protocol for repair of residences damaged by expansive soils," *Proceedings of Sessions on Unsaturated Soils at Geo-Logan '97* sponsored by The Geo-Institute of the American Society of Civil Engineers, Logan, Utah, Geotechnical Special Publication No. 68, 255-276.

Prediction of Heave Using "Effective" Stress

Ronald F. Reed, P.E.[1]

Abstract

Application of "effective" stress principles in the prediction of heave in unsaturated expansive clay requires an evaluation of the swell index and changes in the matric suction. The latter, in turn, requires knowledge of water balance principles. This paper presents a case study to illustrate the practical application of an effective stress approach to heave prediction.

The purpose of this paper is to present application of a method to predict heave proposed by Fredlund and Rahardjo (1993) in typical engineering studies. A brief review of the method is provided, followed by areas of application. A case history with measured movement, swell pressures and matric suction is presented to illustrate the information required and its application in an engineering environment.

Introduction

Prediction of heave generally relies on empirical correlations based on classification tests or direct application of swell tests using laboratory oedometer test devices. Summaries of the typical methods have been provided in various publications to include Johnson and Snethen (1978), and Fredlund and Rahardjo (1993). However, neither the initial nor final stress conditions within the expansive strata are typically quantitatively defined in these methods of analysis. Evaluation of matric suction and application of an effective stress approach provide the geotechnical engineer a theoretical base to evaluate and explain the cause or causes of observed versus predicted heave.

The purpose of this paper is to present application of a method to predict heave proposed by Fredlund and Rahardjo (1993). This procedure varies from the normal direct methods using a standard oedometer by "correcting" the swell pressure

[1] President, Reed Engineering Group, Inc., 2424 Stutz Drive, Suite 400, Dallas, Texas 75235; phone (214) 350-5600; rreed@reed-engineering.com

for sampling disturbance. The corrected swell pressure is then considered to be equal to the sum of the overburden stress plus matric suction.

A brief review of the method is provided, followed by areas of application. A case study is then presented to illustrate application of the method in an engineering environment.

Methodology

Prediction of heave requires:

1. definition of the void volume characteristics of the profile; and

2. an estimate of initial and final stress conditions.

Fredlund and Rahardjo (1993) have proposed modifying the constant volume swell test to account for sample disturbance. The method is presented in ASTM D 4546 and consists of partially consolidating the sample once the absorption pressure is obtained, then correcting the swell pressure by application of a procedure similar to that which Casagrande recommended for consolidation. The resulting void volume curve is shown in Figure 1.

As shown in Figure 1, "correction" of the absorption pressure results in translating the swell portion of the curve to the right. This translation effectively increases the change in void ratio between any two pressures (i.e., increases the magnitude of predicted heave).

Increasing the magnitude of predicted heave from constant volume swell tests may not appear to be initially justified; however, logically, it is reasonable to conclude that an expansive soil would swell upon release of overburden stresses during sampling. The resulting expansion would be restrained by an increase in the matric suction. The effect of this would be to increase the laboratory measured suction while decreasing the absorption pressure because of the limited expansion that occurs prior to restraint by the increase in matric suction.

Historically, others have also recommended modification of the void volume curve to account for differences between laboratory measured swell and observed movement. As early as the mid-50's, Holtz and Gibbs (1956) recommended modifying the constant volume swell curve by construction of a "field" curve that was estimated to lie between the pressure/swell curve using the constant volume swell procedure and the swell/pressure curve from the improved simple oedometer procedure. This modification is shown in Figure 2.

Once the void volume characteristics of the profile are defined, an estimate of the anticipated change in stress is required. Changes in stress include structural and overburden loads and changes in the matric suction pressure. The initial stress condition is taken to equal the "corrected" absorption pressure, P'_s, as shown in Equation 1. These relationships are also illustrated in Figure 1. Pressure line 1 on Figure 1 represents the net overburden pressure plus the matric suction equivalent. Pressure line 3 represents the net overburden pressure with zero matric suction equivalent, i.e., fully saturated.

$$P_o = P'_s = (\sigma_y - u_a) + (u_a - u_w)_e \quad (1)$$

where

σ_y = total overburden pressure

$\sigma_y - u_a$ = net overburden pressure

u_a = pore-air pressure

$(u_a - u_w)_e$ = matric suction equivalent

u_w = pore-water pressure

The final stress condition has the same form as the familiar effective stress equation and is shown in Equation 2.

$$P_f = \sigma_y + \Delta\sigma_y - u_{wf} \quad (2)$$

where

$\Delta\sigma_y$ = change in total stress

u_{wf} = final pore-water pressure

For unsaturated conditions, the pore-water pressure is negative; therefore, decreasing matric suction reduces the magnitude of final pressure. For purposes of this paper, only changes in the matric suction will be considered.

The magnitude of heave is calculated using standard consolidation theory. The swell index is used in lieu of the coefficient of consolidation. The recommended equation and input parameters are:

$$\Delta h = [C_s/(1 + e_o)] \cdot h \cdot \log(P_f/P_o) \quad (3)$$

where

Δh = heave of the layer

C_s = swell index

e_o = initial void ratio

P_o = initial stress state, which is taken to equal P'_s

P_f = final stress state.

This procedure does not account for lateral stress relief; however, it still provides a method for evaluating input data and a means of applying engineering judgment regarding observed or predicted heave.

Application

Application of effective stress in the analysis of building performance provides a base for understanding expansive soil behavior. For example, the method can be used to explain the difference in observed versus predicted heave of ground-supported floors or foundations where extraneous sources of water are present. Some of the forensic engineering applications include evaluating the effect of:

- leaks within utilities or poor drainage;
- changes in the level of ground water; and
- pre-existing vegetation.

Space does not allow for discussion of each of these situations; however, the technique is generally the same. Basically, the initial stress state is taken to equal the corrected swell pressure, P'_s. The final stress state, P_f, is then obtained by calculating the overburden pressure plus any matric suction.

The effect of leaks within utilities and poor drainage on the prediction of heave can be noted by studying Figure 1. If the magnitude of post-construction floor movement was based on a method where 100 percent saturation of the soil profile was not assumed, then leaks within utilities or poor drainage could reduce matric suction resulting in additional movement. This is depicted in Figure 1 by the two pressures indicated as Lines 2 and 3. Assuming that the anticipated stress based on water balance considerations is equal to Line 2, then heave from the initial void ratio, e_o to the void ratio at e_2 would occur. If, however, the final matric suction is reduced to zero (100 percent saturation) and the final stress state is equal to the overburden stress, Pressure line 3 on Figure 1, then additional heave from e_2 to e_3 would be experienced.

Several end conditions can be estimated, to include full or partial saturation (i.e., some residual matric suction). Equilibrium at full saturation is generally conservative; which can result in severely over-estimating the magnitude of predicted heave. An alternative approach is to estimate an equilibrium condition based on the Russam-Coleman (1961) correlation between final matric suction and the prevailing Thornthwaite Climatic Index (I_m). The Russam-Coleman correlation is shown in Figure 3.

Changes in ground water level can also be evaluated using the effective stress approach. For conditions where the water table is less than about 9 meters, the final matric suction below a covered surface is generally taken to equal the negative of the distance above the water table times the unit weight of water. A significant rise in the water table would result in reduction in the matric suction and thus result in heave. The magnitude of heave can be calculated by application of the concepts discussed.

Case Study

The method described in the preceding sections was applied in evaluation of a distressed residence. The residence is supported by a post-tensioned slab-on-ground foundation and has been subject to approximately 140 mm of differential movement.

The site is located with weathered soil of the Upper Cretaceous Eagle Ford Formation. At the subject site, the Eagle Ford Formation consists of soft clay shale that has weathered to a yellowish-brown and light gray CH clay to a depth of approximately 6.1 meters. The weathered profile is highly jointed and fissured and desiccated to the top of unweathered shale at the 6.1 meter depth. The joints and fissures are generally tight. A summary of classification data for the weathered portion of the formation is provided in Table 1. The in situ moisture contents in the drier boring were 5 to 8 percentage points below the plastic limit. This condition is reflected in negative liquidity indices shown in Table 1.

The shape of the deflected foundation and geologic environment indicated heave. An elevation survey indicating the pattern of movement is provided as Figure 4.

Borings were performed as shown in Figure 4 to evaluate soil conditions on both the high and low sides of the house. From this information, it was determined that the soils in the vicinity of the high side of the house (Boring B-2) were relatively moist, whereas they were relatively dry on the low side of the house (Boring B-1). Absorption pressure swell tests (ASTM D 4546, Method C) and matric suction tests (ASTM D 5298) were conducted on selected samples. Plots of the matric suction profiles along with the absorption pressure and swell index are shown in Figure 5.

The initial stress condition was considered to equal the corrected swell pressure within the drier boring (Boring B-1). The final stress condition was taken to equal the overburden pressure plus the matric suction found in Boring B-2. Three conditions for matric suction were used in the analysis:

1. matric suction equal to values measured within the wetter boring (i.e., high side of the house);

TABLE 1. Summary of Classification Data

Liquid Limit L_w (%)	Plastic Limit P_w (%)	Plasticity Index, I_w (%)	Liquidity Index I_l* (%)	-2μ (%)
68 - 80	26 - 33	35 – 53	-0.13 to -0.17	58 - 63

*in situ condition from the drier boring, B-1.

1. matric suction equal to an equilibrium value predicted by the Russam-Coleman correlation; and

2. matric suction dropping to zero, or fully saturated conditions.

Three constant volume swell tests were conducted on samples from the drier boring (B-1). The average swell index, C_s, was measured to be 0.107, with an average initial void ratio, e_o, of 0.935. Swell tests were conducted at the in situ moisture content, which was 5 to 8 percentage points below the plastic limit.

The profile was divided into four layers for calculation of the magnitude of movement as shown in Table 2. The grade beams extended to an approximate depth of 0.8 meters, therefore this portion of the profile was not used in the calculations.

Application of Equation 3, considering initial stress equal to the corrected swell pressure and final stress, P_f, equal to the overburden pressure plus the matric suction values encountered in the wetter boring, B-2, is shown in Table 2. Structural loads from the single-story residence were relatively light and were not used in the calculation.

TABLE 2. Summary of Heave Calculations, Condition 1

Layer	Depth, meters	P'_s,kPa	u_{wf}, kPa(1)	σ_y,kPa(2)	Δh, mm
1	0.8 - 2.3	718	167	30	55
2	2.3 - 3.8	598	167	60	35
3	3.8 - 5.3	718	215	90	26
4	5.3 - 6.1	718	335	112	22
					Total = 138

1. Final pore pressure, u_{wf} based on values measured in B-2.

2. Based on unit dry weight of $\gamma_d = 19.64$ kN/m^3.

TABLE 3. Predicted Heave w/Varying End Conditions

Condition	Initial Stress	Final Stress	Calculated Movement (mm)
1	P'_s	O.B.[1] + 167-335 kPa suction[2]	138
2	O.B. + 167-335 kPa suction	O.B. + 62 kPa suction[3]	91
3	O.B. + 167-335 kPa suction	O.B.	188

1. O.B. - overburden
2. Matric suction measured in Boring B-2
3. Matric suction predicted by Russam-Coleman correlation

Results of similar application of Equation 3 for the three initial and final stress conditions discussed above are shown in Table 3.

Condition 1 is considered applicable for the change in stress from that indicated on the drier side of the house to the wetter side. In other words, Stress Condition 1 should represent the approximate observed heave. The calculated value of 138 mm compares to a measured dis-elevation of about 140 mm.

Condition 2 represents additional heave for an end condition equivalent to that predicted using the Russam-Coleman correlation shown in Figure 3. For the subject site, considering the influence of irrigation, the average I_m is +20. This results in a predicted equilibrium of 2.8 pF (62 kPa). The calculation indicates that additional heave of 91 mm would occur if the suction is reduced from approximately 167 kPa to 62 kPa.

Condition 3 represents full saturation or zero matric suction. This analysis indicates that additional movement of 188 mm is possible (total heave of 326 mm) if fully saturated conditions prevailed throughout the profile. This end condition is possible if a leak developed below the residence, although it is unlikely full saturation of the entire profile would occur. By applying some engineering judgment, or measuring the matric suction profile below the house, an evaluation of the magnitude of heave associated with a leak could be made.

The advantage to this analysis is that some approximation of the magnitude of heave can be obtained for various end conditions. By changing the end stress conditions, the method provides some insight into the effects of external environmental conditions on the magnitude of heave.

As mentioned previously, the geologic formation consists of a dry, weathered clay shale. Limited desiccation cracks are present within the formation, thus limiting horizontal movement or lateral stress relief. No reduction in the calculated heave was applied to account for lateral strain relief associated with closure of desiccation cracks.

Conclusions

Use of the illustrated procedure provides a means of evaluating heave resulting from changes in the matric suction. By estimating an end stress condition, the engineer can evaluate various scenarios and make educated conclusions as to the cause of movement. By comparison of equilibrium matric suction predicted using published correlations, such as that presented by Russam and Coleman, with measured values of matric suction, the effect of leaks or drainage can be quantifiably assessed.

For the case study illustrated, application of the procedure recommended by Fredlund and Rahardjo appears to predict reasonably well the observed movement. Additional studies should be conducted to evaluate the effectiveness in other geologic settings.

References

Fredlund, D., and Rahardjo, H. (1993). "Soil Mechanics for Unsaturated Soils," John Wiley & Sons, Inc., New York, N.Y.

Holtz, W.G. and Gibbs, H.J. (1956). "Engineering Properties of Expansive Clays." Journal of the Soil Mechanics and Foundations Division, ASCE 121, 641-663.

Johnson, L.D., & Snethen, D.R. (1978). "Prediction of Potential Heave of Swelling Soil." Geotechnical Testing Journal, ASTM 1 (3), 117-124.

Russam, K. and Coleman, J.D. (1961). "The Effect of Climatic Factors on Subgrade Moisture Conditions." Geotechnique 1(1), 22-28.

APPENDIX

C_s = Swell Index

e_o = Initial Void Ratio

h = Thickness of Layer

I_m = Thornthwaite Climatic Index

P_o = Initial Pressure

P_f = Final Pressure

P'_s = Corrected Swell Pressure

σ_y = Total Overburden Pressure

U_a = Pore-Air Pressure

$(U_a - U_w)_e$ = Matric Suction Equivalent

U_w = Pore-Water Pressure

U_{wf} = Final Pore-Water Pressure

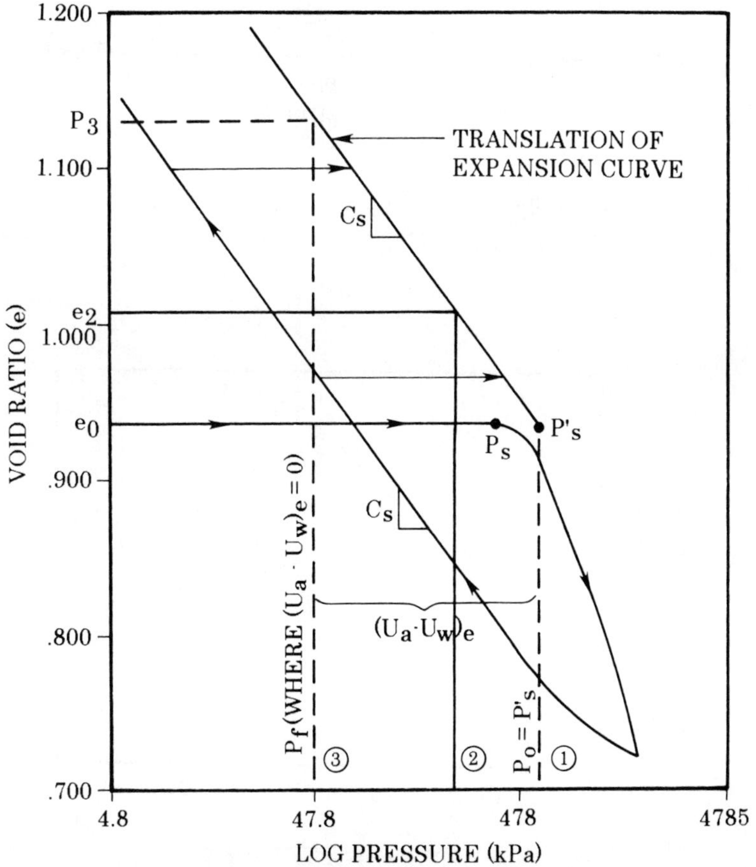

Figure 1. Pressure Swell Curve/Corrected Swell Pressure.

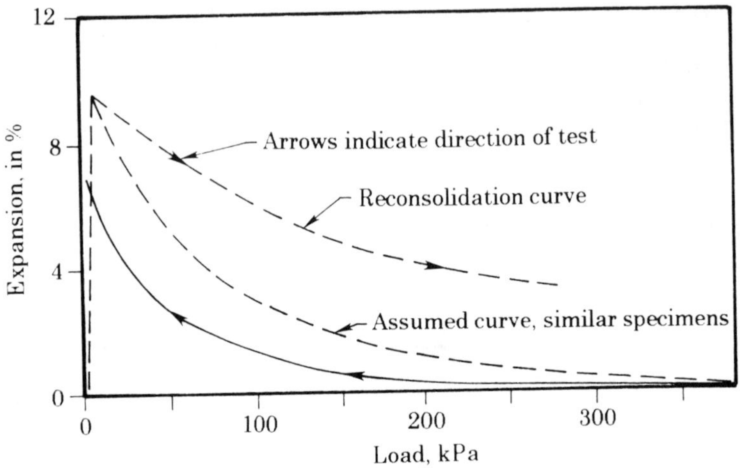

Figure 2. Holtz & Gibbs (1956) Void Volume Curve.

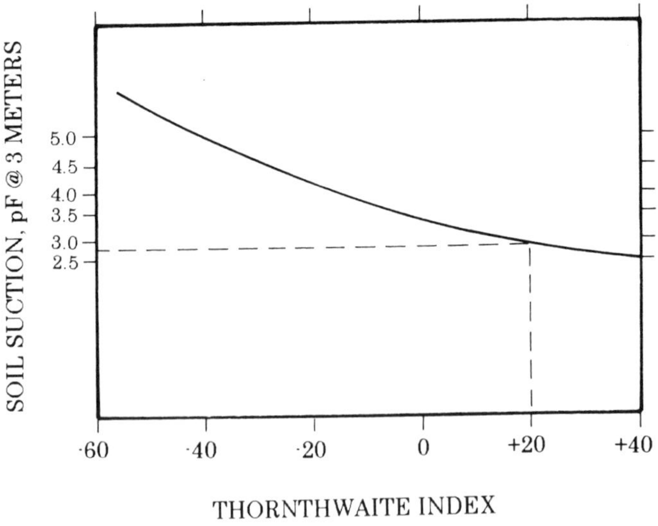

Figure 3. Russam-Colman Correlation (1961).

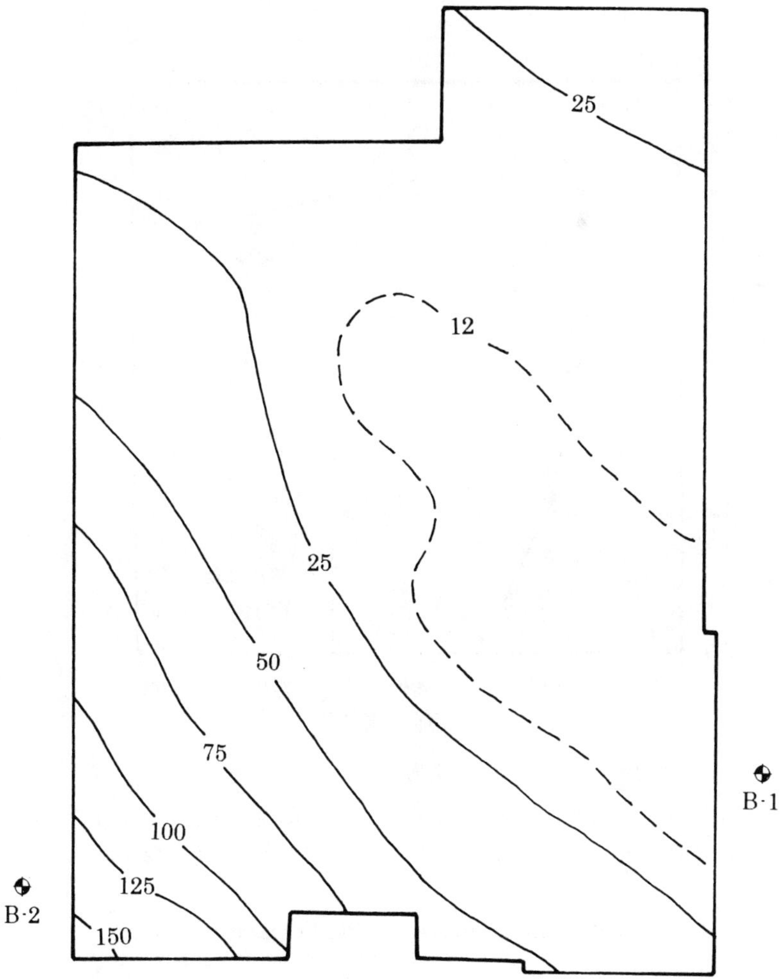

Measurements in millimeters

Figure 4. Plan of Borings & Relative Elevations.

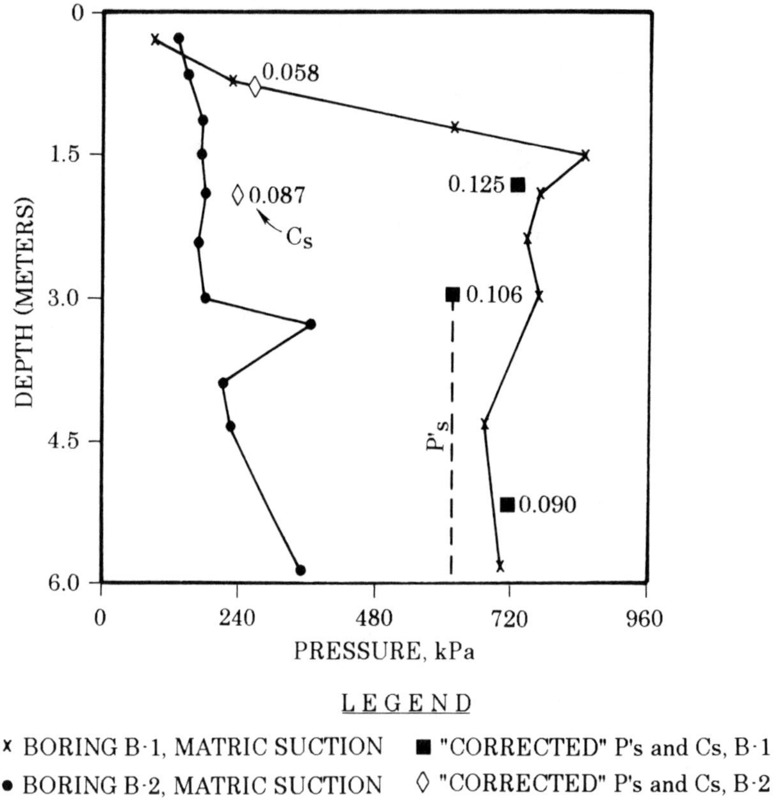

Figure 5. Matric Suction and Corrected Swell Pressure.

Impact Of Climatic Variation On Design Parameters For Slab On Ground Foundations In Expansive Soils

Ronald F. Reed, P.E.[1] and Monica Kelley, P.E.[2]

Abstract

Adequate design of post-tensioned residential slab-on-ground foundations in expansive soils is dependent on understanding the geotechnical and climatic factors inherent to the procedure. A method for design of ground-supported slab foundations was developed by the Post-Tensioning Institute (PTI). The design stiffness factors (edge moisture variation distance, e_m, and differential movement, y_m) are derived in part based on the average Thornthwaite Climatic Index, I_m, for the design area.

The purpose of this paper is to show, by case study, the importance of changes in water balance as measured by the I_m on expansive soil behavior and in particular to the PTI design stiffness parameters, e_m and y_m.

Surveyed differential movements of post-tensioned foundations in two expansive clay formations in the North Texas region, where average I_m, not including the influence of irrigation, is 0, are provided. Data is presented to illustrate that use of the average I_m value results in slab foundations that are under-designed. Re-calculation of the published average I_m value for the design area is recommended to account for changes in water balance caused by development, for example the increase in I_m from landscape watering or extreme dry periods.

The case histories show that the PTI values for e_m and y_m, using the re-calculated I_m, are in closer agreement with observed differential movement. The

[1] Member, A.S.C.E., Principal, Reed Engineering Group, Inc., 2424 Stutz, Suite 400, Dallas, Texas 75235; phone (214) 350-5600, fax (214) 350-0019; rreed@reed-engineering.com

[2] Member, A.S.C.E., Geological Engineer, Reed Engineering Group, Inc., 2424 Stutz, Suite 400, Dallas, Texas 75235; mkelley@reed-engineering.com

studies also show that PTI y_m values derived using average Thornthwaite values can be in error by 100 percent or more.

Introduction

Current industry practice in design of post-tensioned slab foundations is heavily dependent upon application of procedures presented in the PTI manual "Design and Construction of Post-Tensioned Slab-on-Ground" (1980). Included in the manual is Appendix A.3, which outlines procedures for developing soil design parameters used in the PTI analysis. Although the PTI method significantly advanced the art of engineering for slabs-on-ground when introduced in the early 1980's, Appendix A.3 contains data and recommendations that need to be more critically evaluated than currently practiced by the engineering community. Specifically, this paper discusses derivation of the Thornthwaite Climatic Index, I_m, and its influence on estimating the design differential movement, y_m, from the tables in Appendix A.3. Two case studies are also presented illustrating observed failure modes and differential movement.

Familiarity of the reader with the PTI design procedure, and specifically Appendix A.3, is necessary to follow the general discussion. A brief review of the procedures presented in Appendix A.3 of the PTI manual and some of the definitions in the PTI manual are presented in the following section.

Review of PTI Procedure

Design of slabs-on-ground using the PTI manual requires various factors, with two of the more critical being the design movement, y_m, and the edge moisture distance, e_m. These parameters are required for both the "edge" lift and "center" lift condition. The conceptual definition of these two parameters for the "edge" lift and "center" lift condition is shown in Figure 1. Appendix A.3 of the manual presents a means of deriving e_m and y_m for both the edge lift and center lift condition of movement.

Appendix A.3 contains various figures and tables for development of e_m and y_m. Required input is the type and amount of clay. The designer can then proceed through Appendix A.3 and derive the design values. This odyssey begins with a map of the United States or Texas that plots average Thornthwaite values. The map of Texas is reproduced as Figure 2. From the average Thornthwaite value, the designer proceeds to two other figures and obtains the estimated constant suction and values for e_m for both the center lift and edge lift condition. These figures are reproduced as Figures 3 and 4. With the values obtained from Figures 3 and 4, coupled with the percent and type of clay, the designer proceeds to tables and obtains "design" values for y_m. A portion of a table from Appendix A.3 used for obtaining y_m is shown in Figure 5.

Analysis of Figures 3 and 4 indicates the reliance of the PTI method on the Thornthwaite Climatic Index. Variations in I_m directly influence the constant suction and, more dramatically, e_m. Analysis of Figure 5 indicates the influence of these parameters on y_m.

Analysis Of Thornthwaite Index In The Dallas/Fort Worth Region

The Thornthwaite Climatic Index, I_m, is an empirical method developed by C.W. Thornthwaite (1948) to classify climates. The method basically compares potential evapotranspiration and precipitation on a yearly basis, and then averages the values to arrive at a climatic rating. Positive values of I_m indicate humid climates. Negative values represent sub-humid to arid climates. The importance of this statement is to realize that the I_m values presented in PTI Appendix A.3 are averages and do not represent the potential range of values. Focusing on the Dallas area, Fig. A.3.3 of the PTI manual (Figure 2) suggests an I_m of zero centimeters (cm) (0 inches).

Calculation of the I_m was made for the Dallas area for the last 60 years using procedures presented by McKeen and Johnson (1990). The results of the calculations from 1955 to 1973 are presented in Figure 6. While the average does indeed fall around zero cm (-0.95 inch), the extremes range from -33 to 45 cm (-13 to +18 inches), with the positive extreme of 45 cm (+18 inches) occurring in 1957 with 140.1 cm (55.14 inches) of rainfall. This range is consistent with that reported by Thornthwaite.

As can be seen in Figure 6, the Thornthwaite Index does not always fall on an "average" condition. Any year that is significantly wetter or drier than the average will test the integrity of the "average" design.

The addition of irrigation also increases the "average" calculated I_m. Irrigation in the Dallas area can add the equivalent of 5 cm of rainfall per month for the five winter months and 10 cm per month for the remaining portion of the year. Results of the effect of irrigation on the calculated I_m are presented in Figure 7.

Analysis of Figure 7 provides an indication of the moisture range to be expected. For example, a home constructed in 1980 or 1988 (dry years) and irrigated could be expected to undergo significant movement associated with edge lift.

Comparison of PTI design parameters affected by I_M

To illustrate the impact the I_m has on design, a hypothetical case has been developed to compare e_m and y_m values for both average and extreme values of I_m. For this study, values of -25.4, 0 and +50.8 cm (-10, 0 and +20 inches) will be used. The values that will vary with I_m are e_m and constant soil suction, pF. The percent and type of clay will stay constant at 50 percent montmorillonite, values typical of the Dallas area. The maximum depth to constant suction (2.1 meters or 7 feet) will be used in the analysis. The maximum moisture velocity of 1.8 cm/month (0.7 inches/month) will be used. Calculations used a slab length of 12.2 meters (40 feet), with 25.4-cm (10-inch) wide beams spaced at 3.7 meters (12 feet) on-center. The resulting e_m, soil suction and y_m values can be compared in Table 1.

As can be seen from this comparison, the value of I_m can significantly affect the structural design of the slab.

Case Studies

Two case studies are presented. The first is to illustrate that the "edge" lift values obtained from Appendix A.3 can be low by a significant value compared to the magnitude of differential movement in the "edge" lift condition. The second study illustrates the influence of the revised procedure for evaluation of I_m on the design of slab foundation. Both cases are from residential subdivisions where significant claims regarding distress were reported and investigated. The cases represent the average claim, and neither represents an isolated occurrence.

A rigorous structural analysis was not performed to evaluate the required beam depth. In general, beam depths were calculated based on a computer model incorporating the design formulas recommended in the PTI manual. Input variables include perimeter line and floor load; slab length; beam spacing and width; e_m; y_m; and allowable deflection. Output includes perimeter load and beam depth for each lift condition.

It should be noted that the depth to constant suction used for analysis is 2.1 meters (7 feet). This depth was used because it represents the maximum depth in Appendix A.3.

Case 1, Carrollton, Texas

This case consists of a two-story residence within Carrollton, Texas. The foundation system consists of a post-tensioned slab-on-ground reportedly designed in accordance with the PTI procedures. The residence is located within residual soils of the Eagle Ford Group. The Eagle Ford consists of a clay shale. Chemical weathering of the shale produces a highly expansive soil. The weathered profile at this site extends to a depth of 16 feet, below which unweathered shale is present.

Design movements provided by the geotechnical engineer consisted of estimated potential vertical rise of 8.26 cm (3.25 inches) and differential movement of 5.72 cm (2.25 inches). PTI design values were not provided in the geotechnical report.

The home was constructed in the spring of 1990 and was approximately 9.4 meters (31 feet) by 20.7 meters (68 feet) in plan dimension. A limited amount of vegetation was planted around the residence, and a sprinkler system was installed.

The foundation consisted of 25.4-cm (10-inch) wide by 61-cm (24-inch) deep beams spaced at approximately 3.35 meters (11 feet) on-center in both directions.

By May 1992, the foundation had undergone differential movements exceeding 8.9 cm (3.5 inches). The home owner reported severe cracking of sheetrock and mis-aligned doors and windows. There was also evidence of differential movement of the roof frame and cracks within the brick veneer. The foundation was re-surveyed in December 1992, with measured differential movement exceeding 20.3 cm (8 inches). The results of the respective elevation surveys are shown in Figures 8 and 9. With the exception of the garage area, there was no evidence of cracks within the slab, although severe warping of the slab associated with edge lift occurred along with general tilt from the high to low side.

Soil borings were performed around the residence to evaluate subsurface conditions. The location of the borings is shown in Figure 8. Tests were performed to evaluate plasticity, moisture, total suction and constant volume absorption-pressure swell. Constant volume swell tests were performed in accordance with procedures outlined by Johnson and Snethen (1978). Suction tests were performed in accordance with ASTM D-5298. Two boring logs, representing the high and low side of the residence, are shown in Figures 10 and 11. Moisture content, plasticity and suction are shown graphically on the logs. The range of results of nine pressure-swell tests is shown in Figure 12. Previous studies within the subdivision indicated the percent of the weathered zone finer than 2 microns varied from approximately 55 to 65 percent. Based on Activity Ratio (A_c) and Cation Exchange Activity (CEA_c), the clay type was a mixture, falling in the "interstratified" area of Fig. A.3.5 of the PTI manual. The montmorillonite tables were therefore used to determine design values.

Derivation of e_m, using an average I_m of 0 and procedures presented in Appendix A.3, results in the PTI design values shown in Table 2. Values of e_m are derived based on the upper curves shown in Figure 4. Design values of y_m for the center and edge lift condition obtained from Tables A.3.14 and A.3.29 of Appendix A.3 [60 percent montmorillonite, constant suction of 5140 kPa (3.4 pF) at a depth of 2.1 meters (seven feet)] are also shown in Table 2. Analysis of the derived beam depth and spacing using these values is consistent with the constructed condition.

Alternative values of e_m and y_m using a range of Thornthwaite values are provided in Table 3. I_m was varied from -25.4 cm (-10 inches) for the center lift condition (i.e., the driest value leading to shrinkage along the perimeter) to +50.8 cm (20 inches) for the edge lift condition (i.e., the average I_m, including the influence of irrigation leading to edge lift or heave). Analysis using the e_m and y_m values in Table 3 results in beam depths of 99 cm (39 inches), considering a beam spacing of 3.35 meters (11 feet). Clearly, beam depths of 99 cm (39 inches) would have provided a more rigid slab.

It can be noted from Table 3 that the design y_m for edge lift equaled 4.3 cm (1.7 inches); however, differential movement in the edge lift condition exceeded 20.3 cm (8 inches). Resolution of this apparent conflict between observed differential movement and "design" movements is beyond the scope of this paper. However, experience in the geologic formation has shown that beam depths between 91.4 and 101.6 cm (36 and 40 inches), spaced 3.05 to 3.7 meters (10 to 12 feet) on-center, are capable of withstanding movement of 20.3 to 25.4 cm (8 to 10 inches) with limited deflection. However, tilting of the foundation of a magnitude equaling the vertical movement would have taken place. Tilt of this magnitude is generally unacceptable.

Case 2, Mesquite, Texas

This case consists of a one-story residence with a post-tensioned slab. The residence is located within residual soils of the lower Ozan or Taylor Formation. The Taylor Formation tends to be more calcareous, however, than the Eagle Ford

Formation, and swell pressures and potential movements within the Taylor Formation are, in general, less severe in nature than those of the Eagle Ford.

Design movements provided to the structural engineer are not available. The foundation plan indicates beams were spaced at 3.05 to 4.42 meters (10 to 14-1/2 feet) along the long dimension (14.9 meters or 49 feet) and at 4.42 meters (14-1/2 feet) along the short dimension (8.84 meters or 29 feet). Beams were designed to be 25.4 cm (10 inches) wide by 66 cm (26 inches) deep. The perimeter beam is discontinuous along the outside garage wall.

The home was constructed in November 1992. Vegetation around the house is relatively sparse, being concentrated at the front of the residence. Differential movements across the floor slab of 7.6 cm (3 inches) have occurred to date. Cracks are present within the brick veneer and within the sheetrock on the interior of the house. The floor slab is cracked across the garage.

The foundation exhibits movement in an edge lift condition. Differential movement between the center of the slab and edge is approximately 4.4 cm (1-3/4 inches) on the side of the residence with higher soil moisture. The elevation survey is provided in Figure 13.

Soil borings were performed around the residence to evaluate subsurface conditions. The location of the borings is shown on Figure 13. Tests were performed to evaluate plasticity, moisture, total suction and absorption pressure-swell. Two boring logs, representing the high and low sides of the house, are shown in Figures 14 and 15. Moisture content, plasticity and suction are shown graphically on the logs. The range of results of four pressure-swell tests is shown in Figure 16. Hydrometer results indicate the percent clay is on the order of 75 percent.

The derived values of e_m and y_m for I_m of 0 are shown in Table 4. Seventy percent montmorillonite clay with a constant suction of 5140 kPa (3.4 pF) at a depth of 2.1 meters (seven feet) was used in the analysis. Values of y_m are taken from Figures A.3.15 and A.3.30.

Design beam depths considering the values in Table 4 are on the order of 74 cm (29 inches) compared to the 66 cm (26 inches) actually used. Alternative values of e_m and y_m using Thornthwaite values ranging from -10 to +20 are shown in Table 5. Analysis using the e_m and y_m values in Table 5 results in beam depths of 160 cm (63 inches), considering a beam spacing of 4.4 meters (14-1/2 feet). The required beam depth decreases to 94 cm (37 inches) for beams spaced at 3.65 meters (12 feet) on-center.

As noted, the y_m value for edge lift is 5 cm (2 inches) using I_m of +50.8 cm (+20 inches). This compares with the measured differential movement of approximately 4.4 cm (1-3/4 inches).

Conclusions

It would appear that the range of I_m needs to be evaluated for a given locality to include the influence of irrigation, and then "design" values covering 80, 90 or 95 percent of the range used to evaluate e_m and y_m. Using a range of I_m would result in more rigid slabs, which in turn, should result in less foundation distress during the normal variation in yearly rainfall.

Soil parameters used for the PTI analysis based on "average" Thornthwaite values can be significantly different from values calculated using a site-specific I_m, where I_m is calculated to include the influence of vegetation.

References

Johnson, L.D., & Snethen, D.R. (1978). "Prediction of Potential Heave of Swelling Soil." Geotechnical Testing Journal, ASTM 1 (3), 117-124.

McKeen, R. G. and Johnson, L. D. (1990). "Climate-Controlled Soil Design Parameters for Mat Foundations." ASCE, J. of Geotechnical Engineering. Vol. 116, No. 7, pp. 1073-1094.

Post-Tensioning Institute (PTI), (1980). "Design and Construction of Post-Tensioned Slabs-On-Ground." PTI, Phoenix, Arizona.

Thornthwaite, C. W. (1948). "An Approach Toward a Rational Classification of Climate." Geographical Review. Vol. 38, No. 1, pp. 55-94.

Table 1. Comparison of design parameters for different values of Im

Condition (1)	Thornthwaite Index (2)	Suction pF (3)	Edge Moisture Variation Distance (Em) (4)	Soil Movement (Ym) (5)	Estimated Beam Depth (6)
Center Lift	0 cm	3.4	1.2 m	2.46 cm	46 cm
	0 in		4.0 ft	0.97 in	18 in
Edge Lift	0 cm	3.4	1.5 m	1.22 cm	23 cm
	0 in		5.0 ft	0.48 in	09 in
Center Lift	-10	3.6	1.8 m	11.66 cm	76 cm
			6.0 ft	4.59 in	30 in
Edge Lift	+20	3.6	1.5 m	3.45 cm	74 cm
			5.0 ft	1.36 in	29 in

Table 2. e_m and y_m Using I_m Equal to 0, Case Study 1

Mode of Movement	E_m		y_m	
	meters	feet	cm	inches
(1)	(2)	(3)	(4)	(5)
Center Lift	1.5	5.0	3.04	1.20
Edge Lift	1.2	4.0	1.52	0.60

Table 3. e_m and y_m Using I_m Equal to -10 and +20, Case Study 1

Mode of Movement	e_m		y_m	
	meters	feet	cm	inches
(1)	(2)	(3)	(4)	(5)
Center Lift	1.8	6.0	14.43	5.68
Edge Lift	1.5	5.0	4.27	1.68

Table 4. e_m and y_m Using I_m Equal to 0, Case Study 2

Mode of Movement	e_m		y_m	
	meters	feet	Cm	inches
(1)	(2)	(3)	(4)	(5)
Center Lift	1.5	5.0	3.66	1.44
Edge Lift	1.2	4.0	1.80	0.71

Table 5. e_m and y_m Using I_m Equal to -10 and +20, Case Study 2

Mode of Movement	e_m		y_m	
	meters	feet	cm	inches
(1)	(2)	(3)	(4)	(5)
Center Lift	1.8	6.0	17.17	6.76
Edge Lift	1.5	5.0	5.08	2.00

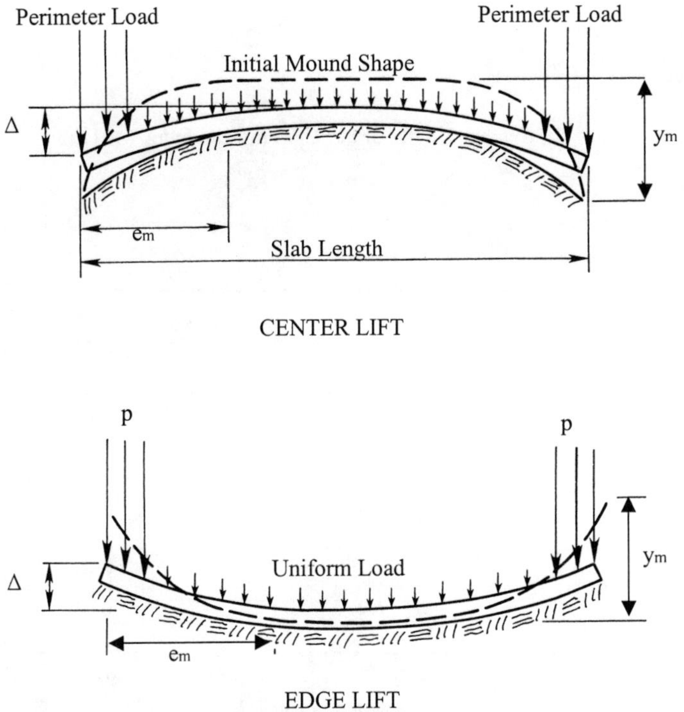

Fig. 1. Illustration of "Edge" and "Center" Lift Conditions.

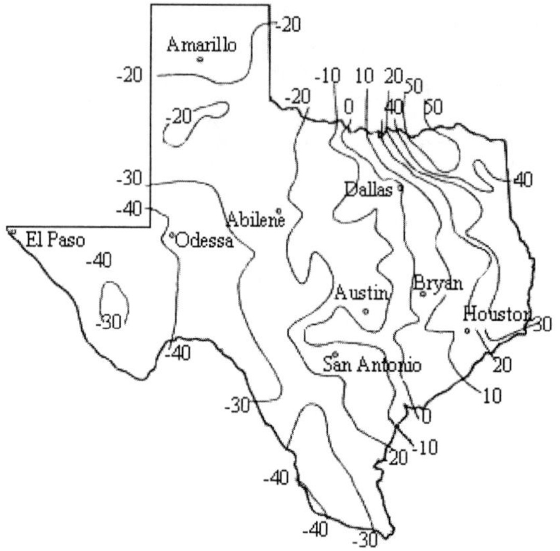

Fig. 2. Average Values of the Thornthwaite Index for Texas.

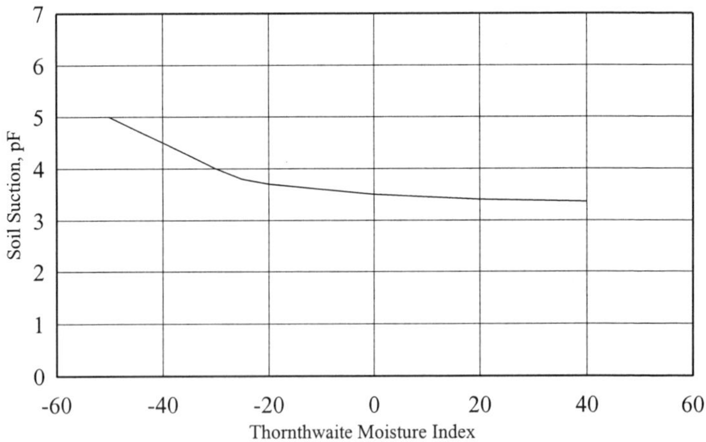

Fig. 3. Values of Constant Suction.

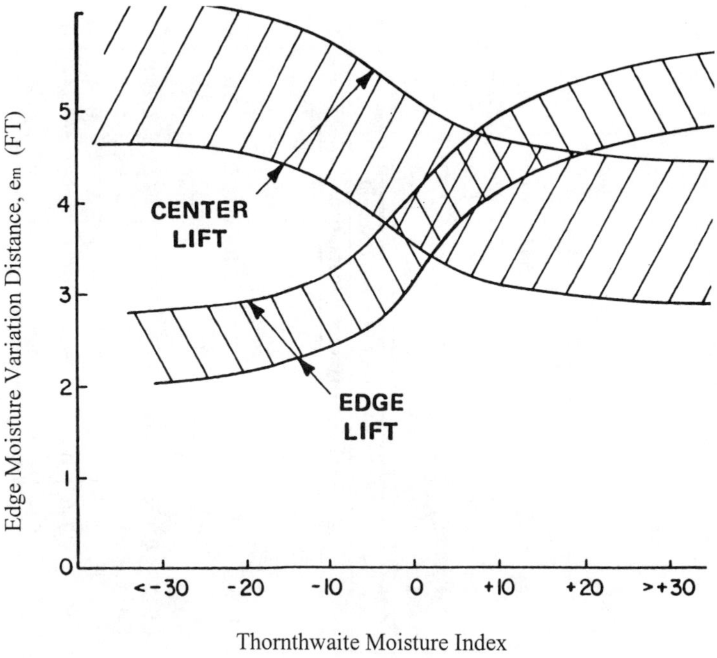

Fig. 4. Edge Moisture Variation Distance em.

Percent Clay (%) (1)	Depth to Constant Suction (m) (2)	Constant Suction (kPa) (3)	Velocity of Moisture Flow (cm/month) (4)	Differential Swell (centimeters) — Edge Distance Penetration (m)							
				1 m (5)	2 m (6)	3 m (7)	4 m (8)	5 m (9)	6 m (10)	7 m (11)	8 m (12)
40	2.1	3240	1.3	0.081	0.161	0.239	0.318	0.396	0.471	0.547	0.620
			1.8	0.113	0.224	0.335	0.444	0.547	0.650	0.753	0.854
		5140	1.3	0.176	0.348	0.514	0.675	0.832	0.985	1.134	1.280
			1.8	0.247	0.484	0.713	0.930	1.142	1.346	1.545	1.736
		8150	1.3	0.444	0.857	1.242	1.605	1.945	2.271	2.578	2.873
			1.8	0.625	1.189	1.704	2.177	2.621	3.034	3.422	3.790

Percent Clay (%) (1)	Depth to Constant Suction (ft) (2)	Constant Suction (pF) (3)	Velocity of Moisture Flow (in/month) (4)	Differential Swell (inches) — Edge Distance Penetration (ft)							
				1 ft (5)	2 ft (6)	3 ft (7)	4 ft (8)	5 ft (9)	6 ft (10)	7 ft (11)	8 ft (12)
40	7	3.2	0.5	0.032	0.064	0.095	0.126	0.157	0.187	0.217	0.246
			0.7	0.045	0.089	0.133	0.176	0.217	0.258	0.299	0.339
		3.4	0.5	0.070	0.138	0.204	0.268	0.330	0.391	0.450	0.508
			0.7	0.098	0.192	0.283	0.369	0.453	0.534	0.613	0.689
		3.6	0.5	0.176	0.340	0.493	0.637	0.772	0.901	1.023	1.140
			0.7	0.248	0.472	0.676	0.864	1.040	1.204	1.358	1.504

Fig. 5. Typical Design Table for y_m.

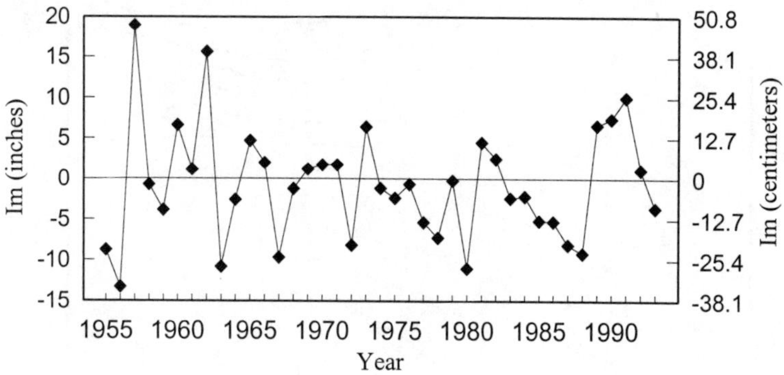

Fig. 6. Variation in Thornthwaite Index, Im Dallas-Fort Worth from 1955 to 1993

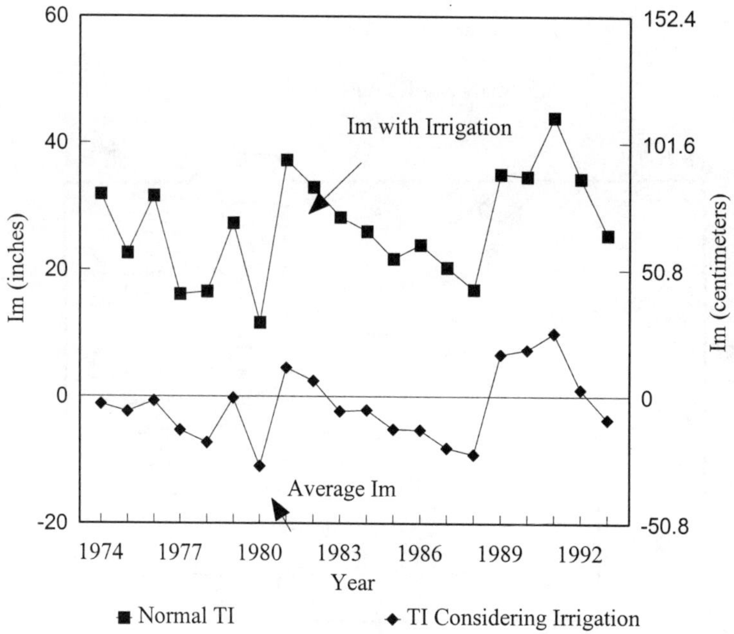

Fig. 7. Variation of Thornthwaite Index, Im D/FW Airport, 1974 - 1993

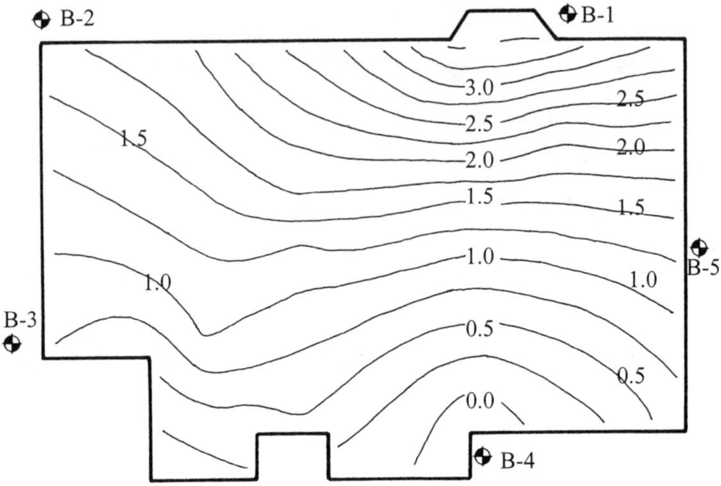

Contour interval = 0.25 inches (1" = 2.54 cm)

Fig. 8. Case Study No. 1, Initial Relative Elevation Survey.

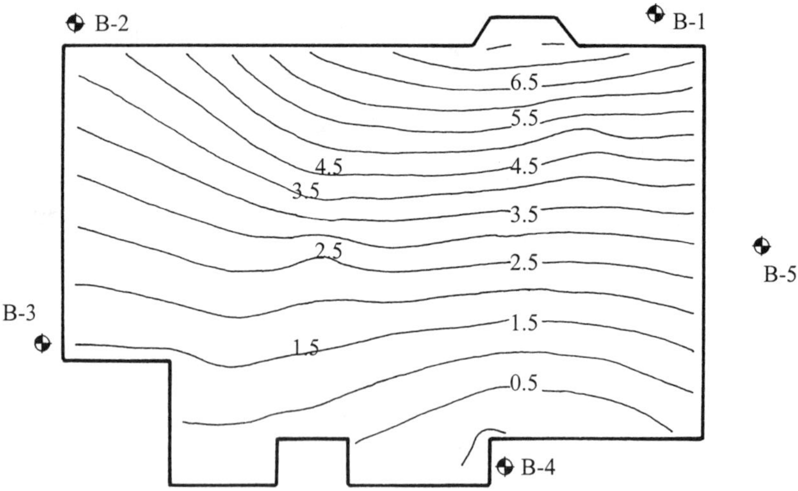

Contour interval = 0.5 inches (1" = 2.54 cm)

Fig. 9. Case Study No. 1, Second Relative Elevation Survey.

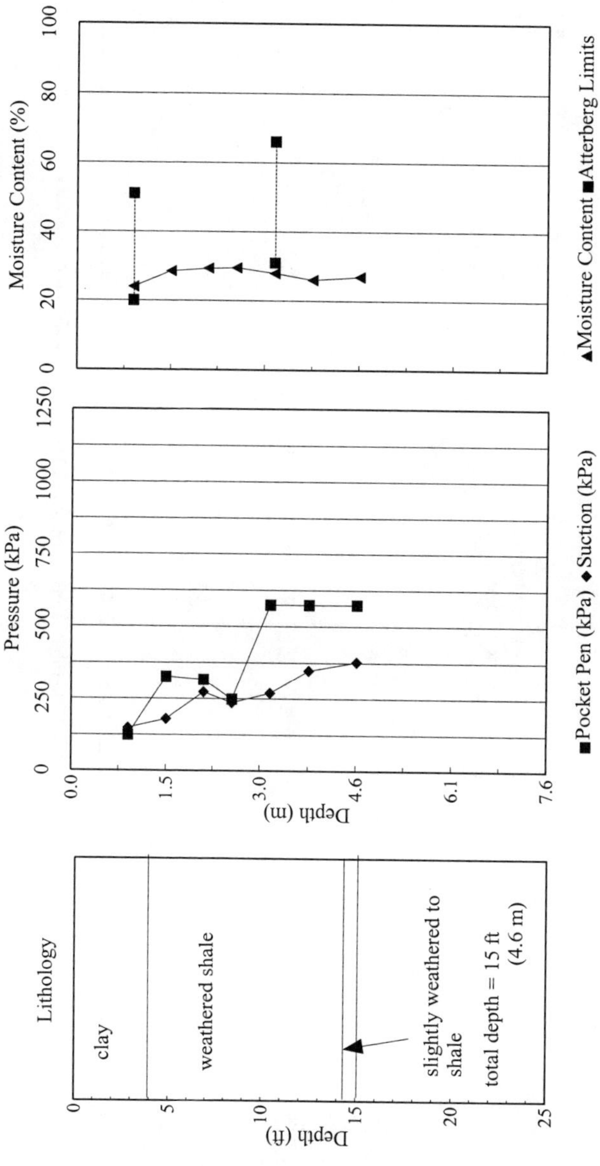

Fig. 10. Case Study No. 1, Boring Log B-1.

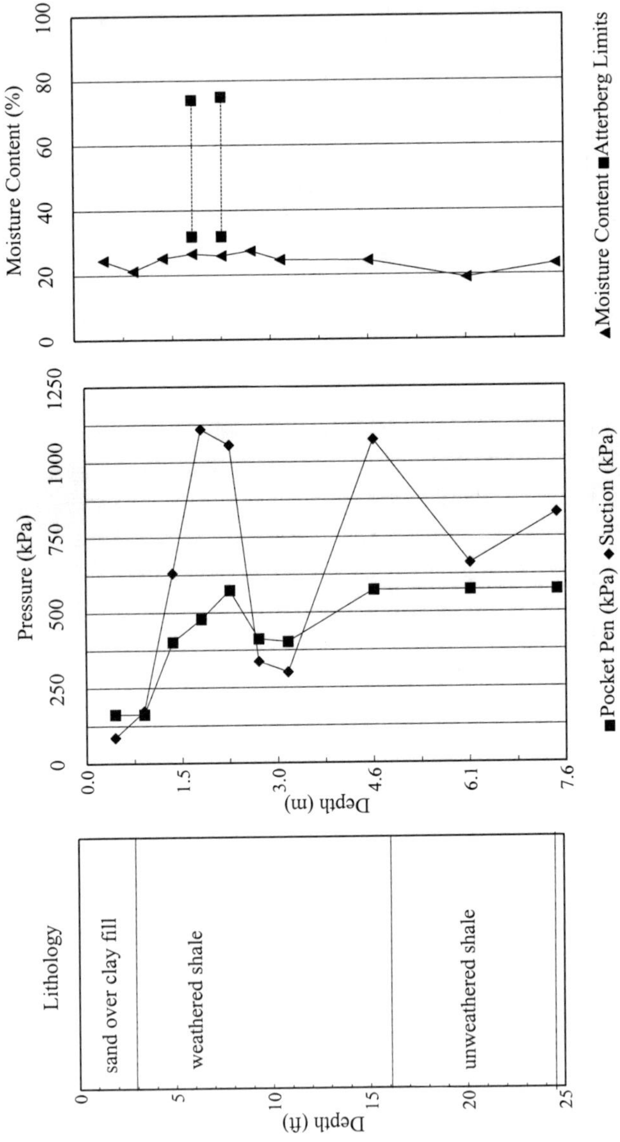

Fig. 11. Case Study No. 1, Boring Log B-5.

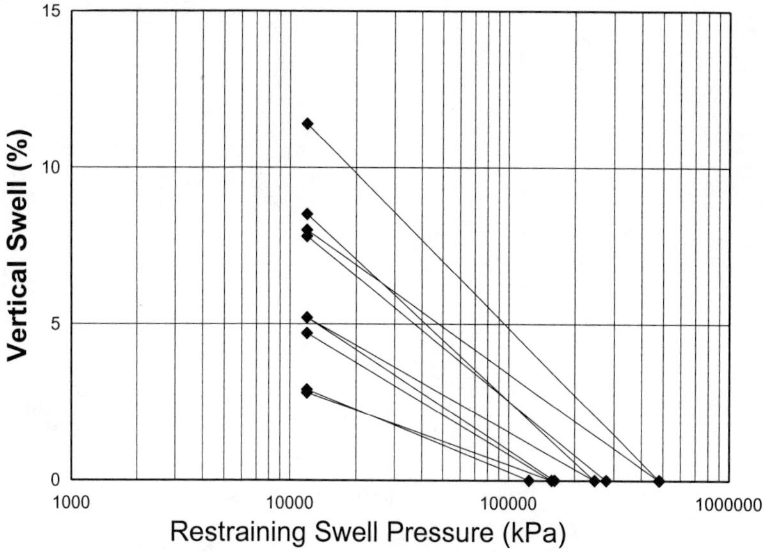

Fig. 12. Case Study No. 1, Summary of Swell Tests.

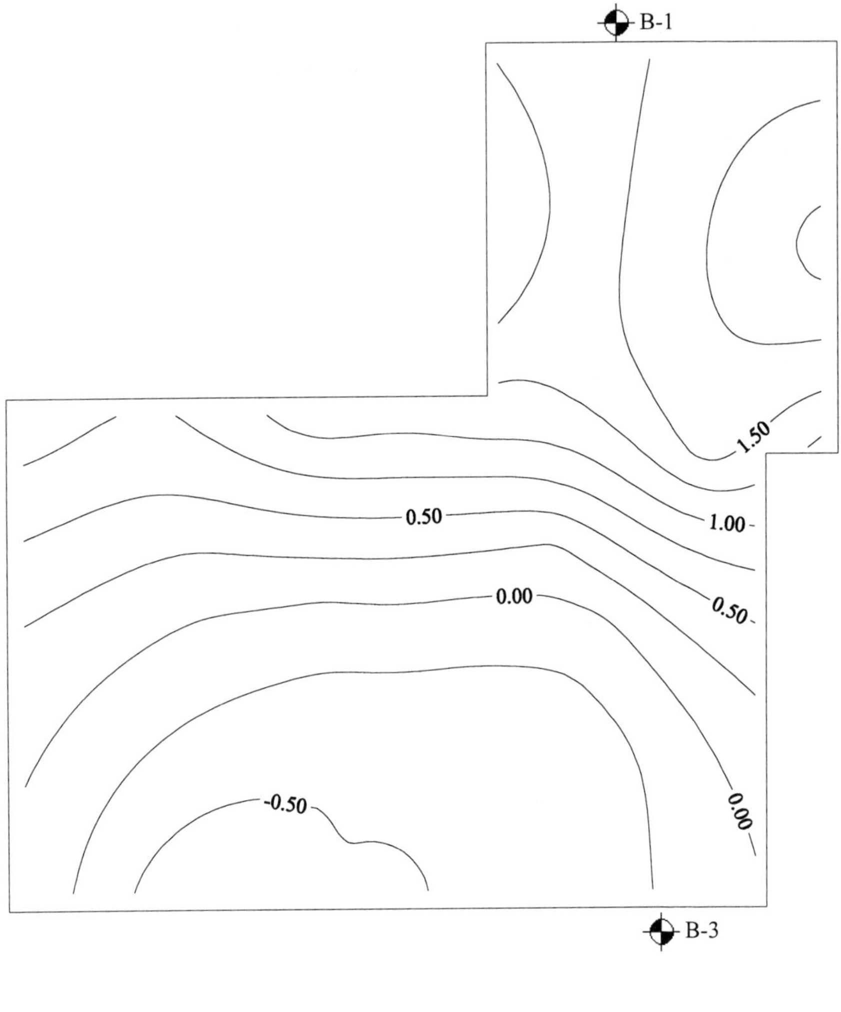

Contour interval = 0.5 inches (1" = 2.54 cm)

Fig. 13. Case Study No. 2, Relative Elevation Survey.

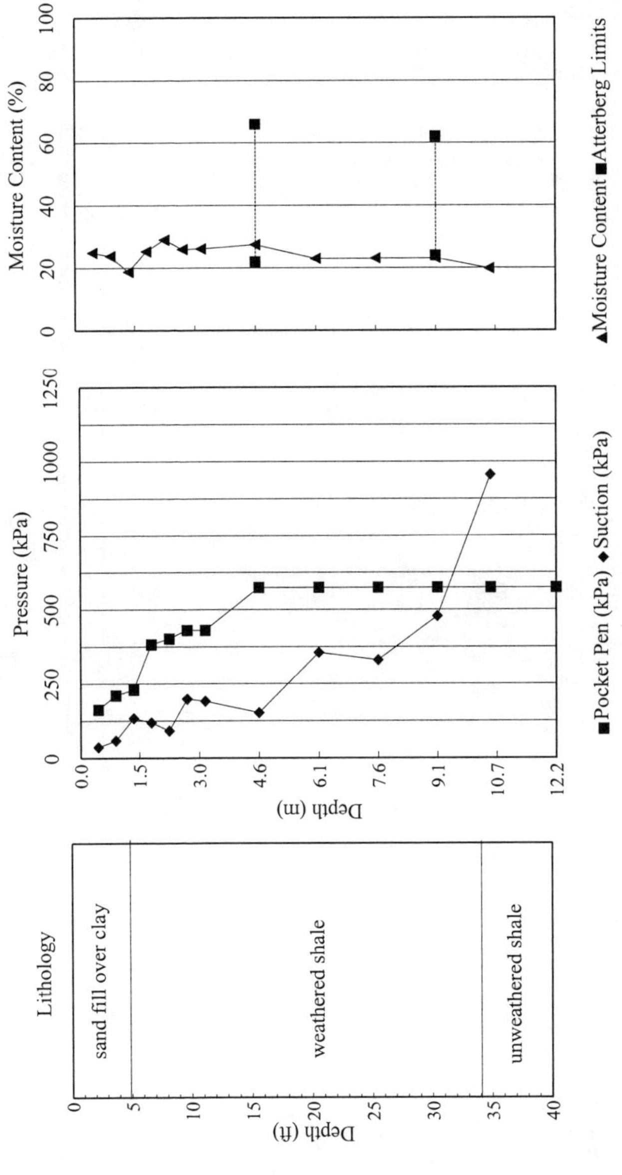

Fig. 14. Case Study No. 2, Boring Log B-1.

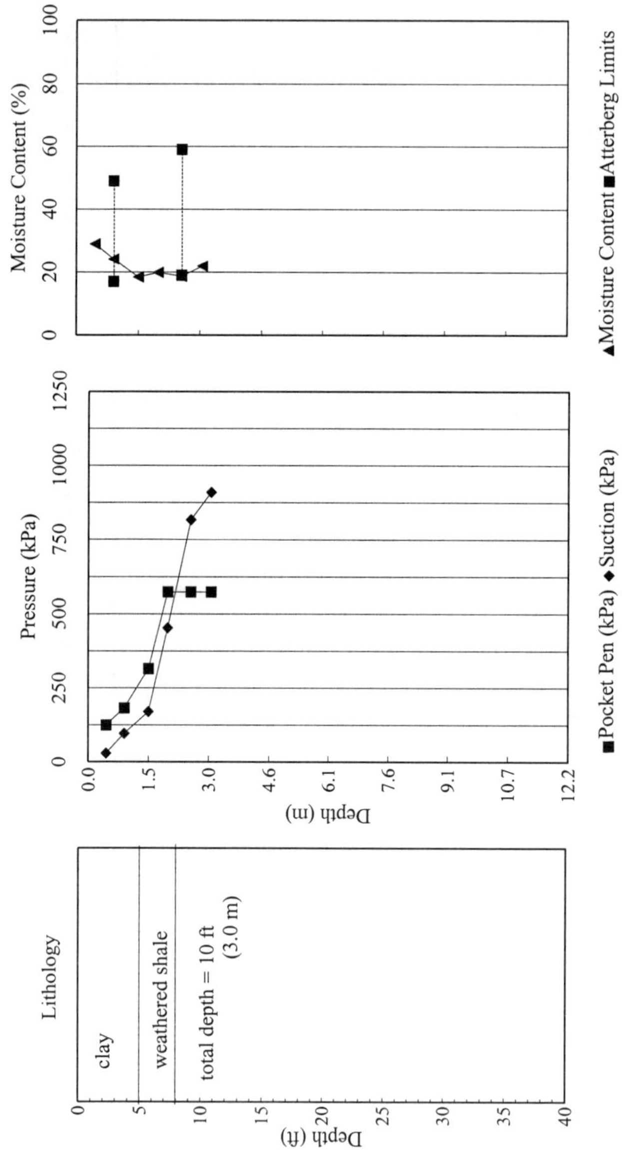

Fig. 15. Case Study No. 2, Boring Log B-3.

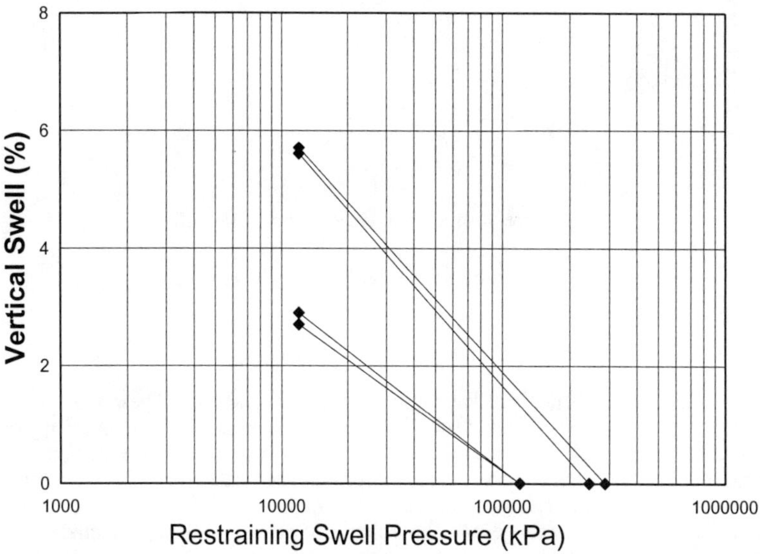

Fig. 16. Case Study No. 2, Summary of Swell Tests.

Expansive Soils and the Geomembrane Remedy

Malcolm Steinberg, P.E., F.ASCE[1]

Abstract

Expansive soils are well known in Denver and Colorado. These soils are also found in 40 of the 50 United States and worldwide in all but the polar regions. Different remedies have been used to meet the expansive soils challenge. One material being used is the geomembrane. This geotextile is treated to make it relatively waterproof. Moisture change will cause the expansive soil to swell or shrink. Minimize the moisture change and the volumetric change will be minimized. The DuPont Company brought their geomembrane to the United States in the 1960's. One of the first geomembrane projects in this country took place in Colorado. The geomembrane was placed horizontally between sand layers. Initial reports were encouraging. The Texas Highway Department used the geomembrane on a San Antonio urban arterial street project in 1976. Testing took place prior to and following construction. The geomembrane was placed horizontally on the subgrade, base dumped and spread, watered, and compacted. Paving and testing followed. The next Texas highway geomembrane project was on Interstate Loop 410 in San Antonio. The fabric was placed 8 feet deep along the shoulders of both northbound lanes. The adjacent southbound lanes of the divided highway received no geomembrane. Post construction testing showed the geomembrane protected lanes provided the smoother ride and required less maintenance. Over twenty additional Texas highway projects used the geomembrane usually as a deep vertical moisture barrier and usually successfully. Wyoming DOT has used the geomembranes on over 50 projects across the state beginning in 1981 totaling more than 4,000,000 square yards at prices of $1.25 per square yard. Their early projects placed geomembrane horizontally. Later, and currently they are placing it horizontally and vertically, and are pleased with the results. Other states have used the geomembranes on their highways. Australia is using geomembranes on their highways. and around some buildings. They have developed a trenching boom capable of cutting 6 to 6-1/2 feet deep, only 3 inches wide, permitting the use of a flowable backfill of cement, sand, water, and fly ash. Several buildings sites in Texas have been similarly protected by using geomembranes. The fabric is placed horizontally and then vertically. The geomembranes have been used on

----

[1] CEO, Steinberg and Associates, 1201 Prospect Street, El Paso, Texas 79902

railroads in France, China, and Russia. Israel and the United States have used them on canals and airport runways. Tests of geomembrane material confirm its decades long durability. The geomembrane costs have significantly decreased in Texas from early bids of $20 per lineal foot down to $6 per lineal foot. Geomembranes can be an effective remedy when dealing with expansive soils and their resultant damages.

Introduction

Expansive soils are a worldwide challenge. Denver is no stranger to these soils and their impact. Expansive soils have been identified on all the world's continents except the polar ones. Expansive soils have been identified in forty of our fifty United States (Table 1).

Table 1. States With Expansive Soils

Eastern States	Central States	Western States
Maine	Ohio	Arizona
Vermont	Michigan	California
New York	Indiana	Oregon
Pennsylvania	Illinois	Washington
New Jersey	Kentucky	Idaho
Delaware	Tennessee	Utah
Maryland	Louisiana	Wyoming
Virginia	Texas	South Dakota
West Virginia	Arkansas	North Dakota
North Carolina	Oklahoma	Montana
South Carolina	New Mexico	Nevada
Florida	Kansas	
Mississippi	Colorado	
Alabama	Nebraska	
Georgia		

Extracted from: "Technical Guidelines For Expansive Soils in Highway Subgrades" by the Geotechnical Laboratory, U. S. Army Corps of Engineers, Waterway Experiment Station, for the Federal Highway Administration.

These soils also appear in North and South America, Spain, South and East Africa, the Middle East, India, Burma, China, and Australia.

Dr. Karl Terzaghi, known as the father of soils mechanics, brought engineering skills to the understanding of the soils on the earth's crust. He built on the work of D. Atterberg who developed the tests bearing his name. The Atterberg Limits provide us to this day with a device for identifying expansive soils. Dr. Terzaghi's work in the early decades of the 20th Century drew the attention of many civil engineers. A 1905 Massachusetts Institute of Technology (MIT) graduate, Willard E. Simpson, Sr., P.E., started correspondence dealing with the impact of these soils on building foundation.

Willard Simpson's work as a San Antonio, Texas consulting engineer led to construction experiences with expansive soils and their reactions to moisture changes. Dr. Terzaghi's summation statement of "without moisture change there'd be no soil mechanics" had further confirmation with these expansive soils experiences. Two papers delivered at the Harvard University 1936 conference on soil mechanics are particularly noteworthy. One by Henry Porter, the Research Engineer for the Texas Highway Department (TxDOT), studied moisture changes in a highway section between San Antonio and Seguin, Texas Bearing capacity and pavement elevations, along with the Atterberg Limits, were correlated with this expansive clay. D. Wooltorton's paper focused on the damages sustained by building in the Mandalay area of Burma. He related damages directly to the swelling Kitti clays the structures were founded upon.

An Expansive Soil or a Swelling Clay

Terzaghi and Peck in "Soil Mechanics in Engineering Practice" state the inherent ability to swell depends on the plasticity index of the clay (Terzaghi & Peck, 1967). The swelling soil may be a clay or a shale or a marl. The damages expansive soils cause were estimated in the proceedings published by the American Society of Civil Engineers in 1973 at $1.2 billion annually in the United States. The Jones and Holtz study concluded these damages exceeded the total costs of all other natural disasters occurring in this country (Jones & Holtz, 1973). Through the years these estimates have been raised. One of the latest ones in 1998 estimated the swelling soil damages at $9 billion annually. Investigations and remedies for these damages continued through the 1930's, the World War II years, and into the 1950's. During those decades Conard Kelly, with the Texas Highway Department, the U. S. Army Corps of Engineers, and then the National Lime Association, developed a method of treating the swelling soil. This frequently resulted in admixing the lime with the top 6 inches of subgrade. Fu Hua Chen reminded us the Chinese had been using lime to stabilize soils for 2000 years (Chen, 1975). Dallas Little's book, "Stabilization of Pavement Subgrades and Base Courses With Lime", brings the treatment into detailed studies (Little, 1995).

Professor Spencer Buchanan of Texas A&M University directed the first international conference on swelling soils at Texas A&M University, College Station, Texas in 1965. Conferences followed at Texas A&M in 1969 (Buchanan, 1969), at Haifa, Israel in 1973, at Denver, Colorado in 1980, and at Dallas, Texas in 1992. Books were published addressing the subject which continued to attract attention. "Pavements on Expansive Clays" by Kassiff, et al. (Kassiff, Livneh & Wiseman, 1969) and "Foundations on Expansive Soils" by Fu Hua Chen focused on the swelling soils during that period (Chen, 1975). The Federal Highway Administration (FHWA) of the U. S. Department of Transportation (USDOT) funded a $700,000 study of these soils. The research was done by the Waterways Experiment Station (WES) of the U. S. Army Corps of Engineers. Generally directed by Drs. Donald Snethen and the late Lawrence Johnson, results were published in "Expansive Soils in Highway Subgrades" in 1979 (Snethen & Johnson, 1979). "Expansive Soils, Problems and Practice in Foundation and Pavement Engineering" by Dr. John Nelson and Debora Miller published in 1992 continued the studies, to name a few of the group (Nelson & Miller, 1992). Universities, federal, and state agencies along with a variety of organizations pursued the expansive soils challenges.

Cures and Remedial Measures

Cures and remedies continued. During World War II, L. Casagrande used electro-osmotic chemical stabilization of a clay. Professor C. E. O'Bannon conducted experiments to use this method to stabilize expansive Montmorillonite and Chinle clays in Arizona (O'Bannon, 1975). Porter's study with ponding led to similar tests during the 1960's in Mississippi and Texas, The test site in the Jackson, Mississippi area directed by Paul Teng, then MissDOT Research Engineer, now with the FHWA, had holes drilled 20 feet deep and backfilled with a pervious material. Ponding with water followed. Twenty years later it was reported to be the only Interstate Highway in the Jackson area not requiring substantial maintenance work. The Texas ponding project on U. S. Highway 90 in west San Antonio was conducted with the University of Texas' (UT) Center for Transportation Research. This study led to a zone of activity determination where most moisture changes in the clay took place. The 6 to 8 foot depth became a factor of geomembrane placement. The reports of this work became parts of the University of Texas' (UT) series of 118 reports. These studies began in 1969 with Professor R. L. Lytton, now at Texas A&M University, then a graduate student at the University of Texas, leading the expansive soils studies (Lytton, 1969).

Options for dealing with the expansive soils, other than ponding or lime treatment, are available. One difficult option might include relocating the facility to a non-expansive soil site. Another option is to replace the expansive material with non-expansive soils. This has been used in a variety of locations including the Denver International Airport, on several toll road projects in Orange County, California, and some Texas highway and commercial building projects. One project in El Paso, Texas, where the Franklin Mountains (part of the Rockies) bisect the city, required a 20 foot replacement of expansive soils.

Chemical treatment of the swelling soils extend beyond the lime admixtures noted previously. Cement, fly ash, and asphalt have been used in some cases to reduce the plasticity of the swelling soils, in others to create a moisture barrier. Laboratory and construction studies have indicated that expansive soils react to moisture changes with volumetric variations. This can result in a cracked building or a distorted transportation facility.

The Geomembrane Option

Colorado's highway department had used asphalt, applied at one gallon per square yard, to create a moisture seal. The DuPont Company's Dr. Harry Tan offered a TYPAR geomembrane to Bud Brakey of the Colorado Highway Department as a alternative. The geomembrane is a manmade hydrocarbon geotextile with an impermeable treatment. The TYPAR was placed between two 6 inch layers of sand. The results were initially viewed as a success and reported at the meeting of the Expansive Soil Technical Advisory Group (ESTAG), part of the WES study for the FHWA, where the geomembrane attracted attention. The Texas Highway Department, now the Texas Department of Transportation (TxDOT) was offered TYPAR for a test project on General McMullen Drive, a San Antonio arterial street. After preconstruction testing, work began and the geomembrane material was placed in 1976. The geomembrane varied between 7 and 20 mils thick. A 600 foot section was placed over the subgrade soil having the highest potential vertical rise (PVR). The PVR index had been developed by Chester McDowell, TxDOT Research Engineer. Base was dumped directly on the geomembrane, spread by maintainers, watered, and roller compacted. No tearing of the TYPAR occurred. Paving, traffic loading, and testing followed. The results were so favorable it was decided to try the geomembrane on another rehabilitation project.

The project was on Interstate Loop 410 in southwest San Antonio. In the Valley Hi Drive underpass area pavement distortion led to the exposure of the base course through the asphaltic concrete pavement despite constant maintenance efforts. A one-half mile test section through the area was chosen for a vertical placement of the geomembrane. The northbound lane received the geomembrane placed 8 feet deep (which was the active zone for the U. S. Highway 90 ponding project) along both shoulders. The geomembrane, a deep vertical fabric moisture barrier (DVFMB), was lapped 2 feet onto the paved shoulders. The adjacent south bound lane had no geomembrane. Both had level ups and a final paving course. In addition to serviceability, spreadability, and photo logging testing, moisture sensors were placed both inside and outside the geomembranes. Test results were encouraging. The soil moisture sensors had a short survival life, but their early readings indicated less moisture change inside the geomembrane protected area. The vertical geomembrane placement was bid at $20 per lineal foot. It included excavating the trench, placing the fabric, and backfilling with sand. After trying to cut the trench with a backhoe and experiencing sliding, the contractor announced it was impossible. The District Engineer did not accept the diagnosis. The contractor subcontracted the work with a utility contractor. Using a backhoe with sliding shoring, 400 feet of geomembrane was placed in a day. Several years later when both lanes were widened, geomembrane was placed along the shoulders.

Buoyed by signs of success on IH 410, when a southeast San Antonio legislator was told IH 37 in her district was going to receive an asphalt level up the response was: “Are you going to do that again?” The six lane concrete pavement freeway, located in a depressed two mile section, had received an asphalt level up before being opened to traffic. The pavement over an expansive clay subgrade had become rough-riding, in one instance reputedly causing a tractor trailer truck to overturn. The 2 mile section required $50,000 to $100,000 in annual pavement maintenance.

The District Engineer hearing the legislator’s response quickly replied that this year they were going to do something different. The section was going to be reshaped with drainage from the median centerline to the outside shoulders. Along the shoulders of both the north and southbound lanes, the geomembrane would be placed 8 feet deep. On this project the Resident Engineer specified a concrete aggregate backfill with a cement stabilized cap of 1-1/2 feet. The successful low bidder chose the same IH 410 utility subcontractor to excavate and place the geomembrane. This time the excavation was done with a trenching machine. Setting a planned placement rate of 600 feet per day, work was often completed by noon. Considerable testing followed, including the use of moisture sensors. Again the sensors had a brief project life. The results were encouraging and two more projects were soon let to contract involving rehabilitation of existing freeways using DVFMBs. No pavement maintenance work was required on the IH 37 project for twelve years.

The two highways, one in east San Antonio, the other on the northside, were constructed with the geomembrane placed vertically. On the IH 10 eastside project the contractor used a trenching machine and a conveyor for the excavated material which was dumped into a truck. When some trench wall sliding took place, the trench was moved away from the shoulder. Horizontal fabric placement was extended between the top of the vertical barrier and the paved shoulders. Several years after the completion of the work the Resident Engineer stated maintenance costs were a thing of the past and that the geomembrane worked well. The second project was on U. S. Highway 281. The contractor used a backhoe to excavate the trench where the DVFMB was placed. The clay was very stiff and no sliding of the trench walls took place. A series of rehabilitation contracts were awarded in the San Antonio area using DVFMB. This included work on two city streets, IH 10 from San Antonio to Seguin, three Farm-to-Market highways northwesterly from IH 10, U. S. Highway 87 southeasterly from IH 10, and State Highway 46 west of Pleasanton. Not all of this work was considered free of challenges. One of the first IH 10 projects east of San Antonio experienced distortion and cracking of the shoulders following placement of the DVFMB. Non-destructive testing revealed voids occurring in the fine base scalping used as trench backfill. A field change had permitted the contractor to use the finer backfill material. The solution was to grout the voided areas. A coarser backfill material was specified on subsequent area contracts.

The DVFMB rehabilitation work on the Pleasanton project, and two of the Farm-to-Market highways did not result in better-riding roads. Post construction investigation on the Pleasanton project revealed considerable cracking and distortions of adjacent structures and roads. It was thought that the cracks on the Pleasanton project resulted from a subsurface fault rather than the fault of the DVFMB. In the case of the Farm-to-Market highways the Resident Engineer felt the fabric was not placed deeply enough.

Two other DVFMB rehabilitations in Texas require special note. One on U. S. Highway 84 in west central Texas, north of Snyder, included experimentation with a variety of trench backfill materials. In addition to the contractor working with a variety of compaction equipment, the backfill material varied from concrete aggregate to the soil excavated for the trench. The Resident Engineer concluded that using the native soil as backfill created no problems and was just as effective as any other material. The moisture sensors again proved to be of short life and limited value.

Another noteworthy project using DVFMB was on IH 45 near Waxahachie. The IH 45 is the main highway between Dallas and Houston. The plan was to widen the reinforced concrete pavement and to place the geomembranes 8 feet deep along the shoulders of the north and southbound lanes. The low bidder on this IH 45 DVFMB geomembrane contract price was $6 per lineal foot for a total cost of over $1 million. This included excavation, fabric, placement, and backfill. During the contract the fine aggregate specified for backfill became unavailable and a field change permitted use of a modified material at $6.25 per lineal foot. Reports have indicated the geomembrane project was a success. Production rates averaged 2,500 feet per day.

Geomembranes, Other States and Other Countries

The Wyoming Department of Transportation (WYDOT) began using the geomembrane to control expansive soil movements in the early 1980's. Their first project used geomembranes placed horizontally. They currently are placing the geomembrane horizontally and vertically on their pavement rehabilitation contracts. These projects now total about 60, extending from the eastern borders of Wyoming westward to the mountains. Vertical placement is generally considered to extend to the underlying rock formations, ranging from 3 to 5 feet in many of the projects discussed. The geomembrane specified has a minimum thickness of 12 mils. A wide number of manufacturers have supplied the geomembranes (Table 2, see page 8, following).

The projects discussed herein are viewed as successful. Michael Hager, WYDOT's Chief Engineering Geologist stated in late October 1999 that he had been recently contacted by engineers from the South Dakota Department of Transportation. They wanted information on WYDOT's geomembrane operations as they were planning to use the DVFMB. Their action would be poetic justice since three decades ago E. B. McDonald of their Department placed perhaps the earliest vertical fabric barriers on a South Dakota highway. McDonald placed the geomembrane 4 feet deep and conjectured that had it been placed twice as deep it would have stopped all pavement distortion over the swelling Mancos shale.

WYDOT is continuing to place the DVFMB. Total placement is estimated to be 6,000,000 square yards. Their horizontal and vertical placement involves covering the geomembrane with base before the close of the day's work. This has tended to limit their rates to one-half mile per day, about the rate that TxDOT reached with its DVFMB. WYDOT's average geomembrane bid price is $1.25 a square yard. Hager says they are pleased with the results of the geomembrane barriers.

The use of the DVFMB has also been proposed in Montana on an Interstate Highway rehabilitation.

The Arizona Department of Transportation is currently placing 150,000 square yards of geomembranes on a highway rehabilitation project in the Flagstaff area. The Chinle clays have long been recognized for their expansive qualities. O'Bannon experiments with electro-osmotic treatment were viewed as too expensive. Jim Demeree, Arizona DOT's Materials Engineer, reported that the state has an estimated 15 highway projects using geomembranes. The geomembranes are placed vertically 8 feet deep along the shoulders and are felt to provide a solution to the challenges of the swelling soil's destructive abilities. ADOT's geomembrane specifications include a minimum membrane thickness of 14 mils.

Geomembranes have been used in a dozen states of the United States. They have also been used on streets in several cities around this country including Las Vegas, Nevada.

Australia has used the vertical moisture barriers extensively in the state of Victoria. Located in the southwestern part of Australia around the metroplex of Melbourne. Victoria Roads Authority (VicRoads) led by their Research Engineer, Dr. J. C. Holden (now Adjunct Professor of Civil Engineering at Swinburne University),

Table 2. Geomembrane Characteristics

	TYPAR		REEMAG	REEMAY			PHILIPS		MCF		LORETEX	
	3401	3153	3353***	3358*	3354	3368	MB	MB11	MC 1212	500	140N	T5500**
Weight - oz./yd.^{2}	3.9	2.4-2.6	4.8-5.1	7.5	4.0	12.3	-	9.0	6.5	4	4.5	7.0
Thickness - Mils	-	7-9	17-20	15.5	-	24	-	-	18	23	60	12.5
Grab Tensile - lbs.	130	85	130	180	140	175	140	150	200	200	120	195
Grab Elongation - 5%	60	67	55	65	-	75	40	40	-	30	55	-
Trapesoidal Tear - lbs.	60	30	60	54	40	51	45	45	80	115	50	41
Mullen Burst - psi	140	70	150	225	-	160	250	260	440	400	210	366
Permeability (perms)	0.1	3	25	-	0	-	0	0	0	0.2	0.2	-
Puncture - lbs./in.^{2}	40	52	130	51	40	51	80	80	120	35	70	-

* 3358 (T-343) no longer available.

** The same as the Midwest canvas product.

*** 3353 replaced with 3354.

has placed the geomembrane 6 to 6-1/2 feet deep, corresponding to the zone of activity. The fabric is tacked to the paved shoulder of the highway or roadway. To reduce the width of the trench for the geomembrane a slim line trencher has been developed. Using a Ditch Witch machine the trenching attachment cuts this depth with a 3 to 3-1/2 inches wide trench. This has permitted use of a flowable backfill material of cement, fly ash, sand, and water. The cost of the vertical moisture barrier is reduced considerably and increased daily production to 2,000 feet. Attention to VicRoads efforts has resulted in the use of the methodology to protect many community buildings in a similar manner.

Currently, VicRoads' Research Engineer Andrew Papacostas reports that the geomembranes are being used on three major projects in their state. The lightweight geomembrane, 0.2 mm thick, is used also as a barrier for pouring concrete on subgrades. VicRoads uses it as a double thickness in its moisture barriers. They note having some durability problems with the slim line trencher. Also one project where the geomembrane trench was not backfilled for several days resulted in sliding of the trench walls. Subsequent projects avoided that delay. Additional geomembrane vertical moisture barrier highway rehabilitation projects have been built in the states of South Australia and Queensland. Some of these projects used the geomembrane horizontally under the shoulders along with vertical placement. The results are considered a success and more are planned in the future.

Israel is another country where geomembranes have been and are being used to minimize damages from expansive soils. There has been an awareness of the problem since the earliest days of the state. Professor George Zeitlin of the Technion, Haifa, Israel followed by Professors Kassiff, Komornick, Wiseman, Livneh, Uzan, and Frydman all of the Technion have addressed this worldwide challenge. Engineer Shimon Nesichi at the Israel Department of Public Works and Uri Ackerbaum at Ben Gurion Airport have put into practice much of the research. In October 1999 conversations with Amir Aurbach indicated as many as 30 highways in that small country have used the geomembranes. They are usually placed vertically along the shoulders. Some have a considerable element of horizontal placement. Most are assessed as a success. Some are not. The question remains as to whether underestimating the projected traffic is the problem.

Buildings Also Benefit With Geomembrane Treatments

Highways and streets are not the only beneficiaries of the geomembrane treatments to reduce the damages of swelling soils. Building uses in Australia's state of Victoria were mentioned. There the slim line trencher minimizes the disruption caused by digging the trench, placing the geomembrane, and backfilling the trench. Several of the town councils in the Melbourne area of Victoria have requested and received assistance from VicRoads in carrying out membrane projects. The geomembrane is generally placed vertically, and is viewed also as a root barrier. The roots are contributors to the transfer of moisture in the soil, frequently leading to volumetric changes and building cracking.

Two Texas engineers addressed efforts to control expansive soil's destructive impact on buildings by using geomembranes. Kirby T. Meyer, P.E., a consulting civil engineer in Austin, recommended its use in addressing the effects of expansive soils on residences. He placed the geomembrane horizontally from the building wall a sufficient distance to permit a trenching machine for use in the excavation. The machine would cut the trench for the vertical placement of the geomembrane. Recently Meyer reported that he is proposing geomembrane placement in conjunction with some new building construction. The geomembrane will reduce the amount of stress and movement that the reinforced slab will experience.

Professor Kent Wray, P.E., then at Texas Tech University and now Dean of Engineering at Ohio University , consulted on an apartment development in Amarillo, Texas. Expansive soils were causing distortion in several buildings in the complex. Wray recommended (after thorough investigations) the use of geomembranes to reduce the moisture variation resulting in the soil movements.

Guy Hoop of DuPont's European geomembrane moisture barrier operations reported several years ago on uses in their marketing area. This involved placement over expansive soils in Saudi Arabia's airports and railroad lines in France and Russia.

Chinese engineers at the 1992 conference on expansive soils reported on their railroad maintenance using geomembranes. They stated 20% of their system was built over expansive soils. The geomembrane was placed horizontally over the subgrade. Ballast, ties, and rails followed. Rail maintenance expenses decreased with the geomembrane use and they were pleased with the results (Wu, Wei & Chen, 1992).

Conclusion

Expansive soils exist around the globe. They can cause costly damages. Geomembranes can be a valuable material in minimizing destructive movements resulting from expansive soils. Membrane placement rates have increased and costs decreased. Their durability has been proven. Expansive soils can frequently be controlled by the geomembranes remedy (Steinberg, 1998).

References

Aurbach, Amir, unpublished data, 1999

Buchanan, Spencer J., Fourth International Conference on Expansive Soils, American Society of Civil Engineers, New York, 1980.

Chen, Fu Hua, "Foundations on Expansive Soils", Elsevier, Amsterdam, 1975.

Demeree, James, unpublished data, 1999.

Hager, G. Michael, unpublished data, 1999.

Hoop, Guy, unpublished date, 1995.

Jones, D. E. and Holtz, W. G., "Expansive Soils - The Hidden Disaster", Journal of Engineering Proceedings of the American Society of Civil Engineers, Volume 43, New York, August 1973.

Kassiff, G., Livneh, M., Wiseman, G., "Pavements on Expansive Clays", Jerusalem Academic Press, Jerusalem, Israel, 1969.

Little, Dallas, "Stabilization of Pavement Subgrades and Base Courses With Lime", Kendall Hunt, Dubuque, Iowa, 1995.

Lytton, R. L., "Theory of Moisture Movement in Expansive Soils", Center for Highway Research, University of Texas, Austin, Texas 1969.

Mathias, Chris, unpublished data, 1999.

Meyer, P.E., Kirby T., unpublished data, 1999.

Nelson, John and Miller, Debora, "Expansive Soils, Problems and Practice in Foundation and Pavement Engineering", Wiley, New York, 1992.

O'Bannon, C. E. and Mancini, Frank P., "Field Stabilization of Chinle Clay by Electro-Osmosis and Base Exchange of Ions", Arizona Department of Transportation, Phoenix, Arizona, 1975.

Papacostas, Andrew, unpublished data, 1999.

Snethen, D. R., and Johnson, L. D., "Technical Guidelines for Expansive Soils in Highway Subgrades", U. S. Army Corps of Engineers, Waterway Experiment Station, Federal Highway Administration, Washington, D.C., 1979.

Steinberg, Malcolm, "Geomembranes and the Control of Expansive Soils in Construction", McGraw Hill Co., New York, 1998.

Terzaghi, Karl, and Peck, Ralph B., "Soil Mechanics in Engineering Practice", Wiley, New York, 1967.

Wu, X. M., Wei, P. X., and Chen, J. H., "Damage to Roadbeds and Remedial Measures", Seventh International Conference on Expansive Soils", Dallas, Texas, 1992.

COULUMBIAN - COHESION APPROACH TO ENGINEERING OF EXPANSIVE SOIL DEPOSITS - FIELD PERFORMANCE

By Ramanath K. Katti[1]

Abstract

Quest for a solution to construct stable civil engineering structure including light structure on expansive soil deposits started for the author in the year 1953, when he was told that the buildings he was supervising will heave and crack. Thus, a programme of acquisition of knowledge and planning a long-term four-stage research commenced. The first and second stage research were directed towards understanding swelling pressure and heave phenomenon, through field simulated large scale and small scale experiments, which lead to identifying role played by Coulumbian cohesion bonds, structure of clay minerals and Newtonian weight, in causing, swelling pressure, heave and self equilibrating phenomenon/ no volume change zone. This helped in putting forward fundamental relations between alteration in cohesion and change in stress-deformation and strength behaviour of saturated expansive soil system. This also led to discovery of CNS technology and its possible application to engineering problems. In the third and fourth stage research prototype experiments were performed with CNS technology in MRBC to evolve material and construction specifications for field application. In 1986 a review of performance of CNS technology was conducted, which resulted in putting forwards simplified guidelines by CWC. The last performance review was conducted on 3rd and 4th of September 1999 to assess long-term performance. Out come of this endeavor covering entire spectrum from concept to construction to long-term performance is briefly described in this paper.

INTRODUCTION

Galileo and Newton were responsible for separating the weight of a body into two components (i) mass and (ii) gravitational acceleration. Thus w = m.g.

Coulomb showed that surfaces of matter possess electrical charges. He indicated that different surfaces exhibit different quantum of charges and he put forward the concept of charge density. He showed that electrical force field between two

[1]Professor Emeritus, Indian Institute of Technology, Bombay, Director and Consultant UNEECS (Pvt.) Ltd., Advisor Simplex Concrete Piles (I) Ltd., 501/D, Poonam Chambers, Mumbai - 400 018, India.

charged bodies is an inverse function of distance and is given by: $F = \frac{Q_1 . Q_2}{r^n}$.

Later work revealed that mass is made up of one or different types of atoms and molecules and the mass may exist in different states such as solid, liquid or gaseous form depending upon, temperature, pressure and external stress conditions. Further, atoms are made up of electrons, protons and neutrons arranged in different configuration and geometry. Thus different atoms can interact with each other and produce different substances held together by internal bonding due to Coulumbian force field and can exhibit varying degree of surface charges. Surface charges can influence structure of polar liquid depending upon proximity of polar liquid, nature of polar liquid and charge density of solid surface. Due to Coulumbian force field the structure of polar liquid may change from a liquid structure to a solid state structure and states in-between (Levin 1946, Baver 1956).

When a clayey soil containing a large specific surface accompanied by a large magnitude of unbalanced surface charges, comes in contact with dipolar nature of water, it exhibits development of cohesive bonds between soil particles and altered water, which is manifested in the form of plasticity. If the density of clay water system is higher, the cohesive bonds are higher which is exhibited in the form of unconfined compression strength (UCS). The state of water may change from liquid to solid (Levin 1946, Handy, et. al. 1987). Index properties are indirect manifestation of cohesive bonds in the clay water system with increase in `r' function. Coulomb and Rankine attempted to develop physical models to evaluate lateral pressure assuming equivalent c and ϕ values.

However, it is Terzaghi who brought soil media under stress in the realm of mechanics. He based his entire work on the following definition of soil mechanics which states that soil mechanics as application of mechanics and hydraulics to sediments and unconsolidated accumulation of solid particles produced by mechanical and chemical disintegration of rocks regardless of whether or not they contain an admixture of organic constituents (Terzaghi 1943). The definition assumes that (i) the solid particles do not develop any internal stresses while interacting with dipolar nature of water due to its structure or due to Coulumbian forces in the mass 'm' of the body, (ii) surface electrical charges resulting from the mass 'm' do not alter the structure of liquid water in the pores or capillaries, (iii) water does not contribute to shear strength under stress and (iv) effective stress develops strength. The soils, which behave according to above definition, are designated as conventional soils.

However, civil engineering structures designed and constructed on expansive soil deposits, using conventional soil mechanics principles experienced large-scale distress to structures due to heave, swelling pressure and loss of bearing capacity, under all weather conditions. The deposits during saturation and under saturated conditions exhibited properties different from corresponding conventional soils (Katti, 1994).

These deposits significantly affected a large river valley project (Malaprabha river valley, India) development in particular and others in general. The damage caused by these soils amounted to several billions of rupees in India alone, where in almost one third of surface area is covered with these deposits. We understand that similar damages were experienced all over the world including U.S.A.

A cursory look at the properties of saturated expansive soil deposit with depth exhibited development of cohesion of high magnitude. This indicated that to solve the problem there is a need to incorporate influence of cohesion causing Coulumbian force due to 'm' properties and structure of solid particles.

Thus, in the year 1958 under the guidance of the author, (who had a strong exposure to soil stabilization research at Iowa State University, Ames, Iowa) a long term research program was planned, initiated and executed, to cover the entire spectrum starting from concept to construction and performance. The work was conducted at IIT, Bombay between the years 1958-1988 with the help of a large number of Masters and Ph.D. students and later on at UNEECS supported by Simplex Concrete Piles (I) Ltd. Constant dialogue with Dr. R.L. Handy of Iowa State University and Dr. G.A. Leonards of Purdue University, gave confidence in directing research. Many organizations joined to make the project a success.

Malaprabha authorities of state government of Karnataka, India joined in the research endeavor to conduct prototype studies in the field and later on develop construction methods and specifications for CNS (cohesive non swelling soil layer) and (mechanically stabilized mix) MSM technologies, which were a spin off from the research studies. MSM is a graded granular mix containing size fraction ranging from boulder size to fines. This material is used in base course and subbase of roads wherein the subbase is weaker. Construction of hundreds of kilometers of canals and cross drainage structures was done using this technology. State government of Rajasthan, India used the technology for the construction of the canal, Indira Gandhi Nahar, in highly expansive bentonite deposits. Simplex promoted use of this technology for industrial structures.

India's Central Board of Irrigation and Power (CBIP) and central water commission (CWC) under the advice of World Bank and U.S.A.I.D. funded the fundamental and basic research component of the programme. The results of this endeavor since 1958 till to date are briefly presented in this paper.

DEVELOPMENT OF COMPREHENSIVE PROGRAMME OF RESEARCH STUDIES

It was reported that all types of civil engineering structures constructed on expansive black cotton soil deposits experienced large-scale distress. Such structures were buildings, sides of canals, canal beds, dams, retaining walls, and roads etc. Some limited studies conducted on these soils showed that presence of predominantly montmorillonite clay mineral and or combination of montmorillonite and illite contribute to swelling pressure and heave. X-ray, D.T.A. and B.E.C. studies also corroborate this.

Some subsurface investigations conducted during hot summer and heavy monsoon rainy season conditions showed a pattern of variation of dry density, moisture content and vane shear strength with depth as shown in Figures 1(a) and (b). The data presented is typical representative pattern. While programming studies it was noted that vane shear strength during rainy season conditions is almost zero near the surface and increases rapidly and curvilinear with depth upto 0.8 to 1.5 m depending upon swelling pressure and then remains constant. The zone where in no change in density, moisture content or shear strength are observed, it is designated as self equilibrating zone or no volume change zone. It was also realized that any model study, if it has to be representative or simulate field conditions, the model size should be at least 2 to 3 times larger than self equilibrating depth. Thus, large-scale models have to be of size greater than 1m x 3m x 4m, in width, height and length respectively.

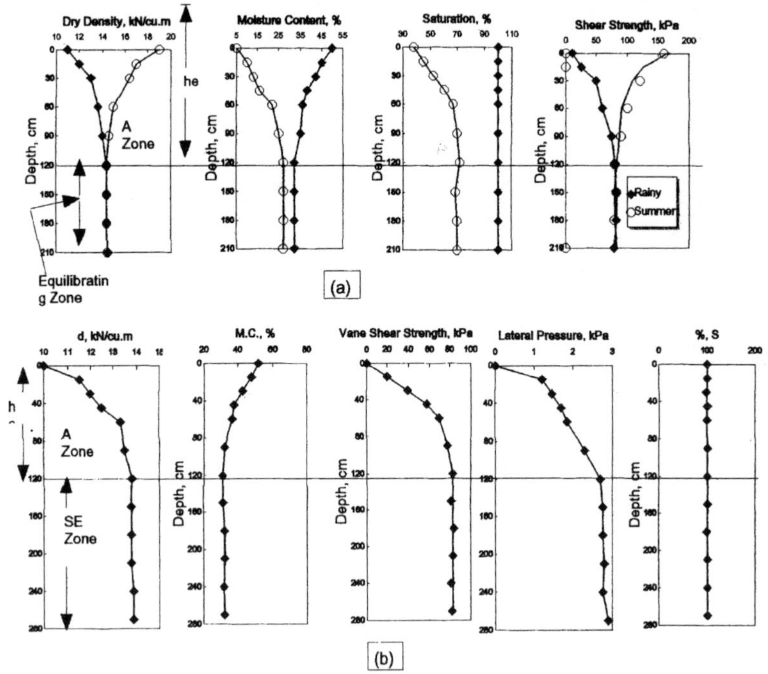

Figure 1. Typical depth vs engineering property results (a) Field situation both for rainy season and hot summer conditions (b) Large scale MRBC-76 sample compacted and saturated under inundation conditions.

First Stage Programme

In the first stage programme, efforts were made to understand factors responsible for heave, swelling pressure, loss of shear strength and self equilibrating phenomenon. In view of this, before planning laboratory simulation and testing programme an extensive field study covering different parts of the country were conducted.

Field Studies

In all, 12 locations were selected. In each of this locations, investigations relating to density, moisture content, and vane shear strength with depth were collected up to three to four times self equilibrating depth or more. Samples were also collected to conduct routine physical, chemical and engineering tests.

Laboratory Study Programmes

Laboratory study under stage 1 programme was conducted in two parts. Under part I, large scale instrumented models of size greater than 1m x 3m x 4m in width, height and length were selected for study to simulate even self equilibrating depth. In part II a large number of laboratory studies were planned using conventional and modified equipment to study samples molded to different moisture content, density, restraint during saturation, admixture, stress application etc.

Large Scale Set Up

In the large scale devices, air dried expansive soil was compacted to specified density and air dry moisture content which remained within 6 to 8%. By providing surrounding sand layer uniformity in saturation was attempted. Saturation was planned by ponding water. Each experiment took around six month or more. Conventional soils were also tested for comparison.

Dry density, moisture content, heave, vane shear strength, lateral pressure with depth etc. was measured, swelling pressure and heave vs. time were also recorded. After saturation tanks were emptied gradually and samples were collected for direct shear, triaxial, consolidation, UCS and other studies. Tests were conducted on five samples having different swelling pressure and locations.

Laboratory Studies

A large number of studies were undertaken with small scale or conventional equipment to study reproducibility of shear strength, heave, swelling pressure etc. corresponding to those in the field under all weather conditions. X-ray, D.T.A, and Base Exchange Capacity and chemical analysis were conducted on various samples to identify clay minerals. The first stage studies were completed in about ten years.

Outcome Of First Stage Programme Of Studies

The outcome of the first stage of studies is described below.

i) The soil deposits, which exhibited swelling pressure and heave, contained predominantly montmorillonite or combination of montmorillonite or illite type of clay minerals. These are three layered phyllosilicates with expanding lattice structure. The expansion takes place in mineralogical axis -c and not in a and b direction (Grim 1950). The magnitude of swelling pressure and heave depends upon index properties in general but also on magnitude of internal and external surface area and nature of isomorphic substitution in particular. This is corroborated by extensive x-ray diffraction, D.T.A. and Base Exchange Capacity studies on various fractions of soils.

Sand fractions and silt fractions did not show swelling pressure but showed settlement during saturation similar to conventional soils. Clay fraction showed swelling pressure much greater than that of the whole soil. The swelling pressure of individual clay particles in c - direction can be as high as 5000 to 10000 kPa as against measured value of 300 to 500 kPa for the whole soil (Katti 1978).

ii) The large scale model studies confirmed the phenomenon of self equilibrating depth and other property including vane shear strength in the field. Heave is confined to active zone of 1 to 1.5 m above no-volume-change zone. Fig. 1(b) shows typical plots of factors varying with depth in an expansive soil compacted to controlled density and moisture content and then subjected to full saturation under impounded conditions. Lateral pressure to vertical stress ratio was as high as 15 to 20 in the volume change zone and it was more than one in the constant volume zone. Thus, such high lateral pressures could not be balanced by vertical stress alone and corresponding ϕ values. Thus, developed cohesion appears to have contributed to the equilibrium.

iii) During unloading it was observed that magnitude of lateral pressures released is much less than corresponding vertical stress released.

iv) Shear strength tests show a bilinear trend in the failure envelope (Katti 1978, 1987 and 1994)

An Approach To Analysis Of First Stage Programme

The effort of the study was to identify the causes of swelling pressure and heave and response of the media under external and internal stresses to the deformation. The effort was also directed to identify which measurable parameters help in predicting stress-deformation and strength aspect of the media. Pattern of variation of cohesion with depth in a saturated expansive soil under impounded water condition for saturation is also reflected in the same pattern of variation of lateral swelling pressure with depth, density and void ratio with depth. This appears to be contributing to over all stability of the system. Suction is widely used as a measure to deal with swelling pressure and heave in expansive soil problems (Mustafa et al. 1993 and Chen 1975).

An expansive soil mass contains a large number of irregular capillaries due to a complex arrangement of clay size particles, silt size particles and sand size particles. These particles of clay fractions possessing internal and external surfaces due to expanding lattice, exhibit considerable Coulumbian electrical charges on the surface due to isomorphic substitution and also due to large specific surface. Other particles also contribute to external surface area of the capillaries (Taylor 1953). In unsaturated state, partially filled up capillaries and voids and in the presence of air can cause capillary pressure or capillary potential in the media due to surface tension of water and radius of capillaries. Thus, they can exhibit matrix suction $\psi = T_{\Delta}\left(\frac{1}{r_1}+\frac{1}{r_2}\right)$ which is a negative quantity. T_{Δ} is the surface tension, ψ is the capillary pressure. It is also expressed as pF wherein, $pF = \log_{10}(hc)$ and hc is the capillary rise in cm. (Taylor 1953, Spangles et al. 1973).

Due to its polar nature, water possesses dipolar properties and can adhere to clay mineral and soil particle surfaces on one hand and also may experience change in its liquid structure to the extent of ice and manifest in the form of cohesion / plasticity, and in expansive soil, the soil can heave / lift up even under stress indicating that water molecules possess such structure which provides shear strength and looses its liquid state behavior. Thus, in the case of saturated expansive soil water in the inter layer capillaries and also in the voids and capillaries may be in the state of solid water (Baver 1956). By saturating the expansive soil, capillary potential caused by air - solid, and air - water interface can be reduced to zero. Thus, cohesion may exhibit a measure of Coulumbian force field influence on the dipolar nature of water.

In expanding clay mineral, when the water is sucked in due to electrical differential potential in between inter layers, the particle may heave. Applying external stresses can arrest part of heave. Cohesion formed on the outer surface of the minerals may counteract expansion (Katti 1978, 1987 and 1994).

Lateral pressures developed with depth in large scale models were measured under saturated condition. By using cohesion approach relations are developed (Katti 1978, Katti & Katti 1994) to predict theoretically lateral pressures. These values closely match with observed experimental values. Mustafa, Warrent K. Wray and G. Vallabhan by using suction approach have predicted lateral pressure in their studies for MRBC-76 soil and other CNS aspects.

As shown in Figure 2 these values are also close to predicted values. This confirmed the inference that cohesion can be a measure of effect of Coulumbian forces in a saturated expansive soil media. By considering structure of expanding clay minerals, clay mineral structure, measured values of cohesion change in cohesion with void ratio and change in void ratio with depth, one may be able to explain self equilibrating phenomenon in saturated expansive soil. Self equilibrating phenomenon can also be used to develop soil improvement methods (Katti 1987, 1994).

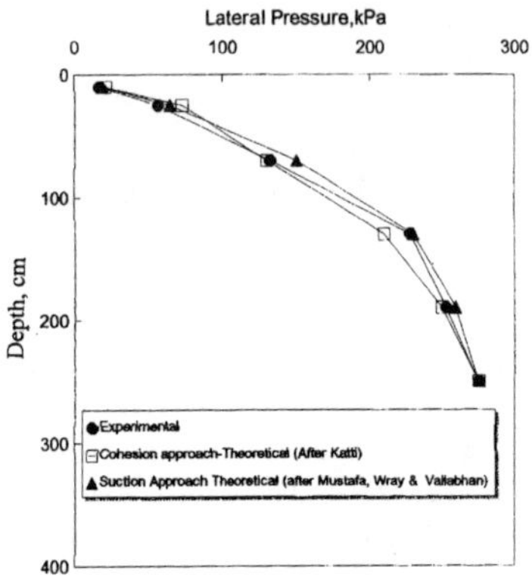

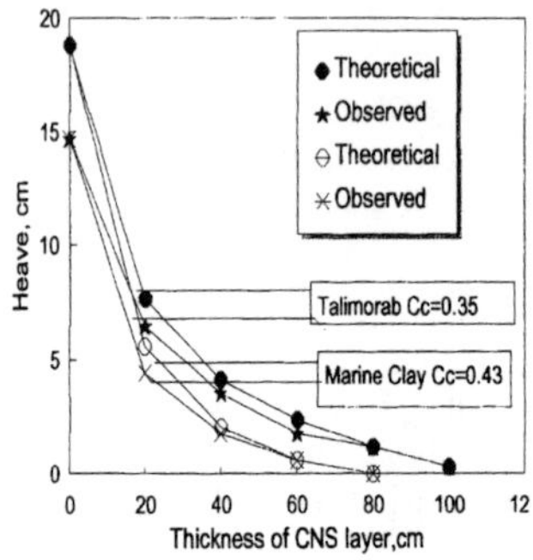

Figure 2. Results of lateral pressure with depth observed in large scale test of MRBC-76 compacted expansive B.C. soil sample under fully saturated condition and comparison with predicted values by Suction Method and Cohesion Approach Method.

Figure 3. Heave vs thickness of CNS selected from different locations on MRBC-76 expansive B.C. soil after full saturation.

Based on cohesion approach, the following relationships were developed:

Q_{swi} - Q_{sw} ; C_u - r_e ; $C_u = C_{uSL}\left[\frac{r_{SL}}{r_e} + C_1 r_e + C_2\right]^n$;

$r_e = \frac{D_s}{2}.\sqrt[3]{e}$; $C_u = C_{cu\ SL}$;

e - log p ; e vs. self equilibrating depth ;

$e = e_o - C_c \log_{10}\left[\frac{\gamma_b.he}{\gamma_b.ho}\right]$; Q_{sw} - C_u ;

he vs. C_u :

$$he = ho \times \frac{1}{e^{0.4343 \times C_c}}$$

$$\times \left[e_o - \left[\frac{-(C_2 - K) - (C_2 - K)^2 - 4C_1 r_{SL}}{C_1.D_s}\right]^3\right];$$

$\Delta h = \frac{e_o - \bar{e}}{2 + e_o + \bar{e}} \times he$; $\bar{H} = he - \Delta h$; $C_u = \frac{\sigma p}{2}$ were developed.

Further details and derivations can be found in the reference (Katti, 1987, 1994).

An examination of self equilibrating depth shows that a small stress equivalent to 1 to 1.5 m overburden (15 to 20 kPa) can counteract swelling pressure of 300 to 500 kPa or more due to cohesion. This is the part which contributes to heave, if the soil contains clay minerals with expanding lattice. Thus, if an equivalent layer of similar Coulumbian and Newtonian weight properties is imposed on the surface of an expansive soil, it may be possible to raise the no volume change zone to surface. Secondly if the imposed soil layer does not possess heaving properties the entire system will have no heave or change in volume. If the equivalent layer is made up of such soil, which contains non-expanding clay minerals it is designated as CNS layer and the concept is called CNS concept.

Second Stage Programme

In the second stage programme, efforts were directed to search for a material system, that by itself it is not exhibiting swelling pressure and heave properties but possessing both Coulumbian cohesion developing properly and Newtonian m•g property such that an appropriate equivalent layer of this material placed on the surface of expansive soil deposit can produce same effect as active zone thickness of expansive soil `h_e', so that the active zone in the underlying expansive soil can be completely converted into no volume change zone under saturated condition and thus, the entire system can be made into a non-expansive system. Such effects can be realized only in a large-scale set up.

The studies were planned to observe alteration in density, shear strength, lateral pressure and heave under various thickness of CNS material placed on the top, on the side and both on the top and side, of a controlled density and moisture content of expansive soil sample in an instrumented large scale set up saturated by impounding water and also without saturation. In the large scale set up facilities were provided to conduct studies under K_o, K_A and K_p conditions. Bearing capacity studies could also be conducted in these large scale tanks. Large scale set up varied in height from 3 to 6 m. The large scale experimental programmes were accompanied by a large number of experiments using conventional small scale equipment with modification to simulate specific conditions. Experiments were also conducted on conventional soils for comparison. Each experiment took around 4 to 6 months. The second stage programme as started in 1967 and was completed in June, 1988, spanning over a period of 21 years.

During part of first stage programme and later during second stage programme experiments were conducted with various chemical additives to suppress swelling pressure and heave of expansive soils and improve compressive strength. Subsequently the results of these studies were used for developing CNS mix design by converting expansive soils artificially.

Large Scale Studies

In all 3 expansive soils were selected for comprehensive studies out of which two were from Malaprabha river valley project out of 5 CNS materials selected, three were from Malaprabha. Some small scale studies were conducted on bentonite deposits of Rajasthan for Indira Gandhi Nahar Canal Project, wherein swelling pressures were of the order of 1500 kPa. By 1970 one set of results regarding effectiveness of CNS to counteract swelling pressure and heave were available and thus, without waiting for the completion of second stage, third stage programme on field trials, and field applications were initiated. The outcome of second stage programme is briefly brought out below.

Out Come Of Second Stage Programme

Following are the significant results of second stage programme:

(i) CNS material is effective in counteracting swelling pressure, lateral swelling pressure, and heave. There exists an optimum thickness of CNS for each expansive soil deposits, and thickness depends upon swelling pressure of expansive soil, and cohesion and grain size properties of CNS. See Figure 3.CNS also improves passive resistance of the expansive soil, which is very low without CNS backing.

(ii) Bearing capacity of expansive soil significantly improves with CNS and MSM.CNS being a fine grained soil normally has a low bearing capacity. To reduce the stresses on CNS, a MSM layer is used whose function is the same as a base course or sub base in a road. CNS preserves the "no volume change bearing capacity" of the underlying expansive soil. Unconfined compressive strength (UCS) of expansive soil under no volume change conditions is comparatively very high.

Approach To Analysis Of Second Stage Programme

Based on the cohesion concept, and incorporating concept of interface equivalent equilibrium void ratio difference approach the following relations were developed to estimate required thickness of CNS for zero heave and estimation of heave if thickness of CNS is less than optimum:

$$\log_{10}\left[\frac{h_e}{ho}\right]^{c_c} = \log_{10}\left[\frac{\overline{h_e}}{\overline{h_o}}\right]^{\overline{c_c}}$$

Theoretical results of lateral pressure with CNS backing closely match with experimental results. See Figure 4.

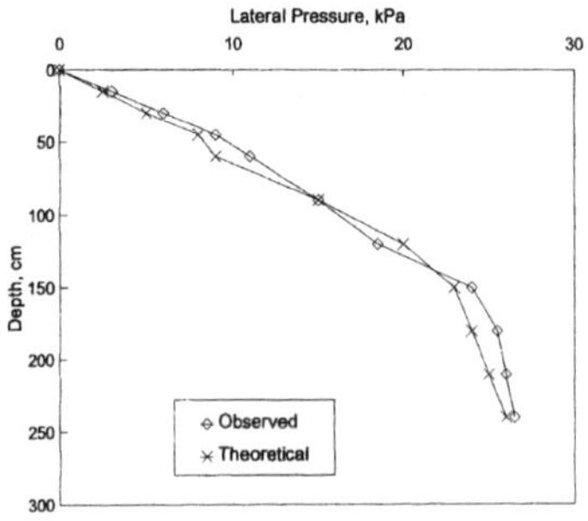

Figure 4. Lateral pressure with depth of MRBC-76 sample in large-scale test with 100 cm thick CNS backing and comparison with predicted

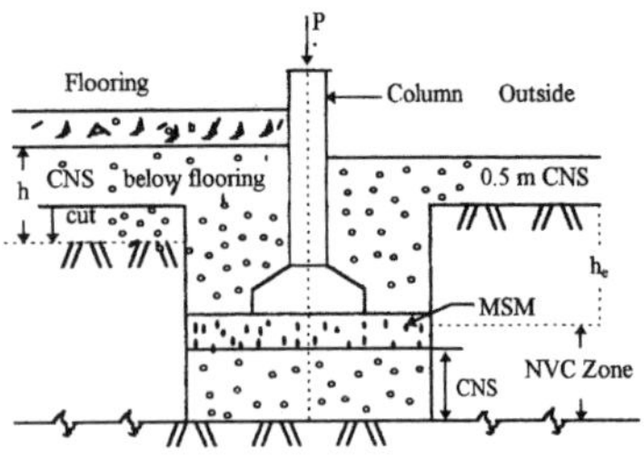

Figure 5. Shallow foundations with CNS and MSM intercepting layers

Third Stage Programme

Primary purpose of the third stage programme was to observe the effectiveness of CNS layer on an expansive soil surface, in-situ or in a compacted condition, with respect to (i) in raising the non-volume change zone to interface between CNS and expansive soil underneath, (ii) in preventing both swelling pressure and heave being transmitted to the surface of CNS layer due CNS phenomenon, (iii) in preserving the high value of shear strength in saturated expansive under no-volume change conditions, (iv) stability of bed and side lining of canal, (v) stability of high embankments and cuts constructed with a layer of covering with CNS, (vi) bearing capacity of foundations with and without CNS layer and with and without MSM layers on top of CNS, (vii)stability of wing walls and abutments of cross drainage structures, where in high lateral pressures were prevented by an intercepting layers of CNS, (viii) stability of pipe culverts and siphons etc. and (ix) foundations of walls, columns and floors of light residential structures etc. under all weather conditions.

Prototype Experiments

These experiments were started in the years 1970 in part of the Malaprabha right bank canal and were completed in 3 to 4 years. Each structure had passed through minimum two summers, two rainy seasons and two winter season conditions, before part of the sections were opened for taking observations of density, moisture content, vane shear strength with depth both in CNS and also in expansive soil below. A large number of undisturbed and disturbed samples were collected to conduct simulated small scale laboratory tests and property tests. Part of the canal sections were bunded and impounded with water upto FSL for at least 6 months.

Instrumentation

To facilitate studies, the experiments were carried out by providing, piezometric gauges, settlement gauges, embedded vane shear equipment, lateral pressure measuring equipment etc. This stage programme was concluded in 2 to 5 years.

Outcome Of Third Stage Programme

Following are the significant results of third stage programme :

i) Behavior of prototype experiments on various structures closely matched phenomenon observed in the large scale tests and that predicted by theoretically by using Coulumbian cohesion approach coupled with Newtonian m•g approach.

ii) The studies indicated that one can go ahead with laying down specifications for selection of CNS, MSM material and identification of parameters for design values for expansive soil.

It is also necessary to lay down procedure for construction, quality control and assessment of long term performance.

Fourth Stage Programme

Studies conducted under the above three programmes indicated that CNS has emerged as a viable construction technique for the construction of stable, (i) canal linings, (ii) foundations of cross drainage structures, (iii) embankments and dams with expansive soil as a construction material and, (iv) to reduce lateral pressure on retaining walls and to improve passive resistance of expansive soils and bearing capacity MSM technology further helped in improving the bearing capacity.

Thus, in the fourth stage programme an effort is made to lay down broad specifications for CNS material, MSM material, identification of expansive soil and broad outline for design and construction of canals in embankments and cuts and cross drainage structures. This was necessary as more than 100 km in length of canal along with cross drainage structures had to be constructed in deep seated expansive soil at Malaprabha river valley project in particular and also in other projects in general.

Identification of Expansive Soil

An expansive soil normally falls under the classification, CH but all CH soils may not be expansive. Thus, it is specified to conduct a no volume change swelling pressure test on disturbed expansive soil at zero moisture content and compacted to a density of 1.4 gm/cc. If the swelling pressure is more than 0.4 kg/cm^2, it may be considered for treatment. If it is less, normal construction with 30 cm murrum filling will be adequate.

Specification for CNS Material

CNS layer has to satisfy dual criteria, (i) it has to counteract swelling pressure and heave and, (ii) it has to possess as high a shear property as possible to satisfy engineering requirements. Thus, inspite of a clayey soil containing non-expanding clay minerals may be best suited to counteract swelling pressure and heave, keeping the engineering criteria, as less a clay content as possible is specified in the specifications. See Table 1. Approach to mix design may be resorted to only if a single CNS material is not available within economic haulage distance (Katti, 1966).

Design

Embankments and Earth Dams

Entire fill is constructed with the expansive soil compacted to conventional criteria and is covered with similarly compacted CNS layers of required thickness. By adopting shear parameters for no volume change conditions for expansive soil, conduct stability analysis taking into consideration level of stress and making use of shear parameters corresponding to rainy season, winter and summer conditions. Similar stability analysis be conducted for CNS zone and contact zones. Settlement and heave analysis was conducted for different seasonal conditions. The separation at contact zone between CNS and expansive soil (Katti 1994) is carefully undertaken.

Table 1 : Specifications for Soil to be considered for CNS Material

Sr. No.	Properties	Range
1.	Grain Size Analysis Clay, (<0.002 mm), % Silt, (0.06 to 0.002 mm), % Sand, (2 t 0.06 mm), % Gravel, (>2 mm), %	 15 - 25 30 - 45 30 - 40 10 - 0
2.	Consistency Limits Liquid Limit, % Plastic Limit, % Plasticity Index Shrinkage Limit, %	 30 - 50 20 - 25 10 - 25 15 and above
3.	(a) Swelling pressure of samples compacted to SDD at zero moisture content under no-volume-change condition, kg/cm^2	Less than 0.1
	(b) Swelling pressure of samples compacted to SDD at OMC under no-volume-change condition, kg/cm^2	Less than 0.05
4.	Clay Minerals	Preferably Kaolinite and Illite
5.	Shear Strength of compacted samples to SDD at OMC after saturation (a) half UCS, kg/cm^2 (b) consolidated direct shear test at 0.0125 mm/min : c, kg/cm^2 ϕ, degrees	 0.15 - 0.35 0.1 - 0.3 8 - 15
6.	Approximate thickness of CNS layer, swelling pressure, kg/cm^2 1.0 to 1.5 2.0 to 3.0 3.5 to 5.0	Thickness, cm 75 - 85 90 - 100 105 - 115

Lining of Canals

Different types of canal lining are in vogue in different parts of the country depending upon the availability of local material and its economics. Lining is provided after providing adequate thickness of CNS backing such that differential heave does not disturb the lining. Adequate care needs to be exercised to prevent uplift, formation of hydraulic channel behind lining, slipping of lining and prevention of chairing effect. Bed is lined with concrete after placing adequate CNS intercepting layer.

Cross-drainage Structures

Normally to improve bearing capacity of CNS with respect to punching, MSM layer of calculated thickness and gradation is provided. Lateral pressures on sides are reduced by providing adequate thickness of CNS with normal drainage arrangements (Katti 1994).

Foundation and Flooring for Light Structures

CNS is provided as an intercepting layer as shown in Figure 5,both for foundation and flooring (Katti, 1996) and Under-reamed Pile (Katti D. R 1988). Due to the Coulumbian surface charge effect between concrete surface of under reamed pile and expansive soil a zone of reduction in cohesion takes place around the pile upto a distance of equivalent active zone. This develops full swelling pressure at the bottom of the pile. This needs to be taken into consideration for design both with respect to uplift and settlement (Katti D R 1988).

Construction

Preferably construction is done during non-rainy season period. Care is taken to immediately cover the expansive soil surface with CNS. Once it is exposed to prevent it from heaving due to water falling or due to rains. Other conventional approach to construction is strictly followed.

Quality Control

In addition to conventional quality control approach, for expansive soil swelling pressure tests are done at fixed intervals for large jobs and minimum three tests for small jobs. In case of CNS great care needs to be exercised with respect to grain size, swelling pressure and heave and thickness. Thickness can be more than the designed value but shall not be less. Permissible tolerance can be upto 2%.

By 1974 above specifications and method of construction were made available to Malaprabha authorities, Government of Karnataka and C.B.I. & P for field use.

Out Come Of Fourth Stage Programme

(i) I.S.I. 9451 (1985), Guide lines of lining of canals in expansive soil was issued by Indian Standards Institute.

(ii) By 1985, Satisfactory performance since 1970 of (i) canals in Malaprabha, (ii) dams at Almatti, Narayanpur and six composite dams in Malaprabha, (iii) upper Krishna and Tungabhadra canal system were reported to Central Water Commission Government of India. At the request of New Delhi office of the world bank, C.W.C. formed a subgroup (committee) to visit the sites and after assessing the performance prepare guide lines which can be used by practicing engineers easily all over the country in their respective projects. In 1987 the committee declared the CNS technology as viable technology for civil engineering construction in expansive soil deposits and issued guide lines.

iii) By 1997 the technology was applied to 22 km of Indira Gandhi Nahar Canal, where in highly expansive bentonite deposits and also to several industrial projects in Gujarat. Performance of these, as of to date is satisfactory. Thus, the four stage research programme initiated in 1958 conceptually is successfully completed by providing a proven viable construction technique for the construction of stable civil engineering structures on highly expansive, deep seated expansive soil deposits.

Long Term Performance

To assess the performance of canals, cross drainage structures constructed since 1970, and some of them upto km 76 are in operation for more than 25 years, the Central Board of Irrigation and Power, suggested to Government of Karnataka to conduct an inspection of the MRBC upto km 108 wherein the last stage of construction is coming to an end.

Karnataka Niravani Nigam Ltd. organized a joint inspection of the project by Prof. R.K. Katti and the Senior Officers of the project who are in charge of maintenance and final stage of construction, on 3rd and 4th September 1999. Some representative photographs taken during inspection are presented in Figure 6 and comparison Figure 7. The committee concluded that CNS technology is a viable technology for the construction of stable and durable civil engineering structures on deep seated expansive soil deposits. MSM helps in improving the bearing capacity.

CONCLUSIONS

Based on the studies conducted under four stages of research programmes spanning over a period of 40 years followed by joint inspection on 3rd and 4th of September 1999 to assess long term behavior of structures the following conclusions are arrived at:

(1) To predict the behavior of expansive soil deposits under all weather conditions under stress, it is necessary to take into account Coulumbian force field generated by charged soil particles, structure of expanding lattice of clay minerals, formation of cohesive bonds of various degree in the soil water system due to charge field interacting with dipolar nature of water, which may be present in the inter layer of clay minerals, around the soil particles and in the voids. In addition to using Terzaghi's Principles of soil mechanics to deal with engineering of expansive soils above properties of expansive soil need to be incorporated.

(2) Unique self equilibrating phenomenon observed in an expansive soils media may be attributed to Coulumbian cohesion phenomenon coupled with structure of clay

FIG. 6 Photographs of Canal, Cross Drainage Structure, and Light Buildings at MRBC. Constructed with CNS Technology taken on 3rd and 4th September, 1999 during Joint Inspection. (a) to (g) Constructed during Period 1972-74. (h) km 108 under Completion Inspected by the Members of Committee. (a) 17m high Bennihalla approach embankment km 62, (b) 19m high wing wall and abutment km 60, (c) Canal in deep cut km 76, (d) Box entrance to road bridge km 34, (e) Combined Structure km 22, (f) Pipe Culvert km 29, (g) Primary School Building at Tirlapur around 80 km. For Comparison (Katti, 1994).

minerals and the Newtonian body weight. CNS technology is a spin off from above phenomenon.

(3) Results of large scale tests conducted simulating field conditions and small scale experiments conducted, helped in developing fundamental relations between cohesion, swelling and swelling pressure, void ratio, index properties and stress-

deformation and strength behavior with and without CNS intercepting layer. Lateral Pressure and heave behavior for the saturated state of expansive soil media under

stress can be closely predicted by either using suction approach or cohesion approach.

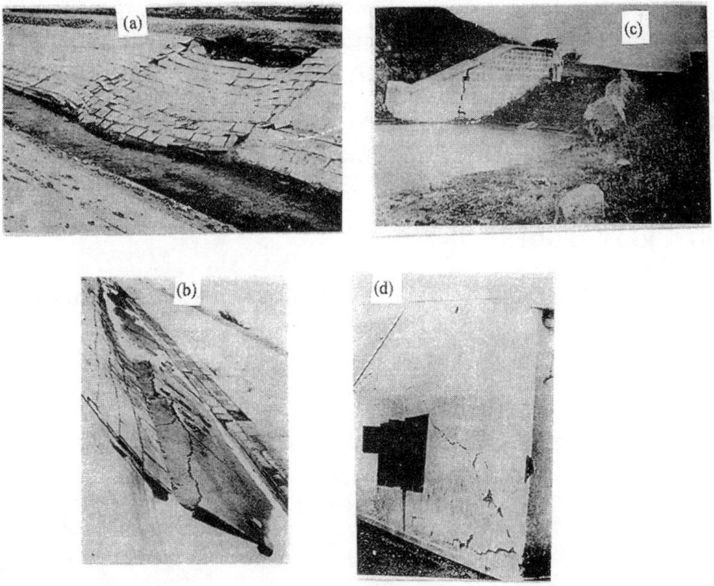

Figure 7. 1970 Photograph of Structures Constructed without CNS and MSM, under Distress, even Prior to Charging of the Canal under Rainy Season Condition, at km 21, (a) Side Slip, (b) Bed Heave, (c) Wing Wall Instability, (d) Light Building.

(4) CNS specifications have to serve duel engineering requirements of (i) arresting swelling pressure and heave and ii) providing adequate shear strength for structure, MSM helps in improving bearing capacity with respect to shear.

(5) Structures designed and constructed with CNS technology based on above fundamental relations are performing satisfactorily even after a long period of 27 years.

(6) CNS technology can be considered as a viable technology for the construction of stable and durable civil engineering structures on deep seated expansive soil deposits.

Acknowledgments

Various organisations, agencies, personalities and students who supported the programme time to time have been acknowledged in the text. Special thank are due to Dr. Dinesh Katti and Dr. Kalpana Katti of North Dakota State University USA for critically reviewing the paper. My co-author of the book Prof. Anand Katti of Datta Meghe College Airoli, on the basis of which the theme for the paper is formatted is acknowledged.

Thanks are due to Mr. Prashant Navalakha, Ms. Bharati Mulik and Ms. Lalita Chavan and other staff of UNEECS and Simplex for their help in the preparation of the paper.

I acknowledge the unstinted support extended by my wife Nalini Katti for participating in conducting the 4 stages of research programme spanning over a period of 40 years.

Appendix I. REFERENCES

Baver, L.U. (1956). Soil Physics. John Wiley and Sons, New York (3rd edition).

Chen, F.H. 1975. "Foundations on Expansive Soils". Elciver Publication, Amsterdam.

Grim, R.E. (1950). Clay Mineralogy. McGraw Hill Book Co., New York.

Handy, R.L. and Turgat, D. (1987). Swelling Pressure Vs. D-spacing of montmorillonite. 6th International Conference on Expansive Soils, New Delhi, India.

Indian Standards Institute (1985). Indian Standards 9451 (1985). Guidelines for lining of canals in expansive soils.

Katti, R. K. Radhakrishnan, N. and Kulkarni, K. R. (1966). Stabilization of black cotton soils without and with inorganic additive, India Road Congress Research Bulletin No. 10, New Delhi, India, pp 1-97.

Katti, R.K. (1978). Search for solutions to problems in black cotton soil, 1st I. G. S. Annual Lecture 1978. Indian Geotechnical Society New Delhi, India.

Katti, R. K. (1987). "Mechanics of saturated expansive soil", a lecture delivered at 6th International Conference on Expansive Clay, Organised by Central Board of Irrigation and Power, New Delhi, India, pp 1-42.

Katti, R. K. and Katti, A. R. (1994), " Behaviour of saturated expansive soil and control methods." Published by I.B.H. Oxford Under the auspices of CBI & P New Delhi, India, pp 1 to 1125.

Katti, R.K. and Katti, A.R. (1996). Procedure for design and construction of shallow foundations in expansive clayey soils with CNS and MSM technology, National Seminar on partially saturated and expansive soils. Terzaghi's birth day celebrations Oct. 2, 1996. Indian Geotechnical Society Kakinada, India, pp 5 to 14

Katti, Dinesh R. (1988). Role of passive resistance phenomenon in expansive soils on equilibrium depth of under reamed piles. Golden Jubilee, Indian Geotechnical Conference, 1998. New Delhi, India, pp 131-134, Vol. I.

Levin, S. (1946). On the interaction of colloidal particles, particular application to parallel plates. In swelling and shrinkage: General discussion, Faraday Society (eds).

Mustafa Autekin, Warren K. Wray and C. V. Girija Vallabhan (1993). Transmitted swelling pressures on retaining structures, Geotechnical special publication No. 39. Edited by Sandra L Houston and Warren K. Wray unsaturated soils A. S. C. E. Convention Dallang, pp 102-117.

Spangles, M.S. and R.L. Handy, 1973 "A soil Engineering". Intext Education Publication, New York.

Taylor, D.W.,1953 Fundamentals of soil mechanics Asia Publishing House, Bombay, India.

Terzaghi, K. (1943). Theoretical Soil Mechanics John Willey and sons. New York (1st Edn.).

APPENDIX II. NOTATION

B.E.C. = base exchange capacity
c = cohesion
C1, C2 = constants
C2 = a constant
Cc = compressibility index
$\bar{c}_c$ = compression index of CNS
C_u = undrained shear strength
CNS = cohesive non swelling soil
CWC = central water commission
D_s = equivalent particle size
Δh = heave
D.T.A. = differential thermal analysis
e = void ratio
γ_b = buoyant weight
he = distance of self equilibrating depth from surface
ho = depth below surface
$\bar{h}_e$ = thickness of CNS @ zero heave
$\bar{h}_o$ = obtained from consolidation test of CNS moulded at m.c. > LL. It is around 1.5 to 2.0 cm
$\bar{H}$ = equivalent initial non swollen thickness
K_o= coefficient of lateral earth pressure at rest
K = a constant
K_A= coefficient of later earth pressures

eo = initial void ratio
F = electrical force field
ϕ = internal friction
g = gravitational accelerations
Mineralogical axis- c, a, b = crystal direction
n = power
Q_{swi} = swelling pressure of individual clay particle
Q_{sui} = swelling pressure of whole soil
Q1 = charged body 1
Q2 = charged body 2
r = distance between the two charged bodies
r_1, r_2 = radius of curvature of air water interface film
under active state condition
Kp = coefficient of lateral earth pressure under passive state condition
m = mass
M.S.M.= mechanically stabilized mix
MRBC = Malaprabha Right Bank Canal
r_{LL} = Void radius at L.L.
S_L = shrinkage limit
σ_p = passive pressure
T_Δ = surface tension
UCS = unconfined compression strength
W = weight
ψ = capillary pressure
h_c = capillary height in cm
T_Δ = surface tension
r_{SL} = void radius at shrinkage limit S.L.
r_e = void radius

Injection Stabilization of Expansive Clays Using a Hydrogen Ion Exchange Chemical

Shondeep L. Sarkar[1], Bruce E. Herbert[1] and Russell J. Scharlin[2], M. ASCE

Abstract

Soil samples treated in the field (Crowley, Texas) with a hydrogen ion exchange chemical (HIExC) solution over two years ago, were analyzed in the laboratory using different techniques, such as X-ray diffraction, zeta potential, transmission electron microscopy/ energy dispersive X-ray analysis and environmental SEM to better understand the mechanism of soil stabilization by the HIExC solution. A limited number of engineering properties such as swell-shrink, strength, and permeability were also tested. Untreated soil samples from the same location were included as a "control," to help delineate the intricate changes in the soil structure and engineering properties that occur as a result of the amendment with HIExC solution.

The principal conclusion of this study is that a change in the shrink-swell behavior of smectite in soils treated with HIExC occurs, which is distinctly beneficial from the standpoint of its stabilization.

[1]Research Scientist, Texas A & M University, College Station, Texas 77843

[2]Vice President, Environmental Soil Stabilization, 4025 Woodland Park Boulevard, Suite 165, Arlington, Texas 76013

Introduction

In January 1997, a roadway subgrade in the City of Crowley, Texas, was treated with a hydrogen ion exchange chemical (HIExC) solution. Earlier attempts to stabilize the upper 6 inches of the failing subgrade beneath the asphalt paving with both lime and cement had proved unsuccessful. The HIExC solution was diluted at the normal 300:1 and injected through the existing paving using single rods on 4-foot centers. The hand-held rods inject at 5.5 gallons per minute.

Two sets of soil samples, one treated with the HIExC solution and the other an untreated sample from the field trial site were obtained from 1-2 and 2-3 feet depths in 1999 for laboratory testing and analyses.

These samples that are referred to as TTI-T and TTI-U (for "treated" and "untreated") respectively, were visually distinct. TTI-T was a dense clayey material colored bright red to orange-brown. TTI-U had a similar texture but was easily distinguishable by its olive-gray-brown color. Both samples had small open vugs, pea-gravel size indurated clasts and black organic matter. Liquid limits and plasticity indices indicated moderate expansion potential before treatment. The Atterberg limits did not change appreciably between untreated and treated samples.

Methods

Testing methods consisted of the following laboratory techniques:

- X-ray diffraction (XRD)
- Zeta potential
- Soluble ions analysis by induced couple plasma (ICP)
- Environmental scanning electron microscopy (ESEM)
- Transmission electron microscopy/energy dispersive X-ray analysis (TEM/EDXA)

The bulk soil samples were prepared according to the conventional method to determine their moisture contents, which are presented in Table 1.

Before size fractionation, several pre-treatments were performed to permit the dispersion of clay-sized particulate matter. Organic matter and carbonates were removed using hydrogen peroxide and dithionate-citrate-bicarbonate solutions, respectively. Following chemical treatment, TTI-U was observed to be thoroughly dispersed whereas the HIExC-treated sample maintained a clear supernatant. The size fractions thus obtained are summarized in Table 2. Sample preparation for XRD analysis was carried out according to the method outlined by Dixon and White, 1996.

Observations made during sample handling and processing are consistent with the activity of the HIExC solution as a shrink-swell stabilizer. The treated sample was much more difficult to disperse, requiring three additional washing/centrifugation steps. The treated sample was much more readily flocculated by excess NaCl solution, whereas the untreated sample was "stickier," making it more difficult to wash out of centrifuge tubes.

X-Ray Diffraction Analysis

Sand and silt size fractions. XRD analysis clearly indicated that quartz dominates the sand and the silt-sized fractions of both the treated and untreated soils. Its prevalence was visually verified under the petrological microscope. The sand was observed to be almost totally comprised of vitreous, translucent, glassy and conchoidally fractured grains that are typically quartz (Drees et. al, 1989). The grains also appeared to be coated with iron and manganese oxides. As these grains constitute less than 1 to 2% of the total sand and silt fractions, the identification of these minerals is beyond the resolution power of the XRD.

Coarse clay fraction. The XRD plots of the clay fractions are presented in Figures 1 and 2 as a nested series of diffractograms that permit comparison of the different cation saturations and treatments upon the X-ray behavior of the minerals. The lowermost pattern is Mg-saturated at 25C. Magnesium occupies the interlayer position of clay minerals along with hydration water and provides a standard measure of the condition of the clay in natural settings. The next diffractogram upwards in the nest represents the pattern for Mg-saturated clays that have been exposed to ethylene glycol solution, which invades the interlayer position of expandable clays like smectite and vermiculite causing them to swell, with an attendant increase in the d-spacing of the (001) lattice plane of the mineral (represented by a peak shift to the left). This behavior is identical to the swelling observed in these clay types by hydration water. The third peak pattern from the bottom is K-saturated at 25C. The interlayer position of K is without hydration water which causes a collapse of expandable clays to approximately 1.00 nm (8.5 2θ) represented by a shift of the peak to the right (relative to Mg-glycol). The uppermost peak pattern is K-saturated at 550C. At this temperature, expandable clays dehydrate and collapse totally to the mica position of 1.00 nm and clay mineral such as the kaolin group decomposes, resulting in obliteration of the peak.

The diffractogram for TTI-U coarse clay reveals a number of peaks. At approximately 12 2θ (0.714 nm), a well-defined peak is present through all the treatments except when heated to 550C, which verifies the presence of kaolinite. A secondary peak for kaolinite appears at approximately 25 2θ (0.357 nm) at half the spacing of 0.714 nm. The relative peak intensities indicate that kaolinite is a dominant phase in this coarse clay sample. A comparison of the diffractograms also signify that treatment with HIExC solution does not affect the mineralogy of the low-expansion kaolinite in any appreciable way.

The strong peak at approximately 27 2θ is a signature of quartz. The broad peak at approximately 6.5 2θ for the Mg-saturated 25C sample, which corresponds to a d-spacing of 14.1 2θ, indicates the presence of chlorite, vermiculite or a smectite. The behavior of the clay, when exposed to various treatments (ethylene glycol, K-saturated at 25C and 550C) permits distinction between the phase identity. When saturated by ethylene glycol, there is a slight shift of the peak toward lower 2θ (higher d-spacing). When saturated by K at room temperature, the peak shifts toward a higher value of 2θ (corresponding to a decrease in the d-spacing resulting from the pull of the unhydrated K^+ in the interlayer position). When heated to 550C the structure of the clay mineral collapses completely to the mica d-spacing value of about 9 2θ. The behavior of the clay minerals, described by expansion and contraction of interlayer or d-spacings across the various treatments is a verification of smectite. Given the intensity of the peaks and the muted shrink-swell behavior, smectite is indicated as a subdominant third mineral phase in this sample. The presence of smectite is also indicated by the second order peak at about 18 2θ. The mineral identities observed in the untreated coarse clay are mimicked by the minerals observed in the treated coarse clay with the exception of smectite, which appears to lose most of its crystalline structure.

Fine clay fraction. Some universal trends are apparent for the fine clay series (Figures 3 and 4) versus the coarse clay series. The fine clay fraction peaks are invariably broader, reflecting the decrease in average crystallite size (Schulze, 1989; Dixon and White, 1996). The fine clays also reflect a greater sensitivity to glycolation, an indication of a greater overall abundance of smectitic clays in this size fraction.

The mineralogy is the same in both the untreated and treated fine clay samples (Figures 3 and 4). Both fine clay samples yield very broad peaks, as a result of much smaller crystallite size. The expansion and contraction of the d-spacings in smectite (the "smectite dance"), while subdued in the coarse clay samples, is clearly evident in the fine clay samples. This reflects the greater overall abundance of smectite in the fine clay fraction (Borchardt, 1989). The broad peaks at 18 2θ and 27 2θ are second and third order smectite peaks, respectively. There are insignificant, broad peaks in the kaolinite (12.4 2θ) and 24.9 2θ positions, suggesting it as a subdominant phase.

Shrink-Swell of treated fine clay. The treated fine clay sample exhibits a striking change in shrink-swell behavior relative to the untreated sample. The treated sample maintains a slight peak at 18 2θ (Figure 4), otherwise, the smectite pattern is completely different from that of the untreated fine clay sample. The well-defined smectite dance (expansion with glycol followed by progressive collapse with K saturation and heating to 550C) observed in the untreated sample is absent in the treated sample.

For the distinct diffraction peaks of smectite in the untreated fine clay sample to disappear, it is necessary either to destroy the smectite lattice or alter it so as to remove the uniform diffraction from the 1.4 nm d-spacings (18 2θ). It is likely that the smectite has been modified by the HIExC solution. It is noteworthy that no additional peaks were observed signifying no new mineralization, although the XRD pattern did not scan past 32 2θ. Since there is no known process that can weather or destroy smectite so completely at surface temperatures and pressures, it is conjectured that smearing or coating of the smectite surface by the HIExC solution, a noncrystalline substance, restricts the diffraction from the characteristic atomic planes of smectite.

Zeta Potential Tests

The velocity of a colloidal particle in an electrolytic liquid with an applied electric field is related to the magnitude of charge density on its surface. An instrument that measures this velocity is able to calculate the electrophoretic mobility and the zeta potential of the particle, thereby providing data on the nature and magnitude of charge on particulate surfaces.

The zeta potential of the soils was determined for samples that had not been subjected to the dispersion treatments necessary to size segregate the soil. Thus, the destruction of organic matter (and possibly, the HIExC treatment) was avoided. Additionally, two samples of reference clay minerals (kaolinite and calcium montmorillonite) were selected for comparison with the TTI-series samples. The reference clay minerals were tested both as untreated samples and after having been exposed to the aqueous HIExC solution. The reference clays were prepared by two similar methods. Measurements of electrophoretic mobility and zeta potential were run at a constant pH and temperature. The data on zeta potential is presented in Table 3.

Negative values for zeta potential indicate that the particles are negatively charged. The values for kaolinite are observed to be less than those for the smectite (and the smectite-rich TTI series). This is expected because kaolinite is well documented to have less surface charge (Dixon, 1989). Smectite, on the other hand, has a much greater tendency toward high surface charge due to enhanced isomorphic substitution in both the tetrahedral and octahedral sheets (Borchardt, 1989). This tendency is reflected by greater zeta potential values for the smectite and smectitic TTI samples.

The addition of the HIExC solution results in a similar effect on all of the clays tested (including the TTI-series). Electrophoretic mobility and zeta potential are diminished in all cases because of the HIExC addition. This decrease in mobility and zeta potential can be interpreted as an overall reduction in the charge density of the surface of the particulate matter, possibly by adsorption and shielding of the surface by the HIExC solution or by collapse of the double layer resulting in a greater tendency of the particulates to aggregate.

Soluble Ions Analysis

1M hydrochloric acid was used to extract aluminum and iron from the treated and untreated samples. The soils were homogenized and split into two replicates. The results shown in Table 4 supports the view that HIExC treatment changes the surface chemistry of the clay through a complex precipitation-dissolution process. The low pH may be a contributory factor to this change.

Environmental Scanning Electron Microscopy (ESEM)

The advantage of using ESEM lies in the ability to examine in-situ samples at the same high resolution as that of a conventional SEM, which requires the sample to be completely dry. As a result of this drying, artifacts such as drying shrinkage, swelling, etc. tend to form, which can be somewhat misleading when interpreting the observational results. The ESEM, on the other hand, allows moist or even slurry samples to be examined. This proves particularly useful in the study of clay samples in which shrinkage-swelling of interlayers is often the pivotal factor.

The ESEM was operated at 20 KV when examining the treated and untreated TTI samples. A fragment of the as-received untreated clay sample was gently fractured so as to obtain a certain degree of interplanar orientation. The morphology of the very fine clay particles is illustrated in Figure 5. The interplanar orientation of clay particles is clearly visible in these micrographs.

An example of HIExC-treated clay sample is shown in Figure 6. The interplanar spacing does not appear to have increased, suggesting that swelling has not taken place. The interlayers are undistorted, and the same peripheral sharpness is present as in the untreated sample. This is in contrast to a plain water-treated sample (Figure 7), in which a distinct swelling was observed, and the periphery of the particles had acquired a certain amount of bulge or edge-rounding, characterizing the swelling of smectitic clay.

From ESEM examination of several untreated, water-treated, and HIExC-treated clay samples, it was evident that HIExC treatment does not alter the morphology of the clay particles or change the interlayer spacing, whereas these changes become quite apparent in the counterpart water-treated clay sample. It must be noted, however, that in the absence of any chemical data of the clay particles being examined under the ESEM, it is difficult to establish with certainty that the particles are smectite and not kaolinite or chlorite. Nevertheless, it is safe to surmise that since XRD revealed smectite to be the dominant clay mineral in the fine fraction, the particles examined are most likely smectitic in composition.

Transmission Electron Microscopy/Energy Dispersive X-Ray Analysis (TEM/EDXA)

The treated and untreated samples prepared for zeta potential, were also used for TEM/EDXA. The TEM is capable of achieving a much higher resolution than the SEM because of its much higher acceleration potential (200 KV), and use of ultrathin specimen for examination. The attached EDXA allows simultaneous analysis of the object being viewed on the TEM specimen stage. The other integral difference with SEM/ESEM is that the electrons transmit through the specimen, which is ultra-thin, whereas in SEM it is reflected from the surface of the specimen.

Under the TEM, it was possible to analyze individual smectite particles. The major difference observed between the untreated and treated sample was the absence of the crystalline electron diffraction pattern of smectite in the treated sample, whereas the atomic reflections appeared quite clearly from the smectite in the untreated sample (Figures 8 and 9). The crystalline diffraction pattern was also prominently displayed in a laboratory-treated sample of clay, which had been saturated with plain water (Figure 10). These results corroborate well with the XRD results in which the smectite peaks were observed to have disappeared in the treated sample. No morphological differences, such as rounding of grain boundary, etc. appeared in the treated sample. The other difference noted was in the chemistry of the two samples. Semiquantitative analyses of selected smectite grains are presented in Table 5. Significant differences in aluminum and iron content are noticeable.

Engineering Properties Tests

Selected undisturbed samples of the treated and untreated soils were tested for swell-shrink characteristics, strength (by unconsolidated, undrained triaxial compression testing) and permeability.

Swell-Shrink Characteristics. Swell-shrink tests were performed on three undisturbed HIExC-treated and untreated soil samples. Free swell tests were conducted in accordance with ASTM D 4546, Method A, but with no overburden pressure. Each undisturbed sample was trimmed into the consolidomter ring and placed in the consolidomter apparatus. A filter stone was placed on top of the sample and the spring of the LVDT measuring device stem was placed in contact with the top of the stone. The sample was then inundated with water. At the completion of the swell tests, shrink tests were conducted on the swell samples using a procedure developed by Briaud (1998). After completion of the swell test, the soil sample was taken out of the consolidometer, removed from the ring, weighed and then placed on a laboratory table to dry. Weight and sample dimensions are measured at regular intervals until the readings become constant for consecutive readings (at the shrinkage limit). The sample is then oven-dried to obtain final weight and dimensions.

A plot of volumetric strain change versus moisture content is in Figure 11. The slope of the shrink-swell line is called E_w, the shrink-swell modulus. It is evident that the E_w values are higher for the treated samples (0.95 and 1.30) than for the untreated sample (0.72), indicating a narrower range of volume change with change in moisture content. This implies that HIExC-treated soils shrink or swell less than the untreated soil with an equal change in moisture content.

Strength Tests. Unconsolidated, undrained triaxial compression tests were conducted on three undisturbed soil samples, one untreated sample (U) and two treated samples (T). The results are plotted on Figure 12. All three samples indicated very stiff consistency, although the two HIExC-treated samples failed at a lower strain. It is evident that the treated samples "acted" more like a brittle or friable material, whereas the untreated soil had a stress-strain curve more indicative of a plastic clay.

Permeability Tests. To evaluate the permeability of HIExC-treated soil, permeability tests were conducted on undisturbed samples. The results are presented in Table 6.

All three test results were within one order of magnitude of each other, ranging from 1.2 to 6.4x 10^{-9} cm/s, indicating no appreciable difference in the coefficient of permeability of the untreated and HIExC-treated soils.

Conclusions

On the basis of the mineralogical and microstructural analyses, such as, X-ray diffraction, zeta potential, environmental scanning electron microscopy, and transmission electron microscopy/energy dispersive X-ray analysis, the following conclusions may be drawn.

1. Mineralogical identification determined that sand and silt fractions of the untreated and treated samples are dominated by quartz. Coarse clay fractions consist of a mixture of kaolinite, quartz and smectite. Fine clay fractions are dominated by smectite with minor amounts of

kaolinite.

2. The identifiable peaks of smectite were noted to be absent in the HIExC-treated coarse and fine fractions.

3. Zeta potential is highest for the untreated field samples (average of 38.11 mv) and is lowered by 32% for the HIExC-treated field samples (average of 26.06 mv). It is interesting to note that the zeta potential for the treated field samples were closer to the values for low-expansion kaolinite (average of 20.8 mv for untreated lab samples and average of 18.0 mv for HIExC-treated lab samples) than to the values for the untreated field samples.

4. Soluble ions analysis of Al and Fe supports the hypothesis that the HIExC treatment changes the surface chemistry of clays through a complex precipitation-dissolution process; the low pH may be a contributory factor.

5. ESEM examination showed that no morphological changes occur when clay sample is treated with the HIExC solution, but swelling becomes evident in the counterpart water-treated sample.

6. TEM examination revealed that the crystalline electron diffraction pattern of smectite in the treated sample is replaced by diffuse halos, implying a change in the clay structure or a smearing of the smectite particles by an amorphous or non-crystalline material.

7. During sample handling and processing, the treated sample was continually recognized as having different behavior compared to the untreated sample. The untreated sample was more "stickier", as opposed to the treated sample that was much more difficult to disperse.

8. The dominant conclusion of this study is that soil samples amended with the HIExC solution exhibit a change in the crystallographic characteristics of the soil, which can be associated with a change in the shrink-swell behavior of smectite.

On the basis of the engineering properties tests, the following conclusions may be drawn.

1. Treatment of expansive clays using the HIExC solution reduces the swell-shrink characteristics of these clays as evidenced by the steeper slopes and higher values of of modulus E_w of the treated samples shown on Figure 11. The HIExC solution treatment reduced the swell-shrink potential an average of 35% over the untreated swell-shrink potential.

2. The HIExC-treated soils exhibit a more friable or brittle nature, indicating some loss in "cohesive" characteristics, as shown on the strength stress-strain plots shown on Figure 12.

3. Treatment of expansive clays using the HIExC solution does not increase the permeability of the soil, and therefore, "bathtub" conditions will not develop in the future when water becomes available.

Acknowledgements

A substantial part of the work was done by Mr. Ronald L. Parker under the supervision of Dr. Bruce Herbert, Department of Geology & Geophysics, Texas A&M University. The engineering tests were performed by Mr. Khaled Choudhury, a graduate student, under the supervision of Dr. Jean Louis Briaud, Department of Civil Engineering, Texas A&M University.

References

Borchardt, Glenn, 1989, "Smectite", in Minerals in Soil Environments, J. B. Dixon and S. B. Weed, Eds. Soil Science Society of America, Madison Wisconsin, SSSA Book Series # 1, pp. 675-727.

Briaud, Jean Louis, 1998, "Shrink Test for Predicting Heave and Shrink Movements", in Proceedings, ASCE Texas Section Fall 1998 Meeting, Dallas, Texas.

Dixon, J. B, 1989, "Kaolin and Serpentine Group Minerals", in Minerals in Soil Environments, J. B. Dixon and S. B. Weed, Eds. Soil Science Society of

America, Madison Wisconsin, SSSA Book Series # 1, pp 913-974.

Dixon, J. B and G. N. White, 1996. Soil Mineralogy Laboratory Manual. 2nd edition. Published by the authors, Department of Soil and Crop Sciences, Texas A & M University, College Station, Texas 77843-2474. 139 pp.

Drees, L. Richard, Larry P. Wilding, Neil E. Smeck and Abu L. Senyaki, 1989, Silica in Soils: Quartz and Disordered Silica Polymorphs, in Minerals in Soil Environments, J. B. Dixon and S. B. Weed, eds. Soil Science Society of America, Madison Wisconsin, SSSA Book Series # 1, pp 913-974.

Schulze, D. G., (1989) "An Introduction to Soil Mineralogy" in Minerals in Soil Environments, J. B. Dixon and S. B. Weed, eds. Soil Science Society of America, Madison Wisconsin, SSSA Book Series # 1, pp. 1-34.

White, J. L., 1971, "Interpretation of Infrared Spectra of Soil Minerals", Soil Science, 112: 22- 31.

Table 1. Air Dried Moisture Contents During Sample Preparation

Sample ID -- (1)	Moisture Content (%) (2)
TTI-T1	8.10
TTI-T2	8.99
TTI-U1	10.62
TTI-U2	10.56

Table 2. Size Fractionation Results

Sample ID -- (1)	Size Fraction -- (2)	Weight (%) (3)
TTI-T1	Sand	38.61
	Silt	14.85
	Coarse Clay	10.42
	Fine Clay	29.52
	Moisture	8.10
	Total	101.50
TTI-T2	Sand	38.13
	Silt	14.90
	Coarse Clay	10.70
	Fine Clay	26.47
	Moisture	8.99
	Total	99.19
TTI-U1	Sand	19.72
	Silt	26.07
	Coarse Clay	8.95
	Fine Clay	27.34
	Moisture	10.62
	Total	92.69
TTI-U2	Sand	19.15
	Silt	25.77
	Coarse Clay	8.10
	Fine Clay	23.13
	Moisture	10.56
	Total	86.71

Table 3. Zeta Potential for Reference Kaolinite, Reference Smectite and TTI-Series

1 KGa-1b w/o HIExC (mv) (1)	2 KGa-1b w/HIExC (mv) (2)	3 STx-1 w/o HIExC (mv) (3)	4 STx-1 w/HIExC (mv) (4)	6 TTI-U w/o HIExC (mv) (5)	5 TTI-T w/HIExC (mv) (6)
-20.76	-17.87	-26.01	-21.95	-38.51	-26.94
-19.98	-17.22	-26.52	-21.16	-38.64	-21.42
-21.55	-18.79	26.50	-21.95	-37.19	-29.83

KGa-1b = Kaolinite Standard
STx-1 = Calcium Montmorillonite Standard

Table 4. Soluble Ions Analysis

Sample (1)	Aluminum (ppm) (2)	Iron (ppm) (3)
Blank	0.27	0.24
TTI-U1	436.1	272.1
TTI-U2	431.9	271.7
TTI-T1	249.7	730.0
TTI-T2	248.5	755.0

Table 5. EDX Analysis of Smectite in Treated and Untreated Samples

Sample ID -- (1)	Na (% mass) (2)	Mg (% mass) (3)	Al (% mass) (4)	Si (% mass) (5)	Cl (% mass) (6)	K (% mass) (7)	Ca (% mass) (8)	Fe (% mass) (9)	O (% mass) (10)
TTI-U1	1.5	0.9	24.7	40.2	3.3	0.8	0.1	6.5	22.0
TTI-U2	3.1	0.8	21.6	42.2	2.5	1.6	0.2	7.0	21.0
TTI-T1	2.6	0	13.1	65.0	0	1.5	1.0	0.7	15.6

Table 6. Results of Permeability (Hydraulic Conductivity) Tests

Sample ID -- (1)	Coefficient of Permeability (cm/s) (2)
U-1	1.2E-09
T-2	6.4E-09
T-3	1.4E-09

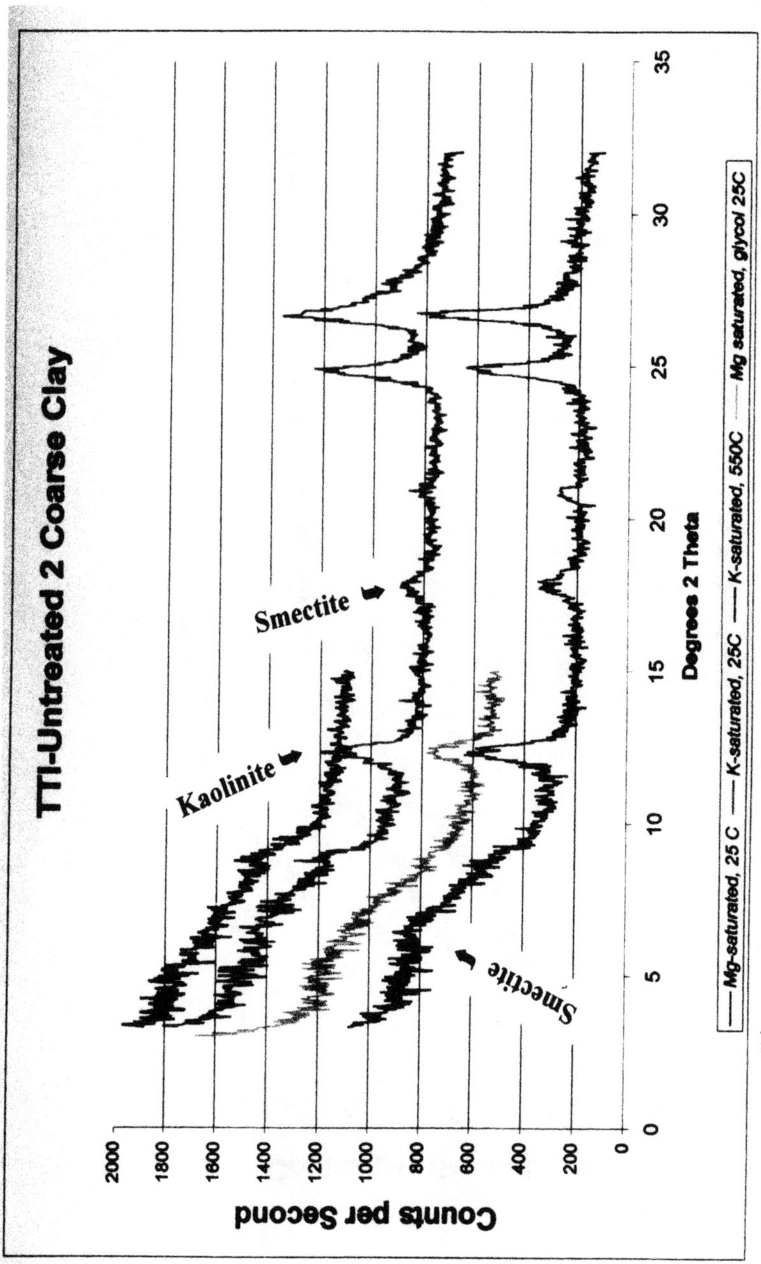

Figure 1 - XRD Patterns of Untreated Coarse Clay

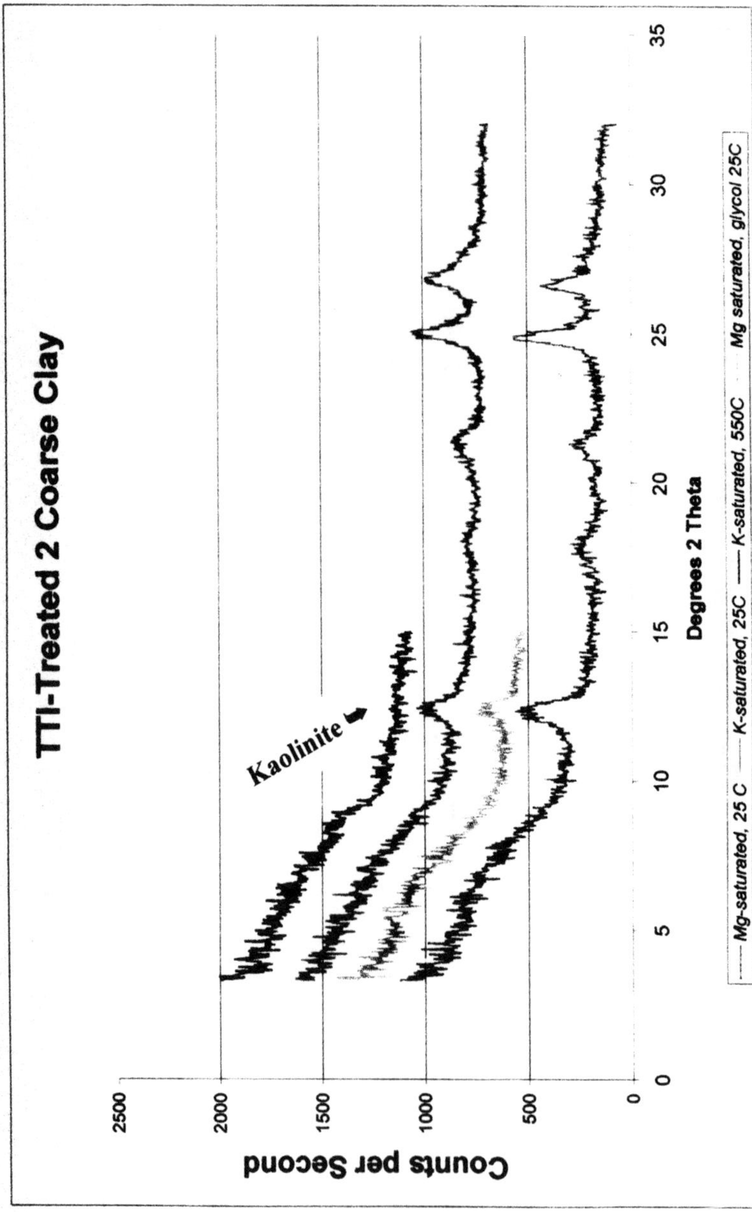

Figure 2 - XRD Patterns of HIExC-Treated Coarse Clay

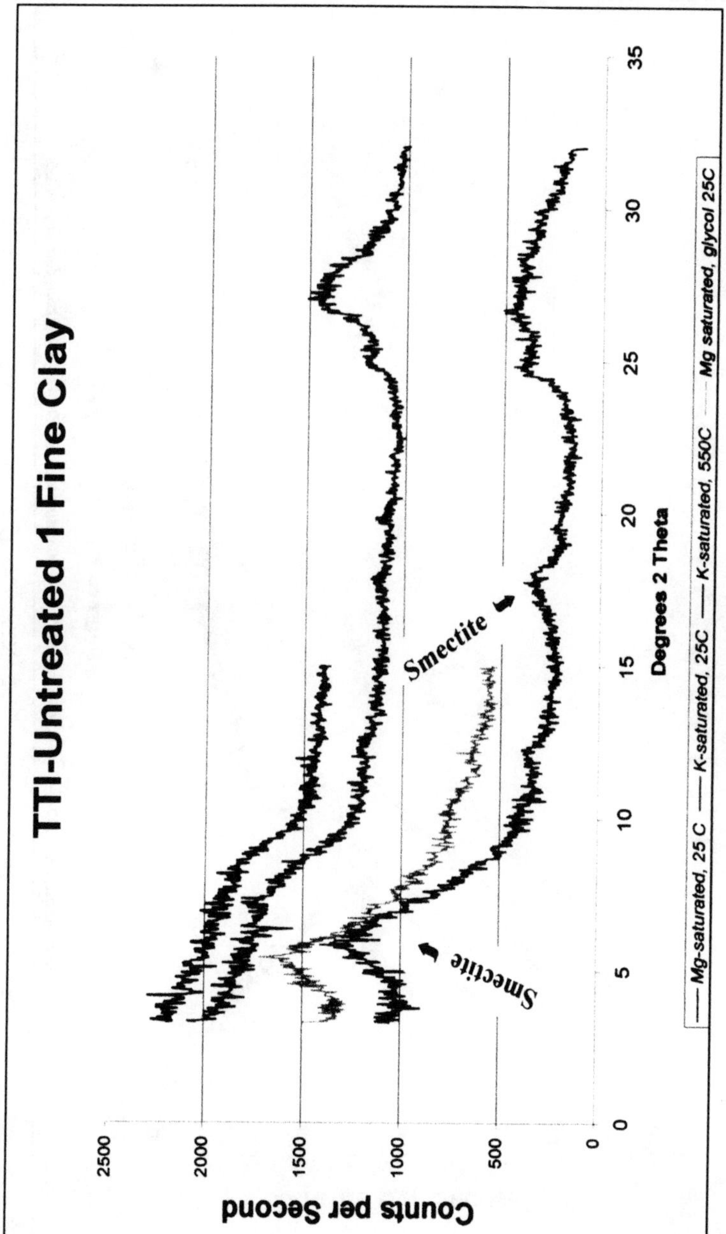

Figure 3 - XRD Patterns of Untreated Fine Clay

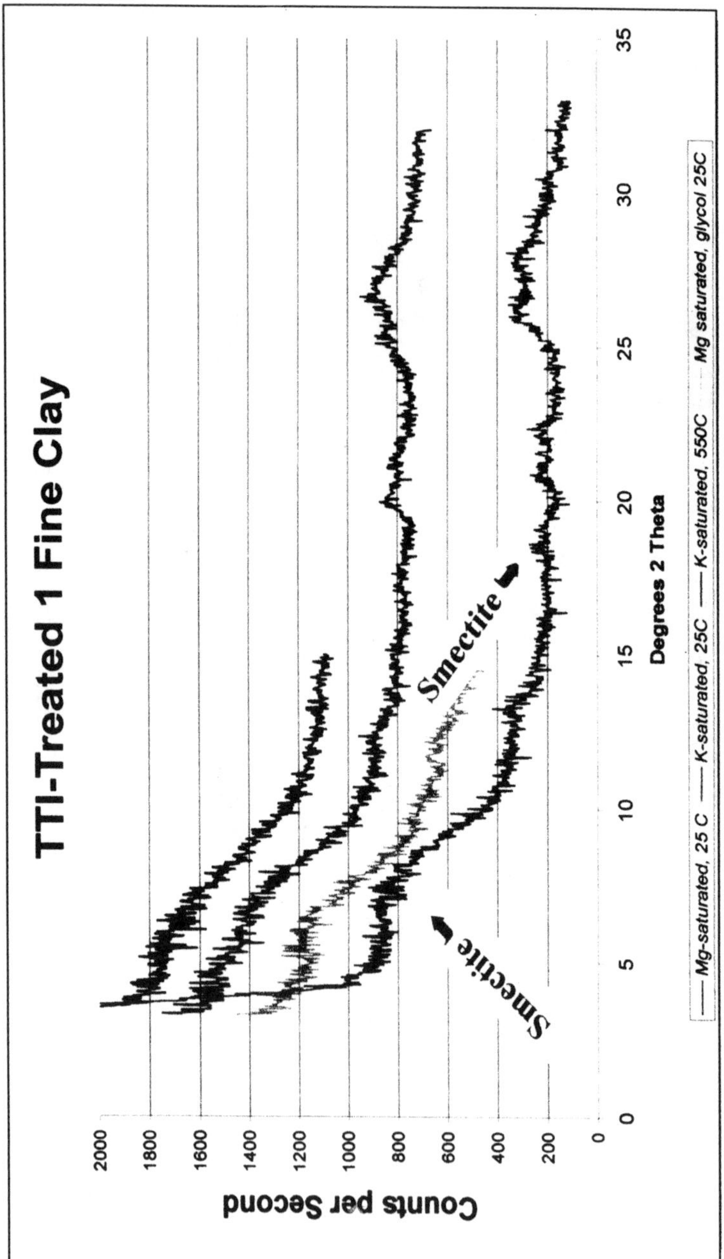

Figure 4 - XRD Patterns of HIExC-Treated Fine Clay

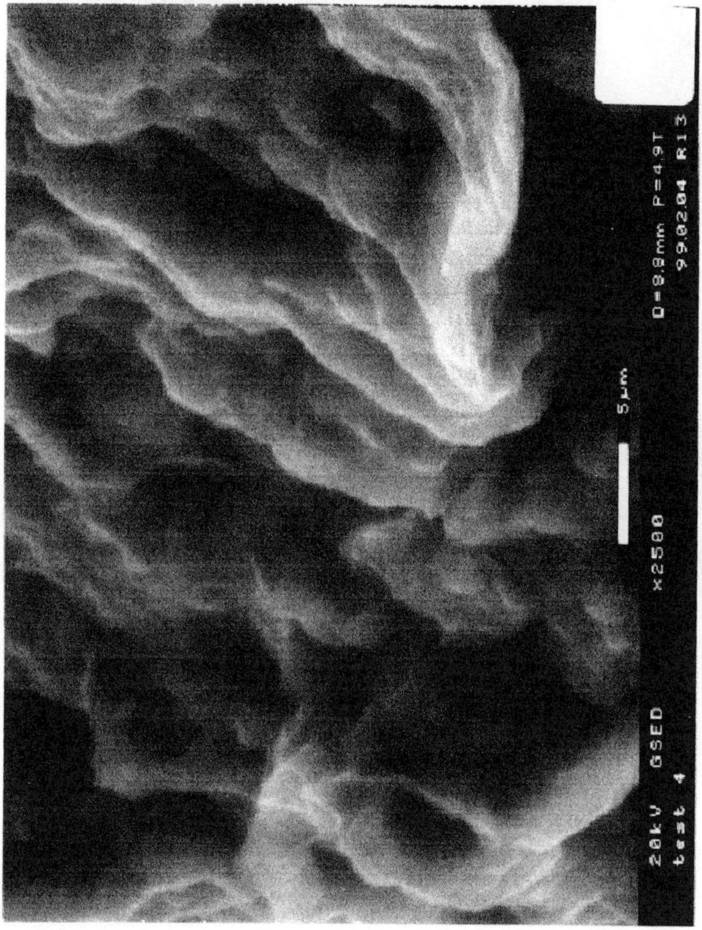

Figure 5 - ESEM of Untreated Clay

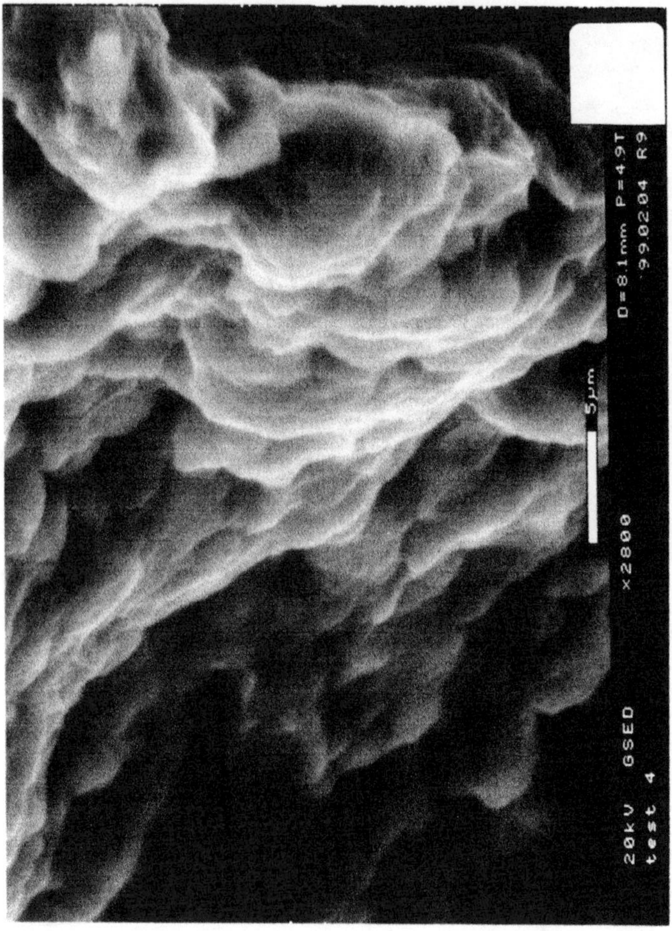

Figure 6 - ESEM of HIExC-Treated Clay

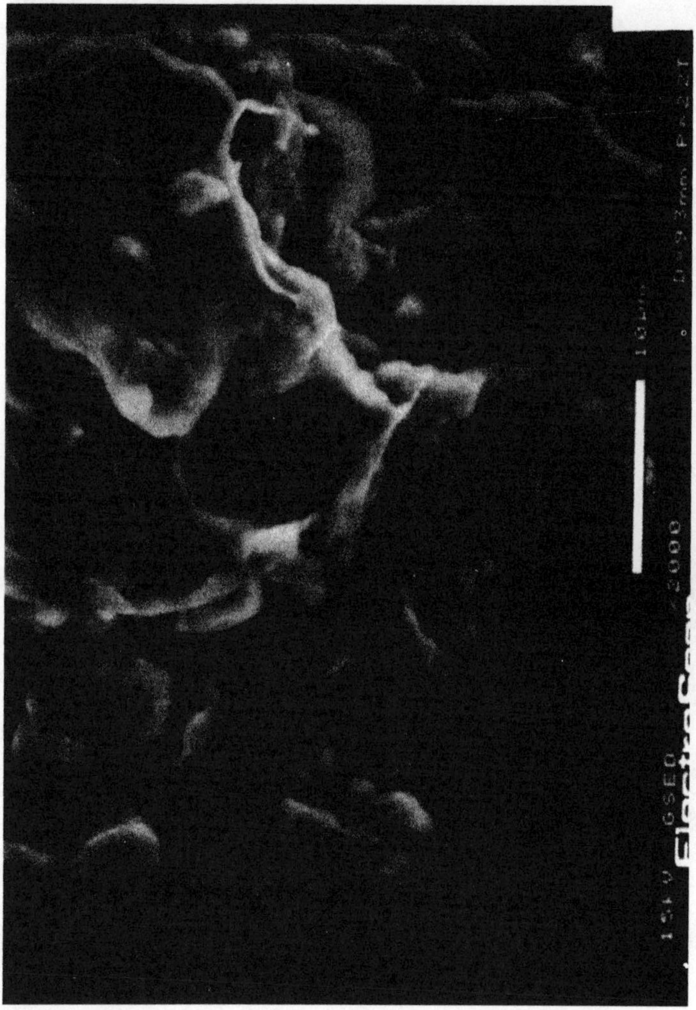

Figure 7 - ESEM of Water-Treated Clay

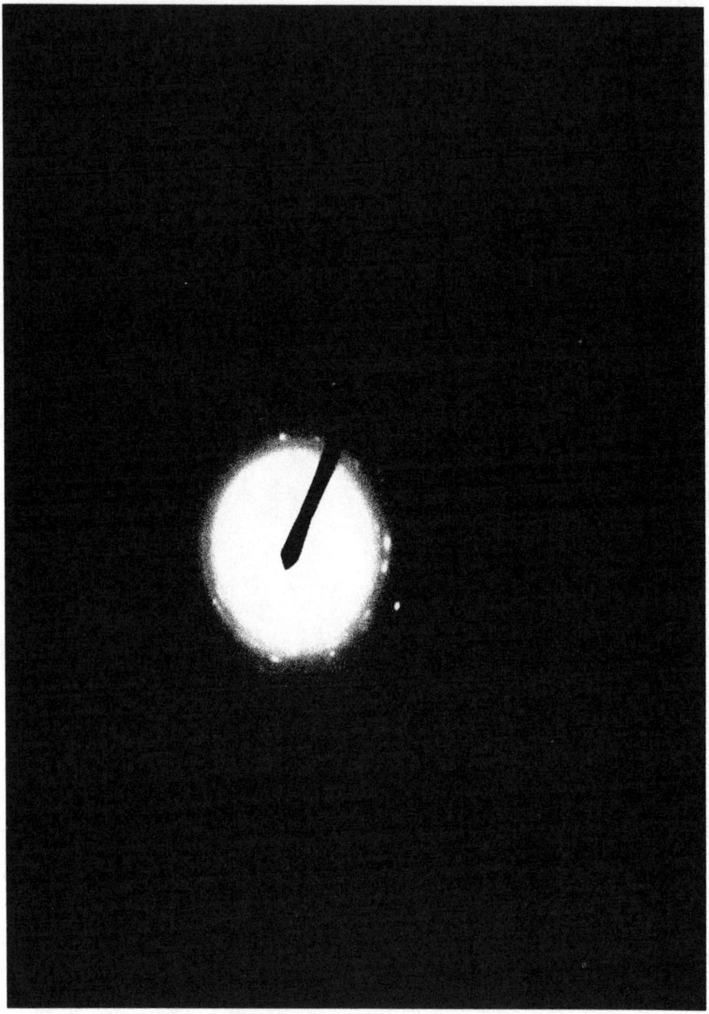

Figure 8 - TEM/EDXA of Untreated Clay

Figure 9 - TEM/EDXA of HIExC-Treated Clay

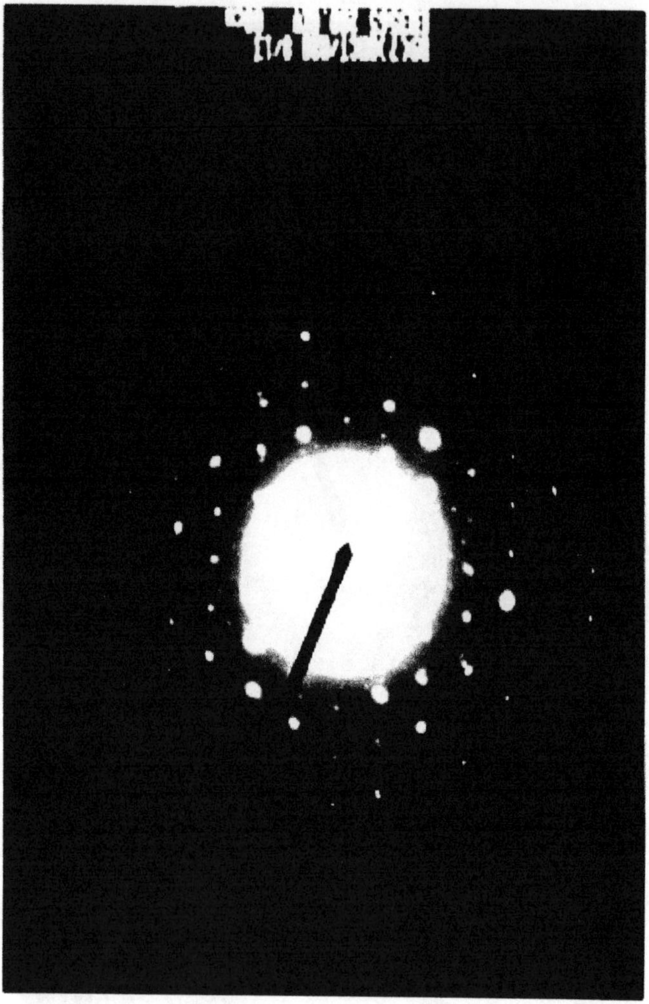

Figure 10 - TEM/EDXA of Water-Treated Clay

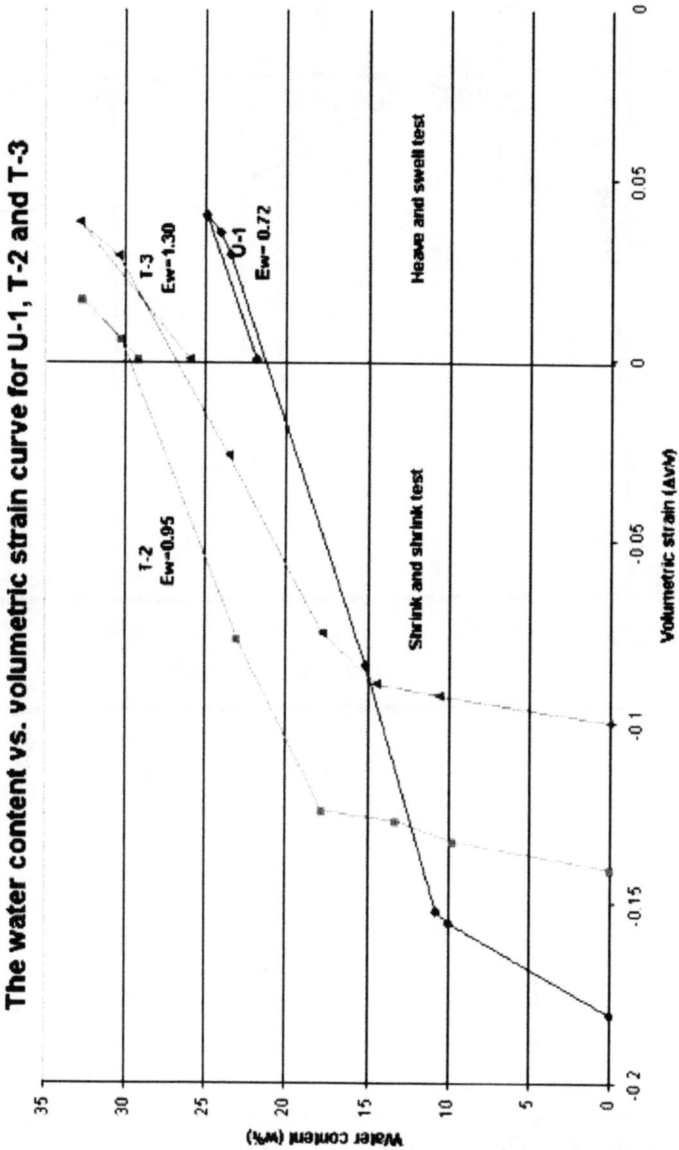

Figure 11 - Moisture Content versus Volume Change for Treated and Untreated Clays

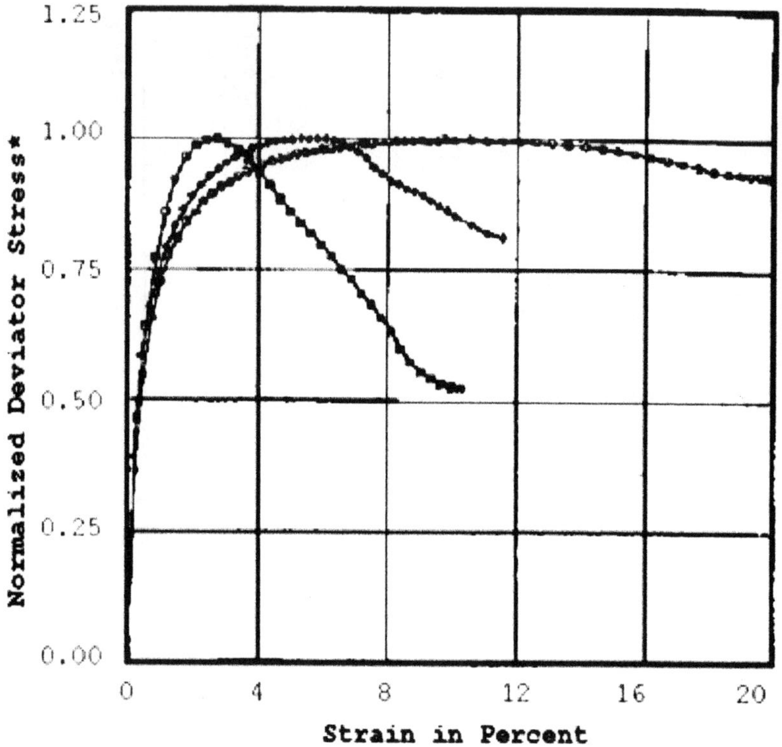

Curve	Sample No.	Depth (ft)	Test Type	Confining Pressure (psi)	Maximum Deviator Stress (ksf)	ε_{50} (%)
O—O	U-1	2.0	UU	4.0	5.29	0.4
□—□	T-3	2.0	UU	4.0	4.14	0.3
◆—◆	T-2	2.0	UU	4.0	7.23	0.3

*Normalized with respect to maximum deviator stress

Figure 12 - Stress-Strain Curves from Unconsol.-Undrained Triax. Compression Tests for Treated and Untreated Clays

Air Entrapment Effects on Hydraulic Properties

Katia Vanessa Bicalho[1], Dobroslav Znidarcic [2], Member, ASCE, and Hon-Yim Ko [2], Member, ASCE

Abstract

The purpose of this paper is to show that the soil-water characteristic and unsaturated hydraulic conductivity functions derived for soils with continuous air channels, in which the air is assumed to exist at constant atmospheric pressure, should not be used for simulating infiltration process. Ongoing experimental work is oriented towards investigating the influence of the entrapped air on the hydraulic constitutive relations of a compacted fine-grained soil imbibing liquid. Testing is being conducted in two phases to determine (1) soil-water characteristic function and (2) unsaturated hydraulic conductivity function for systems involving discontinuous air phase. Hydraulic constitutive functions for soils with discontinuous air phase are proposed using the obtained experimental results.

Introduction

Knowledge of soil-water characteristic and unsaturated hydraulic conductivity functions is essential when modeling unsaturated flow in practical applications, including clay liners in waste containment and compacted clay cores in earth dams. Due to the relative difficulty associated with the measurement of unsaturated hydraulic conductivity, the soil-water characteristic curve is often used as the basis for the prediction of the unsaturated hydraulic conductivity function (Burdine, 1953; Kunze et al., 1968; Mualem, 1976; Fredlund et al., 1994). Therefore, it is important to have a reasonably accurate characterization of the soil-water characteristic curve. The soil-water characteristic curve for a soil is defined as the relationship between matric suction (also known as capillary pressure) of the soil (i.e., the pressure difference between air and water) and volumetric water

[1] Civil Engineering Department, Federal University of Espirito Santo, Centro Tecnologico, Caixa Postal 01-9011, CEP 29060-970, Vitoria, ES, Brazil

[2] Department of Civil, Environmental and Architectural Engineering, University of Colorado, Boulder, CO 80309-0428

content. In geotechnical engineering practice and the petroleum literature, it is customary to plot capillary pressure as a function of the degree of saturation. This relationship exhibits a marked hysteresis, with the wetting (or absorption) curve differing from the drying (or desorption) curve. A number of equations have been proposed to empirically produce the best fit to the soil-water characteristic curve (Brooks and Corey, 1964; van Genuchten, 1980). Modeling efforts have also been made to estimate the soil water characteristic curve from particle size distribution (Arya and Paris, 1981) or pore size distribution data of the soil (Fredlund and Xing, 1994). Although most of the models were developed to represent the drying soil-water characteristic curve, they also are applied routinely to describe wetting soil-water characteristic curves required for simulating an infiltration process (Chiu and Shackelford, 1998). However, at higher degrees of saturation on the wetting cycle the air phase becomes discontinuous with isolated air pockets within the pore space which have no connection over macroscopic distances (occluded air bubbles or entrapped air). Considering that the measurement of pore air pressure in an occluded sample is extremely difficult, the relationship between capillary pressure and degree of saturation for discontinuous air phase is not well defined. To the authors' knowledge, none of the existing predictive soil-water characteristic models has been developed or evaluated for a porous medium containing a discontinuous air phase. It will be shown in the following discussion that whereas an increase in the capillary pressure decreases the saturation values in the existing models for continuous air phase, in the occluded zone the saturation increases in response to the increase in the capillary pressure. Therefore, the values of the capillary pressure, P_c, at higher degrees of saturation on the wetting soil-water characteristic curve can not be replaced simply by decreasing P_c. The assumption of the air pressure being equal to the atmospheric pressure in the occluded zone should be avoided and the predictive models for unsaturated hydraulic conductivity based on drying soil-water characteristic curve should not be used for simulating an infiltration process.

Soil-Water Characteristic Function

Based on analysis of a large number of experimental data, Brooks and Corey (1964) presented the following empirical expression for describing the soil-water characteristic function:

$$P_{c1}(S) = \frac{P_b}{(S_e)^{\frac{1}{\lambda}}}, \qquad P_{c1}(S) > P_b \tag{1a}$$

$$P_{c1}(S) = P_b, \qquad P_{c1}(S) \leq P_b \tag{1b}$$

where $P_{c1}(S)$ is the relationship between P_c and S considering continuous air channels, P_b and λ are characteristic constants of the medium, S_e is the effective saturation, defined as $S_e = \frac{(S - S_r)}{(1 - S_r)}$ and S_r is the residual saturation. Although the function was originally developed for porous media not involving discontinuous air phase (drying soil-water characteristic curve), it is also applied routinely to describe wetting soil-water characteristic curve.

Corey (1994) and Chiu and Shackelford (1998) recommend that the value of S_e should be redefined as $S_e = \frac{(S - S_r)}{(S_m - S_r)}$ in the case of infiltration (wetting) processes where the soil is typically not completely saturated due to the presence of entrapped air. The value of S_m, which is the correspondent degree of saturation where the air starts to be entrapped, is considered the maximum field saturation (Corey, 1994). Corey indicates that $0.80 \leq S_m \leq 0.92$ for infiltration (wetting processes), which is consistent with the range of S_m values reported by Chiu and Shackelford (1998), i.e., $0.84 \leq S_m \leq 0.90$. However, Bloomsburg and Corey (1964) show that the porous medium will eventually become fully saturated by imbibition even though there is no liquid flowing through the porous material. Moreover, compacted soils at great depths below the phreatic surface (i.e., subjected to higher pressures) may have the saturation in the occluded zone increased to full saturation. Therefore, it is necessary to define a relationship between the capillary pressure and saturation in the occluded zone (i.e., $S \geq S_m$).

At higher degrees of saturation on the wetting soil-water characteristic curve, where the air phase is discontinuous, the pore air pressure may differ significantly from the atmospheric air pressure. Because the measurement of pore air pressure, u_a, in an occluded sample is extremely difficult, the air pressure must be predicted. The theoretical relationship between air and water pressures proposed by Schuurman (1966), considering a porous medium containing occluded air bubbles, will be adopted in this study. Schuurman considered an undrained situation and could not obtain an explicit expression for the relationship between capillary pressure and saturation. In this study, consideration of zero variation in the volume of void space (i.e., the volume of the air-water mixture is compensated by adding water) leads to a final expression for the wetting soil-water characteristic function ($S \geq S_m$).

A relationship between the pore-water pressure, u_w, and the degree of saturation, S, involving discontinuous air phase is presented. The theoretical constitutive relationship is determined by using Boyle's law and Henry's law of solubility, and the Kelvin equation for pressure difference across the interface between the air and water phases. Schuurman (1966) also assumes that all spherical air bubbles are of the same size and the number of air bubbles within the pore space is constant. According to Henry's law the amount of dissolved air only depends on the air pressure in the bubbles. Hilf (1956) concluded that a uniform distribution of dissolved air requires an identical air pressure in all the bubbles. Since the interfacial tension is assumed constant, the equilibrium condition demands that free air bubbles in a uniform fluid have the same radius. Therefore, assuming that the volume of the air-water mixture is compensated by adding water (i. e., there is no variation in the volume of void space) the proposed u_w-S relationship is given by (Bicalho, 1999):

$$u_w = \frac{[1 + S_0(H-1)]u_{a0}}{S(H-1)+1} - (u_{a0} - u_{w0})\left(\frac{(1-S_0)}{(1-S)}\right)^{1/3}, \quad S \geq S_m \tag{2}$$

u_w and S are the variables in (2) while u_{w0}, u_{a0} and S_0 are the initial values of air pressure, water pressure and degree of saturation, respectively, H is the dimensionless coefficient of solubility (equals 0.02 at 20 ^{0}C) and S_m is the correspondent degree of saturation where the air starts to be entrapped. The initial air pressure is assumed to be the atmospheric pressure. A compacted cohesive soil contains air that may initially be under pressure due to the compaction process. However, if the soil is not immediately sealed, the air pressure will soon become atmospheric (Hilf, 1956).

This relationship is in terms of absolute pressures. For the fully saturated state (S = 1) the function has a singularity with water pressure tending towards - ∞. This indicates that prior to their disappearance, the bubbles become unstable, collapsing due to the capillary effect. This phenomenon has yet to be explored experimentally. While it may be important at the microscopic scale for individual air bubbles, it is unlikely that it would have a major effect on the macroscopic behavior of soils. Not all air bubbles would collapse at the same time, so the singularity will be smoothed out. Once the bubbles disappear, the difference between the air pressure and the water pressure no longer exists.

Based on (2), the relationship between capillary pressure, P_c, and the degree of saturation, S, at higher degrees of saturation on the wetting cycle is given by

$$P_c = P_{c2}(S) = u_a - u_w = (u_{a0} - u_{w0})\left[\frac{(1-S_0)}{(1-S)}\right]^{1/3}, \quad S \geq S_m \tag{3}$$

where P_{c2} (S) is the relationship between P_c and S considering discontinuous air phase.

Fredlund and Rahardjo (1993) recommend not to apply Kelvin's equation to the relationship between the pore-air and pore-water pressures in the occluded zone due to conceptual difficulties of Kelvin's equation (i.e., a slight increase in total stress could cause a progressive process which would reduce the air bubbles to an infinitesimal size while the capillary pressure goes to infinity). It is suggested that the pore-air and pore-water pressures be assumed to be equal in the occluded zone (i.e., disregarding surface tension, Bishop and Eldin, 1950). The effect of the Kelvin's equation on Schuurman's theory is very small (Bicalho, 1999). Of course, because the air-water interface is curved, the air pressure inside the bubble will be larger than that of the surrounding water. Therefore, it is necessary to take into account the surface tension at the interface between air and water. This effect should not be dismissed in a proper description of the behavior of unsaturated soil (Barends, 1979). Whereas the increase in the capillary values decreases the saturation values in the existing models for continuous air phase, in the occluded zone the saturation increases in response to the increase in the capillary pressure (3). Therefore, the values of the capillary pressure, P_c, at higher degrees of saturation on the wetting curve ($S \geq S_m$) can not be replaced simply by assuming P_c equal to zero. Considering that the measurement of pore-air pressure in an occluded sample is extremely difficult, Schuurman's theory is an attractive tool for defining the soil-water characteristic in the occluded zone. If the Kelvin's equation is omitted from the model, and the air and water pressures are assumed to be equal, the suction at the moment of occlusion cannot be properly taken into consideration.

The ability of a theoretical expression to match experimental results is a useful indicator of the predictive capability of the proposed theory. A new experimental procedure for measurement of the wetting soil-water characteristic curve in the occluded zone has been proposed (Bicalho, 1999). The measurements were accomplished by using a modified triaxial cell connected to a flow pump. The system includes pore-water back pressure facilities, a differential pressure transducer, a differential mercury manometer and the data acquisition system, Figure 1. The standard triaxial cell was designed to accept a high air entry value porous ceramic plate at the bottom platen bellow the soil sample. The flow pump enables a precise control of water flow in and out of the soil sample. The flow pump forces water through the soil sample at a constant predetermined rate and creates a pressure difference across the sample that is continuously recorded with a precision differential pressure transducer connected to a personal computer-based data acquisition system. The experimental procedure proposed by Znidarcic et al. (1991) was used to obtain the soil-water characteristic curve involving continuous air phase. Once water appeared at the top of the sample in the wetting cycle, an impervious top boundary was created for measuring the wetting soil-water characteristic involving discontinuous air phase. The pore-water pressures versus saturation (or volumetric water content) were obtained from the pressure-time data recorded. The assumption is made that the degree of saturation is uniform within the sample for the total duration of the experiment. This is a reasonable assumption since small samples were used (diameter = 71 mm and thickness ≤ 30 mm) and a very low Darcian water velocity was applied in the test ($V_w = 3.3 \times 10^{-8}$ m/s). The validity of this assumption was verified by numerical analysis. A variety of tests were performed for verification of the occluded state theory proposed in this study (Bicalho, 1999). The results of the experimental investigation indicated that the theory is in reasonable agreement with experiment for the soil tested. The experiments conducted indicate that the degree of saturation that marks the change in the condition of the pore fluid (occlusion), S_m, is around 70 % for the soil investigated. This soil was a silt material from a sedimentary formation at the Bonny Dam Site, in eastern Colorado. The index properties for Bonny Silt are LL = 25%, PL = 21%, clay fraction 12%, Casagrande classification CL-ML, specific gravity 2.63, Proctor optimum water content 14.5% and maximum dry density 17.5 kN/m^3 . The samples were compacted at their molding water content in a two-part split metal mold using a static compaction press.

As shown in Figure 2, the measured drying soil-water characteristic curve is fitted with the Brooks and Corey function by using the following values: $P_b = 14.4$ kPa, $\lambda = 0.46$ and $S_r = 0.3$. The term suction is used to indicate the negative pressure of water relative to atmospheric air pressure, i.e., $-(u_w - u_{atm})$. The complete drying curve was not measured because the objective of this study was to investigate the soil-water characteristic function at higher degrees of saturation on the wetting cycle. Therefore, insufficient data were available to define precisely the shape of the soil-water characteristic curve on the drying cycle. Since the complete drying curve was not measured, the main wetting soil-water characteristic curve could not be determined and it is assumed to be defined by the Brooks and Corey function using the value of S_e determined by using S_m (Chiu and Shackelford,

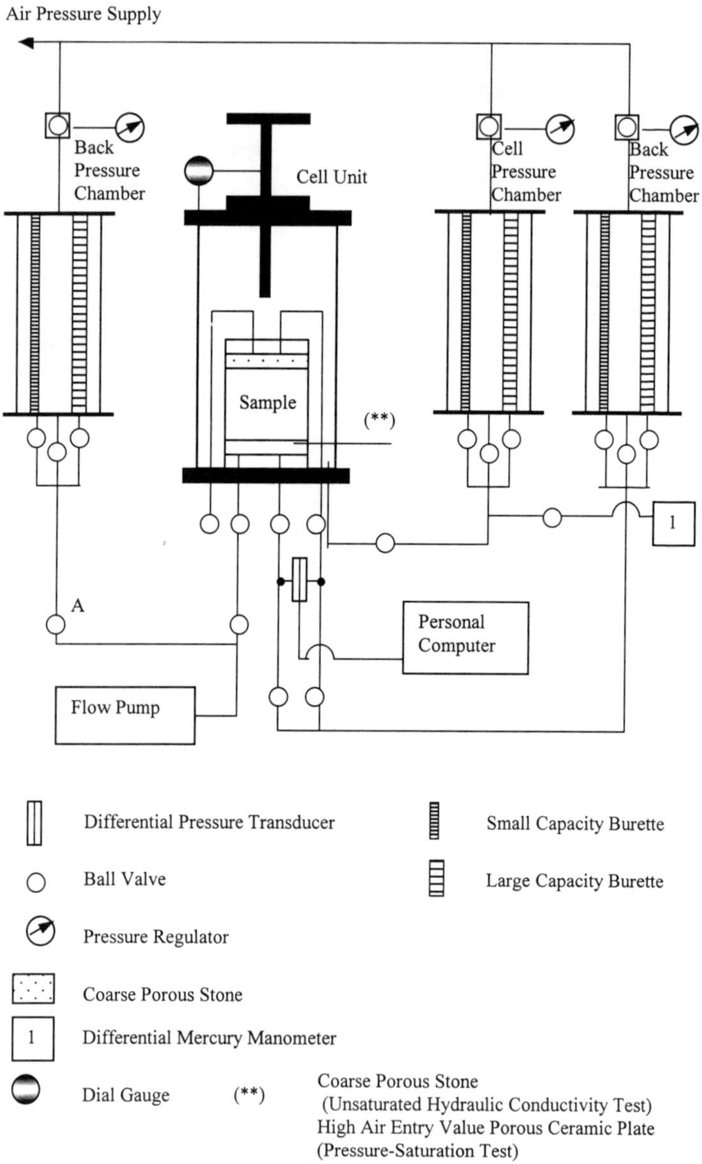

Figure 1. Basic Apparatus – Laboratory Measurements

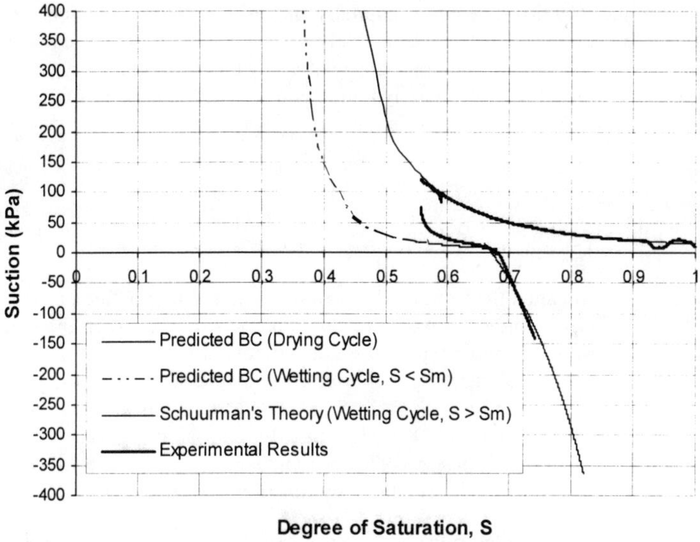

Figure 2. Measured and Fitted Soil-Water Characteristic Curves for Compacted Bonny Silt based on Brooks and Corey Function and Schuurman's Theory

1998). The main wetting curve was defined by the following values: P_b = 7 kPa, $\lambda = 0.46$ and $S_r = 0.3$ and $S_m = 0.7$. The value of S_r was assumed to be the same for drying and wetting and was obtained from the drying branch of the curve. The measured data of suctions at higher degrees of saturation on the wetting curve, $S \geq S_m$, are used for verifying the theoretical relationship between the pore-water pressure and the degree of saturation (2) derived by using the occluded state theory proposed by Schuurman (1966). The occluded state theory adopted in this study has been found to be satisfactory in predicting the relationship between the pore-water pressure and saturation considering the presence of discontinuous air phase as presented in Figure 2.

Unsaturated Hydraulic Conductivity Function

For a given soil, assuming that the fabric does not change significantly, the unsaturated hydraulic conductivity is affected by combined changes in the void ratio (or porosity) and the degree of saturation (or water content) of the soil. Whereas the influence of void ratio on hydraulic conductivity is well documented, there is a strong need to develop a better understanding of how the water and air hydraulic conductivities vary with the saturation (Lloret and Alonso, 1980).

Burdine (1953) derived a capillary model for the prediction of the unsaturated hydraulic conductivity from the basic law of fluid flow in porous media (Poiseuille's law). Burdine's equation for relative hydraulic conductivity, which is defined as the ratio of the effective hydraulic conductivity at a given saturation to the saturated hydraulic conductivity, is given by:

$$k_r(S) = \frac{\dfrac{1}{T_S}\displaystyle\int_{S\min}^{S}\frac{dS}{P_c^2(S)}}{\dfrac{1}{T_1}\displaystyle\int_{S\min}^{1}\frac{dS}{P_c^2(S)}} = = \left(\frac{T_1}{T_s}\right)\frac{\displaystyle\int_{S\min}^{S}\frac{dS}{P_c^2(S)}}{\displaystyle\int_{S\min}^{1}\frac{dS}{P_c^2(S)}} \tag{4}$$

where T_s has been called the tortuosity at saturation S, and it is defined as $(L_e/L)^2$, L_e is the length of the tortuous tube and L the direct distance between the ends of the tube and P_c (S) is the capillary pressure at a particular saturation S.

The possibility of deriving a closed-form expression for the hydraulic conductivity by substituting a soil-water characteristic function into the capillary model proposed by Burdine (1953) makes the Burdine approach very attractive (Brooks and Corey, 1964; Mualem, 1976; van Genuchten, 1980). The Brooks and Corey-Burdine (BC-B) hydraulic conductivity function is given as follows:

$$k_r(S) = (S_e)^{\frac{2+3\lambda}{\lambda}}, \qquad P_{c1}(S) > P_b \tag{5a}$$

$$k_r(S) = 1, \qquad P_{c1}(S) \leq P_b \tag{5b}$$

where P_b and λ are defined as in (1).

Although Burdine's theory is concerned primarily with desorption or drainage of liquid and was derived for systems not involving discontinuous air phase, it is realized that these hydraulic conductivity functions are widely used for modeling infiltration into unsaturated soil as well. At some higher degree of saturation on the wetting cycle the air phase becomes discontinuous and consists of isolated air bubbles. The air pressure is different from the atmospheric pressure when the air is present in the form of occluded bubbles. Therefore, the relationship between the relative hydraulic conductivity, k_r, and degree of saturation, S, defined for continuous air channels may not be adequate for occluded air bubbles. It must be emphasized that whereas the increase in the capillary values decreases the saturation values in the proposed models for continuous air phase, in the occluded zone the saturation increases in response to the increase in the capillary pressure. Therefore, the capillary effects (3) introduced by the Kelvin's equation into the Schuurman's model cannot be used as integral functions in (4).

A new testing technique for measuring unsaturated hydraulic conductivity in the occluded zone is presented in more detailed in Bicalho (1999). The basic apparatus used for measuring unsaturated hydraulic conductivity is the same as used for measuring the soil-water characteristic function. The only modification is in the triaxial cell where a coarse porous stone was used at the bottom of the sample replacing the high air entry value porous ceramic plate used for measuring the soil-water characteristic function. The experimental procedure is based on the concept of saturation of soils by increasing the pressure in the pore fluid. The test consists of application of a back pressure in the pore-water lines concurrently with application of an equal increment of the cell pressure. The increments in chamber pressure and back pressure are applied slowly to allow adjustment of the water content under each increment before the application of the next increment. The hydraulic conductivity measurements are performed for each increment of back pressure applied to the pore-water system by using the flow pump experimental technique as discussed by Aiban and Znidarcic (1989). The flow system is treated as a continuous succession of steady states. This procedure is continued until no change in the hydraulic conductivity is found for an increase in pore pressure. It is assumed that the entrapped air is compressed and dissolved and the measured hydraulic conductivity corresponds to the saturated hydraulic conductivity for the tested sample. The theory proposed by Schuurman (1966), which was verified by comparison with detailed experimental data, was used for predicting the increase in the degree of saturation due to the increase in the pore-water pressure. Laboratory measurements were also conducted to confirm that the assumption of no variation in the volume of void space was appropriate (i.e., experimental results – phase B in Figure 3). During the tests conducted in phase B, several cycles of loading and unloading were applied prior to the hydraulic conductivity measurements to minimize total volume changes during the tests. The specimen's height was measured by using a dial gauge and a negligible change in the height of the sample was observed (i.e., $\Delta H < 0.02$ mm). Therefore, a better control of volume's change was observed on the tests conducted in phase B than on the tests conducted in the preliminary phase (i.e., phase A).

Figure 3 presents a dimensionless plot of relative hydraulic conductivity versus degree of saturation for experiments conducted on 5 compacted samples of Bonny Silt with the porosity ranging from 35% to 39%. It is important to mention that the tests were conducted on the same soil compacted at different molding water contents that may result in different soil structures and therefore different hydraulic conductivity functions. The relative hydraulic conductivity increases sharply when the saturation increases from 0.9 to 1. It has a very small value ($k_r \approx 0.1$) while the saturation is relatively high ($S > 0.8$).

Using the experimental results thus obtained, an empirical expression for defining $k_r - S$ relationship at higher degrees of saturation on the wetting cycle was determined by curve fitting to the experimental results and is given by:

$$k_r(S) = a + (1-a)[\frac{(S-b)}{(1-b)}]^n \,, \qquad S \geq S_m \tag{6}$$

where a, b and n are empirical parameters.

The empirical parameters for the proposed hydraulic conductivity function (6) were obtained by calibrating a numerical simulation of the transient water flow to measurements. The obtained empirical parameters for the hydraulic conductivity function for the experimental results (phase B) are: a = 0.22, b = 0.7 and n = 5.7. The proposed expression (6) has been found to be most satisfactory in predicting the transient water flow measurements as presented in the numerical analysis (Bicalho, 1999). The $k_r - S$ relationship using BC-B functions resulting from fitting the drying and wetting soil-water characteristic curves are the same since the same λ and S_r values are used for both curves. The approach proposed by Corey (1994)

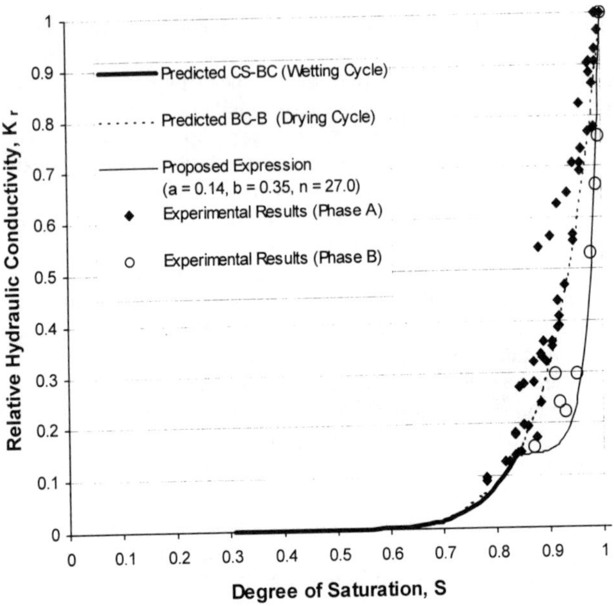

Figure 3. Measured and Predicted Unsaturated Hydraulic Conductivity Functions for Compacted Bonny Silt

and Chiu and Shackelford (1998) (i.e., CS-BC, with the value of S_e redefined as $S_e = \frac{(S - S_r)}{(S_m - S_r)}$) was used for predicting the k_r – S relationship for the wetting cycle. In that approach the value of hydraulic conductivity corresponding to S_m is considered the maximum hydraulic conductivity value for the infiltration processes. Therefore, for comparisons with the other functions where k_r is defined with respect to the saturated hydraulic conductivity, the maximum hydraulic conductivity on the predicted CS-BC (wetting cycle) was selected to coincide with the hydraulic conductivity defined for $S = S_m$ using (6) and the stated parameters. The experimental results (phase B) and the proposed function suggest that the BC-B function tends to overpredict the measured k_r values. The proposed function also shows sharp increases in k_r values when S increases from 0.9 to 1 and small change in the k_r values for $0.80 \leq S \leq 0.90$. However, a broader investigation of the k_r - S relationship for discontinuous air phase is needed before a definite conclusion could be made. Direct measurements of unsaturated hydraulic conductivity function should always be preferred over the prediction using the idealized models.

Summary and Conclusions

Following the occluded state theory proposed by Schuurman (1966), a theoretical soil-water characteristic function involving discontinuous air phase is developed. A new testing technique is developed to provide reliable data of the relationship between the pore-water pressure and saturation involving discontinuous air phase. The experimental results indicated that Schuurman's theory can be used to describe the relationship between the pore-water pressure and degree of saturation considering a porous medium containing occluded air bubbles.

A new method for the measurement of unsaturated hydraulic conductivity where the air is present in the form of bubbles is described. The method is based on the concept of saturation of the soils by increasing the back pressure. The developed relationship between the pore-water pressure and saturation involving discontinuous air phase is used for determining the increase in saturation due to the increase in the back pressure. An empirical hydraulic conductivity function determined by curve fitting the experimental measurements leads to the conclusion that the relative hydraulic conductivity increases sharply when the saturation increases from 0.9 to 1 and it has a very small value ($k_r \approx 0.1$) while the saturation is relatively high ($S > 0.8$).

Acknowledgments

The first author is grateful for sponsorship by the Brazilian government agency CNPq.

References

Aiban, S. A., and Znidarcic, D. (1989). "Evaluation of the Flow Pump and Constant Head Techniques for Permeability Measurements." Geotechnique, 39 (4), 655-666.

Arya, L.M., and Paris, J. F. (1981). "A Physicoempirical Model to Predict the Soil Moisture Characteristic from Particle-Size Distribution and Bulk Density Data." Soil Sci. Soc. Am. J., 45, 1023-1030.

Barends, F. B. J. (1980). "Nonlinearity in Groundwater Flow." Ph. D. Dissertation, Delft, Holland.

Bloomsburg, G.L., and Corey A. T. (1964). "Diffusion of Entrapped Air from Porous Media." Colorado State University, Hydrology papers, No. 5, 1-27.

Bicalho, K. V. (1999). "Modeling Water Flow in an Unsaturated Compacted Soil." Ph.D. dissertation, University of Colorado, Boulder, Colo.

Bishop, A.W., and Eldin, A. K. G. (1950). "Undrained Triaxial Tests on Saturated Sands and Their Significance in the General Theory of Shear Strength." Geotechnique, 2, 13-32.

Brooks R. H., and Corey A. T. (1964). "Hydraulic Properties of Porous Media." Colorado State University, Hydrology papers, No. 3, 1-27.

Burdine, N. T. (1953). "Relative Permeability Calculation size distribution data." Transactions of the American Institute of Mining, Metallurgical and Petroleum Engineers, 198,71-78.

Chiu, T.-F., and Shackelford, C. D. (1998). "Unsaturated Hydraulic Conductivity of Compacted Sand-Kaolin Mixtures." J. Geotech. Engrg, ASCE, 124(2), 160-170.

Corey, A.T. (1994). "Mechanics of Immiscible Fluids in Porous Media." 3rd Ed., Water Resources Publications, Highlands Ranch, Colo.

Fredlund D. G., and Rahardjo H. (1993) "Soil Mechanics for Unsaturated Soils", John Wiley & Sons, Inc., New York, N.Y.

Fredlund, D. G., and Xing, A. (1994) " Equations for the Soil-water characteristic Curve." Can. Geotech. J., Ottawa, Canada, 31 (4), 521-532.

Fredlund, D. G., Xing, A., and Huang S. (1994) "Predicting of the Permeability Function for Unsaturated Soil using the Soil-water characteristic Curve". Can.Geotech. J., Ottawa, Canada, 31(4), 533-546.

Hilf, J. W. (1956). "An Investigation of Pore-Water Pressure in Compacted Cohesive Soils." Ph.D. dissertation, University of Colorado, Boulder, Colo.

Kunze, R. J., Uehara, G., and Graham, K. (1968). "Factors Important in the Calculation of Hydraulic Conductivity." Proc. Soil Sci. Soc. Amer., 32, 760-765.

Lloret, A. M., and Alonso, E.E. (1980). "Consolidation of Unsaturated Soils including Swelling and Collapse Behaviour." Geotechnique, 30 (4), 449-477.

Mualem, Y. (1976). "A New Model for Predicting the Hydraulic Conductivity of Unsaturated Porous Media." Water Resources Research, 12 (3), 513-522.

Schuurman, E. (1966). "The Compressibility of an Air/Water Mixture." Geotechnique, 16 (4), 269-281.

van Genuchten, M. Th. (1980) "A closed-form Equation for Predicting the Hydraulic Conductivity of Unsaturated Soils." Soil Sci. Soc. Am. J., 44 (5), 892-898.

Znidarcic D., Illangasekare T., and Manna M. (1991). "Laboratory Testing and Parameter Estimation for Two-Phase Flow Problems." ASCE, Geotechnical Special Publication, 27, 1089-1099.

Incorporating Near-Surface Processes in Modeling Moisture Movement in Soils

Carl E. Morris[1] MASCE
John C. Stormont[2] MASCE

Abstract

The movement of moisture into and through near-surface soils is of major importance in numerous areas of geotechnical and geoenvironmental engineering, impacting soil strength, contaminant transport, efficiency of waste containment systems, and slope stability. Numerical models are often used to simulate moisture movement and behavior of soil systems due to the cost, complexity, and time required to undertake field studies. However, these simulations often ignore or grossly simplify near-surface processes such as precipitation, snowmelt, run-off, and evapotranspiration. It is not uncommon to see complex systems simulated using a one-dimensional model in which the surface moisture flux for the site is represented by a steady-state infiltration rate that is based on the annual average precipitation. In this paper we examine the impact of incorporating non-steady state models for near surface processes such as evapotranspiration, precipitation, and surface runoff on simulations of soil moisture movement in capillary barrier systems. Simulations with and without transient near surface processes have been conducted for a variety of sites with different climates. These sites include a semi-arid climate, a site where the bulk of the precipitation occurs during the winter months, and a wet, humid climate where rainfall exceeds 2 meters per year. The results demonstrate that simplification or omission of near-surface processes can lead to erroneous results.

Introduction

The infiltration and movement of moisture in near-surface soils is of importance in determining the behaviour of natural and engineered systems.

[1] Senior Lecturer, Faculty of Engineering, University of Wollongong, Wollongong, NSW, 2522, Australia
[2] Associate Professor, Department of Civil Engineering, University of New Mexico, Albuquerque, NM, 87131

Moisture content influences factors such as the strength of the soil, the rate at which water and contaminants will migrate, the remaining storage capacity of the soil system, and the effectiveness of capillary barriers to prevent downward water movement. In systems designed to exclude moisture from buried wastes or to increase soil shear strengths, the management of the soil moisture content is of great importance. In the case of waste containment systems the elimination or reduction of the flux of water through the waste zone eliminates or reduces the associated leaching of contaminants into the vadose zone and groundwater. Similarly, soil shear strength is not reduced in systems that have low moisture contents.

Numerical simulations are often used in the design process to determine the flow of moisture into and through soil systems. Numerical modelling is quick and relatively inexpensive compared to laboratory and field tests, and allows the user to simulate many variables to develop suitable designs. There is a wide range of models that can be used to simulate infiltration and moisture movement, ranging from those that are capable of handling only steady-state, saturated flow in one-dimension, to models that are capable of multi-phase, three dimensional, transient unsaturated flow. In addition to choosing a model suitable for the simulation to be conducted in terms of dimensions and flow characteristics, the modeller must also ensure that near-surface phenomena are correctly accounted for in the modelling process. Major near-surface processes that can strongly influence moisture movement are precipitation, evapotranspiration (ET), runoff, and snowmelt. These transient parameters are often either ignored or average annual values are used in the simulations.

In some locations and for some problems this approach may be reasonable and justifiable, but in many instances can lead to significant errors. In general, ET is usually greatest during the growing season and is significantly reduced during the winter and, sometimes, the early spring months. In some locations, the bulk of the rain falls in a short time interval, while in colder areas where snow accumulates, large volumes of water are released over a short period when the melt occurs. If these phenomena occur during periods of low ET, large amounts of water may infiltrate, causing soil moisture contents to increase significantly. These inputs of moisture over a short time period can lead to failure of capillary barrier systems and saturation of soil storage systems (Hakonson et al. 1994; Khire et al. 1999; Anderson et al. 1993). Similarly, ET can remove large amounts of water from the soil during peak growing seasons, reducing soil moisture contents significantly.

If these near-surface phenomena are ignored when using numerical models, soil moisture contents can be significantly over- or under-predicted. In this paper we examine the influence of ET, the use of average annual precipitation as a steady-state infiltration rate, and the use of average plant dormant season precipitation on the behaviour of a capillary barrier system. To evaluate the differences in distributing precipitation and the inclusion of near surface processes percolate moving into the coarser layer of a capillary barrier and the diversion length of the capillary barrier system was monitored. The simulation was conducted over a period of ten years using a steady-state precipitation rate based on annual average rainfall, steady-state precipitation based on the dormant period rainfall, and daily transient precipitation with and without ET. It should be noted that ET, run-off and snowmelt processes were only included in the transient cases since they cannot be reasonably applied to average rainfall scenarios. These simulations were conducted for a capillary barrier system located at Albuquerque, New Mexico; San Francisco, California; Chicago, Illinois; and Colombia, South Carolina. These locations were chosen to represent a variety of climatic regimes.

Methodology

A model developed to investigate the movement of moisture in sloping capillary barrier systems was used in this study to determine the influence of ET, runoff and the method of distributing precipitation. The same model and soil parameters were used for each scenario and location to allow for direct comparison of results. At each location, five approaches for distributing annual precipitation were investigated. They were:

- Distributing the annual precipitation equally across the year on a daily basis.
- Distributing the precipitation that occurred during the plant dormant period equally across that period on a daily basis.
- Determining the storage capacity of the finer and unsaturated drainage layers of the capillary barrier and subtracting this value from the dormant period precipitation. This remaining precipitation was distributed equally on a daily basis across the period.
- Using the daily available precipitation (rain and snowmelt) as the daily input. ET and surface runoff were also accounted for in the model.
- Using the daily available precipitation (rain and snowmelt) as the daily input, accounting loss of water through surface runoff, while setting ET to zero.

Each of these approaches have been used or suggested as methodologies to distribute precipitation in numerical models or analytical solutions (Ross 1990; Oldenburg and Pruess 1993; Webb et al. 1997).

Model

The computer code used to perform the numerical simulations was TRACER3D, described by Travis and Birdsell (1991). The finite-difference code is capable of simulating 3-D saturated and unsaturated flow behavior in porous media. Minor modifications were made to the code to allow distribution of ET from the surface and within the root zone of the model, to allow large source/sink files to be used, and to provide the output in an easy to use format. The distribution of ET followed the methodology used in the HELP model, which was based on work by Penman, Ritchie, and Knisel and is described in detail in Schroeder at al. (1994). The distribution accounts for both direct evaporation from the soil, and for plant transpiration from throughout the root zone. Previous work by the authors has shown that the program is capable of simulating capillary performance providing reasonable agreement with both analytical and field data (Morris and Stormont 1997b; Morris and Stormont 1998a; Morris and Stormont 1998b). Details of the modelling process used are described in Morris and Stormont (1997a).

The model used in the study simulated two-dimensional flow and had a horizontal extent of 100 m and a vertical extent of 1.03 m. The system had a 5% slope to promote lateral movement of moisture. A uniform grid was used, consisting of 50 cells in the horizontal direction and 31 cells in the vertical direction to simulate the cover system. The top 21 cells of the model were designated to be the fine soil, with the next six cells being an unsaturated drainage layer (UDL). The remaining 4 cells were gravel, which formed the capillary break. The properties of each of the soils used in the model are given in Table 1. The top cell of the system was used only as a source of infiltration, with the ET sink being allocated to the next 20 cells in the model (cells 2-21) when it was used. Since each cell was 0.033 m in depth, ET took place in the top 0.66 m of the system. The ET term, when used, was distributed through the cover to a depth of 0.66 m (the chosen rooting depth) using the algorithm described in the HELP engineering manual (Schroeder et al. 1994).

The model boundary conditions were no-flow on the right, front, back and top, and ambient on the left and bottom of the model. The ambient boundary condition (analogous to a constant head boundary) maintains the designated cells at their initial condition throughout the simulation. A schematic of the model is shown in Fig. 1. The lower boundary is analogous to a landfill system where the gravel layer of the capillary barrier would be underlain by a layer of unsaturated soil, which is placed over the compacted waste. In effect, the boundary condition prevented moisture from accumulating in the gravel layer and did not alter the system behavior at the interface. The model was sloped to the left allowing lateral drainage out of the system. The area of study was 50 m in length, leaving 50 m to eliminate the influence of the downslope boundary on lateral flow. Moisture flow into the second layer of cells in the coarse layer of the capillary barrier system was tracked and tabulated on a two-week basis for the 10-year simulation period. The moisture flow into the monitored cells was designated to be the amount of breakthrough (inflow) produced for the lateral diversion distance and scenario being simulated. It should also

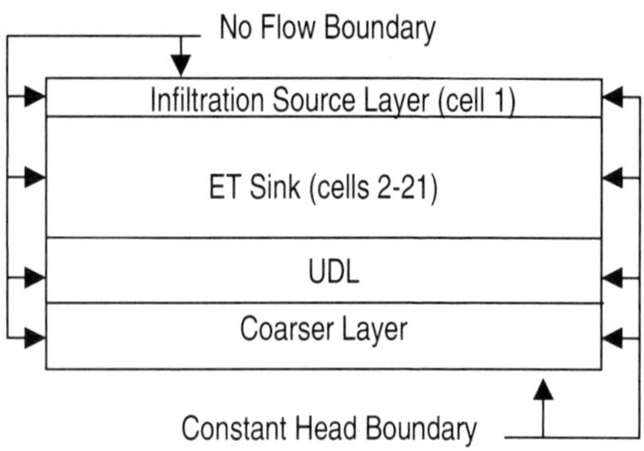

Fig 1. Schematic of the numerical model.

be noted that when the ET term was used, it became inactive when the soil moisture in a cell was below the "wilting point" (15 bar matric potential), with the demand being passed to the next deeper cell in the system. If the entire root zone area was dried to the wilting point, the ET term was ignored. This scenario was again based on the method used in the HELP code.

The initial conditions used for the simulations were identical for all simulations and were as follows: vegetative and unsaturated drainage layers 10% saturation and the gravel layer 1% saturation. Since the focus of the research was a comparison between different methodologies for distributing precipitation, the authors felt that using the same initial conditions for each scenario was important. The initial condition used for the gravel layer (1% saturation) was chosen based on the assumption that water would readily drain from the layer and that the moisture content near the interface would be very low. Increasing the gravel layer saturation by 10% in the simulations increased breakthrough (inflow) by less than 1%, which was considered to be insignificant. Similarly, increases in the initial saturation of the finer soil also led to insignificant increases in breakthrough.

Material Properties

As described above in the model section, three materials were included in the capillary barrier system simulation. They were a finer soil, an unsaturated drainage layer soil, and a gravel. The finer layer (also called the vegetative layer) was a fine-grained, uncompacted soil commonly found on the Sandia National Labs site. The UDL material was a 100 mesh sand, while the coarser layer was a medium gravel. An uncompacted finer layer was used since there is no need to minimise the hydraulic conductivity of this soil.

Table 1. van Genuchten parameters for the cover system soils

Soil/ Layer (1)	**k_s (m/s) (2)**	**θ_s (3)**	**θ_r (4)**	**α (mm^{-1}) (5)**	**n (6)**
Vegetative	$1.4x10^{-6}$	0.442	0.077	0.015	2.03
100 Mesh Sand	$2.1x10^{-4}$	.39	0.031	0.0038	4.95
Gravel	0.1	.42	0.005	0.493	2.19

The properties of the soils used in this study are given in Table 1 using the van Genuchten functions (van Genuchten 1980). The moisture content, θ, and the matric potential, h, are related by means of:

$$\Theta = \frac{\theta - \theta_r}{\theta_s - \theta_r} = \left[1 + (\alpha h)^n\right]^{-m} \tag{1}$$

and the hydraulic conductivity is given as:

$$k = k_s \Theta^{1/2}\left(1-\left(1-\Theta^{1/m}\right)^m\right)^2 \tag{2}$$

where Θ = dimensionless water content,
k_S = saturated hydraulic conductivity,
θ_S = saturated moisture content,
θ_r = residual moisture content,
m = 1-1/n, and,
n and α are fitting parameters.

The soil properties for the fine layer were based on soil samples taken from Sandia National Laboratories as determined by McTigue (1994). The gravel layer properties were taken from Stormont (1995), while the properties 100-mesh sand were measured and reported by Pease (1995).

Input Variables

Three major parameters used in the simulations are functions of location and time; daily precipitation, length of the dormant season, and daily ET. The range of total precipitation over the 10-year study period ranges from a low of 2.19 m at Albuquerque, to a high of 13.12 m at Columbia. Similarly, ET varied from a high of over 18 m over the ten years at Albuquerque, to a low of under 12 m at Chicago.

Though cumulative precipitation and ET over the simulation period is important, the timing of the precipitation periods in relation to ET should also be examined. This relationship is shown in figs 2-5. In Albuquerque (fig 2) the ET is higher than most days of precipitation, which results in moisture being rapidly removed from the soil if suitable plant growth is in place. San Francisco (fig 3) provides a contrast to this scenario since the bulk of the rain occurs during the winter

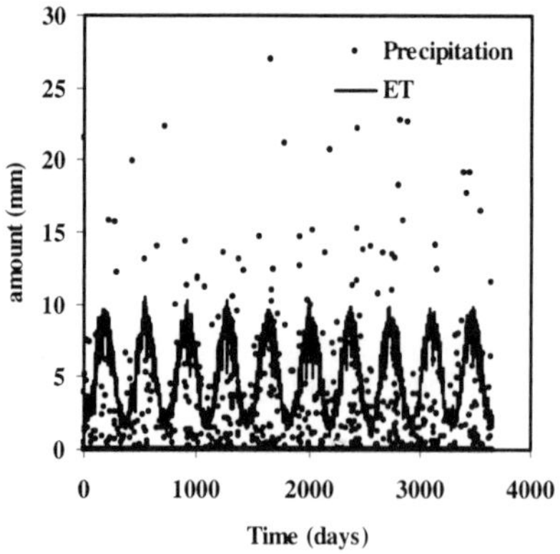

Fig 2. Albuquerque precipitation and ET

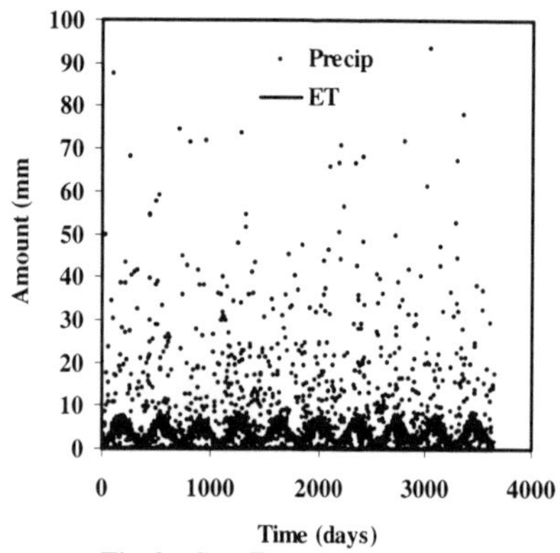

Fig 3. San Francisco precipitation and ET

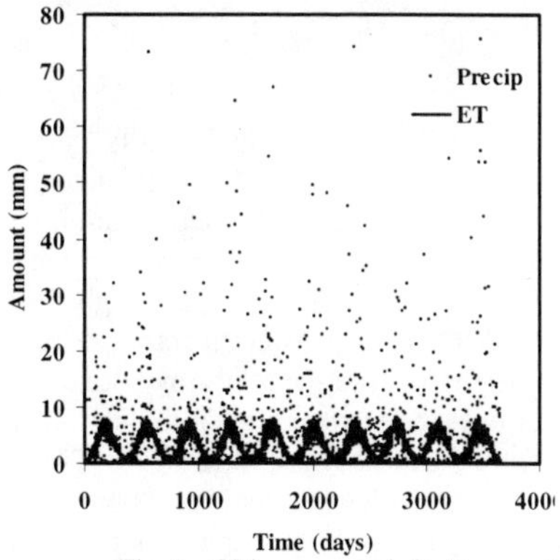

Fig 4. Chicago precipitation and ET.

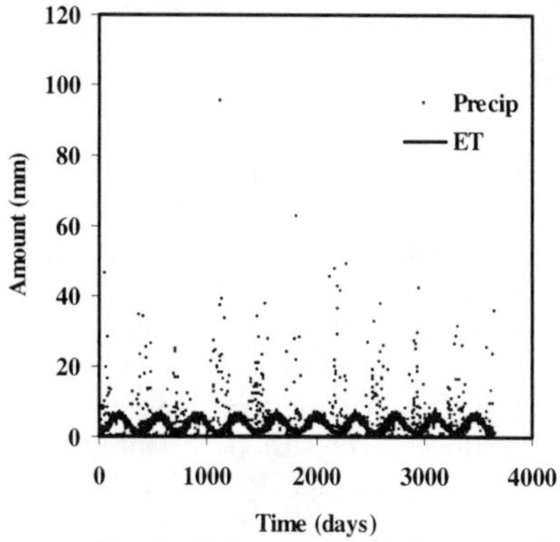

Fig 5. Columbia precipitation and ET.

systems.

The climate and precipitation parameters used in the simulations are shown in Table 2. They were derived from the data provided in the HELP model for each of the locations. The 10-year precipitation data used in the transient simulations and upon which the average annual and dormant period data is based upon, was chosen from the 100-year record. The 5-year period of highest precipitation was found for each location to provide a worst-case scenario. Three years prior to the period and two years after the period were included to allow stabilisation of the model and to determine if there were carry-over effects. The dormant period precipitation rate was determined for each location on a yearly basis from the daily rainfall record. This value was summed and applied uniformly to the model over the dormant season as daily precipitation.

Table 2. Climate parameters.

	Albuquerque	San Francisco	Chicago	Columbia
Average Precipitation (m)	0.22	0.52	0.92	1.31
Dormant Period	164	115	192	111
Storage (m)	0.276	0.276	0.276	0.276

Results

Albuquerque

The results from the different modelling scenarios varies significantly by location due to differences in precipitation patterns, dormant periods and ET rates. Albuquerque, New Mexico has a semi-arid climate due to its high desert location, and, therefore, has low annual precipitation. Much of this precipitation comes in the form of intense periods of rain during the late summer months when ET is high. Based on this pattern, evapotranspiration is expected to have a significant influence on soil moisture levels. This behavior is seen in fig 6 where the total flow into the underlying coarse layer of the capillary barrier over the 10-year simulation period is plotted versus downslope distance (distance from the left edge of the model). The figure illustrates that for the capillary barrier system modelled, there is failure of the system (flow into the coarse layer) only when ET is ignored in the transient case. The data for dormant period precipitation and dormant period precipitation minus the system storage capacity are not shown because they also prevented system failure. Therefore, in Albuquerque, the only method that does not provide reasonable results for the system modelled is the transient case when ET is ignored. This is an expected outcome based on the climate. However, it should be noted that this is the case only for the capillary barrier simulated. If a 100-m long system were modelled the results would likely be different.

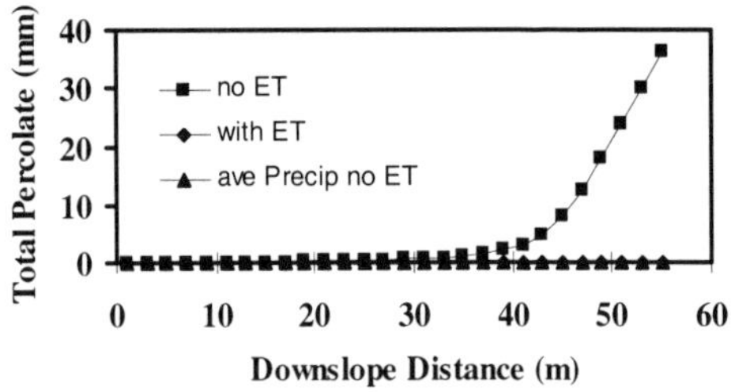

Fig 6. Albuqerque total percolate and diversion lengths.

San Francisco

San Francisco has a significantly different climate, and is often classed as Mediterranean. The bulk of the rainfall occurs during the fall and winter months, with little to no rain during the summer months. In this scenario the heaviest rainfall occurs during the period that plants are dormant and ET is low. The effects of four methods to represent precipitation at San Francisco are shown in fig 7. Again there is significant differences between the methodologies used, ranging from no failure of the capillary system to over 2.5 m of water flowing

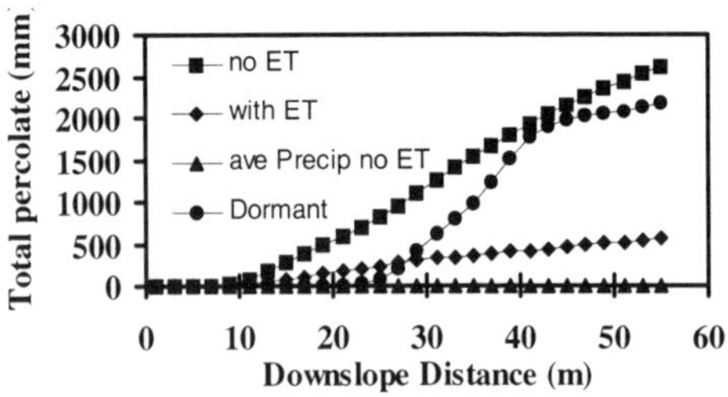

Fig 7. San Francisco total percolate and diversion lengths.

into the underlying layer of the barrier. As expected, the worst case occurs for the transient case when ET is ignored. The point of failure occurs about 10 m downslope, with flow increasing rapidly from that location. Since ET was ignored, the only mechanism for moisture removal was by lateral flow downslope. When ET was allowed, the soil of the system was very dry before the rain season began, providing more storage capacity in the system.

Failure also occurs in the system at San Francisco when ET is included in the transient case, though the total amount of percolate is only about 20% of that when ET is ignored. Breakthrough into the coarse layer of the barrier also takes place slightly further down slope. This result is expected since the bulk of the precipitation falls during the dormant season when ET is low.

If precipitation only during the dormant season is considered, failure of the system again occurs, but at a distance of about 25 m downslope compared to 10-12 m for the transient cases. This greater diversion length is due to the distribution of the rainfall over the entire dormant period, which results in decreasing the magnitude of the individual storm events. The overall flow into the capillary barrier at the downslope end of the system is much greater than the transient case with ET and nearly as great as the transient case without ET. Since the bulk of the rainfall at San Francisco occurs during the winter (dormant) months, it is reasonable that the dormant case and the transient case without ET would provide similar overall results.

The final simulation shown in fig 7 is the averaging of the precipitation over the year and applying it daily. Using this process, there is no failure of the capillary barrier over the 50-m length modelled. This methodology smooths the precipitation pattern over the entire year, reducing the magnitude of the winter events, which in turn allows the infiltrating moisture to be diverted laterally within the barrier and flow downslope out of the system. (Note: the dormant period precipitation minus the storage capacity of the system was also modelled, with the amount of percolate produced near zero.)

Chicago

Chicago is the third location simulated in this study, and again the methodology employed to distribute the precipitation strongly influences the behavior of the cover system. The results of the simulation are shown in fig 8. As in the previous two locations, the transient simulation without ET produces the largest amount

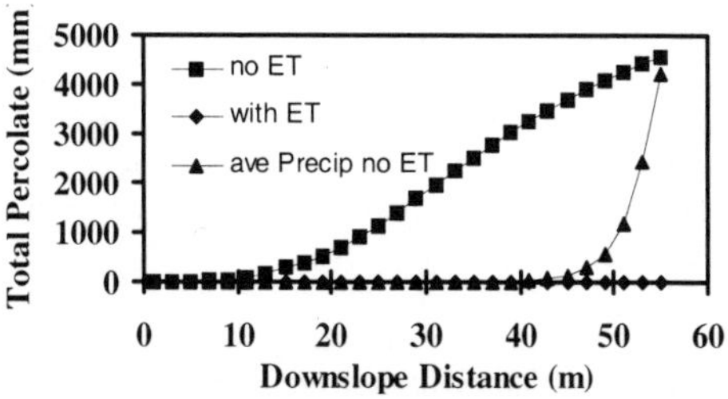

Fig 8. Chicago total percolate and diversion lengths.

of percolate. However, when ET is included in the transient simulation, the capillary barrier system does not fail. Since precipitation tends to fall in greater amounts during the growing season, and during the winter it falls as snow, the ET process is able to remove significant volumes of soil moisture as it infiltrates.

The third modelling methodology illustrated in fig 8 is spreading the annual precipitation over the entire year and ignoring ET. Failure of the system occurs at about 46 m downslope, with a very rapid increase in breakthrough after this point when using this approach. This failure of the system is not surprising since there is a large amount of moisture infiltrating the soil with no removal mechanism other than lateral diversion. Failure did not occur for this scenario at San Francisco and Albuquerque due to lower annual precipitation rates. (Note: The dormant and dormant –minus storage methodologies were modelled for Chicago, but no failure of the system occurred. Therefore, the data was not included in fig 8 for clarity.)

<u>Columbia</u>

Columbia is the location with the highest rainfall and shortest dormant season of the four locations and contrasts greatly with Albuquerque. It is expected that a capillary barrier would not work well at this location due to the high rainfall levels encountered throughout the year. Fig 9 confirms this expectation. In all cases, failure of the system occurred within the 50 m barrier length. In the transient case with ET, breakthrough into the underlying coarse layer occurred at about 32 m downslope with total percolate exceeding 1 m over the 10 year period at the diversion distance of 50 m. The transient case in which ET was ignored produced about 8 times as much percolate over the simulation period, with breakthrough occurring at 10 m downslope. This difference in behavior is consistent with the high rainfall and high ET conditions found at Columbia.

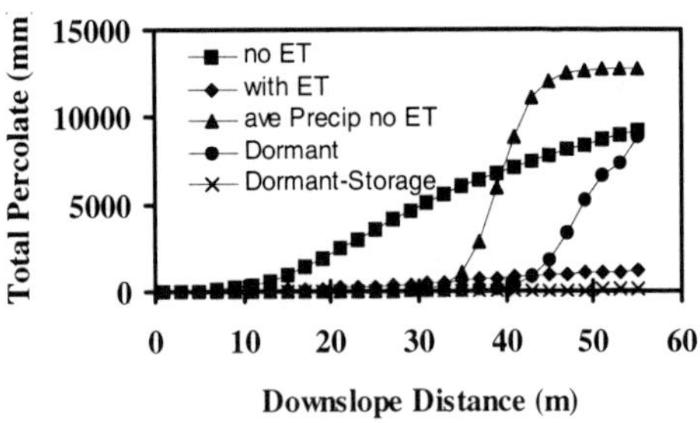

Fig 9. Columbia total percolate and diversion lengths.

The third methodology shown in fig 9 is the annual average precipitation approach. Failure of the system also occurs using this approach at a downslope distance of approximately 35 m. It should be noted that the total percolate produced using this methodology is over 12 m in the 10-year simulation period, which is 50% greater than the 8 m seen in the transient case with no ET. This 50% increase is due to the ignoring of surface run-off in this methodology, which removes a large portion of precipitation at this location. Similar behavior is seen for the dormant case in which only rain that fell during the 111-day period of plant dormancy is input into the model. Breakthrough again occurs, with the total percolate that flows into the coarse layer of the capillary barrier equivalent to the amount from the transient case with no ET. Again surface runoff is ignored in this methodology, leading to greater amounts of percolate than for cases with greater precipitation.

The final methodology used at Columbia is the dormant minus storage approach. In this case, breakthrough occurred at the downslope end of the barrier with only about 0.5 m of percolate produced over the modelling period. This is significantly lower than for the previous four methodologies. This low amount of percolate and long diversion length is based on the assumption that the fine and unsaturated drainage layers are completely dry at the start of each dormant period. This is definitely not the case at Columbia when heavy amounts of rain fall year round and in amounts greater than can be removed by ET.

Discussion

A significant variation in results was seen at each of the four locations at which simulations were conducted, depending on the methodology employed for representing near-surface processes. This variation is not surprising considering the assumptions made with each approach and the varying importance of each near-surface process at each location. If rainfall is uniform over the year at a location, then using an average rainfall may well provide realistic results. However, at a location like San Francisco, this approach will likely lead to inaccurate levels of soil moisture at any given time.

To provide a comparison between the 5 methodologies employed in this study and their impact, diversion lengths for each approach are plotted by location in fig 10. It is seen in the figure that there is a wide range of diversion lengths for each location that are determined by the methodology employed for distributing precipitation and representing near-surface processes. At each location, the methodology best representing actual processes is the transient approach that includes evapotranspiration. At Albuquerque, which is a relatively benign climate, the transient methodology with ET produces a diversion length in excess of 55 m. The only methodology that produces a shorter diversion length is the transient approach without ET. This would suggest that ET is a major process that cannot be ignored in this climate.

At San Francisco we see that the diversion lengths for the two transient cases and for the dormant case are very similar, indicating that ET does not play a major role in preventing system failure. However, if we refer back to fig 7, we see that the total amount of percolate produced at the far end of the barrier is significantly different for the two transient cases. This difference is likely due to the much lower

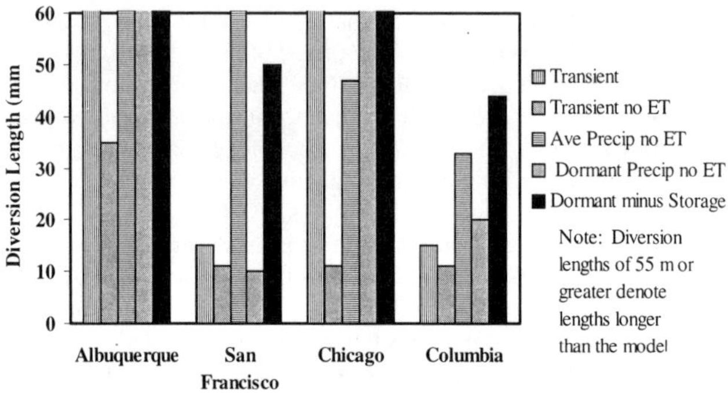

Fig 10. Comparison of diversion lengths for the methodologies of distributing precipitation.

moisture content in the barrier system before the onset of the rains, allowing significant amounts of infiltrating moisture to be stored in the upper layers of the system. Also, though ET is low during the rainy period, there is some removal of moisture by this process between storms. Also worth noting is that the diversion lengths and total percolate produced at the downslope end of the capillary barrier are similar for dormant and transient without ET cases. This indicates that the bulk of the breakthrough occurs during the dormant period.

The results from Chicago are somewhat different than San Francisco, with the transient with ET, dormant and dormant minus storage cases all having diversion lengths greater than 55 m. The transient without ET and average annual precipitation cases both had shorter diversion lengths of approximately 10 m and 45 m, respectively. These results illustrate that ET is a significant process at Chicago and that it cannot be ignored. The lack of breakthrough in the dormant case indicates that failure of the capillary barrier system does not tend to occur during these months. This observation agrees with rainfall distribution for the location, where the major storms tend to occur during the summer months. Winter precipitation tends to be lighter and in the form of snow.

The final location studied was Columbia, which has a significantly wetter climate than do the other three sites. This is seen in the results from the simulations for all cases. Diversion lengths tend to be short and total percolate much greater. In all cases, the diversion lengths are less than 55 m, with the transient case with ET approximately 15 m. This is equivalent to San Francisco. However, the total percolate produced is approximately double that of San Francisco. The average annual approach results in a longer diversion length, but much greater percolate at the downslope end of the barrier than the transient case with ET. Similar results are also seen for the dormant case. The dormant minus storage case yields the longest diversion length and produces the least amount of percolate of the methodologies employed.

Conclusions

Based on the work presented in this study there does not seem to be a methodology by which one can approximate the near-surface processes and employ simple steady-state modelling routines. Results from transient models based on daily climatic parameters yield significantly different results for three of the four locations studied. Similar results are likely if a longer model were used at Albuquerque that would define the system diversion length. In all cases, it is seen that ET is a major process that should not be ignored in simulating near surface moisture fluxes, especially if plants are present. Therefore, the use of simplifying approaches and steady-state models should be avoided and results based on them used with extreme caution. Though the process of developing and using transient models that incorporate significant near-surface processes is more time consuming and more difficult, the results provided are much more likely to reflect actual field conditions.

References

Anderson, J. E., Nowak, R. S., Ratzlaff, T. D., and Markham, O. D. (1993). "Managing Soil Moisture on Waste Burial Sites in Arid Regions." *Journal of Environmental Quality*, 22(1), 62-69.

Hakonson, T. E., Maines, K. L., Warren, R. W., Bostick, K. V., Trujillo, G., Kent, J. S., and Lane, L. J. (1994). "Hydrologic Evaluation of Four Landfill Cover Designs at Hill Air Force Base, Utah." *LAUR-93-4469*, Los Alamos National Laboratory, Los Alamos, NM.

Khire, M. V., Benson, C. H., and Bosscher, P. J. (1999). "Field data from a capillary barrier and model predictions with UNSAT-H." *Journal of Geotechnical & Geoenvironmental Engineering.*, 125(6), 518-527.

McTigue, D. (1994). "Estimation of unsaturated hydraulic properties by inversion of moisture content data from a field experiment." , Sandia National Laboratories, Albuquerque, NM.

Morris, C. E., and Stormont, J. C. (1997a). "Capillary Barriers and Subtitle D Covers: Estimating Equivalency." *Journal of Environmental Engineering*, 123(1), 3-10.

Morris, C. E., and Stormont, J. C. (1997) "Numerical simulations of capillary barrier field tests." *International Containment Technology Conference*, St. Petersburg, Florida, USA, 275-281.

Morris, C. E., and Stormont, J. C. (1998a). "Discussion and closure for 'Capillary Barriers and Subtitle D Covers: Estimating Equivalency'." *Journal of Environmental Engineering*, 124(5), 483-484.

Morris, C. E., and Stormont, J. C. (1998b). "Evaluation of numerical simulations of capillary barrier field tests." *Geotechnical and Geological Engineering*, 16(3), 201-214.

Oldenburg, C. M., and Pruess, K. (1993). "On numerical modeling of capillary barriers." *Water Resources Research*, 29(4), 1045-1056.

Pease, R. E. (1995). "Increasing the Diversion Length of Capillary Barriers," Master of Science, University of New Mexico, Albuquerque.

Ross, B. (1990). "The Diversion Capacity of Capillary Barriers." *Water Resources Research*, 26(10), 2625-2629.

Schroeder, P. R., Aziz, N. M., Lloyd, C. M., and Zappi, P. A. (1994). "The Hydrologic Evaluation of Landfill Performance (HELP) Model: Engineering Documentation for Version 3." */600/9-94/xxx*, U.S. Environmental Protection Agency Risk Reduction Engineering Laboratory, Cincinnati, OH.

Stormont, J. C. (1995). "The Performance of Two Capillary Barriers During Constant Infiltration." Landfill Closures ...Environmental Protection and Land Recovery, R. J. Dunn and U. P. Singh, eds., American Society of Civil Engineers, New York, 77-92.

Travis, B.J., and Birdsell, K.H. (1991). "TRACR3D: A Model of Flow and Transport in Porous Media." Los Alamos National Laboratories LA-11798-M (UC-814), Los Alamos, NM.

van Genuchten, M. T. (1980). "A closed-form equation for predicting the hydraulic conductivity of unsaturated soils." *Soil Science Society of America Journal*, 44, 892-898.

Webb, S. W., McCord, J. T., and Dwyer, S. F. (1997). "Prediction of tilted capillary barrier performance." *International Containment Technology Conference*, St. Petersburg, Florida, USA, 296-301.

Movement of Moisture and Dissolved Salts from Soils into Concrete Foundations

by D.H. Gray[1]

Abstract

The flow regime in a concrete foundation (stem wall and slab) resting on an unsaturated soil was analyzed to determine where and how water moves into and through the foundation in response to water pressures (or moisture suctions) acting at the boundaries. Flow nets were constructed using an iterative, finite difference technique. Steady state conditions and isotropic hydraulic conductivities were assumed in the analyses. Various cases were examined including effects of moisture barriers beneath the foundation and different degrees (or rates) of evaporation at foundation boundaries.

An increasingly greater proportion of the total, advective water flow through the foundation occurred into and out of the sides of the stem wall as either 1) upward, water flow into the slab was prevented because of the presence of a moisture barrier beneath the slab, or if 2) a higher amount of evaporation occurred at the outside surface of the stem wall/grade beam compared to the upper surface of the slab. In every case water entering the concrete foundation from irrigated soil adjacent the foundation moved upward and into (and in some cases out of) the grade/beam stem wall. In no case did any of this irrigation water enter the slab.

These illustrations and finding are significant with regard to the problem of possible sulfate damage to concrete foundations. The sulfates may occur naturally in the native soil or be introduced in the irrigation water or as sulfates added to the soil from agricultural amendments, e.g. gypsum or sulfur. The illustrations are consistent with the nature of minor damages (e.g., surface flaking and spalling) observed in concrete foundations next to irrigated plantings where sulfates are present in the soil or irrigation water.

Introduction

The movement of water and dissolved salts into concrete foundations resting or imbedded in the ground, either slabs and/or stem walls, can adversely affect the foundation system and superstructure. The movement of moisture is driven largely by capillary action or suction in which water is transferred from areas of higher soil moisture potential in the underlying ground to areas of lower potential (or higher suction) in the overlying foundation. Evaporation at the top or outer surfaces of concrete members perpetuates this cycle of moisture transfer (Gray, 1969). If salts are present in the

[1] Professor, Dept. of Civil & Envl. Engr. *The Univ. of Michigan*, Ann Arbor, MI 48109. Phone (734) 764-4354. E-MAIL: dhgray@engin.umich.edu

capillary liquid they tend to precipitate out and concentrate at or near the surface of evaporation. This phenomenon is sometimes referred to as rising damp and/or salt damp. The problem tends to be more severe in arid regions or maritime climates characterized by hot, dry summers and cool, wet winters. Severe problems with rising damp and salt attack have been reported in Southern California (Yen, 1990) and Alidade, Australia (Salt Damp Research Committee, 1976).

Dissolved salts in the rising damp make the problem much worse. Several attack mechanisms are associated with the presence of dissolved salts in the liquid water that is imbibed into a concrete foundation. Two of the principal mechanisms are briefly described below:

Salt crystallization. Salts present in the imbibed water precipitate at evaporative surfaces where they can effloresce and discolor the concrete and any plaster course that might be applied to the surface. As the salts precipitate out of solution they also crystallize. Salt crystallization causes minor scaling and fretting of surface layers somewhat similar in appearance to freeze-thaw damage. Scaling occurs due to volume changes associated with salt crystallization. In most cases this salt induced scaling is a minor cosmetic nuisance and not a structural problem because of the limited depth of crystallization. Salts known to cause salt crystallization attack at the surface included sodium sulfate, sodium carbonate and sodium chloride.

Sulfate attack. Sulfate attack is a chemical reaction between certain cement minerals (calcium aluminates) and sulfates in the pore liquid of hardened concrete. Sulfate attack problems with concrete have been widely studied and are fairly well recognized (ACI, 1998; DePuy, 1994). Sulfate attack occurs when calcium, alumina, and sulfate combine to form ettringite ($3CaO \cdot Al_2O_3 \cdot 32H_2O$). The volume of ettringite is over 200% that of the original constituents, which can result in massive swelling and cracking when sufficient ettringite forms by sulphation of alumina. Ettringite is expansive only when it is formed by chemical reaction; once formed it is no longer expansive. Ettringite can be recognized by the presence of needle-like crystals in electron micrographs of degraded concrete that has been exposed to sulfates. Sulfate attack may result from an external sulfate source, e.g., groundwater or irrigation water that contains sulfate compounds.

Sulfate attack is potentially the most serious of the two degradation mechanisms as it can lead to crumbling and disintegration of concrete with major attendant structural problems. Sulfate attack has even caused the demise of roads constructed over cement stabilized base courses (Rollings et al., 1999). Sulfate attack of concrete is minimized or mitigated by using sulfate-resistant cements (type II or V depending on the severity of the problem) and by limiting the water cement ratio.

Source of Sulfates

A sulfate attack of concrete foundations requires an external sulfate source. There is not sufficient residual sulfate in the cement to cause a significant problem. A sulfate reaction requires that dissolved sulfates be imbibed or drawn into the concrete foundation by the capillary action-evaporation mechanism described previously. The sulfates may already be present in the underlying natural soil and groudwater or they could be supplied from amendments, e.g., gypsum, that are added to the soil for either agronomic or engineering purposes. Water used to irrigate foundation plantings is another likely source of sulfates.

Any attempt to ascertain the nature and cause of possible sulfate damage to a concrete foundation requires that the source and pathway of sulfates into the foundation be determined properly. Sulfate can only migrate into the foundation in an aqueous liquid phase from moist soil in contact with the foundation beneath or at its sides. A damp course barrier or polymeric membrane is normally placed between a concrete slab and the underlying ground to limit or prevent such moisture transfer. Another possible source of dissolved sulfates is the moist or wet soil adjacent the outside surface of the short concrete stem wall that is connected to the slab at its sides. This soil which supports foundation plantings typically is amended with gypsum ($CaSO4 \cdot nH_2O$) to improve its tilth and plant support status. In arid or water scarce regions landscape irrigation is frequently done with "grey water" or reclaimed water that normally contains some dissolved sulfates.

Flow Pathways

The flow regime in a typical concrete foundation resting on partially saturated ground can be analyzed to determine where and how water moves through the foundation in response to water pressures (or moisture suctions) acting at the boundaries. These analyses will reveal the transit path and destination of any solutes or dissolved solids that might be present in the water. The solutes must be transported in the liquid phase within a porous medium by a process of either advection or by diffusion (or both).

Flow nets for concrete foundations resting on (or imbedded in) a partially saturated soil can be constructed using an iterative, finite difference technique. This procedure allows one to construct the flow net within a porous medium based on known (or assumed) conditions along the external boundaries. The method can handle quite complicated flow cross-sections or geometries such as those of a concrete footing. The analyses can be run under a variety of assumed external boundary conditions. A parametric variation approach should be followed in order account for possible uncertainties (or variations) in the external boundary conditions and to examine explicitly the effect of such variations or uncertainties on the flow regime in the concrete foundation. The following cases were examined in the present study:

- presence of a moisture barrier beneath the slab portion of the foundation
- absence of a moisture barrier beneath the slab portion of the foundation
- evaporation of moisture at the top surface of the slab only
- equal rates of evaporation of moisture at the top of the slab and on the outside surface of the grade beam/stem wall respectively
- greater evaporation on the outside of the grade beam/stem wall compared to the top surface of the slab.

Additional flow nets can be constructed (or updated) for other specified boundary conditions as needed. The general approach outlined herein would still be valid.

Advective vs. Diffusive Transport

The analyses described herein pertain to solute (salt) transport under advection, i.e., the movement of water and solute under a hydraulic gradient. Water moves from a region of high to low energy potential.... or from a region of low to high moisture suction at a rate proportional to the hydraulic conductivity. Suction driven gradients in porous media are important where evaporation occurs. Evaporation results in high suctions (or negative pore water pressures). Water will tend, therefore, to flow from wetter (or saturated) regions with low moisture suctions into drier areas with high moisture suctions.

Dissolved solids (salt) can also be transported by diffusion, i.e., the movement of solute under a chemical or concentration gradient. Solute tends to move from areas of high concentration to low concentration at a rate proportional to the diffusion constant. Diffusion and advection can occur simultaneously---and in either the same or opposite directions. When evaporation occurs at a flow boundary any salt present tends to precipitate out. Accordingly, the salt concentration increases at the evaporative boundary and the diffusive flux of solute will be opposite to the advective flux. This counter diffusion acts as a "brake" or self limiting control on the accumulation of salts at the evaporative boundary.

Flow Nets - General

Theoretical Basis. A flow net is simply a graphical representation of flow lines and equi-potential lines in a porous medium through which water is flowing. The potential (or hydraulic head) distribution in a net under steady state flow must obey the LaPlace equation as follows:

$$k_x \, \partial^2 h/\partial x \; + \; k_y \, \partial^2 h/\partial y \; + k_z \, \partial^2 h/\partial z = 0 \qquad (1)$$

where: h = the hydraulic head
x,y,z = Cartesian co-ordinates
k_x, k_y, k_z = hydraulic conductivity in the x, y, and z directions

For flow in two dimensions in isotropic media ($k_x = k_y = k_z$) Equation 1 reduces to the following simpler form:

$$\partial^2 h/\partial x \; + \; \partial^2 h/\partial z = 0 \qquad (2)$$

In the present analyses the hydraulic conductivity was assumed to remain constant throughout the whole flow regime in the concrete foundation. In actual fact the liquid hydraulic conductivity will decrease with decreasing degrees of saturation near surface boundaries where evaporation is occurring. The flow nets cannot be used to calculate liquid transfer or discharges under these conditions. They are intended solely, however, to indicate the direction and destination of liquid flows not rates of transfer.

A flow net can be used to calculate the hydraulic head and gradient at any point. The gradient (i) is given by the following equation:

$$i = dh \, / \, dl \qquad (3)$$

where: dh = the head difference (or drop) between equipotentials
dl = the distance or spacing between adjacent equipotential lines

When equipotential lines are spaced closed together the gradient is high and vice versa.

Flow Net Construction. Flow nets are most easily drawn and constructed by using numerical methods. The region of interest is "discretized" or subdivided into small elements (nodes) and the hydraulic head computed at the center of each element. The hydraulic head assigned or computed for each element must satisfy the LaPlace equation (Equation 1).

The elements(nodes) of the "discretized" two-dimensional region of interest are represented by a grid of cells in a spread sheet. The hydraulic head for each cell is calculated using an iterative, finite difference calculation technique. In other words, a computer is used to repeatedly calculate and update the head at each node(or cell) until the LaPlace equation if satisfied everywhere.

Various rules (Kleiner, 1985) can be formulated to determine the head at any point in the grid of nodal points (or cells) based on the value of the hydraulic head at surrounding points and conditions at adjacent boundaries. For example, the hydraulic head (h_o) at an interior point in a region of uniform hydraulic conductivity is given by the following equation:

$$h_o = 0.25 \times [h_1 + h_2 + h_3 + h_4] \qquad (4)$$

where: $h_1, h_2, h_3, h_4 =$ the hydraulic heads at surrounding points

A grid of cells conforming in shape or outline to the cross section of the flow region of interest is laid out on a spreadsheet. Known or assumed hydraulic heads at external boundaries, or alternatively, appropriate equations are entered into the cells of the spreadsheet. The hydraulic heads at all interior nodes are calculated by using a finite difference, numerical iteration procedure and entered into the cells of the spreadsheet. The equipotential lines (contours of equal head) are then plotted using a graphical interpolation routine. Electronic spread-sheets such as EXCEL have the capability to calculate the hydraulic heads at discrete points and to plot the equipotential lines.

The final step is to draw in the flow lines in such a way that the elements in the net are "square", i.e., so that the flow lines and equipotential lines intersect at right angles and so that the elements in the net are equi-dimensional. Some distortion of the elements will invariably occur near boundaries, but the majority should be reasonably "square."

Flow Nets - Concrete Foundations

Procedural Approach. The general procedure outlined in the previous section was followed. The flow region of interest in this case consists of a concrete foundation with a cross section or outline as shown in Figure 1. Essentially, the foundation consists of a slab and stem wall/grade beam combination. Water flows by capillary action from the below-ground side or base of the foundation were soil moisture suctions are low to the top or above-ground surfaces of the foundation where soil moisture suctions are high because of evaporation. Soil moisture suctions will tend to be highest at the above-ground, exterior surfaces of the stem wall and slab that are directly exposed to the atmosphere. In many cases, evaporation will be even more intense, and soil moisture suctions accordingly higher, at the above-ground, exterior surfaces of the stem wall particularly if exposed to the drying action of wind and sun. Conversely, soil moisture suctions will tend to be lowest at the below-ground, exterior surfaces of the stem wall because of irrigation in flower beds and lawns adjacent the foundations.

Parametric Variation Analysis. A parametric variation or sensitivity analysis was conducted. Various assumptions were made about the hydraulic heads (soil moisture suctions) along the boundaries of the foundation and about flow constraints along any particular bounding surface. The effect, for example, of having an impermeable barrier (polymeric sheet) below the slab portion of the foundation can be simulated by making

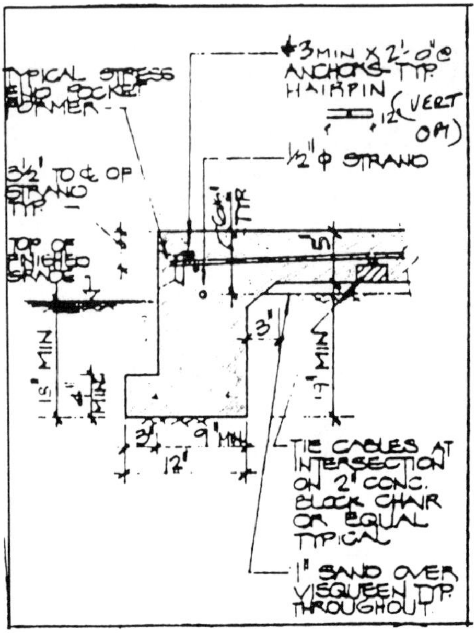

Figure 2. Typical exterior concrete grade beam/stem wall detail

this surface a "no flow" or impermeable boundary. Similarly, the effect of more intense evaporation at the exterior, above ground portion of the stem wall relative to the interior, top of the slab can be simulated by assigning higher negative pressures (suctions) along this surface. The effect of these assumptions or "scenarios" on the flow regime (flow net) was determined by the procedures described in the previous section.

An example of the geometry, assumed boundary conditions, and calculated distribution of hydraulic heads in a concrete foundation.... for one these scenarios is shown in Figure 2. Frequent irrigation in the adjacent lawns and flower beds is assumed to keep the soil at or slightly wetter than the field capacity of the soil along the exterior surface of the below-ground portion of the stem wall.

Flow Net Results

Case i) Evaporation at Top Only. No Moisture Barrier Beneath Slab. In this case evaporation was assumed to occur uniformly along the top of the slab and stem wall. The soil moisture suction along the top was assigned a value of -2.6 bars everywhere. It was also assumed that frequent irrigation in the adjacent lawns and flower beds keeps the soil at or slightly wetter than the field capacity of the soil along the exterior surface of the below-ground portion of the stem wall. Accordingly, the hydraulic head along this exterior surface was assigned a value of -0.2 bars over a depth of 1 foot. The rest of the below-ground portion of the concrete foundation was assumed to reach and remain at a residual value equal to or slightly drier than the field capacity, and assigned a value of -0.8 bars. No moisture barrier was assumed to be in place beneath the slab; accordingly, soil moisture could enter the concrete from any below-ground surface.

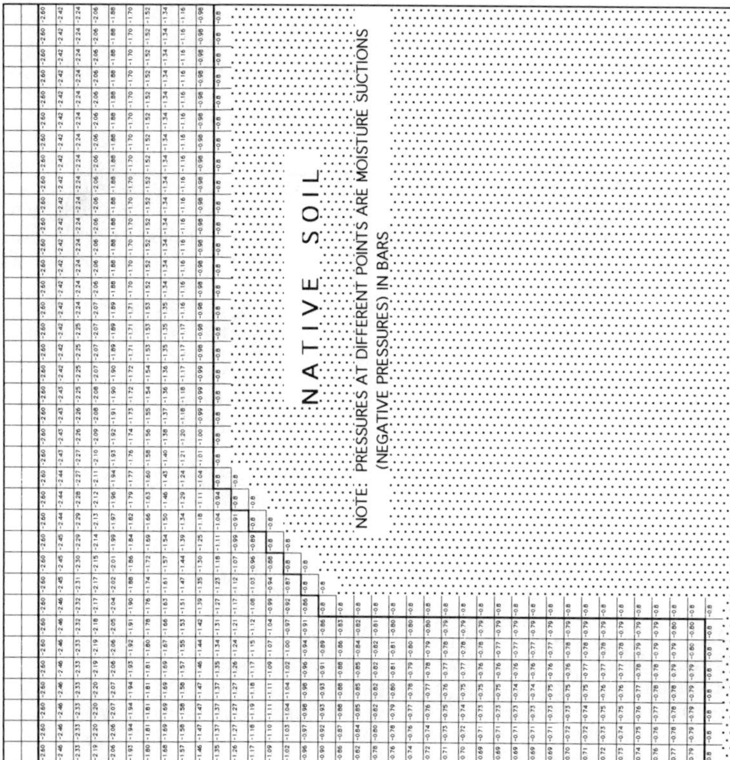

Figure 2. Cross-section of concrete foundation showing "discretized" flow region, assumed boundary conditions, and calculated distribution of hydraulic heads (pressures) for one of several cases considered.

The flow net for this case is plotted in Figure 3. The flow lines are solid and have arrow heads which show the direction of flow. The equipotential lines are plain solid lines; they intersect the flow lines at right angles. Virtually all the flow is upwards and originates from either the irrigated soil adjacent the below-ground, exterior surface of the stem wall or from beneath the slab. A very small fraction of the flow originates from the irrigated zone and flows down and across the stem wall.

Case ii) ***<u>Evaporation at Top Only. Moisture Barrier Beneath Slab.</u>*** This case is similar to the previous one (Case i), viz., evaporation was assumed to occur uniformly along the top of the slab and stem wall. As before the soil moisture suction along the top was assigned a value of -2.6 bars. However, in this case a moisture barrier (polymeric or plastic sheet) was assumed to be in place beneath the slab that prevents moisture flow from the ground upward along this interface.

The flow net for this case is plotted in Figure 4. Approximately half the flow lines now originate from the irrigated zone adjacent the exterior, below-ground surface of the stem wall and the other half from the below-ground, inclined interior surface of the stem wall (below the level of the moisture barrier). A small fraction of the flow also originates from the irrigated zone and flows down and across the stem wall as in Case i. Most of the flow passes vertically upward through the stem wall.

Case iii) ***<u>Evaporation at Top and Side. No Moisture Barrier In Place.</u>***

In this case evaporation was assumed to occur along the exterior, above-ground portion of the stem wall as well as at the top. Accordingly, moisture suction values of -2.6 bars were assigned along most of the above-ground, exterior portions of the stem wall in addition to the top. Like Case i, no moisture barrier was assumed to be in place beneath the slab; thus, soil moisture could enter the concrete from any below-ground surface.

The flow net for this case is plotted in Figure 5. As before, most of the water transfer occurs vertically upward; however, a substantial fraction now flows in a semi-circular pattern from the below-ground portion of the stem wall to the exterior, above ground portion where evaporation is occurring. *It is important to note that virtually all the flow that is emanating from the above-ground, exterior portion of the stem wall now originates from the irrigated soil zone adjacent the wall.* This means that the composition and quality of the irrigation water together with the addition of any soluble salts or agricultural amendments (e.g., gypsum or other sulfate compounds) in the soil adjacent the stem wall will determine to a very large extent the type, transfer and accumulation of salts in the above-ground, exterior surface of the stem wall. This finding has significant implications with regard to the question of possible sulfate attack of the concrete.

Case iv) ***<u>Evaporation at Top and Side. Moisture Barrier In-Place.</u>***

Like Case iii evaporation was assumed to occur along the exterior, above-ground portion of the stem wall as well as at the top. Accordingly, moisture suction values of -2.6 bars were assigned along most of the above-ground, exterior portions of the stem wall in addition to the top. However, unlike Case iii), a moisture barrier was assumed to be in place beneath the slab to prevent flow of moisture through the slab from the ground below.

The flow net for this case is plotted in Figure 6. Most of the flow lines originate from the irrigated zone adjacent the exterior, below-ground surface of the stem wall; less than half originate from the inclined, interior surface of the stem wall (below the level of the moisture barrier). A substantial fraction of the flow (approximately 25%) now emanates or exits from the above-ground, exterior portion of the foundation or stem wall.

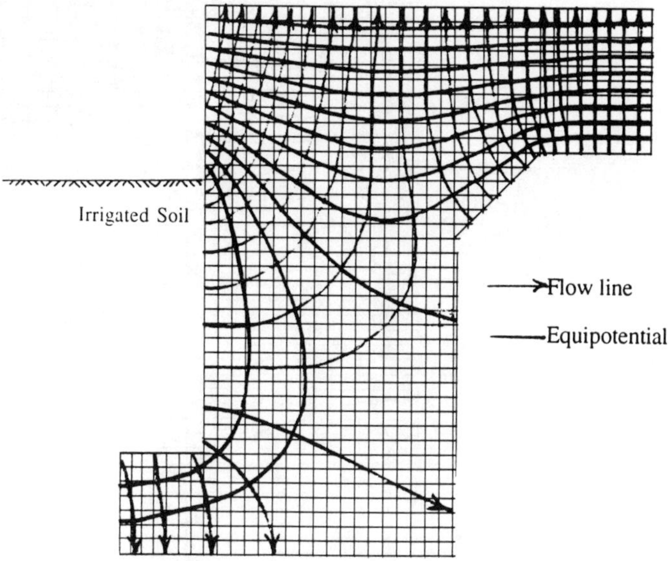

Figure 3. Evaporation at top only. No moisture barrier beneath slab *(Case i)*

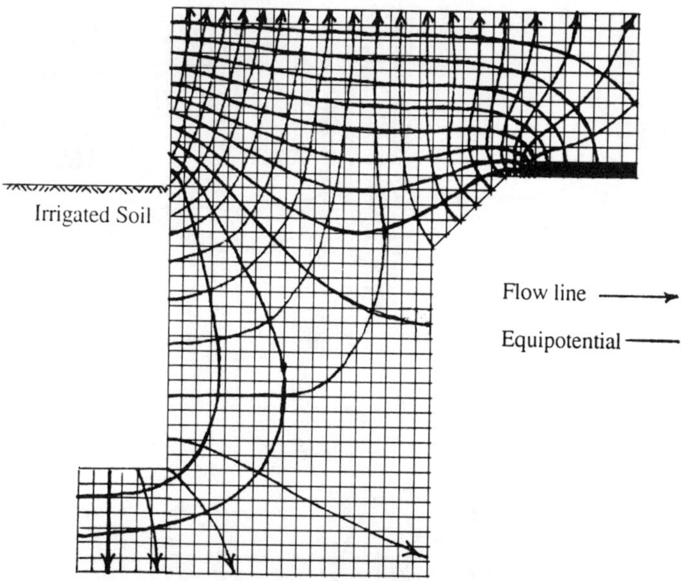

Figure 4. Evaporation at top only. Moisture barrier in place beneath slab *(Case ii)*.

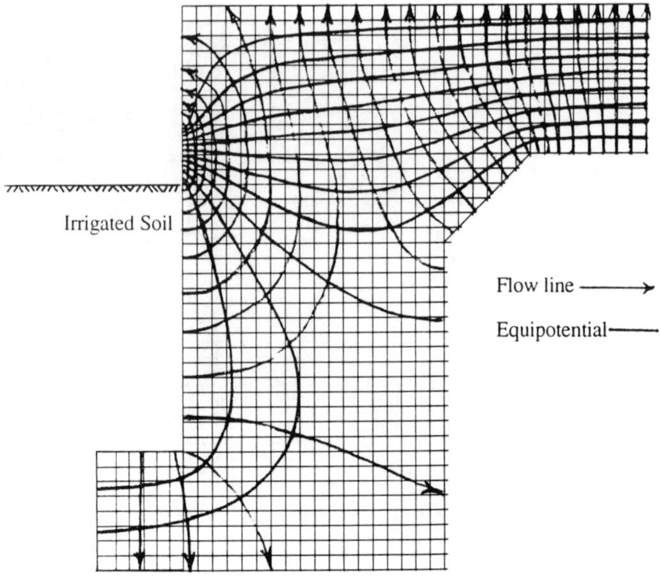

Figure 5. Evaporation at top and side. No moisture barrier beneath slab *(Case iii)*.

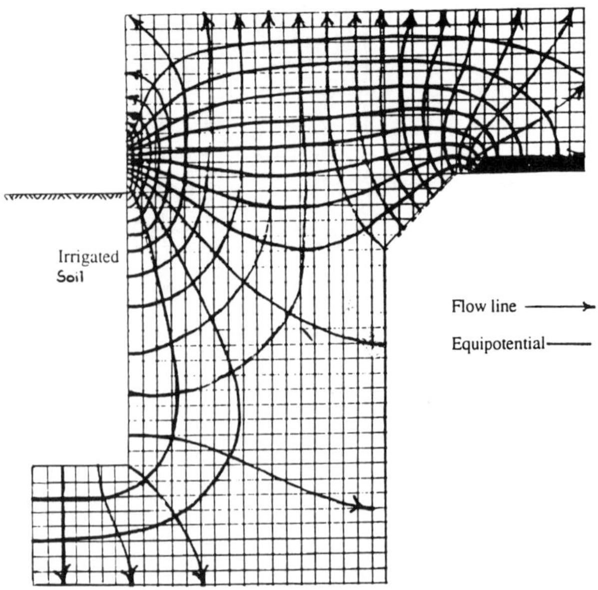

Figure 6. Evaporation at top and side. Moisture barrier in place beneath slab *(Case iv)*.

Case v) Evaporation at Top and Side (Higher Suctions at Side). Moisture Barrier In Place Beneath Slab.

This case examines the possibility and consequences of having much higher soil moisture suctions in the above-ground, exterior portions of the stem wall relative to the top of the slab portion of the foundation. This situation can arise as a result of exposure to severe drying effects of sunshine and/or wind on outside surfaces. Accordingly, a hydraulic head (suction) of -2.6 bars was assigned to most of the above-ground, exterior surface of the stem wall vs. a lower value of -1.8 bars to the top of the slab. A moisture barrier (polymeric or plastic sheet) was assumed to be in place beneath the slab that prevents moisture flow from the ground upward along this interface.

The flow net for this case is shown plotted in Figure 7. Case i) and Case v) represent quite different conditions/assumptions and their flow nets likewise differ substantially. Most of the flow (discharge) in Case v) is exiting or emanating from the above-ground, exterior portion of the stem wall. The source of this flow comes entirely from the irrigated zone adjacent the below-ground portion of the stem wall.

This means that the composition/quality of the irrigation water together with the addition of any soluble salts or agricultural amendments (e.g., gypsum or other sulfate compounds) in the soil adjacent the stem wall will determine to a very large extent the type, transfer and accumulation of salts in the above-ground, exterior surface of the stem wall. As noted previously this finding has significant implications with regard to the question of possible sulfate attack of the concrete and its manifestations.

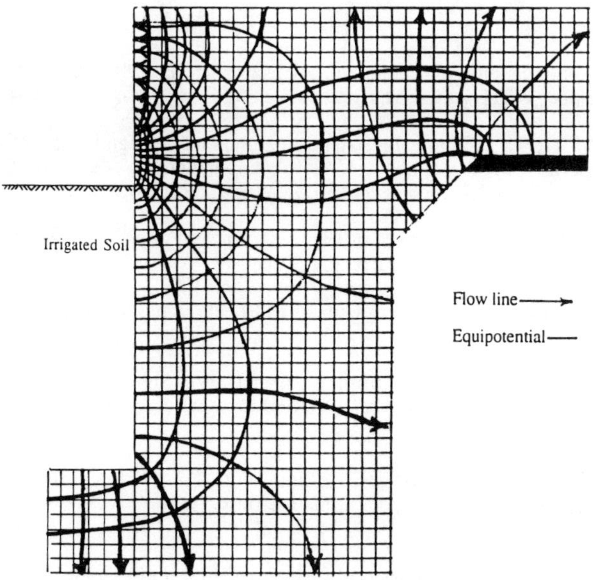

Figure 7. Evaporation at top and side. Moisture barrier in place beneath slab. Higher suction at sides *(Case v)*

Discussion and Conclusions

The flow nets were constructed for a range of possible soil moisture boundary conditions or scenarios. The flow nets show the origin, direction, and destination of flow in the concrete foundations under different scenarios.

Cases iii and iv, which assume evaporation at the above-ground, exterior portion of the stem wall, show that a significant portion of the flow (and any dissolved or soluble salts) simply originates in the irrigated soil adjacent the below-ground portion of the stem wall. Evaporation at the side in turn causes much of this water to exit or emanate from the above-ground, exterior portion of the foundation or stem wall. The proportion of the flow emanating in this manner increases further if evaporation rates (and soil moisture suction) are assumed to be higher (Case v) at the sides of the above-ground exterior portion of the stem wall relative to interior, top surface of the slab. This assumption seems quite reasonable since the above-ground exterior surface of the stem wall is much more likely to be exposed to severe drying effects of the sun and wind during much of the year in the dry, arid climates like Southern California.

If most of (or a substantial proportion of) the flow originates from the irrigated soil adjacent the below-ground portion of the stem wall, this means that the composition and quality of the irrigation water together with the addition of any soluble salts or agricultural amendments (e.g., gypsum or other sulfate compounds) in the soil adjacent the stem wall largely determine the type and accumulation of salts in the above-ground, exterior surface of the stem wall. This in turn has far reaching implications for the likely manifestation and cause of any sulfate attack and damage to the concrete foundation.

References

ACI (1998). Guide to durable concrete. ACE 201.2R-92, ACI Manual of Concrete Practice, Vol. 1, American Concrete Institute

DePuy, G.W. (1994). Chemical resistance of concrete. In Significance of tests and properties of concrete and concrete-making materials. ASTM 169C, P. Klieger and J. Lamond, eds. ASTM, West Conshohocken, PA, pp. 263-281.

Gray, D.H. (1969). Frevention of moisture rise in capillary systems by electrical short circuiting. *Nature* **223**: 371-374.

Kleiner, D.E. (1985). Engineering with spreadsheets. *Civil Engineering* (ASCE) **55**(10): 55-57

Rollings, R.S., Burkes, J.P. and Rollings M.P. (1999). Sulfate attack on cement-stabilized sand. *J. of Geotechnical and Geoenvl. Engineering* (ASCE) **125**(5): 364-372.

Salt Damp Research Committee (1976). Salt Damp: First Report of the Salt Damp Research Committee, Alidade House, Alidade, Australia, October 1976.

Yen, B.C. (1990). Residential foundation deterioration study for the cities of Lakewood, La Palma, and Cypress, California. Report prepared by California State University, Long Beach, pursuant to Senate Bill 2016, Chapter 1438 of the California State Statutes of 1988. 117 pp.

Unsaturated Performance Comparison of Compacted Clay Landfill Liners

Carol J. Miller[1], Member, ASCE, Saad Merayyan[2], Student Member, ASCE, and Nazli Yesiller[3], Associate Member, ASCE

Abstract

Tests were conducted to determine the variation in volumetric water content and pore water suction for a variety of compacted clay soils used in the construction of landfill liners. The fit of the experimental data to an existing parametric model was investigated for two different fitting techniques. The first technique involves the use of the retention curve computer program (RETC) developed for the U.S. EPA. The second technique employs the Solver subroutine included in Microsoft Excel. The parametric models resulting from either technique correlated well to the experimental data. However, the individual curve fit parameters varied significantly. The effect of these variations on the unsaturated behavior of compacted clay liners was evaluated using the Hydrologic Evaluation of Landfill Performance (HELP) model. The curve fit parameters resulting from both the RETC and the Solver techniques were used as input to the HELP routine for simulation of variably saturated flow through a cover liner. There were no significant differences in the volume of leakage or rate of leakage predicted using the input from the two curve-fitting techniques. However, there was significant soil-dependent variation in the HELP output. Examination of the HELP output provides information regarding variation in the volumetric water content of the cover liner soil. This information can be used to predict pore water suction variations and susceptibility to desiccation cracking.

[1] Professor, Department of Civil and Environmental Engineering, Wayne State University, Detroit, MI 48202.

[2] Graduate Teaching Assistant, Department of Civil and Environmental Engineering, Wayne State University, Detroit, MI 48202.

[3] Assistant Professor, Department of Civil and Environmental Engineering, Wayne State University, Detroit, MI 48202.

Introduction

The primary function of a cover liner at a landfill is to limit the migration of moisture to the underlying waste, thereby reducing the volume and rate of leachate production. Cover liners are constructed at volumetric water contents less than the saturated value, and during the life of the containment system moisture is added to and removed from the cover liner. Therefore, it is expected that, for at least a portion of it's life, the cover liner will be operating in an unsaturated mode. The governing equation for flow through a variably saturated liner, assuming one-dimensional, vertical flow is:

$$C(\psi)\frac{\partial \psi}{\partial t} = \frac{\partial}{\partial z}\left(K(\psi)\left(\frac{\partial \psi}{\partial z} + 1\right)\right) \quad (1)$$

where t is the time (sec); z is the elevation assumed positive upward; ψ is pore water suction head (m); $K(\psi)$ is unsaturated hydraulic conductivity (m sec^{-1}); and $C(\psi)$ is specific water capacity (m^{-1}). $C(\psi)$ relates pore water suction to volumetric water content and is equivalent to the slope of the soil-water characteristic curve (SWCC). The data necessary to construct a SWCC is typically determined using pressure plate apparatus (ASTM, 1994). The resulting data can be fit to existing parametric models of the SWCC (Brooks and Corey, 1964; Fredlund and Xing, 1994). One common model is the van Genuchten (van Genuchten, 1980) equation:

$$\frac{\theta_w - \theta_r}{\theta_s - \theta_r} = \frac{1}{\left[1 + \left(\frac{\psi}{a}\right)^n\right]^m} \quad (2)$$

where the optimized parameters are θ_s, θ_r, a, n, and m. The effect of the parameters a, n, and m on the shape of the SWCC is provided by Leong and Rahardjo (1997). Van Genuchten (1980) suggested a simplifying relation between the parameters m and n (i.e., m = 1-1/n). However, in the present investigation the parameters m and n are assumed to be independent for more flexibility in the curve-fitting procedure.

The SWCC can be used to provide insight to the potential for desiccation cracking of a cover liner soil. During desiccation, the saturation of a liner is reduced, and the remaining pore water is held at increasingly large suctions. If the volumetric water content is known, or has been predicted, the SWCC can be used to determine the pore water suctions corresponding to the known volumetric water content. Previous research (Miller et al., 1998) has shown that the onset and resulting amount of cracking is correlated to a soil-specific critical suction level. Hence the maximum pore water suction expected in the cover liner can be compared to the critical suction level to estimate susceptibility to desiccation cracking.

The landfill cover liner problem is not the only environmental application that requires modeling of unsaturated conditions. A similar problem involves surface spills that migrate through the unsaturated zone towards an unconfined aquifer. In general, the solution of pollutant migration through the unsaturated zone is a much more difficult problem than contaminant transport through the aquifer. In the aquifer, the pore water pressure is linearly related to depth below the free surface and the hydraulic conductivity is assumed constant. However, in the vadose zone, the presence of the air phase impacts the relationship between pore water pressure and volumetric water content, as well as hydraulic conductivity and volumetric water content.

This study was conducted to investigate the impact of variations in the characterization of the unsaturated behavior of a cover liner soil on the performance of that liner. The liner performance was evaluated using water balance output from a landfill simulation model.

Materials and Methods

The materials used in this study consist of six clay soils used as landfill cover liners at five municipal solid waste landfills in the metropolitan Detroit vicinity. One of the sites had two active excavation regions, yielding two distinct clay soils. Samples of the liner materials were retrieved from the excavation areas used as borrow sources for liner construction at each site. Table 1 provides a summary of the geotechnical characterization of each soil. In subsequent discussions, the six soils will be referred to using the numerical designation provided in column one of this table, i.e. the clay liner from Auburn Hills will be designated Soil 1.

Table 1 - Geotechnical Characterization of the Clays

Soil	Saturated Hydraulic Conductivity (m/sec)*10^{-10}	Distribution of Fines (% of total)		Max. Dry Density (kN/m^3)	Opt. Moisture Content (% by wt)	Plasticity Index	Liquid Limit
		Clay	Silt				
1	6.32	32	31	21.1	8.6	10	22
2	3.40	28	43	20.4	10.2	10	26
3	1.80	36	41	20.2	9.4	14	30
4	4.83	31	47	20.1	10.3	13	35
5	6.40	36	32	20.3	9.1	16	32
6	0.51	64	34	16.7	22.5	60	83

Note: Soil designations 1-6 refer to soil from the following locations: Auburn Hills, Woodland Meadows (brown clay), Woodland Meadows (gray clay), Evergreen, Canton, and Smith's Creek (with 25% by weight bentonite added), respectively.

Although mineralogical analysis was not included in the scope of this research, previous investigations (Salim et al., 1996) suggest that the mineralogical make-up of clay soils in southern Michigan are fairly similar and that site-specific mineralogical analysis may be unwarranted. A typical mineralogical composition shows the clay fraction to be dominated by three minerals with 60% illite, 12% kaolinite, and 9% chlorite.

Volumetric pressure plate and pressure membrane extractors were used to determine the soil-water characteristic curve (SWCC) during the drying cycle for remolded samples of each of the six soils. The methodology is described in ASTM Standards D2325 (ASTM, 1993) and D3152 (ASTM, 1994) for the plate and membrane apparatus, respectively. The samples were remolded using modified Proctor compactive effort at moisture conditions approximately 2% wet of optimum. The volumetric water contents were used together with the applied suctions to develop the SWCCs.

Data collected was fit to the models described by Eq. 2. An optimization routine was used to fit the parametric models to the measured data by altering the fitted parameters using an iterative process until the difference between the predicted SWCC and the measured data was minimized. The quantity to be minimized is the Sum of the Squared Residuals (SSR):

$$SSR=\sum_{i=1}^{n} w_i\left(\theta_{wi}-\theta_{ci}\right)^2 \tag{3}$$

where w_i is the weighting factor which is equal to 1.0, θ_{wi} is the measured volumetric water content, which is defined as θ_{wi} = nS (n=soil porosity and S=saturation corresponding to θ_{wi}), and θ_{ci} is the calculated volumetric water content from each model. The minimization process for SSR was performed using two techniques: a standard statistical package included with Microsoft Excel® and an optimization package developed specifically for this purpose by van Genuchten et al. (1991). Detailed information regarding the curve fit for each soil and each optimization technique is presented in Table 2. In all cases, the SSR is less than 10^{-3}, which is similar to the SSR values obtained by Leong and Rahardjo (1997). Each of the optimization approaches provides an acceptable fit to the experimental data, with similar values of the SSR. The individual curve fit parameters, however, vary significantly between approaches.

Table 2 – van Genuchten Curve Fit Parameters

Soil	Technique	a (kPa x 10^{-2})	n	m	θr	θs	SSR
1	Solver	0.0035	0.7131	31.6122	0.1449	0.2419	$9.60X10^{-6}$
	RETC	0.2856	1.0050	1.5980	0.1408	0.2304	$10.22X10^{-6}$
2	Solver	0.5809	56.9445	0.0185	0.2424	0.3351	$23.54X10^{-6}$
	RETC	0.5828	150.7430	0.0067	0.2408	0.3351	$23.77X10^{-6}$
3	Solver	0.1125	0.8311	0.3465	0.0003	0.3106	$121.59X10^{-6}$
	RETC	0.1525	1.0050	0.4697	0.1186	0.3065	$10.47X10^{-6}$
4	Solver	0.0102	0.8711	11.8146	0.1962	0.3231	$3.12X10^{-6}$
	RETC	0.1560	1.0050	1.0896	0.1702	0.3202	$3.31X10^{-6}$
5	Solver	0.0121	1.6752	166.8758	0.2700	0.3485	$42.67X10^{-6}$
	RETC	0.0117	1.6768	178.7306	0.2700	0.3485	$49.06X10^{-6}$
6	Solver	0.3668	2.0737	0.5178	0.3488	0.5525	$36.75X10^{-6}$
	RETC	0.3456	2.0848	0.5203	0.3525	0.5525	$49.06X10^{-6}$

It was observed that the best-fit saturation volumetric water content (θ_s^*) was relatively independent of the approach used, although there was significant variation in the value between the six soils. The maximum variation in θ_s between optimization techniques was 5% for Soil 1. The maximum variation in θ_s among the soils was greater than 100%, representing the variation between Soil 1 and Soil 6. For a saturated soil, the saturated volumetric water content, θ_s, is equal to the soil porosity. Thus, it was expected that there would be significant variation in θ_s among the soils.

The optimized value of the irreducible volumetric water content, θ_r, showed greater variation between techniques than the θ_s variation. The variation between optimization methods was on the order of 15% for all soils, with the exception of Soil 3 which showed a much greater dependence on optimization technique and a variation on the order of 100%. The variation in θ_r between soils was on the order of 50%, except for Soil 3 which exhibited a θ_r much less than the others.

The other curve fit parameters: a, n, and m represent the shape and offsets of the SWCC. These parameters, especially the latter two, show significant variation between soil types and between optimization approaches. However, the curve shapes that result from the two approaches appear very similar, as shown in Figure 1. The parameter set defining the shape of the SWCC is not unique. The similarities observed in the SWCCs from the two techniques suggests that the added effort in using the RETC program is unwarranted. Operation of the RETC requires knowledge in DOS and Fortran formatting of input files. On the other hand, the Solver subroutine does not require formatted input and has a Windows interface.

Modeling of flow in the unsaturated zone requires input of the SWCC, usually in a form similar to that in Eq. 2, along with the associated curve fit parameters (θ_s, θ_r, a, n, and m, in this case). Therefore, it is expected that alternative data sets for these parameters will result in variable model output. This hypothesis is tested using the HELP model for prediction of landfill cover liner performance.

Leakage Predictions

The HELP model (Schroeder et al., 1994) was used to simulate the variably saturated conditions in a hypothetical landfill cover liner of one acre surface area. HELP is a computer program developed by the U.S. Army Engineer Waterways Experiment Station for the analysis of a landfill water budget. The model includes precipitation, surface runoff, evapotranspiration, and snow melt in calculation of the site water budget. Excess moisture acts as additions to soil moisture storage in the cover soil while moisture deficits reduce soil moisture storage in the cover soil. The model simulates four different types of soil layers, subdivided based on their functions: vertical percolation layers, lateral drainage layers, barrier soil liners, and combination geomembrane/barrier soil liners. The transport process for each layer is one-dimensional. The model simulates quasi-steady state and quasi-two-dimensional flow (Khanbilvardi et al., 1995).

The landfill cover system was modeled simplistically as a two-component system, comprised of a 60 cm thick upper sandy layer above a 90 cm thick compacted clay layer. The upper sandy layer was modeled as a *vertical percolation layer* of 60 cm thickness with good vegetative cover, a surface slope of 3% and slope length of 305 meters. The run-off curve number corresponding to these surface layer conditions was determined internally in HELP to be 67. The climatological conditions (precipitation, temperature, and solar radiation) simulated were those included in the HELP database as 100 years synthetic data for Detroit, MI.

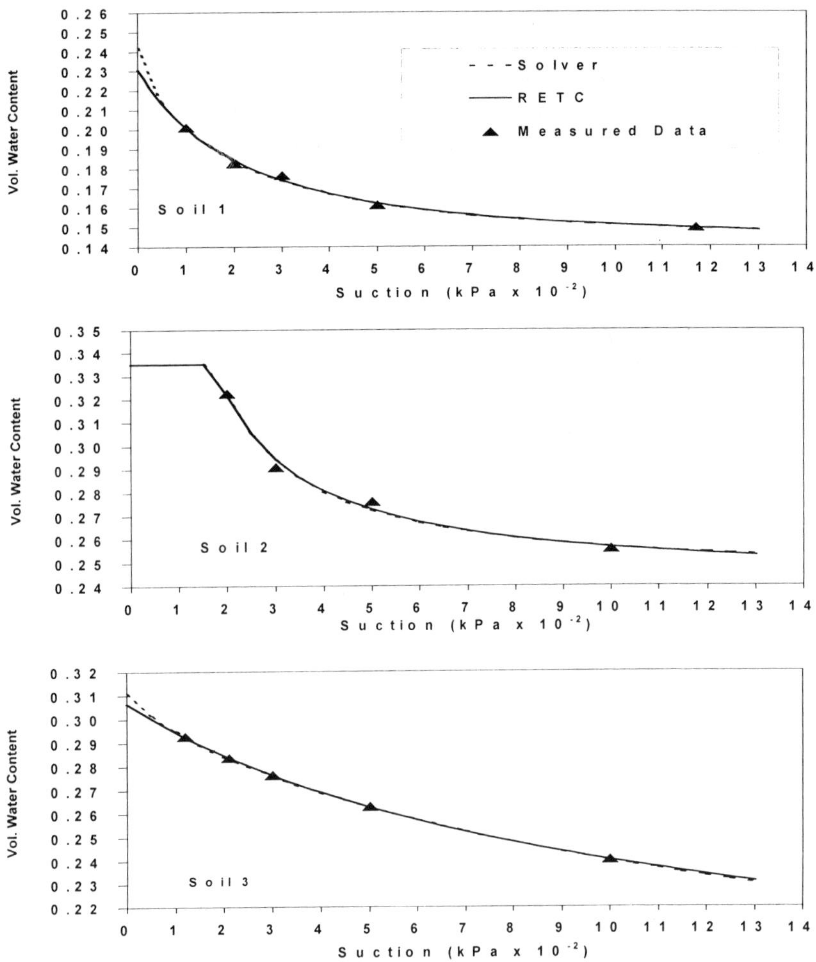

Figure 1a. Van Genuchten Curve-Fit for SWCC for Soils1-3

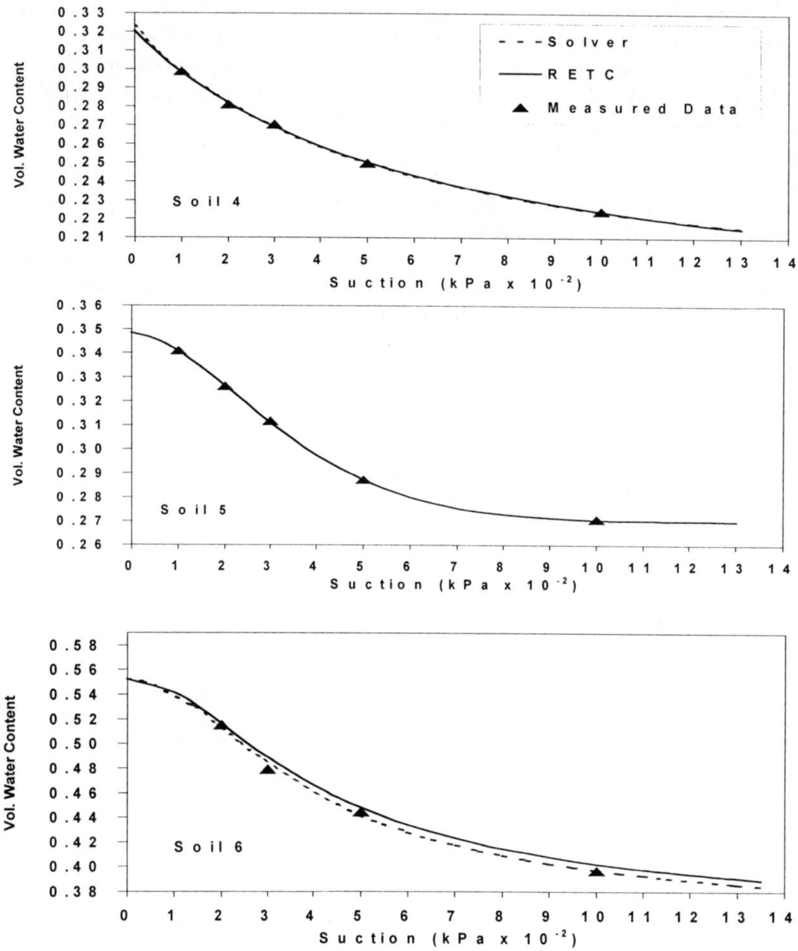

Figure 1b. Van Genuchten Curve-Fit for SWCC for Soils 4-6

It is typical to designate a compacted clay liner as a *barrier soil layer* for HELP modeling. However, the barrier soil layer module assumes the cover liner to be saturated for the entire simulation. Therefore, for all runs performed, the 90 cm thick clay layer was simulated as a vertical percolation layer which allows for variably saturated conditions. This approach allowed the use of the SWCC parameters for the cover liner and permitted the simulation of variably saturated conditions within the cover liner.

The saturated hydraulic conductivity and the volumetric saturated water content (or porosity) used for each soil was that provided in Tables 1 and 2, respectively. The initial moisture condition for all simulations corresponded to a constant moisture profile throughout the clay liner, with the volumetric water content equal to the field capacity. The field capacity is defined as the volumetric water content at a pore water suction of 0.33 Bar while the wilting point is the volumetric water content at 15 Bar (Schroeder et al., 1994). The volumetric water contents corresponding to these pore water suctions were determined using the RETC and Solver best-fit techniques for the van Genuchten formulation of the SWCC (Eq.1) for each of the six cover liner soils. These values are provided in Table 3. The unsaturated characteristics determined using the two curve fitting techniques were used to test the sensitivity of model results to the curve-fit procedure adopted. In each comparative run, the only input parameters varied were the field capacity and wilting point. All other quantities remained at the base run values.

Table 3 – Input Parameters for HELP Simulation

Soil	Technique	FC	WP
1	Solver	0.2203	0.1475
	RETC	0.2185	0.1470
2	Solver	0.3351	0.2519
	RETC	0.3351	0.2514
3	Solver	0.3039	0.2248
	RETC	0.3023	0.2259
4	Solver	0.3130	0.2116
	RETC	0.3123	0.2104
5	Solver	0.3480	0.2700
	RETC	0.3472	0.2700
6	Solver	0.5512	0.3810
	RETC	0.5514	0.3855

Results and Discussion

Figure 2 shows the cumulative leakage over a five year period for each of the cover liners simulated using the HELP model with the Solver parameters, while Figure 3 shows similar information for HELP runs using the RETC parameters. Table 4 presents the cumulative output at the end of the five year simulation for each of the six cover liners and both curve-fitting techniques.

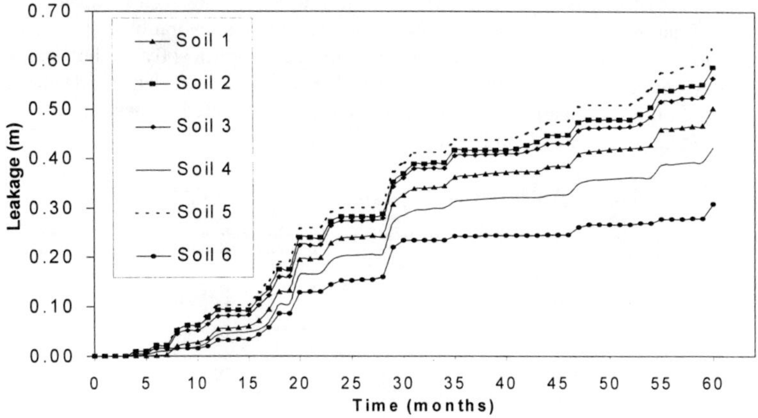

Figure 2. Leakage from each Soil Liner Using the Solver Parameters

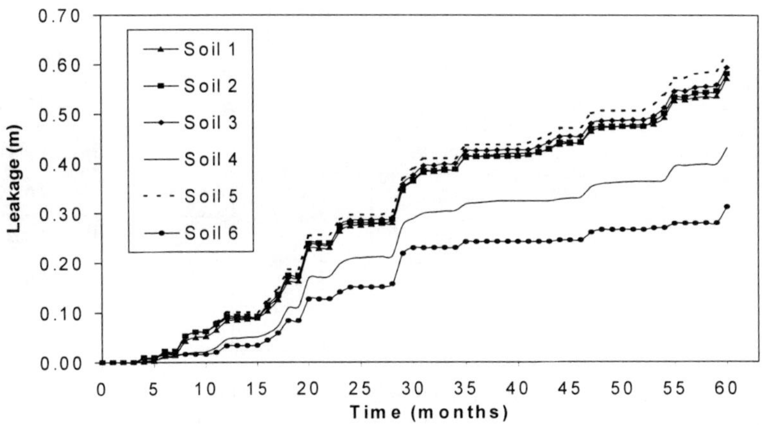

Figure 3. Leakage from each Soil Liner Using the RETC Parameters

Variation due to soil type

Although the cumulative leakage varies between soils (Table 4), the monthly patterns are similar (Figures 2 and 3). For all soils, the leakage was greatest during months 17, 19, and 28. The leakage did not reach steady state conditions during the 60 months simulated. For the entire period of the simulation, leakage through the cover liner constructed with Soil 6 was significantly less than that of the other cover liners. At the end of the simulation period, the leakage through Soil 6 was approximately 60%

less than any of the other cover liners. The diminished leakage can be attributed to a variety of factors, including the decreased value of saturated hydraulic conductivity for Soil 6 and the shape of the SWCC (suctions increase significantly as the volumetric water content is reduced). The leakage through Soil 5 exceeded that of all other soils considered in the simulation. The variation in leakage predicted in these simulations is similar to that reported by Khire et al. (1998).

Table 4 – Water Balance Components from HELP Simulation

Soil	Technique	Cumulative Output (m)			
		Leakage	ET	Runoff	Δ Storage
1	RETC	.5712	2.097	1.471	-0.0358
	Solver	.5038	2.0971	1.4714	0.0379
2	RETC	.5811	2.0567	1.4706	-0.0016
	Solver	.5865	2.0567	1.4706	-0.0075
3	RETC	.5941	2.0693	1.4789	-0.0387
	Solver	.5630	2.0693	1.4789	-0.0047
4	RETC	.4313	2.2078	1.4730	-0.0057
	Solver	.4238	2.2078	1.4730	0.0026
5	RETC	.6227	2.0435	1.4434	-0.0030
	Solver	.6274	2.0435	1.4434	-0.0081
6	RETC	.3127	2.3900	1.4172	-0.0142
	Solver	.3072	2.3946	1.4179	-0.0139

The cumulative evapotranspiration and run-off also are soil-dependent. The evapotranspiration component of the water balance was greatest for Soil 6. Soil 6 was most successful at preventing moisture movement through the cover liner, which increased the moisture available for evapotranspiration. Surface runoff was almost identical for all soils. This is expected due to the identical runoff curve numbers and precipitation history that were used in simulating all six conditions.

Variation due to optimization technique

Differences in the predicted leakage due to the optimization technique adopted for the curve-fitting routine were not significant. The maximum variation in the predicted leakage between the two techniques occurred for Soil 1 and was approximately 13%. The evapotranspiration and runoff components of the water balance were not affected by the optimization technique.

Moisture fluctuations

The HELP model does not provide direct output of the calculated volumetric water content changes in the cover liner soil, except on an annual basis. Monthly changes in soil moisture storage were calculated using the water balance equation:

$$\Delta \text{ Soil Moisture} = P - RO - ET - L - \Delta \text{ SW} \quad (4)$$

where the quantities on the right hand side of the equation reflect the cumulative monthly values of precipitation, runoff, evapotranspiration, leakage, and change in snow water, respectively. The snow water term accounts for that portion of the precipitation occurring when local temperatures are below freezing. Changes in the volumetric water content of the clay cover liner were calculated for Soil 1 and 6. The variation on a monthly basis is shown in Figure 4. Although both soils were subjected to identical climatological conditions, the resulting moisture states were very different. The volumetric water content spans significantly different ranges for the two soils. The volumetric water content for the cover liner using Soil 1 varies between 0.15 and 0.24 while this range for Soil 6 is between 0.50 and 0.55. Soil 6 was much more effective at preventing leakage, and thus a significantly larger fraction of the available moisture was stored above the liner to meet the evapotranspiration demands. Therefore, the liner constructed with Soil 6 maintained comparatively higher values of volumetric water content than Soil 5, which allowed greater leakage.

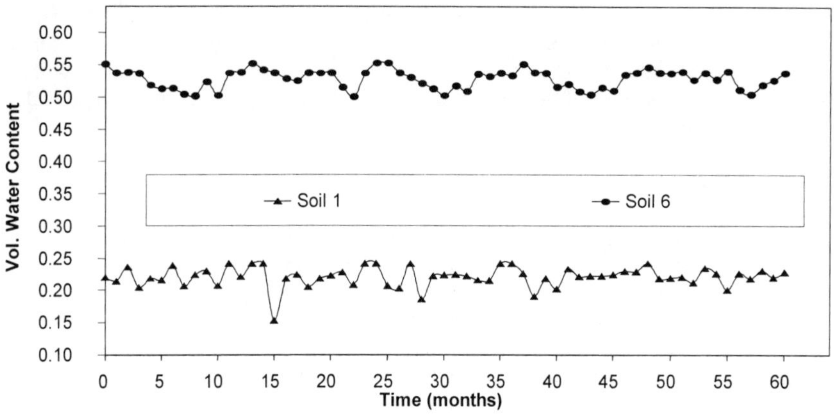

Figure 4. Volumetric Water Content Fluctuations in the Soil Cover Liner

Using the SWCC, the volumetric water contents of Fig. 4 can be converted to values of pore water suction. Previous research by the authors (Miller et al., 1998) has shown that the onset of desiccation cracking may be related to a soil-specific critical suction value. If that value is known for the cover soil under investigation, it would be possible to use the HELP output to determine the likelihood that the cover liner will experience suctions equal to or greater than the critical value. With this information, it would be possible to predict the onset of desiccation cracking. For example, if the critical suction is determined to be 200 kPa for Soils 1 and 6, the HELP simulation suggests that Soils 1 and 6 are subjected to desiccation cracking 2% and 12% of the time, respectively.

Further details regarding moisture conditions within the cover liner soil are provided in Table 5. Although Soil 1 experienced significantly lower volumetric water contents than Soil 6, it did not exceed the critical suction as frequently as did Soil 6. Therefore, it is expected that Soil 6 is more susceptible to desiccation cracking than Soil 1. However, this limited analysis provides no methodology to predict the extent of cracking.

Table 5 - Details Regarding Moisture Conditions Encountered in Clay Cover Liner

Soil	Minimum Volumetric Water Content	Maximum Pore Water Suction (kPa)	% of Time Saturated	% of Time Critical Suction Exceeded
1	.1572	880	18.34	1.7
6	.5004	240	10.00	11.7

Table 5 indicates that the cover liners constructed with Soil 1 and Soil 6 are expected to reach saturation during 18.3% and 10.0% of the 5 year simulation duration, respectively. This prediction suggests that analyses of clay cover liners that rely on the assumption of saturated conditions may be inappropriate. Such assumptions likely result in over-estimates of the amount of leakage through the cover liner, while under-estimating the potential for desiccation cracking.

Summary and Conclusions

The variably saturated performance of six compacted clay soils as cover liners was compared using the HELP model. The relationship between the unsaturated volumetric water content and pore water suction was determined in the laboratory for each soil using the pressure plate apparatus. The data was fit to the van Genuchten model of the soil-water characteristic curve (SWCC) using two different optimization techniques. The shape of the two resulting SWCCs were very similar for each of the six soils, although the values of the individual curve-fit parameters varied significantly (exceeding 100% in some cases). The variation in the shape of the resulting SWCCs for the six soils was much greater than the variation due to optimization technique.

The cover liner was modeled as a *vertical percolation layer* rather than the typical approach of using the HELP option for a *barrier soil layer*. This was done to overcome the assumption of continuous saturated conditions inherent in the *barrier soil layer* specification. The design of each cover liner simulated was identical, the only variable being the input data set used for specification of the saturated and unsaturated soil characteristics. Two simulations were completed for each of the cover liner soils that included the parameter sets resulting from both types of SWCC curve-fitting routines.

The results from the HELP simulations showed that the monthly patterns of leakage were similar for all cover liner soils. In addition, leakage through the cover liner was relatively unaffected by the SWCC curve-fitting routine adopted. However, the leakage varied significantly from one soil to another. The cover liner constructed with Soil 5 exhibited the greatest leakage, while the cover liner constructed with Soil 6 exhibited the least leakage. Although it is well known that the relationship between volumetric water content and pore-water pressure is hysteretic, only the drying limb of the soil water characteristic curve was used in these simulations. There may be great variation in the soil water characteristic behavior on the wetting and drying sides of the hysteretic curve. Therefore, it may be unrealistic to adopt the drying curve for simulation of the infiltration process.

Evapotranspiration and runoff were much less variable than the leakage component. There was essentially no variation attributed to the different optimization approaches, with only minor variations due to cover liner soil specification.

Changes in soil moisture storage were calculated on a monthly basis for Soils 1 and 6. The range in volumetric water contents predicted for each soil was very different. The volumetric water contents were converted to pore water suctions using the SWCC and compared to an assumed critical value of 200 kPa. Critical conditions were exceeded 1.7% and 11.7% of the time for Soil 1 and Soil 6, respectively. Both soils were predicted to be unsaturated during a significant portion of the five year simulation period (82% and 90% of the time for Soils 1 and 6, respectively). Simulation of the variably saturated nature of the flow through the cover liner is important to realistically predict leakage behavior.

Acknowledgement

Partial funding for this study was provided by the National Science Foundation under Grant CMS-9713922. Additionally, the Wayne State University Faculty Graduate Research Assistantship award program supported student involvement on the project.

References

ASTM (1993). "Standard Test Method for Capillary-Moisture Relationships for Coarse -and Medium-Textured Soils by Pressure-Plate Apparatus," Designation: D 2325-68, American Society for Testing and Materials, Annual Book of ASTM Standards.

ASTM (1994). "Standard Test Method for Capillary-Moisture Relationships for Coarse -and Medium-Textured Soils by Porous-Plate Apparatus," Designation: D 3152-72, American Society for Testing and Materials, Annual Book of ASTM Standards.

Brooks, R. H. and Corey, A. T. (1964). "Hydraulic Properties of Porous Medium," *Hydrology Paper No. 3*, Civ. Engrg. Dept., Colorado State University, Fort Collins, CO.

Fredlund, D. G., and Xing, A. (1994). "Equations for the Soil-Water Characteristic Curve," *Canadian Geotechnical Journal*, Vol. 31, 521-532.

Khanbilvardi, R. M., S. Ahmed, and P. J. Gleason (1995). "Flow Investigation for Landfill Leachate (FILL)," J. Env. Engrg., ASCE, 121(1), 45-57.

Khire, M., Benson, C. and Boscher, P. (1996). "Water Balance Modeling of Earthen Final Covers," Jnl. Geotech. And Geoenv. Engrg. ,123 (8), 744-754.

Leong, E. C. and Rahardjo, H. (1997). "Review of Soil-Water Characteristic Curve Equations," *J. Geotech. and Geoenv. Engrg.*, ASCE, 123(12), 1106-1117.

Miller, C. J., Mi, H., and Yesiller, N. (1998). "Experimental Analysis of Desiccation Crack Propagation in Clay Liners," *Jnl. Amer. Water Res. Ass.*, AWRA, 34(3), 677-686.

Schroeder, P. R., Dozier, T. S., Zappi, P. A., McEnroe, B. M., Sjostrom, J. W., and Peyton, R.L. (1994). "The Hydrologic Evaluation of Landfill Performance (HELP) Model: Engineering Documentation for Version 3," EPA/600/9-94/168b, U.S. Environmental Protection Agency Risk Reduction Engineering Laboratory, Cincinnati, OH.

Salim, I., Miller, C. and Howard, J. (1996). "Sorption Isotherm-Sequential Extraction Analysis of Heavy Metal Retention in Landfill Liners," *Soil Science Society of America Journal*, 60(1), 107-114.

van Genuchten, M.T. (1980). "A Closed Form Equation for Predicting the Hydraulic Conductivity of Unsaturated Soils," *Soil Sci. Soc. Am. J.*, 44, 892-898.

van Genuchten, M., Leij, F., and Yates, S. (1991). "The RETC Code Quantifying the Hydraulic Functions of Unsaturated Soils," Rep. No. EPA/600/2-91/065, U.S. EPA, Ofc. Of Res. And Devel., Washington, D.C.

Laboratory and field tests for studying gas flow through MSW landfill cover soil

Felipe Jucá Maciel[1]
Research Assistant - Department of Civil Engineering, UFPE
José Fernando Thomé Jucá
D.Sc. Associated Professor, ASCE member-Department of Civil Engineering, UFPE

ABSTRACT

The impermeabilization of final cover layer of any municipal solid waste landfill (MSWL) is considered an important matter in reducing rain water infiltration as well as gas flow generated by waste organic matter decomposition process. Landfill gas (LFG) liberation may cause a number of environmental impacts, basically due to the presence of methane and carbon dioxide in its composition. An extensive program of laboratory and field tests were carried out in order to investigate the properties of the final cover layer at two municipal solid waste facilities in Recife-PE, Brazil. This paper presents experimental techniques developed to study gas flow through unsaturated soils. These studies included the use of a laboratory permeameter as well as the development of a simple device for field tests. The results allowed to determine the air permeability parameters and estimate of LFG flow to the atmosphere.

INTRODUCTION

The anaerobic decomposition process of putrescible materials in municipal solid waste landfills (MSWL) produces a flamnable and asphyxiating gas called landfill gas (LFG). The presence of methane, carbon dioxide, and similar gases in landfill gas affects the environment by polluting the atmosphere, among others hazardous consequences. Thus, a complete impermeabilization of the landfill final cover layer to LFG is extremely important to prevent harmful consequences.

The study is part of a program on environmental recuperation of the two major sanitary landfills of the Metropolitan Region of the city of Recife, Brazil: Muribeca's Landfill and Aguazinha's Landfill. In both landfills, the refuse was isolated to restrict the leachate and gas flow. Muribeca's Landfill, which has eight cells, receives daily 2,800 tons of domestic, industrial and hospital waste, while Aguazinha's Landfill receives about 400 tons of solid waste disposed in four cells (Jucá *et al.*, 1998).

----

[1]Researcher, Environment Geotechnics Group, Federal University of Pernambuco-Brazil, Av. Beira Mar 1400 Apt. 101, Piedade –Jaboatão (PE), CEP 54410-001.

The aim of this paper is to present laboratory and field techniques developed to study gas flow through unsaturated soils. Laboratory tests were performed in order to obtain the soil air permeability as a function of the degree of saturation. The laboratory tests were performed using the soil of the cover layer of Aguazinha`s Landfill. The field tests investigation, at Muribeca`s Site, allowed the determination of the LFG flow to the atmosphere and the cover soil capacity in retaining the methane gas. Some aspects of gas generation as well as composition in sanitary landfill are also described in this study.

MSW LANDFILL GAS (LFG)

Gas in sanitary landfills is generated, initialy, by an aerobic decomposition process during waste deposition and continues until there is insufficient free oxygen to sustain the process. After covering the cells completely with soil, an anaerobic decomposition process takes place.

During the first phase of anaerobic decomposition, the presence of organic acids reduce the pH to 4 or 5, making it toxic for the methane production bacteria. A small amount of methane is produced during this period. As the process continues, this bacteria will be more predominant (second anaerobic phase) and will be faster the transformation of volatile acids to methane and carbon dioxide. As a result, the pH increases to neutral range values of 7-8. After this long period, the production of methane will be reduced and more aerobic conditions will be established by means of oxygened water percolation through cover layers into the mass of waste.

Some important factors are responsible for the composition and amount of gases produced in landfills (Waste Management Paper No.27, 1989). Table 1 presents some of these factors as they relate to LFG production.

Table1.Factors controlling the LFG generation.

Physical dimension of the site	Anaerobic processes normally dominate in depth greater than 5m.
Waste type	The composition of waste affects the rate, quality, and Quantity of gas generated.
Site operations	Redution of waste volume by compactation and rapid infilling of a small area of a site will shorten the aerobic degradation.
Waste density	At greater waste density, higher the production of gas per unit volume of void space.
Moisture content	Increase in moisture content by recirculation of leachate in the cells will acelerate gas generation.
pH within the landfill	Methane production will proceed optimally between a pH range of 6.5 to 8.5.
Waste temperature	The optimum temperature range for maximizing the methane production is between 35°C and 45°C.
Ingress of oxygen	Undesirable presence of oxygen during the anaerobic phase will delay gas generation.

The LFG is a mixture of different gases and its properties may vary according to composition. Its composition depends on the waste characteristics, cover, age and collected sample locale. Some typical values were taken from the literature: CH_4

(45-65%), CO_2 (35-55%), N_2 (2.4%), O_2 (0.16%), H_2 (0.05%), CO (0.001%), among others.

The main constituents of the LFG, as described earlier, are the methane and the carbon dioxide, called "Greenhouse Gases" and considered responsible for Earth-heating. In addition, the LFG has other hazardous characteristics that need to be well known: methane is a flamnable gas when in open atmosphere, explosive in a confined space, and asphyxiant either alone or mixed with air. Some of minor LFG constituents could have toxic effects, especially in industrial waste disposal landfills. Therefore, these dangerous properties need to be minimized and one of the ways to prevent such problems is to transform and utilize LFG as a combustible gas.

The landfill gas passage from the cells to the atmosphere is done through a compacted soil layer. The compacted cover soil is usually on unsaturated conditions. In this conditions, the air and water permeability are functions of the soil degree of saturation. The air coefficient of permeability increases when the degree of saturation decrease. On the other hand, the water coefficient of permeability is directly proportional to the increase of the degree of saturation of the soil. The air phase for an unsaturated soil generally becomes continuos as the degree of saturation is reduced to around 85%. For higher degrees of saturation, the air permeability coefficient is apprximately zero, with air flow reduced to a diffusion process through the pore-water (Fredlund and Rahardjo, 1993).

LABORATORY TESTS

Equipment.

The present methodology was modified from models developed by Ignatius (1997), Springer *et al.* (1998), and Stylianou and DeVantier (1995). The tests have been carried out using a flexible membrane permeameter system (Tri-flex 2 - Soil Test - ELE). The system allows the application of confining, inflow, and outflow pressures. The inflow and outflow pressures can be applied using any type of gas. The schematic of the system is shown in Figure 1.

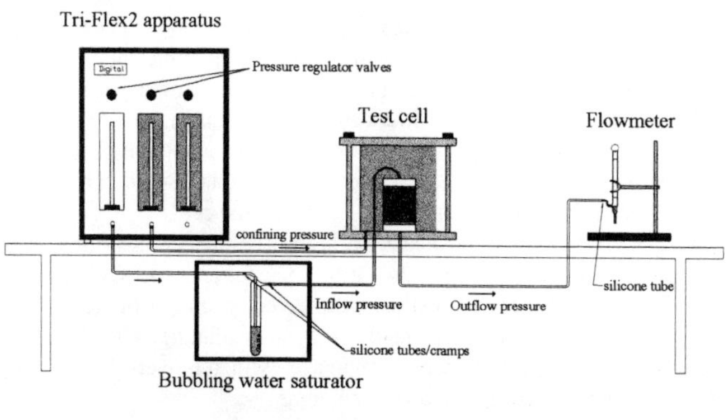

Figure 1. Laboratory test system

In order to eliminate the technical difficulties in working with toxic and explosive gas, the laboratory study has been peformed using air as the permeating fluid. Pressure is delivered to the inlet of the Tri-flex 2 apparatus by means of an air compressor and valve regulated before its injection into the system. Before going into the sample, the air is forced to pass through a glass bubbling air saturator (Fig.1). The saturator is partially filled with deionized water ensuring an increase of saturation of the inflowing air in order to prevent the air from taking water, in vapor form, out of the soil.

A mechanical flowmeter, which is made of glass and graduated to 20 ml of volume, was attached to the outflow end for a volumetric flow rate measurement. A ordinary liquid detergent is used for forming bubbles. It is necessary that the internal walls of the "bubblemeter" had been previouslly cleaned and wetted with detergent to avoid the bubbles "breaking". A stopwatch is also utilized to measure the time of a bubble passage through two indicator marks that correspond to exactly 20 ml of volume. The connections between tubes, the pre-saturator, and the flowmeter were made of silicone tubes. For a complete guarantee of tube connections under high pressures, a "cramp" was added to avoid possible leaks or disconnections.

Soil Studied

The soil used in this research was obtained from the borrow area that supplies materials for the cells` cover of Aguazinha's Landfill. The granulometric soil composition is: 30 % of clay, 8 % of silt, and 59 % of sand (37 % of fine sand). According to the Unified Soil Classification System (USCS), the soil is classified as a sandy clay (SC). The Atterberg limits are: LL= 45 % and PL= 30 %. An optimum moisture content of 14% for a maximum dry density of 17.95 kN/m^3 was obtained from the Standard Proctor compaction test (ASTM D698). The water permeability was determined in saturated condition in the Tri-Flex Permeameter, where the value found for κ_w was 1.8x10^{-8} m/s.

Testing Procedure.

The soil sample tested was compacted using Standard Proctor energy, near the maximum dry density and optimum moisture content. The dimensions of the sample were: 10cm diameter and 5cm height. The lateral surface of the soil was covered with a high viscosity sealing glue (silicone) and two rubber membranes. The fine silicone zone was used to help ensure the absence of sidewall flow between the soil and the rubber membranes in order to avoid preferencial flows. The sample was placed between dry porous stones and filter-papers in the cell.

The system was previously checked with high pressures applications, in order to detect possible air-leakage losses and tube disconnections. Only for applied pressures higher than 250 kPa, anomalies were detected during the test: leaching of soil fine particles through the percolated water and silicone tube disconnections. Therefore, in this study the tests were conducted with relatively low air pressures (up to 120kPa) in order to eliminate these problems.

A low value of net stress was applied in order to obtain a perfect adhesion of the membranes on the specimen. Air flow was oriented vertically downward through the sample. The inflow pressure applied on the top of the specimem varied from 5 to 120kPa (related to atmospheric) depending on the degree of saturation of the sample. The confining pressure was maintained at 10kPa exceeding the inflow pressure for each test. The outflow valve was connected to the flowmeter at atmospheric pressure.

The same sample was used in two series of air permeability tests. In the first series, the sample was peformed with a degree of saturation of 85 %. In this series, it was necessary to apply different air gradient pressures ranging from 10 to 120 kPa. Finishing the first seriers, the sample was saturated and allowed to dry in the laboratory for the next series of tests. The second series consists of 19 air permeability tests conducted in different saturation conditions following a drying path. The range of the degree of saturation varied from approximately 90% to 6%. The first test was conducted with the sample at a degree of saturation of 90,3%. At the end of each air permeability test, the sample was removed from the cell and allowed to dry again. The gradient pressures applied across the specimen were: 5 or 10 kPa. Low pressure (5kPa) was applied to dried samples, while a pressure of 10 kPa was used in order to reduce the time of testing of the wetter samples.

In order to determine the value of κ_a, the air flow rate was measured at steady state condition. The time for reaching the steady state condition varies from 5 (dryier condition) to 30 minutes (wetter condition). The duration of each test depends on the degree of saturation of the sample. For the highest saturation condtion, the test did not exceed two hours.

Results and discussions

The intrinsic permeability was calculated using Equations 1 and 2 which consist of modifications to Darcy's law (Langfelder *et al.*, 1968):

$$\mathrm{K}_{ia} = \frac{q \times \mu \times \Delta x}{\Delta P} \tag{1}$$

$$\mathrm{K}_{a} = \frac{2 \times q \times \mu \times \Delta x \times Ps}{(Pe^2 - Ps^2)} \tag{2}$$

Where: K_{ia} =gas intrinsic permeability considering a non-compressible fluid (m^2),
K_a = gas intrinsic permeability considering a compressible fluid (m^2),
q = apparent flow velocity (m/s),
Δx = sample's height (m),
ΔP = sample's gradient pressure (Pa),
Pe = absolute inflow pressure (Pa),
Ps = absolute outflow pressure (Pa),
μ = dynamic viscosity of air at ambient temperature (Pa.s).

The determination of κ_a could be calculated considering or not air as a compressible fluid using Eqs.(1) and (2) respectively. The influence of gas compressibility on the air permeability determination can be verified through Equation 3. Remembering that $Pe = Ps + \Delta P$ and $p_m = (Pe + Ps)/2$.

$$\frac{K_a}{K_{ia}} = 1 - \frac{\Delta P}{2p_m} \quad (3)$$

The gradient pressure (ΔP) and the average pressure (p_m) applied will determine the relation between the intrinsic permeabilities. For low values of (ΔP) and high values of (p_m) applications, the difference to determine the air coefficient of permeability (κ_a) using Eqs. (1) and (2) will be irrelevant. For gradient pressure applications higher than 10 kPa, the relation (ΔP)/(p_m) is higher than 10% and the difference between K_a and K_{ia} is significant as shown on Table 2. This table presents the gradient pressures used in the first series of tests and also the average pressure calculated, remembering that the outflow pressure was alwallys maintained equal to the atmospheric pressure (Ps=Patm= 101.32 kPa).

Table 2. First series testing program and results

Test	ΔP (kPa)	p_m (kPa)	$\Delta P/p_m$	K_a / K_{ia} (Eq.3)	Apparent flow velocity (q) (m/s)	$K_a / 10^{-15}$ (m^2)	$K_{ia} / 10^{-15}$ (m^2)
1	10	106.32	0.09	0.953	0.5×10^{-4}	4.68	4.91
2	20	111.32	0.18	0.910	1.2×10^{-4}	4.84	5.32
3	40	121.32	0.33	0.835	2.5×10^{-4}	4.78	5.74
4	80	141.32	0.57	0.717	5.6×10^{-4}	4.60	6.44
5	120	161.32	0.74	0.628	9.4×10^{-4}	4.48	7.16

The apparent flow velocity (q) measured for the first series of test and the values of K_a and K_{ia} obtained from Eqs. (1) and (2) respectively, are also presented on Table 2. The results allow to conclude that the equation of K_a is correct used to determine the air permeability for all range of pressures. In the present study, the air coefficient of permeability (κ_a) was calculated in both series of tests using Equation 4:

$$\kappa_a = \frac{K_a \times g}{\upsilon} \quad (4)$$

Where: κ_a = air coefficient of permeability (m/s)
K_a = gas intrinsic permeability considering a compressible fluid (m^2),
g = acceleration of gravity (m/s^2)
υ = cinematic viscosity of air at ambient temperature (m^2/s)

The first series of tests was peformed in order to confirm the laboratory test procedure for determining air permeability from the validy of Darcy's Law. Figure 2 shows the result of the apparent flow velocity as a function of the gradient pressure applied as well as a linear regression of the data. The proportionality between the apparent flow velocity and the gradient pressure is true only for the low range of air pressures used in this tests. The validation of Darcy' Law was experimentally verify by Yoshimi and Osterberg (1963). For high air pressure (over 1MPa) some researchers question the use of Darcy' Law and they suggest the Law of Fick to study the air permeability on unsaturated soils (Blight, 1971; and Dakshanamurthy and Fredlund, 1981).

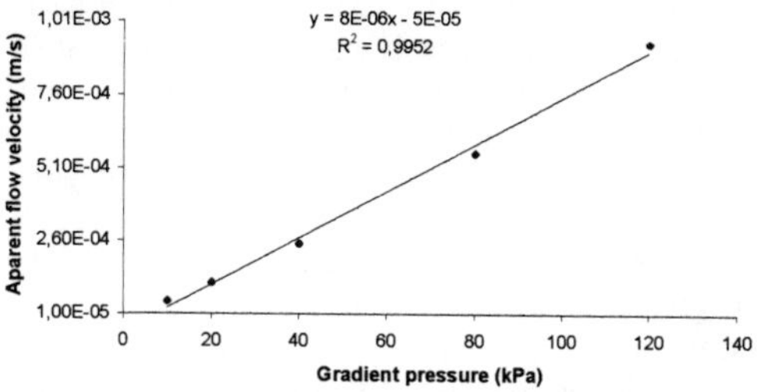

Figure 2. Apparent flow velocity depending on the applied gradient pressure

The second series of tests was peformed in order to evaluate the decrease of the soil air permeability with the increase of the degree of saturation. This evaluation could be representative to understand the mechanism of flow through a compacted unsaturated soil as presented in the cover layer of Muribeca`s landfill. Table 3 provides the second series testing program and results.

Table 3. Second series testing program and results

Test	Gradient Pressure (kPa)	*Confining Pressure (kPa)	Saturation Degree (%)	Aparent flow velocity (m/s)	Air intrinsic permeability K_a (m^2)	Air permeability coefficient (κ_a) (m/s)
1	10.0	20.0	90.3	$4.86x10^{-6}$	$4,06x10^{-16}$	$2,57x10^{-10}$
2	10.0	20.0	88.0	$1.27x10^{-5}$	$1,06x10^{-15}$	$6,71x10^{-10}$
3	10.0	20.0	85.7	$2.64x10^{-5}$	$2,20x10^{-15}$	$1,40x10^{-9}$
4	10.0	20.0	85.2	$6.46x10^{-5}$	$5,39x10^{-15}$	$3,41x10^{-9}$
5	10.0	20.0	82.8	$7.41x10^{-5}$	$6,19x10^{-15}$	$3,92x10^{-9}$
6	10.0	20.0	81.4	$9.88x10^{-5}$	$8,25x10^{-15}$	$5,22x10^{-9}$
7	10.0	20.0	80.4	$9.61x10^{-5}$	$8,02x10^{-15}$	$5,08x10^{-9}$
8	10.0	20.0	79.0	$1.13x10^{-4}$	$9,44x10^{-15}$	$5,97x10^{-9}$
9	10.0	20.0	72.0	$2.31x10^{-4}$	$1,93x10^{-14}$	$1,22x10^{-8}$
10	10.0	20.0	69.9	$2.62x10^{-4}$	$2,19x10^{-14}$	$1,38x10^{-8}$
11	5.0	15.0	56.8	$2.22x10^{-4}$	$3,80x10^{-14}$	$2,40x10^{-8}$
12	5.0	15.0	44.2	$3.34x10^{-4}$	$5,71x10^{-14}$	$3,62x10^{-8}$
13	5.0	15.0	42.4	$3.52x10^{-4}$	$6,02x10^{-14}$	$3,81x10^{-8}$
14	5.0	15.0	31.6	$3.97x10^{-4}$	$6,79x10^{-14}$	$4,30x10^{-8}$
15	5.0	15.0	28.0	$3.54x10^{-4}$	$6,05x10^{-14}$	$3,83x10^{-8}$
16	5.0	15.0	15.1	$4.51x10^{-4}$	$7,71x10^{-14}$	$4,88x10^{-8}$
17	5.0	15.0	13.4	$2.72x10^{-4}$	$4,65x10^{-14}$	$2,94x10^{-8}$
18	5.0	15.0	10.3	$3.51x10^{-4}$	$6,00x10^{-14}$	$3,80x10^{-8}$
19	5.0	15.0	6.3	$3.97x10^{-4}$	$6,79x10^{-14}$	$4,30x10^{-8}$

*related to atmospheric pressure; Patm=outflow pressure = 101,32 kPa

As may be seen the sample follows only a drying path starting at 90.3% of degree of saturation. The hysteresis between the drying and wetting paths was not verified in this program of tests. The maximum value of the air intrinsic permeability was 6.79 x 10^{-14} m^2 for a sandy clay soil. This value is similar to some previous determinations. Fleureau and Taibi (1995) present air intrinsic permeabilities of 6 x 10^{-14} m^2 and 1.5 x 10^{-12} m^2 for a mixture sand/kaolinite and a fine sand respectively. Springer *et al.* (1998) reported the air intrinsic permeabilities depending on the soil water content, the values ranged from 10^{-13} m^2 to 10^{-12} m^2 for a silty sand.

Figure 3 shows the soil air permeability as a function of degree of saturation. The average value of air permeability is about $3x10^{-8}$ m/s for degrees of saturation between 6.3% to 60 %. Above this interval, the air coefficient of permeability is significantly reduced with a corresponding increase in the degree of saturation. The air flow through the soil is extremely low when the soil is reaching the saturation condition (above S = 90%). Some authors has observed this behaviour ranging from 65% to 95% of degree of saturaion depending on the soil type, energy, and moisture content of compaction. This behaviour is probably due to the presence of a discontinuos air phase in the soil. In this phase, the air flows only through the pore water (as entrapped air) and small interconnected air channels. (Langfelder *et al*, 1968, and Fleureau and Taibi, 1995).

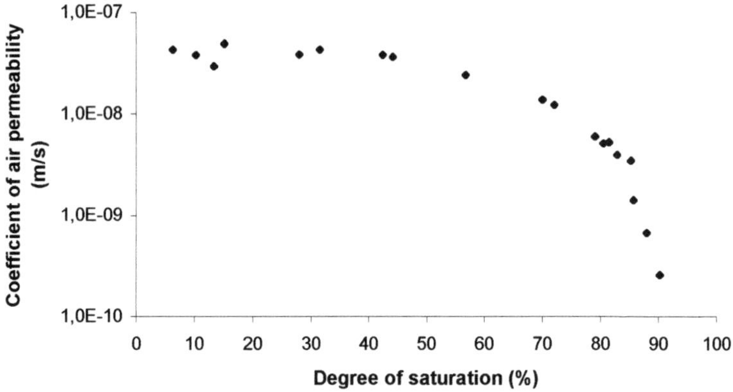

Figure 3. Air permeability coefficient as a function of degree of saturation

FIELD TESTS

Equipment.

Estimation of landfill gas flux from the final cover layer of the cells to the atmosphere is the main objective of the field tests executed at Muribeca's Landfill. In order to obtain this flux rate, an apparatus was designed as shown in Figure 4.

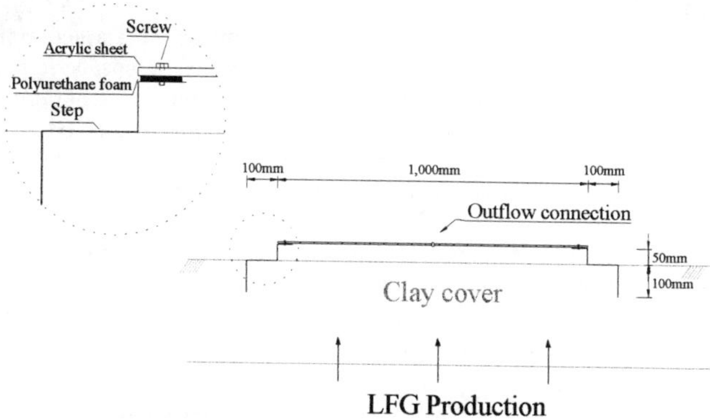

Figure 4. Flux-Box scheme

This equipment has a chamber form made lateraly of a galvanized metal sheet (2 mm thick) with the top made of a clear acrylic sheet (8 mm thick). The acrylic sheet is attached to the metal sheet by five screws on each side of the box. A high-density polyurethane foam between the acrylic and the metal sheets was used to ensure a good seal and prevent gas from leaking. A step of 100 mm height and 100 mm width around the box perimeter was used in order to minimize wind-induced movement of air into and out of the box (see detail). This step has also an important function for setting the Flux-Box into the cover which is to force the box into the soil. The step is placed into the soil thus guaranteeing the flux passage only through an area of 1 m^2, otherwise any empty space between the step and the soil could alter the results.

An outflow gas connection (6,35 mm diameter) was placed in the center of the acrylic sheet, where a fine polyethylene tube was used to convey the landfill gas to a gas detector equipment (Dräger – Multiwarn II) that establishes, by means of sensors, the concentration of methane, carbon dioxide, and oxygen with time for the LFG confined in the Flux-Box. A flux meter is also utlilized on the same outflow connection of the chamber.

Local conditions

The field tests were performed at the final cover layer of Cell 4 at Muribeca`s Landfill. This cell occupies an area of 40.000 m^2 and has an average height of 28 m of waste. The waste age ranged from 12 to 18 months and was completely covered with soil in January 1999. The waste composition in the cell is 60 % of organic matter, 15 % of paper, 8 % of plastic, and 2 % of both metal and glasses. Other materials like woods, stones, and soils are disposed separately and represents 13 % of the total. The soil used in this cover is classified as a sandy clay (SC) and its composition is 17% of clay, 26% of silt, and 57% of sand. The Atteberg limits are: LL = 40 % and PL = 20 %. An optimum moisture content of 17% for a maximum dry density of 16,7 kN/m^3 was obtained from the Standard Proctor compaction test. The water permeability coefficient (κ_w) was determined to be 2.6 x 10^{-8} m/s in saturated condition.

Meteorological conditions may cause significant impact on surface emissions monitoring events. Factors such as precipitation, wind speed, atmospheric pressure, and other seasonal changes in site conditions can alter not only the LFG production but also soil characteristics that are fundamental in controlling LFG emissions. These factors need to be studied and evaluated, along with the results of the tests to determine the influence of meteorological conditions in emissions at any given site.

In this context, data were obtained from a meteorological station at Muribeca`s Landfill on the date of tests. Figure 5 presents the wind speed and atmospheric pressure during the test. The tests lasted for approximately 9 hours from 9 a.m to 18 (6 p.m). Despite the high wind speed values measured during the test, the influence of the wind in the equipment was prevented.

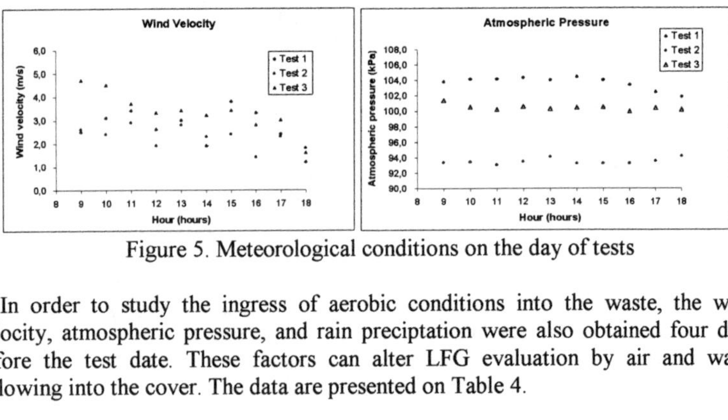

Figure 5. Meteorological conditions on the day of tests

In order to study the ingress of aerobic conditions into the waste, the wind velocity, atmospheric pressure, and rain preciptation were also obtained four days before the test date. These factors can alter LFG evaluation by air and water inflowing into the cover. The data are presented on Table 4.

Table 4. Meteorological conditions four days before tests execution

Test number	Test date	Average wind velocity* (m/s)	Average atmospheric pressure* (kPa)	Rain precipitation (accumulated values) (mm)
1	Aug.19	2.0	102.4	5.2
2	Sept.14	1.6	93.2	2.5
3	Sept.27	2.2	99.5	1.8

Testing Procedure

In order to estabilish the area to be tested, the following aspects were considered: (i) The degree of saturation of the soil must be lower then approximately 80%. The flux rate of gas could be very low or simply non-existent for high values of degree of saturation; (ii) It should be avoided areas near the cells slopes to prevent any influence of the lateral winds or gais drainage, because the gas will hardly percolate through the cover layer in the upwards direction; (iii) It is also recommended to choose a plane area with no cracks on the confined area, otherwise the tests estimate could be mistaken by the soil non-uniformity. In tests with the soil cover presenting cracks, specific studies will be needed such as mapping width and depth of cracks.

After the place is choosen the equipment is set up following some recommendations: (i) The equipment should be positionated with its corner aligned to the wind direction in order to minimize the influence of the wind velocity; (ii) The penetration of the equipment must be done with care to avoid disturbance of the soil; (iii) The outflow gas tube must be already connected to the equipment during its setting in order to avoid any gas accumulation in its interior. This accumulation can alter the upwards direction of the gas percolation through the cover.

Immediately after its setting, a timer should be started and the gas detector equipment must be connected so the reading procedure may be obtained. Atmospreric air must not be allowed to get into the box during the test. Figure 6 shows the Flux-Box set into the cover layer and the gas detector equipment.

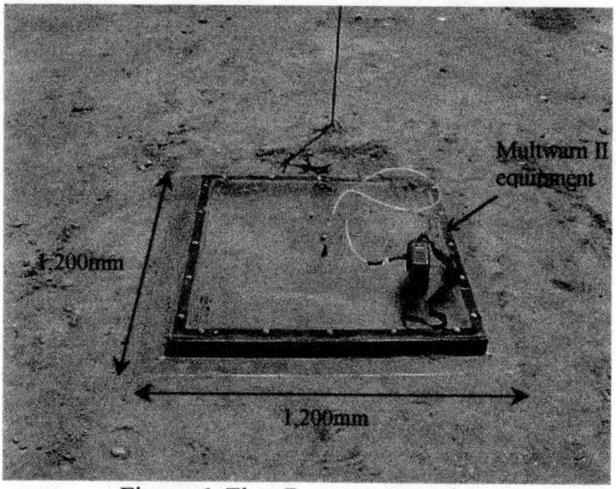

Figure 6. Flux-Box test procedure

The outflowing gas concentration is measured several times, in order to plot concentration as a function of time and finally calculate the gas flux rate. The time intervals for readings are dependent on the flux of gas through the soil. In this case, an interval of 15 minutes was used during the first 2 hours of test and then increased to 30-45 minutes intervals. This change for a longer interval is explained by the decrease in flux and flattening of the curve (CH_4 concentrion versus time). In order to confirm the LFG flux rate, a flowmeter can also be used in this test. It must be connected, on the outflow gas tube, at the beginning of the test before the flux decreases. The flowmeter is the same used in the laboratory tests.

Finished the reading prodedure, which extented for approximately eight hours, the Flux-Box was carefully removed from the cover soil where some parameters of the confined area were investigated. A hole was made up to the waste using a hand auger and its height was measured in order to determine the cover soil height. The soil water content was obtained by weighing three soil samples at the site and oven drying them for 24 hours at a temperature of 110°C in the laboratory. Oedometer rings with a diameter equal to 102 mm and a height of 30 mm were used prior to determine the cover soil natural density. Three soil specimens were retrieved from the middle of the cover by pressing the metallic rings against the soil. Care was taken to remove and trim the samples. The specimens were weighted at the site and had its density determined.

A secondary test was conducted simultaneously with the Flux-Box test in order to determine the methane gas concentration under the cover layer (on the waste). In this test, a hole slightly smaller than 100 mm in diameter was made in the cover soil (up to the waste) using a hand auger, where a plastic PVC tube (100 mm diameter) was placed as shown in Figure 7.

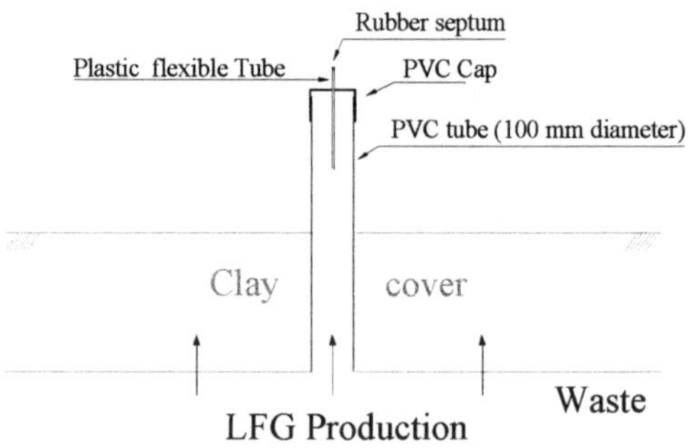

Figure 7. PVC tube test scheme

A plastic flexible tube was fixed on the cap of the PVC tube allowing an evaluation of LFG composition before its passage through the cover. Except when readings are done, this flexible tube must be closed during the test in order not to alter the flux rate in the Flux-Box. Also in this context, this hole should not be done very close to the box. A distance of five meters was used in these experiments.

Results and discussions

Three field tests were peformed on different occasions at Cell 4 of Muribeca`s Landfill. This cell was taken experimentaly in order to estimate the LFG flux to the atmosphere and also determine the cover soil efficiency in minimizing methane surface emissions.

Figure 8 shows the concentration of methane percolating through the cover soil as a function of time. In this figure, a high increase of CH_4 concentration during the first 100 minutes of tests can be observed, followed by the increase of CH_4 concentration at a decreasing rate and finally after approximately 350 minutes, the flattening of the curve which corresponds to the highest value for CH_4 concentration passing through the cover. Table 5 presents the cover soil conditions on the date of each test and its capacity in retaining methane emissions as well as the PVC tube test results.

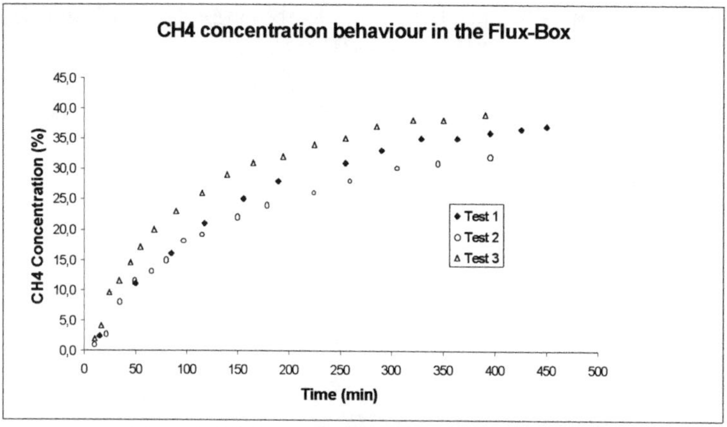

Figure 8. Methane concentration as a function of time

Table 5. Data of the field tests

Test	Cover soil natural density* (kN/m^3)	Water content* (%)	Degree of Saturation (%)	Cover soil height (m)	CH_4 concent. (PVC tube) (%)	Maximum CH_4 concent. (Flux-Box) (%)	Methane retention capacity (%)
1	17.1	15.5	50.2	0.30	42.0	37.0	5
2	16.5	14.5	44.5	0.30	49.0	32.0	17
3	16.8	11.0	38.5	0.30	48.0	39.0	9

* average value of three determinations

The possible reason for the increase of CH_4 concentration at decreasing rate after about 100 minutes of test, is the presence of methane-oxidising bacteria. Although, low rates of methane oxidation was observed in the field tests, large populations of this bacteria and high rates of CH_4 oxidation have been reported in such soils exposed to CH_4 fluxes. This oxidation occurs at ambient temperature in the layer where the concentrations of CH_4 and O_2 are optimal for the micro-organisms involved (Fischer et al., 1995). These low values of oxidation can be explained by the cover soil height of 0.30m. For higher height of soil, the presence of methane-oxidising bacteria would be more predominant causing the increase in the methane retention capacity.

The volume of methane and carbon dioxide in the Flux-Box as a function of the testing time are presented in Figure 9. In order to determine the methane flux rate, inclined lines were fit through the initial points. These lines determined the highest methane flux rate of the tests. On the other hand, a linear regression of the carbon dioxide data allowed the determination of its flux rate. The flux linearity was observed only for CO_2 percolation. The methane non-linearity could be possibly explained by the presence of methane-oxidising micro-organism in the soil. Table 6 presents the CH_4 and CO_2 flux rate for each test. The values determined for CH_4 flux rate were higher than the CO_2 flux due to the slightly difference between the methane and carbon dioxide viscosities ($\mu_{CH4} \approx 0.73\mu_{CO2}$). In Test 3, a slight increase in these gases flux rate is explained by the lowest degree of saturation of the cover soil found in all experiments (see Table 5).

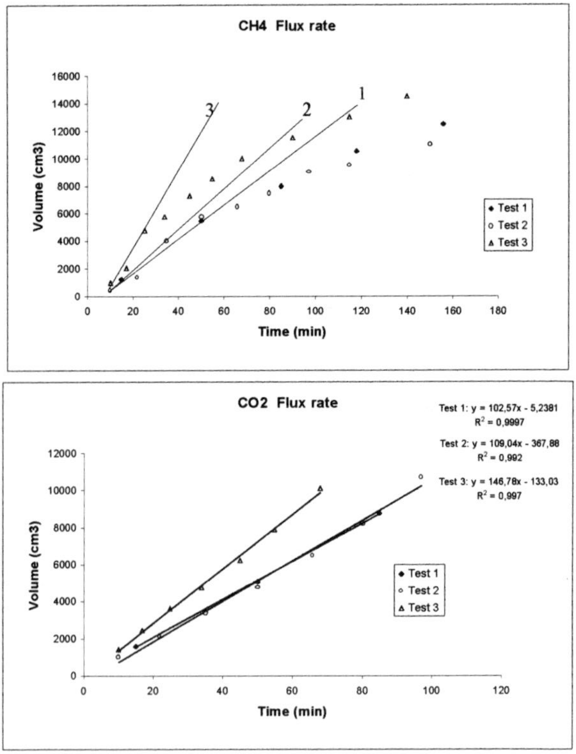

Figure 9. Methane and carbon dioxide flux rate

Table 6. Methane and carbon dioxide flux rates

Test	CH_4 flux rate (m^3/s)	CO_2 flux rate (m^3/s)
1	2.0×10^{-6}	1.7×10^{-6}
2	2.4×10^{-6}	1.8×10^{-6}
3	4.2×10^{-6}	2.4×10^{-6}

The LFG flux rate was also measured using the flowmeter. The average value of 9.8×10^{-7} m^3/s was very similar to the previous determination. This result confirms the estimation of LFG surface emissions to the atmosphere.

In an effort to estimate the pressure gradient across the cover, a non-deformable sample was taken from the same location as the Flux-Box experiment to be tested in the laboratory. The pressure gradient (ΔP) applied across the sample was just enough to represent the same order of flux rate measured in the field: 10^{-6} m^3/s. The values found for ΔP were below 3 kPa. This estimation could be representative to understand mechanism of air flowing into the waste affected by fluctuations in atmospheric pressure. A slight increase of atmospheric pressure will reduce the pressure gradient across the cover causing a decrease in the flux rate. Atmospheric pressure increases higher than the pressure at the bottom of the cover, will reverse the direction of the flow from upward to downward. As a result of this change of directions, the air inflowing will be mixed with the LFG altering its composition by dilution. Atmospheric air may also act reducing the LFG production by the ingress of aerobic conditions.

CONCLUSIONS

The permeability of soil to gas has received a little attention in geotechnical engineering so far plays an important environmental problem, noticeably in the storage of municipal and industrial solid waste. The results from laboratory studies allow the development of a soil gas flow test in a new apparatus adapted from the Tri-Flex Permeameter. The methodology described in this paper allows to determine the soil air permeability for a wide range of degrees of saturation. In this case, a series of 19 tests were performed during a week. This time is much shorter than the time spent for permeability tests performed under controlled suction. In general, controlled suction permeability tests spend this time only to equilibrate the moisture content and obtain a point on the curve of air permeability versus saturation.

The data obtained revealed that the apparent flow velocity is direct and linearly dependent of the gradient pressure, confirming the validity of Darcy's Law for the gaseous fluid utilized in the experiments. It was also concluded that the air permeability is significantly reduced with a corresponding increase in the degree of saturation, especially above 70%. The intrinsic permeability values varied from an order of 10^{-14} m^2 ($\kappa_a = 10^{-8}$ m/s) for samples with 6% of degree of saturation to 10^{-16} m^2 ($\kappa_a = 10^{-10}$ m/s) for samples with approximately 90% of degree of saturation.

The field technique developed and presented in this paper appears to be well suited to provide useful information of the permeability of clay liners cover to gas. Field tests determined a non-efficient final cover layer of Cell 4 at Muribeca's Landfill. The cover layer's capacity in reducing LFG flux rate as well as in retaining methane emissions were investigated. A gas liberation of 86,4 liters/day/m^2 to the atmosphere, or 345,6 m^3/day from Cell 4 (area of 40,000m^2) was observed. The high flux rate measured in the field tests is explained by a poorly compacted cover soil, where low values of natural densities were determined. Low degrees of saturation determined in all the tests contributed to explain the high flux of gases.

Although the simplicity of the Flux-Box methodology presented in this study, the results could provided a satisfactory estimation of the LFG surface emissions at Muribeca`s Landfill site. This parameter is extremelly important to comprehend the mechanism of gas liberation and its pollution at any sanitary landfill site. The future development of this research will futher investigate the final cover layer characteristics such as height, compaction density and cracks, in order to predict the influence of each parameter on the total LFG liberation to the atmosphere.

ACKNOWLEGMENTS

The experimental program described previously was performed with the support of CNPq and EMLURB/PCR. The authors would like to thank Mr. Antônio Brito for his invaluable technical assistence and also Dr. F. Marinho for his suggestions in the review of this paper.

APPENDIX I. REFERENCES

Blight, G.E. (1971). "Flow of air through soils." Journal of the Soil Mechanics and Foundation Division, ASCE, Vol. 97, SM4, pp.604-624.

Dakshanamurthy, V. and Fredlund, D.G. (1981). "A mathematical model for predicting moisture flow in an unsaturated soil under hydraulic and temperature gradients." Water Resource Research 17 (3), pp.714-722.

Fischer, C., Gandolla, M., Aragno, M. (1995). "Heat-generating process around an old sanitary landfill: causes of the phenomenon and possible consequences for landfill management and security." ISWA 25th anniversary congress in Vienna.

Fleureau, J.M. and Taibi, S. (1995). "Water-air permeabilities of unsaturated soils." Proceedings of the First International Conference on Unsaturated Soils, volume 2, pp.479-484, Paris/France.

Fredlund, D. G., Rahardjo, H. (1993). *Soil mechanics for unsaturated soils*. pp.117-143, John Wiley & Sons, Inc. New York, NY, USA.

Ignatius, S. G., Kessler, T., Vicenzetto, F. (1997). "Estimation of quantity of biochemical gas flow through a compacted soil by means of air permeability tests (in portuguese)." 6° Congresso Nacional de Geotecnia – Lisboa/Portugal.

Jucá, J.F.T., Mariano, M.O.H., Melo, V.L.A., Bastos, E.G. (1998). " Monitoring of environmental recuperation of the Muribeca Municipal Solid Waste Landfill." Proceedings of the Third International Congress on Environmental Geotechnics, volume 2, pp.485-489 - Lisboa/Portugal.

Langfelder ,L.J.; Chen,C.F. and J.A.Justice (1968). "Air permeability of compacted cohesive soil." Journal of the Soil Mechanics and Foundation Division, ASCE, Vol.94, SM4, pp.981-1001.

Springer, D.S., Loaiciga, H.ª, Cullen, S.J., Everett, L.G. (1998). "Air permeability of porous materials under controlled laboratory conditions."Ground Water, Ground Water Publisinhg CO.,Westerville, Vol.36, no.4, July-August.

Stylianou, C., DeVantier, B. A. (1995). "Relative air permeability as function of saturation in soil venting." Journal of Environmental Engineering, ASCE, Vol.121, no.4 April.

Waste Management Paper No.27 (1989). "The control of Landfill Gas." Her Majesty`s Inspectorate of Pollution, London.

Yoshimi,V. and J.O.Osterberg (1963). "Compression of partially saturated cohesive soils." Journal of the Soil Mechanics and Foundation Division, ASCE, Vol.89, SM4, pp. 1-24.

USING SOIL DIFFUSION TO DESIGN RAFT FOUNDATION ON EXPANSIVE SOILS

B. M. El-Garhy,[1] **W. K. Wray,**[2] **Fellow, ASCE, and A. A. Youssef**[3]

ABSTRACT: This paper describes the development of a two-dimensional soil-structure interaction model based on the finite element method for calculating the structural design parameters (i.e., moments, shears, and deflections) in a raft foundation (stiffened or constant thickness) resting on expansive soil. The model is capable of estimating the distorted mound shape by considering the soil suction distribution differences throughout the supporting soil mass and the associated volume changes (shrink/heave) with respect to time under a set of different edge conditions. The model requires only a statement of the initial soil suction conditions in the supporting expansive soil mass and the changes in the boundary conditions to predict the response of the raft foundation to these boundary condition changes. The model has been shown to yield results that agree well with reported field measurements of surface volume changes (shrink/heave).

INTRODUCTION

As the soil suction decreases (i.e., the soil water content increases) the ground surface typically moves upward and as the soil suction increases (i.e., the soil water content decreases) the ground surface typically recedes. These cyclic movements (shrink/ swell) in the supporting expansive soils create design and performance problems, particularly for raft foundations.

When a raft foundation is constructed over expansive soils, the effect of evaporation and precipitation on the supporting soil mass beneath the raft foundation is greatly changed.

---

[1] Lecturer, Civil Engineering Department, Faculty of Engineering, Menoufia University, Shebin El-Koom, Egypt.

[2] Professor and Dean, Russ College of Engineering and Technology, 157 Stocker Engineering and Technology Center, Ohio University, Athens, Ohio 45701-2979.

[3] Professor and Vice Dean, Faculty of Engineering, Menoufia University, Shebin El-Koom, Egypt.

The climatic conditions still affect the surrounding soil and cause non-uniform suction or moisture distribution in the supporting soil mass beneath the raft foundation. Consequently, differential soil movements (shrink/heave) occur between the edge and the center of the raft foundation. It is generally accepted that differential rather than total movement of expansive soils is more responsible for major structural damage.

The performance of a raft foundation on expansive soil is largely dependent on the variation of moisture conditions in the supporting expansive soil. Therefore, if an accurate moisture distribution in expansive soil can be predicted, the volume changes (shrink/heave) in these soils and the behavior of a raft foundation resting on them can be reasonably estimated. Moisture condition changes in expansive soils may be caused by different sources such as climate, adjacent trees, and drainage.

Most researchers since about 1965 describe moisture condition variation in unsaturated expansive soils in terms of soil suction (e.g., Richards, 1965; Lytton and Kher, 1970; Mitchell, 1979; Pufahl and Lytton, 1992; Fredlund, 1997). Richards (1974) reported that soil suction could be used to represent the state of the soil water in soils much more effectively than the water content for two reasons. First, soil suction is primarily controlled by the soil environment and not by the soil itself, and it tends to be uniform or change uniformly or at least show continuous trends. Also, the soil suction profile is easier to predict because it tends towards an equilibrium value at a particular depth under particular climatic conditions. On the other hand, water content is highly sensitive to the soil material variables (i.e., soil type, clay content, density, and soil structure). The water content profile is difficult to predict because it can show wide variations and abrupt discontinuities. Additionally, the correlation of soil parameters (i.e., permeability, diffusivity, and shear strength) with water content is poor unless other soil properties (e.g., density and clay content) are considered but, with soil suction these parameters can be most conveniently correlated.

LIMITATIONS IN EXISTING DESIGN METHODS

The available practical design procedures for a raft foundation (stiffened or constant thickness) resting on expansive soil suffer from some limitations on which the approach described herein should improve. These limitations have been discussed by Wray (1978), Holland et al., (1980), and Mitchell (1988).

The major shortcoming is that the design methods currently favored assume the soil-structure interaction to be between the raft foundation and an already distorted soil mound (or dish) shape, which requires an estimate of the initial distorted mound shape. The initial distorted soil surface shape is one of the important variables affecting analysis of raft foundations resting on expansive soils in all of the design methods (Mitchell, 1988). The initial distorted soil surface shape is defined by the edge moisture variation distance, e_m, the maximum differential heave, y_m, and (in some methods) a mound shape exponent, m. The determination of the edge moisture variation distance is considered to be a problem by some persons. Also, estimation of the initial mound shape has the disadvantages of not considering the moisture movement through the supporting expansive soils and the resulting volume change when the soil mass is subjected to different edge effects that commonly cause moisture changes such as climate, trees, and ponded water.

MODELING THE PROBLEM

The analysis method developed in this study considers two problems that are solved independently: (1) three-dimensional moisture movements through the unsaturated expansive soils and the associated soil volume changes, and (2) analysis of structural interaction between the raft foundation and the vertical movements in the supporting expansive soils. For computational simplicity, each problem is solved separately in the approach described here.

In this paper a moisture diffusion and volume change model called SUCH was developed. SUCH is able to determine the distribution of soil suction throughout the unsaturated expansive soil mass and the associated volume changes with respect to time when the soil mass is subjected to different types of edge effects such as climate, trees, ponded water, and moisture barriers. To complete the framework, a soil structure-interaction model called SLAB97 was developed by coupling between the moisture diffusion and volume change model and the structural model to analyze the interaction between the raft foundation and the vertical movements in the supporting expansive soils. The details of SUCH are described completely by El-Garhy (1999).

Water Movement Model

Mitchell (1979) suggested that the movement of moisture through unsaturated expansive soils can be adequately represented by a transient diffusion equation in terms of soil suction expressed in units of pF, Eq. (1). Computer solution of the transient suction diffusion equation is simple and rapid.

$$\frac{\partial^2 u}{\partial x^2}+\frac{\partial^2 u}{\partial y^2}+\frac{\partial^2 u}{\partial z^2}+\frac{f(x,y,z,t)}{p}=\frac{1}{\alpha}\frac{\partial u}{\partial t} \quad (1)$$

where

u = soil suction expressed as a pF (kPa = 0.1×10^{pF})

α = diffusion coefficient (cm^2 / sec)

p = unsaturated permeability (cm / sec)

t = time (sec)

x, y, z = space coordinate

$f(x, y, z, t)$ = internal source of moisture

The diffusion coefficient, α, can be measured in the laboratory (e.g., Mitchell, 1979) or calculated from empirical equations (e.g., McKeen and Johnson, 1990; Bratton, 1991; Lytton, 1994). The different methods used to determine the value of diffusion coefficient, α, for expansive soils are summarized and discussed by El-Garhy (1999).

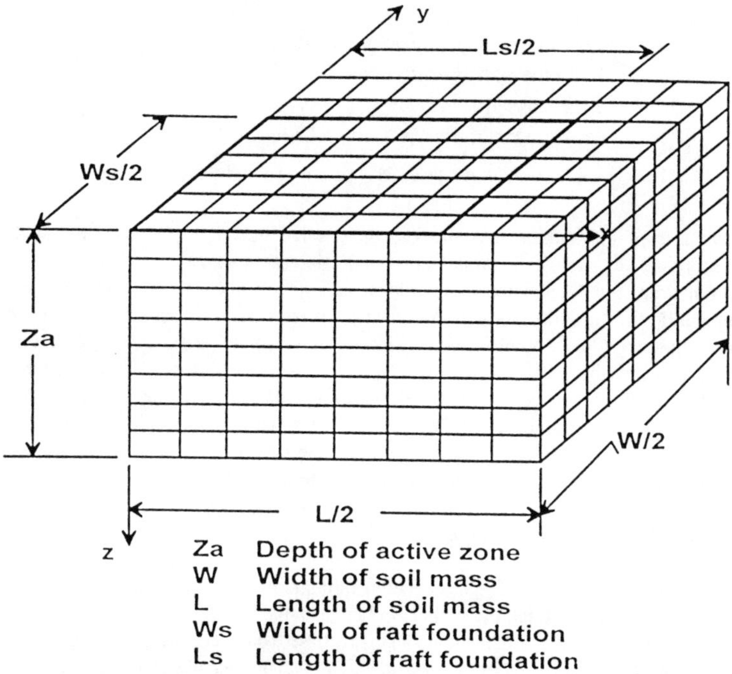

Figure 1. Idealization of raft foundation and supporting soil mass for moisture movement and soil-structural interaction analysis

Method of Analysis

The soil mass is descretized and represented as a grid of points as shown in Figure 1. The finite difference equation representing the change of soil suction with time at each nodal point takes the form:

$$u_{i,j,k}^{t+1} = R\left[R_x^2\left(u_{i+1,j,k}^t + u_{i-1,j,k}^t\right) + R_y^2\left(u_{i,j+1,k}^t + u_{i,j-1,k}^t\right) + \left(u_{i,j,k+1}^t + u_{i,j,k-1}^t\right)\right] + \left[1 - 2R\left(1 + R_x^2 + R_y^2\right)\right] u_{i,j,k}^t \quad (2)$$

where

$$R_x = \frac{\Delta z}{\Delta x}, \quad R_y = \frac{\Delta z}{\Delta y}, \text{ and } R = \frac{\alpha \Delta t}{\Delta z^2}$$

For convergence and stability, the ratio R must be:

$$R \leq \frac{1}{2\left(1 + R_x^2 + R_y^2\right)} \quad \text{or} \quad \Delta t \leq \frac{\Delta z^2}{2\alpha\left(1 + R_x^2 + R_y^2\right)} \quad (3)$$

Δx = division length in x-direction
Δy = division length in y-direction
Δz = division length in z-direction
α = diffusion coefficient ($cm^2 / \sec$)
$u_{i,j,k}^{t}$ = soil suction at time step, t , at grid point (i, j, k)
$u_{i,j,k}^{t+1}$ = soil suction at time step, $t+1$, at grid point (i, j, k)

Two sets of information must be known to solve Eq. (1): (1) the initial suction conditions, i.e., specify the initial values of suction at each node in the soil mass at the initial time, and (2) boundary conditions, i.e., specify the values of suction on the boundaries of the soil mass at each time step. If the values of suction, u^t, at time step t are known at each point in the soil mass, the values of suction, u^{t+1}, at time step $t+1$ can then be calculated by Eq. (2). This procedure, called forward difference method, was used successfully by Richards (1965) to solve the two-dimensional problem of water movement through expansive soils. The value of suction u^{t+1} at time step $t+1$ at each point depends on the values of suction u^t at time step t at that point and the six surrounding points at the previous time step.

Vertical Volume Change Model

After determining the suction distribution throughout the unsaturated expansive soil mass with respect to time, the vertical volume change can be calculated by one of the volume change prediction methods that are based on the concept of soil suction. Equation (4) (Wray, 1997) was used in this model to calculate the vertical volume change at each nodal point in the soil mass:

$$\Delta H_{i,j,k} = \Delta z \, \gamma_{h_{i,j,k}} \left[\Delta pF_{i,j,k} - \Delta pP_{i,j,k} \right] \tag{4}$$

where

$\Delta H_{i,j,k}$ = incremental change in elevation (shrink or heave) at grid point (i, j, k) over the vertical distance Δz
Δz = distance (depth) in z-direction over which shrink or heave occurs
$\gamma_{h_{i,j,k}}$ = suction compression index at grid point (i, j, k)
$\Delta pF_{i,j,k}$ = change of soil suction expressed in pF units (kPa = 0.1×10^{pF}) at grid point (i, j, k)
$\Delta pP_{i,j,k}$ = change of soil overburden over vertical distance (depth) Δz at grid point (i, j, k)

Surface shrink or heave of each nodal point at the surface of the soil mass is calculated by summing the calculated incremental vertical movements of the nodal points on the vertical line passing through that point and extending from the top to the selected bottom of the soil mass. The maximum and minimum expected surface movements (shrink/heave) beneath the raft foundation is determined by comparing the selected bottom of the soil mass. The maximum and minimum expected surface movements (shrink/heave) beneath the raft foundation is determined by comparing the magnitudes of the expected surface movements of nodal points at the surface of the soil mass. Differential movement at each nodal point at the surface of the soil mass can then be calculated by finding the difference between the expected surface movement at that point and the minimum surface movement, usually defined by the edge moisture variation distance.

The different methods used to calculate the value of the suction compression index, γ_h, are summarized and discussed by El-Garhy (1999). McKeen's chart and McKeen's equations are the most frequently cited methods in the literature to estimate the value of γ_h (e.g., McKeen, 1980; Lytton, 1994; and Wray, 1997).

Boundary Conditions

All of the engineering boundary conditions can be expressed as a boundary value or a boundary gradient. Figure 2 demonstrates typical problems that cause a change in soil moisture in the supporting expansive soils and the various types of boundary conditions.

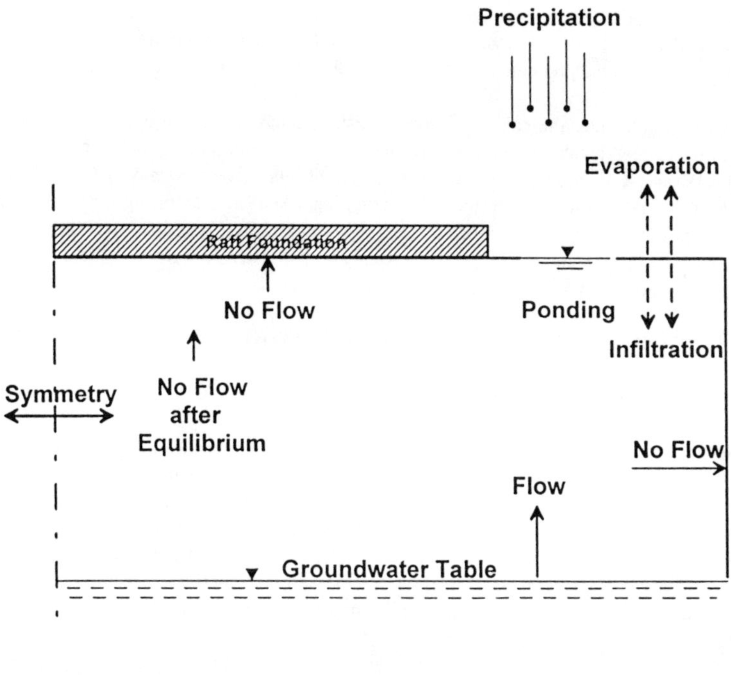

Figure 2. Some types of boundary conditions

Soil-Structure Interaction Model

A FORTRAN computer program called SUCH was written to model the suction distribution problem (El-Garhy, 1999). The program utilized the finite difference technique to solve Eq. (1). SUCH is able to predict the distribution of soil suction throughout the unsaturated expansive soil mass with respect to time and the associated volume changes beneath a raft foundation when the soil mass is subjected to one or some combination of seven different edge effects. These edge effects are: (1) climate only (i.e., no tree, no ponded water, and no barriers), (2) vertical moisture barrier around the perimeter of the foundation, (3) horizontal moisture barrier around the perimeter of the foundation, (4) both vertical and horizontal moisture barriers together around the perimeter of the foundation, (5) tree roots adjacent to the edge of the foundation, (6) trees and vertical moisture barrier at the edge of the foundation, and (7) ponded water adjacent to the edge of the foundation (no barriers).

To complete the framework, a soil-structure interaction model called SLAB97 couples the moisture diffusion and volume change model SUCH and the Finite Element Method to analyze a raft foundation (stiffened or constant thickness) resting on expansive soil. The program SLAB97 was based on the classical theory of thin plates, which considers a plane before bending to remain plane after bending. The raft foundation is assumed homogeneous, isotropic, and linearly elastic and supported on an elastic half space. The genesis of SLAB97 is described by Wray (1978) and is detailed by El-Garhy (1999). The general method for formulating the soil stiffness matrix follows the procedures developed by Cheung and Zienkiewicz (1965).

SLAB97 calculates the values of moments, shears, and deflections at each node in the finite element mesh, as well as the maximum and minimum values of moments, shears, and deflections, which are needed for design. The new model (SUCH plus SLAB97) is able to analyze the raft foundation when the supporting expansive soil mass is subjected to one or more of seven different edge effects that commonly cause moisture changes in the supporting expansive soils which were considered in the moisture diffusion and volume change model SUCH.

COMPARISONS WITH FIELD MEASUREMENTS

Validation of the Model SUCH

SUCH was used to determine the distribution of soil suction through the unsaturated expansive soil mass and the associated volume changes (shrink/heave) with respect to time at a field test site near College Station, Texas (Wray, 1989), for the purpose of program validation. The site was established in 1985 to study the shrink/swell characteristics occurring in the expansive soils beneath a covered surface. Surface elevations and suctions were measured monthly, the suction measurements were monitored by using thermocouple psychrometers to a maximum depth of 2.75m (9.0 ft).

The surface boundary suction may be considered to change as a sinusoidal form with frequency number, n, equal to 1 cycle/year (Mitchell, 1979). Figure 3 shows a comparison of measured and predicted suction values by Mitchell's (1979) cosine wave equation. The boundary suction values at the bottom of the soil mass are fixed and equal to the equilibrium soil suction of 280 kPa (3.45 pF). Figure 4 shows the initial soil suction profile assumed for August, 1985 along with the measured values for the same profile made at the time of site construction (Wray, 1989). The value of the diffusion coefficient, α, of $1.97 \times 10^{-4} cm^2 / \sec$ for the site was calculated by Bratton (1991) and is used in the present study.

The available progressively measured soil suction data at the College Station site is limited due to many of the psychrometers failing to function for a time as a result of soil moisture conditions becoming too wet (Wray, 1989; Bratton, et al, 1990). Therefore, only comparison with the measured values of vertical volume changes (shrink/heave) at the surface of the soil mass for several points were made for each month for the period of five years from August, 1985 through July, 1990. The selected points are located outside the slab edge, at the slab edge, or beneath the slab. Values of suction compression index, γ_h, were calculated for each foot of depth by using McKeen's chart (McKeen, 1980). Comparisons of predicted and measured values for these points are illustrated in Figs. 5 through 8. Referring to these figures, it is observed that: (1) outside and at the slab edge, in direct contact with precipitation and climate, the predicted values match the measured values over approximately the first 1½ years. However, after the first 1½ years, the predicted values are higher than the measured values as shown in Figs. 5 and 6, and (2) inside the slab edge, the measured values are greater than the predicted values for the first approximately three years. The difference between the predicted and the measured values decreases with time and after approximately the first three years the predicted values match the measured values reasonably well under the covered surface as shown in Figs. 7 and 8.

Validation of the Model SLAB97

SLAB97 was tested by comparing the predicted and observed displacements of an actual raft foundation resting on expansive soils. Holland et al., (1980) reported the results of observations of a lightly loaded stiffened raft foundation constructed over a highly expansive soil at the end of a dry season (April, 1976) at Sunshine, Melbourne, Australia.

The 10 cm thick reinforced concrete raft was 7.4 m square, and stiffened by 35 cm wide $\times$ 30 cm depth stiffening beams with spacing between the stiffening beams of 3.7 m in each direction. The raft was loaded by a 7.1 kN/m line load on the perimeter and a 4.7 kN/m load on the center, giving an average uniformly distributed load of 8 kPa over the entire slab area.

The soil profile at the site consists of very stiff gray-brown and gray clays from the surface to at least 5m. Figure 9 shows the measured wet and dry suction profiles at the site. The active zone depth and the equilibrium soil suction are taken to be 2.0m

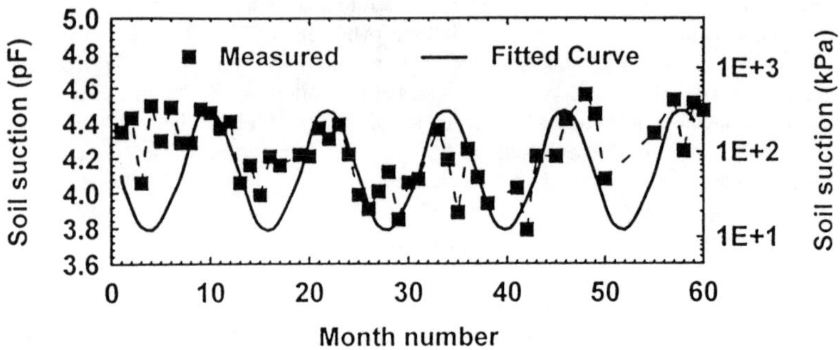

Figure 3. Comparison of measured and predicted suction by Mitchell's cosine wave equation at 0.9m depth (after Bratton, 1991)

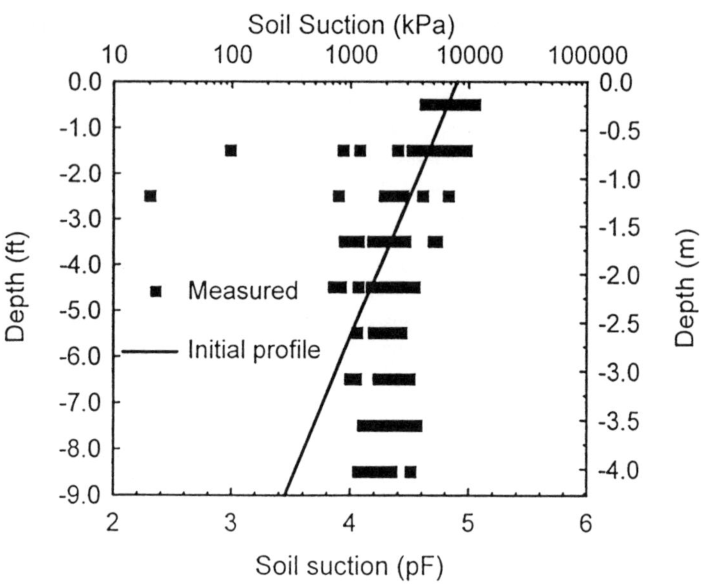

Figure 4. Comparison of assumed initial suction profile and the measured values at the time of site construction at College Station test site

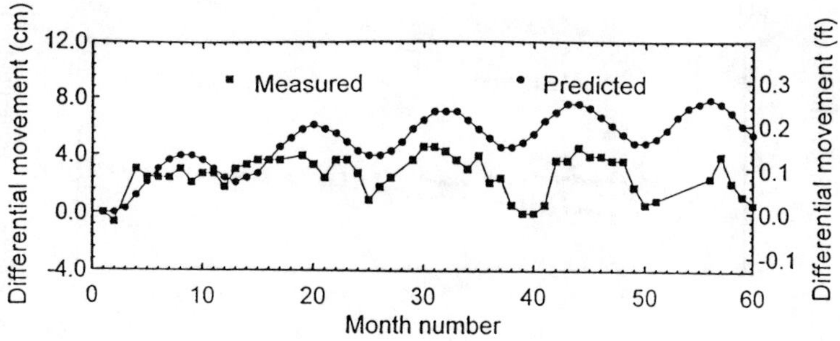

Figure 5. Comparison of predicted and measured monthly surface movements at 1.83 m outside slab edge

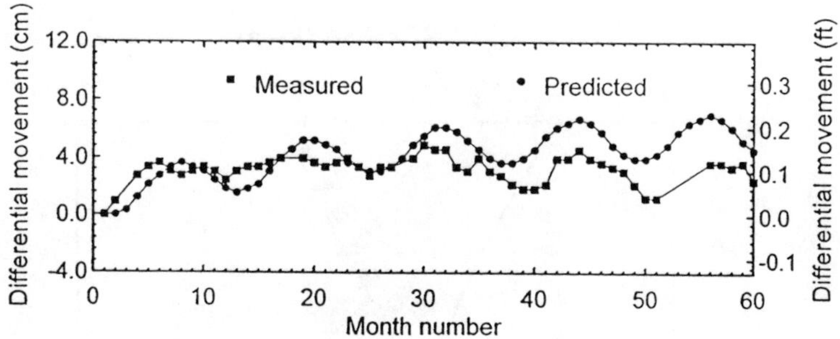

Figure 6. Comparison of predicted and measured monthly surface movements at slab edge

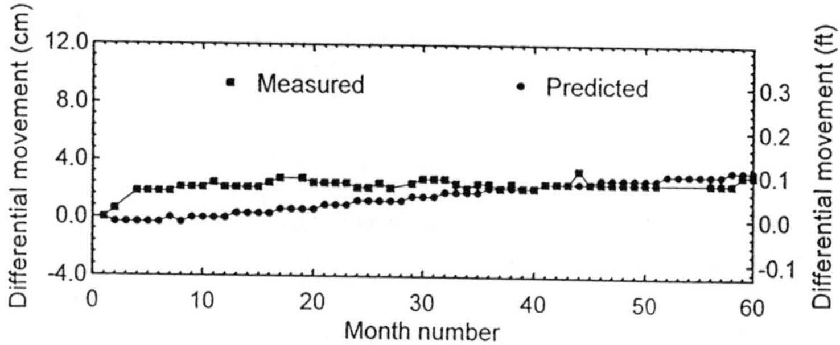

Figure 7. Comparison of predicted and measured monthly surface movements at 1.83 m inside slab edge

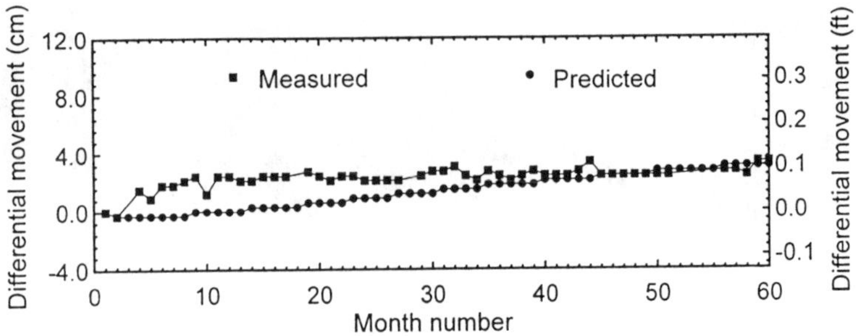

Figure 8. Comparison of predicted and measured monthly surface movements at 2.74 m inside slab edge

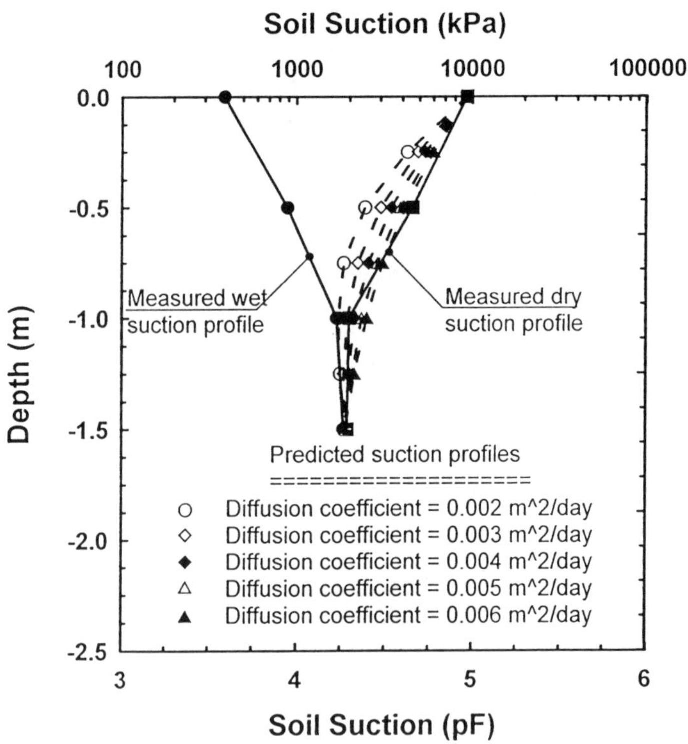

Figure 9. Comparison of predicted and measured soil suction profiles at Sunshine Site, Melbourne, Australia

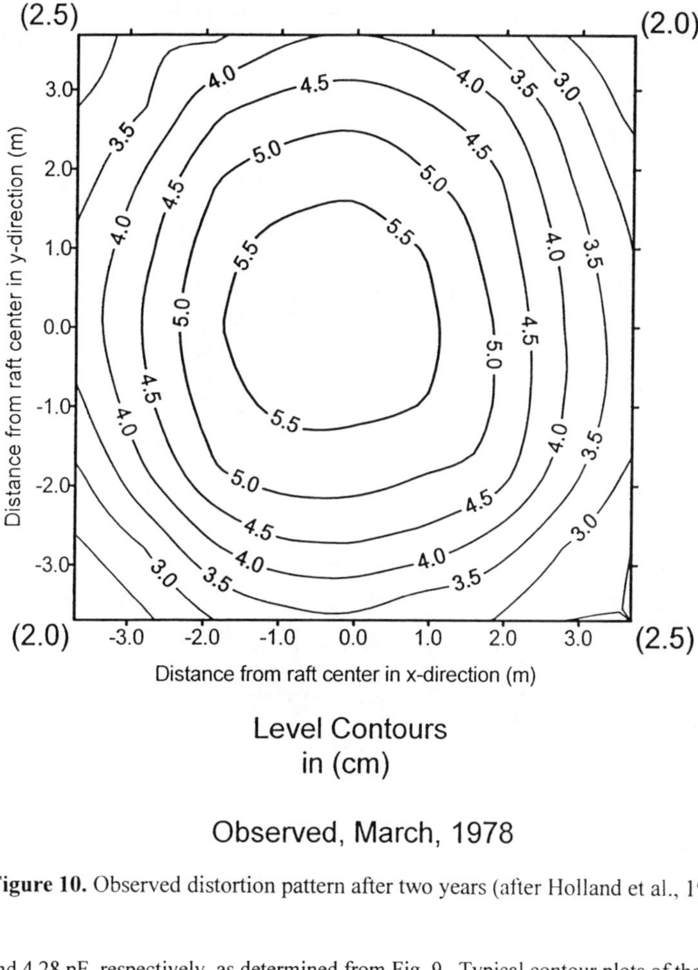

Figure 10. Observed distortion pattern after two years (after Holland et al., 1980)

and 4.28 pF, respectively, as determined from Fig. 9. Typical contour plots of the raft movements after two years from the date of the site construction are presented in Fig. 10.

Unfortunately, no measurements of the diffusion coefficient and suction compression index were made. However, a differential soil movement of 6.3 cm was reported by Holland, et al, as a result of the difference in wet and dry suction profiles shown in Fig. 9. These measurements were used in the model SUCH to estimate the in-situ diffusion coefficient and suction compression index of the soil by back calculation using different values of diffusion coefficient to approximate the suction distribution profile and differential soil movements measured in the field. In these computer runs, the measured wet suction profile was assumed to be the initial suction profile and the

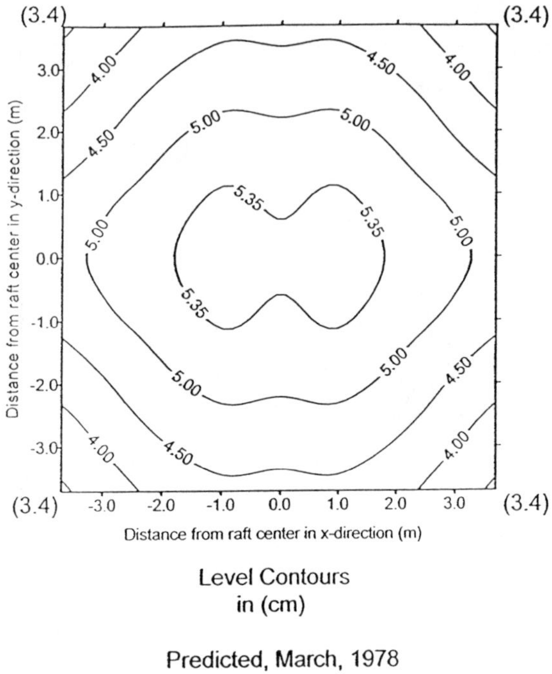

Figure 11. Predicted distortion pattern after two years

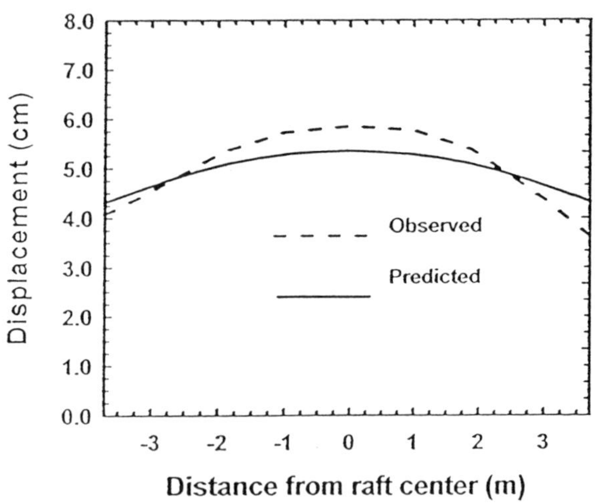

Figure 12. Comparisons of predicted and observed raft displacements after two years

predicted dry suction profile is compared to the measured dry suction profile as shown in Fig. 9. The values of diffusion coefficient and suction compression index of $6.9 \times 10^{-4}\ cm^2 / \sec$ and 0.07, respectively, were selected as providing the best fit for the expansive soil at this site.

A contour plot of the raft deformation pattern after two years as predicted by the model is presented in Fig. 11. This pattern compares closely with the observed distortion pattern of the raft shown in Fig. 10. The predicted raft displacements through the raft's x-axis are compared with the observed values as shown in Fig. 12.

CONCLUSIONS

A two-dimensional soil-structure interaction model incorporating a three-dimensional soil suction distribution model based on soil diffusion has been developed for calculating the structural design parameters in a raft foundation resting on expansive soils. The model improves the principal shortcoming present in design procedures currently used, namely the estimation of the supporting soil distortion shape. It is capable of estimating the distorted soil shape by considering the changes of soil suction distribution throughout the supporting expansive soil mass and the associated volume changes with respect to time under a set of different edge conditions that commonly cause moisture changes (e.g., climate, trees, and ponded water). The model requires an estimation of the most severe boundary conditions which for design purposes may cause the most severe soil suction distribution throughout the soil mass under the raft foundation during the lifetime of the structure. The analysis method introduced in this study appears to be an improvement on other design methods presently available for the design of raft foundations resting on expansive soil because of the direct link between the determination of the distorted soil shape. The results of the analysis show reasonable agreement with the field measurements of both surface elevation changes (shrink/heave) and the behavior of an experimental raft foundation.

APPENDIX I. REFERENCES

Bratton, W.L., C.S. Lee, P.S. Pomper, and Wray, W.K., (1990), "Field Monitoring of Moisture Conditions in Expansive Soils," Proceedings, 1990 Symposium on Engineering Geology and soils Engineering, Pocatello, ID, April 4-6, 1990. pp 36/1-36/14.

Bratton, W. L., (1991), "Parameters for Predicting Shrink/Heave Beneath Slab-on-Ground Foundations Over Expansive Clays," Ph.D. Dissertation, Department of Civil Engineering, Texas Tech University, Lubbock, Texas.

Cheung, Y. K., and Zienkiewicz, O. C., (1965), "Plates and Tanks on Elastic Foundations-An Application of Finite Element Method," Int. Journal of Solids and Structures, Vol. 1, pp. 451-461.

El-Garhy, B. M., (1999), "Soil Suction and Analysis of Raft Foundation Resting on Expansive Soils, Ph.D. Dissertation, Civil Engineering Department, Menoufia University, Shebin El-Koon, Menoufia, Egypt.

Fredlund, D. G., (1997), "An Introduction to Unsaturated Soil Mechanics," Unsaturated Soil Engineering Practice, Geotechnical Special Publication No. 68, ASCE, pp. 1-37.

Holland, J. E., Pitt, W. G., Lawrance, C. E., and Cimino, D. J., (1980), "The Behavior and Design of Housing Slabs on Expansive Clays," Proc. 4th Intl. Conf. on Expansive Soils, Colorado, Vol. 1, pp. 448-468.

Lytton, R. L., (1994), "Prediction of Movement in Expansive Clay," Vertical and Horizontal Deformations of Foundations and Embankments, Geotechnical Special Publication No. 40, A. T. Yeung and G. Y. Felio, Editors, ASCE, Vol. 2, pp. 1827-1845.

Lytton, R.L., and Kher, R.K. (1970), "Prediction of Moisture Movement in Expansive Clays," Research Report 118-3, Center for Highway Research, University of Texas at Arlington, Austin, Texas.

McKeen, R. G., (1980), "Field Studies of Airport Pavements on Expansive Clay," Proc. 4th Int. Conf. on Expansive Soils, Denver, Colorado, Vol. 1, pp. 242-261.

McKeen, R. G., and Johnson, L. D., (1990), "Climate-Controlled Soil Design Parameters for Mat Foundations," Journal of Geotechnical Engineering, ASCE, Vol. 116, No. 7, pp. 1073-1094.

Mitchell, P. W., (1979), "The Structural Analysis of Footings on Expansive Soil," Research Report No. 1, Kenneth W. G. Smith and Associates, Adelaide South Australia.

Mitchell, P. W., (1988), "The Design of Footings on Expansive Soils," Engineering Problems of Regional Soils, Proc. Int. Conf., Beijing, China, pp. 127-135.

Pufahl, D.E., and Lytton, R.L., (1992), "Temperature and Suction Profiles Beneath Highway Pavements: Computed and Measured," Transportation Research Record 1307, pp. 268-276.

Richards, B. G., (1965), "An Analysis of Subgrade Conditions at the Horsham Experimental Road Site Using Two-Dimensional Diffusion Equation on a High-Speed Digital Computer," Moisture Equilibria and Moisture Changes in Soils Beneath Covered Area, Butterworths, Australia, pp. 243-258.

Wray, W. K., (1978), "Development of a Design Procedure for Residential and Light Commercial Slabs-on-Ground Constructed Over Expansive Soils," Ph.D. Dissertation, Texas A & M University, College Station, Texas.

Wray, W. K., (1989), "Mitigation of Damage to Structures Supported on Expansive Soils," Final Report, National Science Foundation, Grant No. EKE-8320493, Vol. 1, 325pp.

Wray, W.K., (1997), "Using Soil Suction to Estimate Differential Soil Shrink or Heave," Unsaturated Soil Engineering Practice, Geotechnical Special Publication No. 68, ASCE, pp. 66-87.

APPENDIX II. NOTATION

The following symbols are used in this paper:

$f(x,y,z,t)$ = internal source of moisture

p = unsaturated permeability ($cm/\sec$)

R = $\dfrac{\alpha \Delta t}{\Delta z^2}$

R_x = $\dfrac{\Delta z}{\Delta x}$

R_y = $\dfrac{\Delta z}{\Delta y}$

t = time (sec)

u = soil suction expressed as a pF (kPa = 0.1×10^{pF})

$u^t_{i,j,k}$ = soil suction at time step, t, at grid point (i, j, k)

$u^{t+1}_{i,j,k}$ = soil suction at time step, $t+1$, at grid point (i, j, k)

x, y, z = space coordinate

$\gamma_{h_{i,j,k}}$ = suction compression index at grid point (i, j, k)

α = diffusion coefficient (cm^2 / sec)

$\Delta H_{i,j,k}$ = incremental change in elevation (shrink or heave) at grid point (i, j, k) over the vertical distance Δz

$\Delta pF_{i,j,k}$ = change of soil suction expressed as a pF (kPa = 0.1×10^{pF}) at grid point (i, j, k)

$\Delta pP_{i,j,k}$ = change of soil overburden over vertical distance Δz at grid point (i, j, k)

Δt = time interval

Δx = division length in x-direction

Δy = division length in y-direction

Δz = division length in z-direction

Subject Index

Page number refers to the first page of paper

Author Index

Page number refers to the first page of paper